Information Theory and Coding - Solved Problems

Predrag Ivaniš · Dušan Drajić

Information Theory and Coding - Solved Problems

 Springer

Predrag Ivaniš
Department of Telecommunications, School
 of Electrical Engineering
University of Belgrade
Belgrade
Serbia

Dušan Drajić
Department of Telecommunications, School
 of Electrical Engineering
University of Belgrade
Belgrade
Serbia

ISBN 978-3-319-84147-2   ISBN 978-3-319-49370-1   (eBook)
DOI 10.1007/978-3-319-49370-1

© Springer International Publishing AG 2017
Softcover reprint of the hardcover 1st edition 2016
This work is subject to copyright. All rights are reserved by the Publisher, whether the whole or part of the material is concerned, specifically the rights of translation, reprinting, reuse of illustrations, recitation, broadcasting, reproduction on microfilms or in any other physical way, and transmission or information storage and retrieval, electronic adaptation, computer software, or by similar or dissimilar methodology now known or hereafter developed.
The use of general descriptive names, registered names, trademarks, service marks, etc. in this publication does not imply, even in the absence of a specific statement, that such names are exempt from the relevant protective laws and regulations and therefore free for general use.
The publisher, the authors and the editors are safe to assume that the advice and information in this book are believed to be true and accurate at the date of publication. Neither the publisher nor the authors or the editors give a warranty, express or implied, with respect to the material contained herein or for any errors or omissions that may have been made.

Printed on acid-free paper

This Springer imprint is published by Springer Nature
The registered company is Springer International Publishing AG
The registered company address is: Gewerbestrasse 11, 6330 Cham, Switzerland

Preface

The aim of the book is to offer a comprehensive treatment of information theory and error control coding, by using a slightly different approach then in existed literature. There are a lot of excellent books that threat error control coding, especially modern coding theory techniques (turbo and LDPC codes). It is clear that understanding of the iterative decoding algorithms require the background knowledge about the classical coding and information theory. However, the authors of this book did not find other book that provides simple and illustrative explanations of the decoding algorithms, with the clearly described relations with the theoretical limits defined by information theory. The available books on the market can be divided in two categories. The first one consists of books that are specialized either to algebraic coding or to modern coding theory techniques, offering mathematically rigorous treatment of the subject, without much examples. The other one provides wider treatment of the field where every particular subject is treated separately, usually with just a few basic numerical examples.

In our approach we assumed that the complex coding and decoding techniques cannot be explained without understanding the basic principles. As an example, for the design of LDPC encoders, the basic facts about linear block codes have to be understood. Furthermore, the functional knowledge about the information theory is necessary for a code design—the efficiency of statistical coding is determined by the First Shannon theorem whereas the performance limits of error control codes are determined with the Second Shannon theorem. Therefore, we organized the book chapters according to the Shannon system model from the standpoint of the information theory, where one block affects the others, so they cannot be treated separately.

On the other hand, we decided to explain the basic principles of information theory and coding through the complex numerical examples. Therefore, a relatively brief theoretical introduction is given at the beginning of every chapter including a few additional examples and explanations, but without any proofs. Also, a short overview of some parts of abstract algebra is given at the end of the corresponding chapters. Some definitions are given inside the examples, when they appear for the first time. The characteristic examples with a lot of illustrations and tables are chosen to provide a detail insight to the nature of the problem. Especially, some limiting cases are given to illustrate the connections with the theoretical bounds. The numerical values are carefully chosen to provide the in-depth knowledge about the described algorithms. Although the examples in the different chapters can be considered separately, they are mutually connected and the conclusions in one considered problem formulates the other. Therefore, a sequence of problems can be considered as an "illustrated story about an information processing system, step by step". The book contains a number of schematic diagrams to illustrate the main concepts and a lot of figures with numerical results. It should be noted that the in this book are exposed mainly the problems, and not the simple exercises. Some simple examples are included in theoretical introduction at the beginning of the chapters.

The book is primarily intended to graduate students, although the parts of the book can be used in the undergraduate studies. Also, we hope that this book will also be of use to the practitioner in the field.

Belgrade, Serbia

Predrag Ivaniš
Dušan Drajić

Contents

1	**Introduction**	1
2	**Information Sources**	5
	Brief Theoretical Overview	5
	Problems	15
3	**Data Compression (Source Encoding)**	45
	Brief Theoretical Overview	45
	Problems	54
4	**Information Channels**	91
	Brief Theoretical Overview	91
	Problems	112
5	**Block Codes**	153
	Brief Theoretical Overview	153
	Problems	164
	Brief Introduction to Algebra I	222
6	**Cyclic Codes**	237
	Brief Theoretical Overview	237
	Problems	250
	Brief Introduction to Algebra II	313
7	**Convolutional Codes and Viterbi Algorithm**	327
	Brief Theoretical Overview	327
	Problems	344
8	**Trellis Decoding of Linear Block Codes, Turbo Codes**	385
	Brief Theoretical Overview	385
	Problems	405

9 Low Density Parity Check Codes 447
Brief Theoretical Overview 447
Problems 459

References 509

Index 515

Chapter 1
Introduction

We are living in the era of digital communications (transmission and storage). As an important feature of such systems, the transmission errors can be detected and corrected (Error control coding), while in analogue transmission practically only signal-to-noise ratio was considered. Classical communication theory started using Fourier analysis, to be latter enriched by the probabilistic approach. However, it considered only signals—the carriers of information. But Shannon, publishing (1948) his paper "A Mathematical Theory of Communication" (some scientists think that the article should be "The" and not "A"), founded Information theory. The Shannon approach was totally different from the classic one. One should say it was at a higher level. He did not consider the signals, but the information. The information is represented (encoded) by signals, which are the carriers of information. Therefore, it is possible that the transmitted signals do not carry any information at all (of course, these signals may be needed for proper functioning of the communication system itself). A simplified block-scheme of communication system from the information theory point of view is shown in Fig. 1.1.

The *encoder* can be divided into *source* encoder (for the corresponding *data compression*) and *channel* encoder (for the corresponding *error control coding*). The encriptor can be added as well. Communication channel is, by its nature, a *continuous* one. Using such an approach the noise and other interferences (other signals, fading etc.) can be directly taken into account. However, it can be simplified introducing the notion of a *discrete* channel, incorporating the signal generation (modulation), continuous channel and signal detection (demodulation). In contemporary communication systems, very often the encoding is connected with the choice of the corresponding (modulation) signals to obtain better system performances. Generally speaking, the fields of interest in information theory are sources, source encoding, channels and channel encoding (and the corresponding decoding).

The next three chapters are dealing with the sources, source encoding and the channel. A few elementary examples are given at their beginnings, followed later with more difficult problems. However, the main body of this book is dealing with the error control coding (last five chapters).

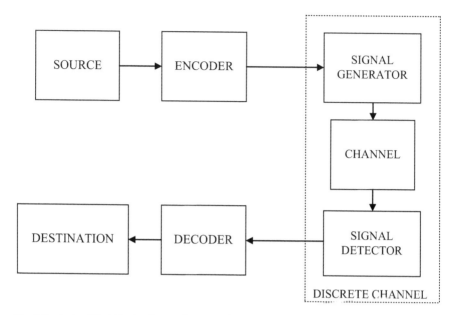

Fig. 1.1 A simplified block-scheme of communication system from the information theory point of view

The second chapter deals with the information sources. Mainly discrete sources are considered. The further division into memoryless sources and sources with memory is illustrated. The corresponding state diagram and trellis construction are explained, as well as the notions of adjoint source and source extension. Further, the notions of quantity of information and entropy are introduced. At the end an example of continuous source (Gaussian distribution) is considered.

In the next chapter source encoding (data compression) is considered for discrete sources. The important notions concerning the source codes are introduced—nonsingularity, unique decodability and the need for an instantaneous code. Notions of a code tree and of average code word length are introduced as well. The short discussion from the First Shannon theorem point of view is included. Further Shannon-Fano and Huffman encoding algorithms are illustrated with corresponding problems. Adaptive Huffman algorithms (FGK, Vitter) are discussed and illustrated. At the end LZ algorithm is considered.

Information channels are considered in the fourth chapter. As the sources, the channels can be as well discrete and continuous, but there exists also a mixed type (e.g. discrete input, continuous output). Discrete channels can be with or without memory. They are described using the corresponding channel matrix. A few discrete channels without memory are analyzed in details (BSC, BEC etc.). Transmitted information and channel capacity (for discrete and continuous channels) are defined and the Second Shannon theorem is commented. The decision rules are further analyzed (hard decoding and soft decoding), and some criteria

(MAP, ML) are considered. At the end Gilbert-Elliott model for the channels with memory is illustrated.

In the fifth chapter block codes (mainly linear block codes) are illustrated by using some interesting problems. At the beginning the simple repetitions codes are used to explain FEC, ARQ and hybrid error control procedures. Hamming distance is further introduced, as well as Hamming weight and distance spectrum. Corresponding bounds are discussed. The notion of systematic code is introduced. Hamming codes are analyzed in many problems. The notions of generator matrix, parity-check matrix and syndrome are explained. Dual codes and McWilliams identities are discussed. The notion of interleaving is illustrated as well. At the end of the chapter arithmetic and integer codes are illustrated. At the end of chapter a brief overview of the corresponding notions from abstract algebra is added (group, field, vector, space).

The sixth chapter deals with cyclic codes, a subclass of linear block codes obtained by imposing on an additional strong structure requirement. In fact, cyclic code is an ideal in the ring of polynomials. The notions of generator polynomial and parity-check polynomial are introduced. The usage of CRC is illustrated in a few problems. BCH codes are as well illustrated. RS codes are analyzed in details, especially decoding algorithms (Peterson, Berlekamp-Massey, Gorenstein and Zierler, Forney). At the end of chapter a brief overview of the corresponding notions from abstract algebra is added (Galois field, primitive and minimal polynomial, ideal).

Convolutional codes and decoding algorithms are analyzed in the next chapter. The corresponding notions are explained (constraint length, transfer function matrix, state diagram and trellis, free distance). Majority logic decoding, sequential decoding and especially Viterbi algorithm are illustrated as well as the possibilities of hard (Hamming metric) and soft (Euclidean metric) decision. Punctured codes are explained as well. TCM is considered in details.

Trellis decoding of linear block codes and turbo decoding are explained in the chapter eight. It is shown how from suitable transformed (trellis oriented) generator matrix, the corresponding parity-check matrix can be obtained suitable for trellis construction. The corresponding decoding algorithms are analyzed (generalized Viterbi algorithm, BCJR, SOVA, log-MAP, max-log-MAP). At the end turbo codes are shortly described.

The last chapter deals with LDPC codes. They provide an iterative decoding with a linear complexity. Tanner interpretation of LDPC codes using bipartite graphs is explained. The various decoding algorithms are analyzed with hard decision (majority logic decoding, bit-flipping) and with soft decision (belief-propagation, sum-product, self-correcting min-sum). At the end the algorithms are compared using Monte Carlo simulation over BSC with AWGN.

For reader convenience, a relatively brief theoretical introduction is given at the beginning of every chapter including a few additional examples and explanations, but without any proofs. Also, a short overview of some parts of abstract algebra is given at the end of the corresponding chapters. This material is mainly based on the textbook *An Introduction into Information Theory and Coding* [1] (in Serbian) from the same authors.

Chapter 2
Information Sources

Brief Theoretical Overview

Generally, the information sources are ***discrete*** or ***continuous***. The discrete source has finite or countable number of messages, while the con source messages are from an uncountable set. In this book will be mainly dealt with discrete sources, especially with those having a finite number of symbols.

The further subdivision of sources is according to the memory they may have. The sources can be ***without memory*** (zero-memory, memoryless) (Problems 2.1, 2.2, 2.3, 2.5, 2.6 and 2.8) emitting the symbols (messages) s_i according only to the corresponding probabilities $P(s)$. Therefore, the zero-memory source is completely described by the list of symbols S (***source alphabet***) $_i$

$$S\{s_1, s_2, \ldots, s_q\}$$

and by the corresponding symbol probabilities.

$$P(s_i)(i = 1, 2, \ldots, q).$$

It is supposed as well that the symbols (i.e. their emitting) is a complete set of mutually exclusive events, yielding

$$\sum_{i=1}^{q} P(s_i) = 1.$$

For the sources ***with memory***, where the emitting of the next symbol depends on m previously emitted symbols (m is the ***memory order***) (Problems 2.4 and 2.7), the corresponding conditional probabilities are needed.

$$P(s_j/s_{i_1}, s_{i_2}, \ldots, s_{i_k}, \ldots, s_{i_m}) \quad (j = 1, 2, \ldots, q) \quad (i_k = 1, 2, \ldots, q)$$
$$(k = 1, 2, \ldots, m),$$

where s_{i_1} is the oldest and s_{i_m} the youngest symbol, proceeding the symbol s_j. The other name for such source is **Markov source** (**Markov chain** is the emitted sequence). For the source with memory-order m, m previously emitted symbols are called the **source state**. Generally, there are q^m states. For a binary source (two symbols only) the number of states is 2^m. For every state the following must be satisfied

$$\sum_{j=1}^{q} P(s_j/s_{i_1}, s_{i_2}, \ldots, s_{i_k}, \ldots, s_{i_m}) = 1 \quad (i_k = 1, 2, \ldots, q) \quad (k = 1, 2, \ldots, m).$$

From every state the source can emit any of q possible symbols, and there are totally $q^m q = q^{m+1}$ conditional probabilities (some can be equal to zero!).

The **state diagram** (Problems 2.4, 2.6 and 2.7) can be drawn, comprising the states and the corresponding conditional probabilities. Instead of a state diagram, **trellis** (Problems 2.5, 2.6 and 2.7) can be constructed being some kind of a dynamic state diagram.

In further considering only the **ergodic** sources will be taken into account. For example, the source having at least one **absorbing state** (Problem 2.4), i.e. where from such a state the source can not pass into the other states, is not ergodic. Loosely speaking, the source is ergodic, if observed for a long time, it will pass through every of its possible states. If the elements of the matrix containing transition probabilities do not change in time, the source is **homogenous**. The source is **stationary** if the **steady state probabilities** can be found by solving the corresponding matrix equation (Problems 2.4 and 2.5)

$$\pi = \pi \times \Pi,$$

where π is steady state probabilities vector and Π is matrix of conditional probabilities. Of course, the additional condition is that the sum of the steady state probabilities equals 1. In this book the convention will be used that element p_{ij} in ith row and jth column in the matrix of conditional probabilities denotes the probability of the transition from the state S_i into the state S_j. Having in view that the elements in transition matrix are probabilities (non negative) as well as that from any state the source must pass to some other state or stay in the same state, the sum of elements in every row equals 1. Such a matrix is called **stochastic**.

$$\sum_{j=1}^{q} p_{ij} = 1 \quad (i = 1, 2, \ldots, q).$$

Some authors for transition matrix use the transposed version of the above matrix and in this case the sum of the elements in every column equals 1.

If the sum of elements in every column also equals 1, the matrix is *doubly stochastic* and all the states are equally probable. For any Markov source the probabilities of emitted symbols can be found as well. The *adjoint source* (Problems 2.4 and 2.7) to the source S, denoted by $\bar{S}$ is a zero-memory information source with source alphabet identical to that of S, having the same symbol probabilities. The nth *extension* (Problems 2.3 and 2.7) of any source S is the source denoted by S^n whose symbols are all possible different sequences (combinations) of the symbols of source S of the length n. If the original source has q symbols, source S^n has q^n different symbols.

Consider binary ($q = 2$) Markov source having memory order 2 ($m = 2$). There are $q^m = 2^2 = 4$ states and $q^{m+1} = 2^3 = 8$ transition probabilities. Let these probabilities are

$$P(0/00) = P(1/11) = 0.7$$
$$P(1/00) = P(0/11) = 0.3$$
$$P(0/01) = P(0/10) = P(1/01) = P(1/10) = 0.5$$

Therefore, the corresponding matrix of transition probabilities is

$$\Pi = \begin{matrix} & \begin{matrix} 00 & 01 & 10 & 11 \end{matrix} \\ \begin{matrix} 00 \\ 01 \\ 10 \\ 11 \end{matrix} & \begin{bmatrix} 0.7 & 0.3 & 0.0 & 0.0 \\ 0.0 & 0.0 & 0.5 & 0.5 \\ 0.5 & 0.5 & 0.0 & 0.0 \\ 0.0 & 0.0 & 0.3 & 0.7 \end{bmatrix} \end{matrix}$$

The states are **00**, **01**, **10** and **11**. It should be noted that the source cannot go directly from any state into the any other state. E.g. if source is in the state **00**, it can pass only to the same state (i.e. stay in it) (**00**) if in the emitted symbol sequence after 00 symbol 0 appears (**000** → **000**), or into the state **01** if after 00 symbol 1 appears (**001** → **001**). It can be easily visualized as the state of the corresponding shift register, where new (emitted) symbols enter from the right side. Equivalently, the "window" of the corresponding width ($m = 2$) sliding over the sequence of emitted symbols can be conceived, showing the current state. The corresponding state diagram is illustrated in the Fig. 2.1.

Steady state probabilities can be calculated as follows

$$P(00) = 0.7P(00) + 0.5P(10)$$
$$P(01) = 0.3P(00) + 0.5P(10)$$
$$P(11) = 0.7P(11) + 0.5P(01).$$

The first equation is the consequence of the fact the entering into the state **00** is a result of two mutually exclusive events—either the source was in state **00** and 0 is emitted, either the source was in state **10** and 0 was emitted. The second and the

Fig. 2.1 State diagram of a considered source

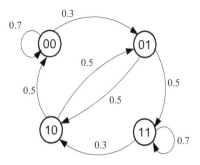

third equation are obtained in a similar way. However, if the corresponding (fourth) equation based on the same reasoning was added for the state **10**, a singular system would be obtained, because this equation is dependent of the previous three. Instead, the following equation will be used

$$P(00) + P(01) + P(10) + P(11) = 1,$$

because the source must be in one of the possible states. The corresponding solution is

$$P(00) = P(11) = 5/16$$
$$P(01) = P(10) = 3/16.$$

The corresponding steady state probabilities vector is

$$\pi = [5/16 \quad 3/16 \quad 3/16 \quad 5/16].$$

Taking into account the fact that state diagram is symmetrical, one should expect as well that the symmetrical states are equally probable. The stationary probabilities of the symbols in emitted sequence can now be calculated. The probability to find 0 in the emitted sequence is equal to the sum (the states are mutually exclusive) of the probability that the source is in the state **00** and the half of the probability that source is in the state **01** or **10**

$$P(0) = P(00) + 0.5(P(01) + P(10)) = 5/16 + 0.5(3/16 + 3/16) = 5/16 + 3/16$$
$$= 0.5.$$

Similarly, one can calculate $P(1) = 0.5$. However, this value can be easily obtained from

$$P(0) + P(1) = 1.$$

Further, because the state diagram is symmetrical (i.e. the values of the corresponding steady state probabilities) the obtained result should be expected as well

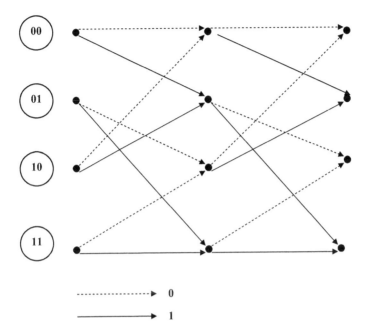

Fig. 2.2 Trellis for the source from the Fig. 2.1

—in the emitted sequence 0 and 1 are equally probable. The corresponding trellis is shown in Fig. 2.2. States are denoted at the trellis beginning only. Further, only points are used. The corresponding emitted bits are marked using arrows.

The state transition probabilities can be marked also. It is easy to note that trellis has a periodical structure. After the first step, the trellis repeats. If the initial source state is defined in advance (e.g. **00**) the trellis starts from this state.

The information is represented (encoded) by signals, which are the carriers of information. From the information theory point of view, the information depends on the probability of the message emitted by the source

$$Q(s_i) = f[P(s_i)].$$

If the probability of some message is smaller, the user uncertainty about this event is greater. By receiving the message (symbol) s_i, the user is no more uncertain about this event. In other words, if the uncertainty is greater, it can be concluded that by receiving the corresponding message, the greater quantity of information is received. Therefore, the quantity of information of some message is equal, or at least proportional, to the corresponding uncertainty at the receiving part. It can be easily concluded that the quantity of information of some message is inversely proportional to its (a priori) probability. A step further, if the user is sure that some message will be emitted ($P(s_i) = 1$), then it does not carry any information, because there was not any uncertainty. That means as well that an information source must have at least two messages at its disposition, to be capable to emit information. It should be noted that the emitting and not emitting of one message correspond to the two different messages.

On the basis of such approach from the definition of the quantity of information carrying by some message one expects that its numerical value is in inverse proportion to its probability, and equals to zero if it is a sure event. The next reasonable condition is that quantity of information is a continuous function of probability, because a small change in probability can not substantially change the quantity of information. Of course, the quantity of information can not have a negative value. There are many functions satisfying these conditions. However, from all these function, as a ***measure of the quantity of information*** the logarithm of the probability reciprocal value was chosen

$$Q(s_i) = \log\left(\frac{1}{P(s_i)}\right) = -\log P(s_i).$$

It is easy to verify that the above conditions are satisfied. Of course, some other function, satisfying these conditions could have been chosen, e.g.

$$Q(s_i) = \frac{1}{P(s_i)} - 1$$

However, the logarithmic function is a unique one making possible the fulfillment of one more necessary condition for the quantity of information—the additivity of information quantities of independent messages. Let the probabilities of messages s_i and s_j are $P(s_i)$ i $P(s_j)$. Then the joint message obtained by successive emitting of these independent messages should have the information quantity equal to the sum of their information quantities. Using logarithmic function, one obtains

$$Q(s_i, s_j) = \log\left(\frac{1}{P(s_i, s_j)}\right) = \log\left(\frac{1}{P(s_i)P(s_j)}\right)$$
$$= \log\left(\frac{1}{P(s_i)}\right) + \log\left(\frac{1}{P(s_j)}\right) = Q(s_i) + Q(s_j).$$

It can be proved that the logarithmic function is the unique one satisfying all mentioned conditions. Logarithm to any base can be used, however, it is suitable to use a logarithm to the base 2. It will be denoted as ld(.).

If a binary source is considered, its output signals (symbols) are known as ***bits*** (usually denoted as 0 and 1). If the bits are equally probable, i.e. $P(s_1) = P(s_2) = 0.5$, one obtains

$$Q(s_1) = Q(s_2) = \text{ld}(2) = 1 \quad [\text{Sh}].$$

This definition gives a very suitable etalon for a unit to measure the quantity of information, especially having in view the digital transmission. Therefore, here

Brief Theoretical Overview

every bit can carry a unit of information. To avoid the confusion between the bits (signals) and the units of information in this book the information unit using a logarithm to the base 2 will be called **shannon** (denoted by Sh). Therefore, one bit can carry up to one shannon.

For a source having q symbols, the corresponding quantities of information for the symbols are

$$Q(s_i) = \operatorname{ld}\left(\frac{1}{P(s_i)}\right) \quad [\text{Sh}] \quad (i = 1, 2, \ldots, q).$$

The average quantity of information emitted (per one symbol) is

$$H(S) = \operatorname{E}[Q(s_i)] = \overline{Q(s_i)} = \sum_{i=1}^{q} P(s_i) Q(s_i) = \sum_{i=1}^{q} P(s_i) \operatorname{ld}\left(\frac{1}{P(s_i)}\right)$$

$$= -\sum_{i=1}^{q} P(s_i) \operatorname{ld} P(s_i) \left[\frac{\text{Sh}}{\text{symb}}\right].$$

The average information rate of the source is

$$\Phi(S) = vH(S) \left[\frac{\text{Sh}}{\text{symb}} \frac{\text{symb}}{\text{s}} = \frac{\text{Sh}}{\text{s}}\right].$$

where v the source symbol rate [symb/s].

The average amount of information per source symbol is called the **entropy** (Problems 2.1 and 2.2), being for the zero-memory source

$$H(S) = \sum_{i=1}^{q} P(s_i) \operatorname{ld} \frac{1}{P(s_i)} = -\sum_{i=1}^{q} P(s_i) \operatorname{ld} P(s_i) \, [\text{Sh/symb}]$$

This expression is analogue to the Boltzmann formula for the entropy of the ideal gas, where $P(s_i)$ is the probability of the state in the phase space. It is some measure of the disorder and according to the Second law of thermodynamics can not be become smaller in a closed system. In information theory it is an average measure (per symbol) of the user uncertainty about the emitted source symbols. Shannon (1948) introduced the name entropy for $H(S)$.

Entropy of the nth extension of zero-memory source is $H(S^n) = nH(S)$ (Problem 2.3). By taking at least m extensions of an mth order Markov source the first-order Markov source is obtained. There will be always the dependence at the borders of of the extension "symbols". The property $H(S^n) = nH(S)$ also holds in this case. The entropy of the mth order Markov source with q symbols $\{s_1, s_2, \ldots, s_q\}$ having the transitional probabilities

$$P(s_j/s_{i_1}, s_{i_2}, \ldots, s_{i_k}, \ldots, s_{i_m}) \quad (j=1,2,\ldots,q) \quad (i_k = 1,2,\ldots,q)$$
$$(k=1,2,\ldots,m).$$

is

$$H(S) = \sum_{j=1}^{q}\sum_{i_1=1}^{q}\sum_{i_2=1}^{q}\cdots\sum_{i_m=1}^{q} P(s_{i_1}, s_{i_2}, \ldots, s_{i_m}, s_j) \mathrm{ld}\left(\frac{1}{P(s_j/s_{i_1}, s_{i_2}, \ldots, s_{i_m})}\right)$$
$$= \sum_{S^{m+1}} P(s_{i_1}, s_{i_2}, \ldots, s_{i_m}, s_j) \mathrm{ld}\left(\frac{1}{P(s_j/s_{i_1}, s_{i_2}, \ldots, s_{i_m})}\right).$$

In the last row the summation over all possible symbols of $m+1$th extension (S^{m+1}) of the original source is symbolically denoted. The number of elements of the sum is q^{m+1}. Further, the entropy of the corresponding source adjoined to the extended Markov source is not equal to the entropy of the extended Markov source (Problems 2.4, 2.5 and 2.7).

The entropy is an important characteristic of the information source. It is interesting how the entropy changes when the symbol probabilities change. If the probability of one symbol equals 1, then the probabilities of the all others are equal to zero, and the entropy is equal to zero as well, because

$$1 \cdot \mathrm{ld}(1) = 0, \quad \lim_{x \to 0} x \cdot \mathrm{ld}\frac{1}{x} = 0.$$

The expression for the entropy is a sum of non negative quantities, and the entropy cannot have the negative values. The next question is do the entropy is limited from the upper side. By using some inequalities, it can be easily proved

$$H(S) \leq \mathrm{ld}\, q.$$

yielding finally

$$0 \leq H(S) \leq \mathrm{ld}\, q.$$

Furthermore, it can be proved as well that the entropy will have the maximum value ($\mathrm{ld}\, q$) if $P(s_i) = 1/q$ for every i, i.e. when all the symbols are equally probable. The "physical interpretation" is here very clear. If the entropy is a measure of uncertainty of the user about the source emitting, then it is obvious that this uncertainty will have the maximal value when all the symbols are equally probable. If these probabilities are different, the user will expect the more probable symbols, they will be emitted more frequently, and at the average, the smaller quantity of information will be obtained per symbol.

The sequence of emitted symbols can be conceived as well as some kind of a **discrete random process**, where all possible sequences generated by a source form an **ensemble**. Sometimes, it is also said that the **time-series** are considered. It will be supposed that the ensemble is **wide sense stationary**, i.e. that the average value and autocorrelation function do not depend on the origin. Further, the process can be **ergodic** for the mean value and autocorrelation function if they are the same if calculated by using any generated sequence, or averaged over the whole ensemble for any fixed time moment. If the symbol duration is T, instead of $x(t)$ symbol $x(nT)$ can be used, or $x(n)$ or simply x_n. The **average value** is

$$E\{x_n\},$$

where $E\{\cdot\}$ denotes mathematical expectation. The **discrete autocorrelation function** (Problem 2.6) is

$$R_x(l) = E\{x_n x_{n+l}\}.$$

If the symbols are complex numbers, the second factor should be complemented. On the base of the Wiener-Khinchin theorem for discrete signals (z-transformation) the corresponding discrete **average power spectrum density** (APSD) (Problem 2.6) can be found.

A special kind of discrete random process is **pseudorandom** (PN—Pseudo Noise) **process** (*sequence*). In most applications binary sequences are used. They are generated by **pseudorandom binary sequence generator (PN)**, i.e. by the shift register with linear feedback (Problem 2.6). PN sequence is periodical and autocorrelation function is periodical as well. If the number of register cells is m, the corresponding maximal sequence length (period) is $L = 2^m - 1$. Of course, it will be the case, if the feedback is suitably chosen.

For a continuous source (Problem 2.8), emitting the "symbols" s from an uncountable set, having the probability density $w(s)$, the entropy can be defined as follows

$$H(S) = \int_{-\infty}^{\infty} w(s) \mathrm{ld}\left(\frac{1}{w(s)}\right) ds.$$

Of course, the following must be satisfied

$$\int_{-\infty}^{\infty} w(s) ds = 1,$$

However, one should be careful here, because in this case the entropy does not have all the properties of the entropy as for discrete sources. For example, it can depend on the origin, it can be infinite etc. Some additional constraints should be added. If the probability density is limited to a finite interval (a, b), the entropy will be maximum for a uniform distribution, i.e.

$$w(s) = \begin{cases} \frac{1}{b-a}, & a \leq s \leq b \\ 0, & \text{otherwise} \end{cases},$$

and a corresponding entropy value is

$$H(S)|_{\max} = \operatorname{ld}(b-a).$$

This result is similar to the result obtained for a discrete source having a finite number of symbols.

If the stochastic variable can take any value from the set of real numbers, but if its variance is finite, i.e.

$$\sigma^2 = \int_{-\infty}^{\infty} s^2 w(s) ds < \infty,$$

the maximum entropy will have Gaussian distribution

$$w(s) = \frac{1}{\sqrt{2\pi\sigma^2}} \exp\left[-\frac{s^2}{2\sigma^2}\right].$$

It is

$$H(S)|_{\max} = \frac{1}{2}\operatorname{ld}(2\pi e \sigma^2).$$

According to the Central limit theorem, the sum of more independent random variables, having limited variances, when the number of variables increases, converges to Gaussian distribution. Therefore, otherwise speaking, Gaussian distribution in the nature is a result of the action of a large number of statistically independent causes. Therefore, it is to be expected that the measure of uncertainty (entropy) in such case has a maximal value.

Some authors consider that the variables whose probability density is limited to a finite interval (a, b) correspond to the "artificial" signals (speech, TV etc.), while the variables with unlimited interval of values correspond to the "natural" signals (noise).

It is also interesting that it is possible to generate Gaussian random process having a predefined autocorrelation function (Problem 2.7).

One more problem is how to define the information rate for continuous sources. In this case the finite frequency band (limited to f_c) of power density spectrum should be supposed. The corresponding signal should be sampled with the rate $2f_c$ samples per second (these values are statistically independent!). The corresponding information rate is

$$\Phi(S) = 2f_g H(S) \left[\frac{\text{sample}}{\text{s}} \times \frac{\text{Sh}}{\text{sample}} = \frac{\text{Sh}}{\text{s}}\right],$$

i.e., the same units are used as for discrete sources (Sh/s). In such a way it is possible to compare discrete and continuous sources.

Problems

Problem 2.1 Zero-memory source is defined by the table

s_i	s_1	s_2	s_3	s_4	s_5	s_6	s_7	s_8
$P(s_i)$	0.3	0.21	0.17	0.13	0.09	0.07	0.01	0.02

(a) Find the entropy and the source information rate if the symbol rate is $v_s = 100$ [symb/s].

b) How the entropy and information rate will change in the case of equiprobable symbols?

Solution

(a) *Source entropy* is the average entropy per symbol

$$H(S) = \sum_{i=1}^{q} P(s_i) \operatorname{ld} \frac{1}{P(s_i)} = -\sum_{i=1}^{q} P(s_i) \operatorname{ld} P(s_i).$$

For this source ($q = 8$) the entropy is

$$H(S) = -0.3 \operatorname{ld}(0.3) - 0.21 \operatorname{ld}(0.21) - 0.17 \operatorname{ld}(0.17) - 0.13 \operatorname{ld}(0.13)$$
$$- 0.09 \operatorname{ld}(0.09) - 0.07 \operatorname{ld}(0.07) - 0.01 \operatorname{ld}(0.01) - 0.02 \operatorname{ld}(0.02)$$
$$= 2.5717 \text{ [Sh/symb]},$$

where, for the easier calculation of logarithms to the base 2 (denoted by ld(x)), the following relation can be used

$$\operatorname{ld}(x) = \frac{\log_a(x)}{\log_a(2)}.$$

Information rate, the average information quantity emitted by the source per second, is

$$\Phi(S) = H(S)v_s = 2.5717 \left[\frac{\text{Sh}}{\text{symb}}\right] \times 100 \left[\frac{\text{symb}}{\text{s}}\right] = 257.17 \left[\frac{\text{Sh}}{\text{s}}\right],$$

where v_s is the symbol rate.

(b) In the case of equiprobable symbols, the expressions become

$$H(S) = \sum_{i=1}^{q} \frac{1}{q} \operatorname{ld} \frac{1}{1/q} = q \times \frac{1}{q} \times \operatorname{ld}(q) = \operatorname{ld}(q), \quad \Phi(S) = v_s \times \operatorname{ld} q,$$

and the corresponding numerical values are

$$H(S) = 3 \left[\frac{\text{Sh}}{\text{symb}} \right], \quad \Phi(S) = 300 \left[\frac{\text{Sh}}{\text{s}} \right].$$

As expected, the zero-memory source has the maximum entropy when the symbols are equiprobable, because the uncertainty about the next emitted symbol is maximal.

Problem 2.2 Zero-memory source is defined by the table

s_i	s_1	s_2	s_3	s_4
$P(s_i)$	0.5	0.25	0.125	0.125

(a) Find the entropy and the information rate if the symbol rate is $v_s = 400$ [symb/s].
(b) Find the entropy if the source emits q symbols according to the following expression

$$P(s_i) = \begin{cases} 2^{-i}, & i = 1, 2, \ldots, q-1 \\ 2^{-i+1}, & i = q \end{cases}$$

and draw the values of entropy depending on the number of symbols for $q = 2, 3, \ldots, 20$.

Solution

(a) Direct using of the corresponding formula for entropy yields

$$H(S) = \sum_{i=1}^{4} P(s_i) \operatorname{ld} \frac{1}{P(s_i)} = \frac{1}{2} \operatorname{ld} 2 + \frac{1}{4} \operatorname{ld} 4 + \frac{1}{8} \operatorname{ld} 8 + \frac{1}{8} \operatorname{ld} 8$$

$$= 1.75 \left[\frac{\text{Sh}}{\text{symb}} \right].$$

In this case, the logarithms can be easily calculated, because the probabilities are inverses of the integer powers of 2. Information rate is

$$\Phi(S) = 1.75 \left[\frac{\text{Sh}}{\text{symb}}\right] \times 400 \left[\frac{\text{symb}}{\text{s}}\right] = 700 \left[\frac{\text{Sh}}{\text{s}}\right].$$

(b) When the probability of one symbol is twice smaller than that of the previous one, and when they are ordered according to the probabilities (while the last two have equal probabilities), the entropy of the source which has q symbols can be found in the form

$$H(S) = \sum_{i=1}^{q} P(s_i) \operatorname{ld} \frac{1}{P(s_i)} = \sum_{i=1}^{q-1} \frac{i}{2^i} + \frac{q-1}{2^{q-1}},$$

after the substitution $k = i - 1$ one obtains

$$H(S) = \frac{q-1}{2^{q-1}} + \sum_{k=0}^{q-2}(k+1)\left(\frac{1}{2}\right)^{k+1} = \frac{q-1}{2^{q-1}} + \frac{1}{2}\sum_{k=0}^{q-2} k \left(\frac{1}{2}\right)^k + \frac{1}{2}\sum_{k=0}^{q-2}\left(\frac{1}{2}\right)^k.$$

The obtained sums can be calculated as the geometrical and arithmetical-geometrical progressions

$$\sum_{k=1}^{n} x^{k-1} = \frac{x^n - 1}{x - 1}, \quad \sum_{k=0}^{n-1} k x^k = -\frac{(n-1)x^n}{1-x} + \frac{x(1-x^{n-1})}{(1-x)^2},$$

yielding

$$H(S) = (q-1)2^{1-q} - (q-2)2^{1-q} + (1 - 2^{2-q}) + (1 - 2^{1-q}) = 2 - 2^{2-q}.$$

Therefore, the entropy here is always limited $1 \leq H(S) \leq 2$, grows monotonously with the number of symbols, approaching asymptotically to $H_{\max}(S) = 2$. In Fig. 2.3 the corresponding entropy is shown (for $2 \leq q \leq 20$) as well as the comparison to the entropy of source having **equiprobable** symbols (Problem 2.1b).

This figure verifies also the known fact that the entropy has the maximal value when the symbols are equiprobable. The difference is very notable when the number of symbols is greater.

Problem 2.3 Consider the binary zero-memory source

(a) Find the entropy of the source, if probability of one symbol equals $p = 0.1$, as well as the entropies for the second and the third source extension.
(b) Find the expression for the entropy of the source nth extension, if the corresponding probabilities are $p = 0.01, p = 0.1$ and $p = 0.5$. For $n \leq 5$ show the results in a diagram.

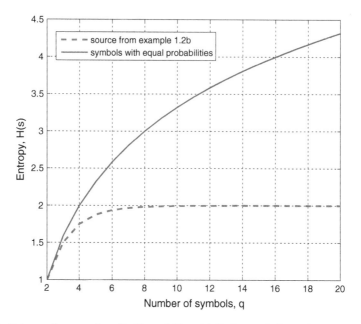

Fig. 2.3 Entropy of sources from Problems 1.1b and 1.2b for $q = 2, 3, \ldots, 20$

Solution

Binary source emits only two symbols with the probabilities $P(s_1) = p$ and $P(s_2) = 1 - p$. As its extension the sequences of original binary symbols as new (compound) symbols are considered.

(a) From $p(s_1) = 0.1$ follows $P(s_2) = 1 - p = 0.9$ yielding the entropy

$$H(S) = -0.1 \text{ ld } (0.1) - 0.9 \text{ ld } (0.9) = 0.469 \text{ [Sh/symb]}.$$

Second extension has $2^2 = 4$ symbols $\sigma_1 = s_1 s_1$, $\sigma_2 = s_1 s_2$, $\sigma_3 = s_2 s_1$ and $\sigma_4 = s_2 s_2$. Having in view that the source is zero-memory (memoryless), the corresponding probabilities are

$$P(\sigma_j) = P(s_{i_1}, s_{i_2}) = P(s_{i_1})P(s_{i_2}), \; i_1, i_2 \in \{1, 2\}, \; j \in \{1, 2, 3, 4\},$$

and can be easily calculated $P(\sigma_1) = 0.81$, $P(\sigma_2) = 0.09$, $P(\sigma_3) = 0.09$ and $P(\sigma_4) = 0.01$, yielding the corresponding entropy

$$\begin{aligned} H(S^2) &= -0.81 \text{ ld } (0.81) - 0.09 \text{ ld } (0.09) - 0.09 \text{ ld } (0.09) \\ &\quad - 0.01 \text{ ld } (0.01) \\ &= 0.938 \text{ [Sh/symb]}. \end{aligned}$$

In the case of nth extension the probabilities are

$$P(\sigma_j) = P(s_{i_1}, s_{i_2}, \ldots, s_{i_n})$$
$$= P(s_{i_1}) \times P(s_{i_2}) \times \ldots \times P(s_{i_n}), \ i_1, i_2, \ldots, i_n \in \{1, 2\}, \ j \leq 2^n,$$

for the **third extension**, which has eight compound symbols, given in Table 2.1 the corresponding entropy is

$$H(S^3) = \sum_{i=1}^{8} P(\sigma_i) \operatorname{ld} \frac{1}{P(\sigma_i)} = 1.407 \left[\frac{\mathrm{Sh}}{\mathrm{symb}}\right],$$

confirming for $n = 2$ and $n = 3$, the known relation [1,2]

$$H(S^n) = nH(S).$$

(b) For zero-memory binary source the entropy is given by

$$H(S) = p \operatorname{ld}\left(\frac{1}{p}\right) + (1-p) \operatorname{ld}\left(\frac{1}{1-p}\right),$$

while for **the entropy of nth extension of zero-memory binary source** one obtains

$$H(S^n) = nH(S) = np \operatorname{ld}\left(\frac{1}{p}\right) + n(1-p) \operatorname{ld}\left(\frac{1}{1-p}\right).$$

The entropy of nth extension of zero-memory binary source for $n = 1, \ldots, 5$ for some values of parameter p, is shown in Fig. 2.4. It can be easily seen that entropy grows linearly with n, yielding the larger differences for greater n. Of course, the entropy is maximal in the case of equiprobable symbols.

Problem 2.4 First-order (Markov) memory source whose source alphabet is (s_1, s_2, s_3, s_4), is defined by the transition matrix

Table 2.1 Symbols of the third extension and its probabilities+

I	1	2	3	4	5	6	7	8
σ_i	$s_1 s_1 s_1$	$s_1 s_1 s_2$	$s_1 s_2 s_1$	$s_1 s_2 s_2$	$s_2 s_1 s_1$	$s_2 s_1 s_2$	$s_2 s_2 s_1$	$s_2 s_2 s_2$
$P(\sigma_i)$	0.729	0.081	0.081	0.009	0.081	0.009	0.009	0.001

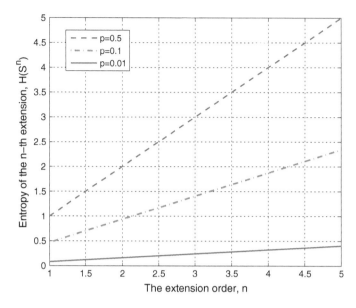

Fig. 2.4 Entropy of the nth extension of zero-memory binary source, one binary symbol with the probability p

$$\Pi = \begin{bmatrix} 0 & 0.7 & 0.3 & 0 \\ 0 & 0 & 0.7 & 0.3 \\ 0.1 & 0 & 0 & 0.9 \\ 0.7 & 0.3 & 0 & 0 \end{bmatrix}.$$

(a) Draw the source state diagram and find the state probabilities and the symbol probabilities. Whether the source is a homogenous one and a stationary one?
(b) If the probabilities in the matrix are changed as follows $P(s_1/s_3) = 0.3$ and $P(s_4/s_3) = 0.7$, draw the state diagram and find stationary probabilities of source symbols.
(c) Repeat the procedure from (b) for $P(s_1/s_3) = 0$ and $P(s_4/s_3) = 1$, the other probabilities being unchanged.
(d) How the state diagram changes for $P(s_1/s_3) = P(s_4/s_3) = 0$ and $P(s_3/s_3) = 1$?
(e) Find the entropies of the sources defined previously. For which transition matrix the entropy would achieve the maximum? What are the symbol probabilities in this case?

Solution

(a) Transition matrix describing **the first-order memory source** (first order Markov source) contains the conditional probabilities—the element in ith row and jth column corresponds to the conditional probability $P(s_j/s_i)$. This matrix

Problems

is always stochastic because the sum of elements in every row must be equal to 1 [3].

In this problem every symbol corresponds to one source state and transition probabilities are determined by the transition matrix. The corresponding state diagram is shown in Fig. 2.5. The source is stationary if the corresponding stationary symbol probabilities can be calculated, for which the following relation holds

$$[P(s_1)\ P(s_2)\ P(s_3)\ P(s_4)] = [P(s_1)\ P(s_2)\ P(s_3)\ P(s_4)] \times \begin{bmatrix} 0 & 0.7 & 0.3 & 0 \\ 0 & 0 & 0.7 & 0.3 \\ 0.1 & 0 & 0 & 0.9 \\ 0.7 & 0.3 & 0 & 0 \end{bmatrix}$$

$$\Rightarrow \pi = \pi \times \Pi,$$

where π is the stationary probabilities vector.

From the previous matrix equation, the following set of equations can be easily found

$$P(s_1) = 0.1P(s_3) + 0.7P(s_4),$$
$$P(s_2) = 0.7P(s_1) + 0.3P(s_4),$$
$$P(s_3) = 0.3P(s_1) + 0.7P(s_2),$$
$$P(s_4) = 0.3P(s_2) + 0.9P(s_3),$$

The source should be always in one of the possible states, yielding

$$P(s_1) + P(s_2) + P(s_3) + P(s_4) = 1.$$

It can be easily verified that the system consisting of the first four equations has not a unique solution. If the fourth equation is substituted by the fifth, the following matrix equation can be obtained

Fig. 2.5 State diagram of a first-order Markov source— (a)

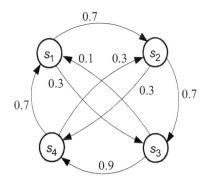

$$\pi = \begin{bmatrix} 0 & 0 & 0 & 1 \end{bmatrix} + \pi \times \begin{bmatrix} 0 & 0.7 & 0.3 & -1 \\ 0 & 0 & 0.7 & -1 \\ 0.1 & 0 & 0 & -1 \\ 0.7 & 0.3 & 0 & 0 \end{bmatrix},$$

yielding

$$\pi = \begin{bmatrix} 0 & 0 & 0 & 1 \end{bmatrix} \times \begin{bmatrix} 1 & -0.7 & -0.3 & 0 \\ 1 & 1 & -0.7 & -0.3 \\ -0.1 & 0 & 1 & -0.9 \\ 1 & 1 & 1 & 1 \end{bmatrix}^{-1}$$

finally, the *stationary probabilities* are obtained

$$\pi^{(a)} = \begin{bmatrix} P^{(a)}(s_1) & P^{(a)}(s_2) & P^{(a)}(s_3) & P^{(a)}(s_4) \end{bmatrix}$$
$$= \begin{bmatrix} 0.2263 & 0.2451 & 0.2395 & 0.2891 \end{bmatrix}$$

The source is *homogeneous* because the probabilities in the transition matrix do not depend on time. Further, it was verified that, in this case, the state probabilities fulfilling the relation $\pi = \pi \times \Pi$ can be found. Therefore, the source is *stationary*. It is obvious that for the first-order Markov source state stationary probabilities are equal to the emitted symbols probabilities.

(b) The corresponding state diagram is shown in Fig. 2.6a. It is symmetric and the corresponding transition matrix

$$\Pi^{(b)} = \begin{bmatrix} 0 & 0.7 & 0.3 & 0 \\ 0 & 0 & 0.7 & 0.3 \\ 0.3 & 0 & 0 & 0.7 \\ 0.7 & 0.3 & 0 & 0 \end{bmatrix}$$

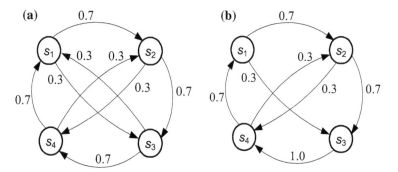

Fig. 2.6 State diagrams for (b) (a) and (c) (b)

is double stochastic, because the sum of elements in every row equals 1, as well as the sum of elements in every column. In the case of **double stochastic transition matrix** all states have equal probabilities, and the symbol probabilities here are

$$\pi^{(b)} = [0.25 \quad 0.25 \quad 0.25 \quad 0.25].$$

(c) For $P(s_1/s_3) = 0$ and $P(s_4/s_3) = 1$, the other probabilities remaining unchanged, state diagram is shown in Fig. 2.6b. The source from state s_3 goes deterministically to the state s_4. However, the source is stationary, because the solution can be found

$$\pi^{(c)} = \begin{bmatrix} P^{(a)}(s_1) & P^{(a)}(s_2) & P^{(a)}(s_3) & P^{(a)}(s_4) \end{bmatrix}$$
$$= [0.2152 \quad 0.2429 \quad 0.2346 \quad 0.3074]$$

(d) For $P(s_1/s_3) = P(s_4/s_3) = 0$ and $P(s_3/s_3) = 1$, the other probabilities remaining unchanged, state diagram is shown in Fig. 2.7. In this case the equation

$$P(s_3) = 0.3P(s_1) + 0.7P(s_2) + P(s_3)$$

has not solution except for $P(s_1) = P(s_2) = 0$ (being in contradiction to the other relations, i.e. not yielding the unique solutions for $P(s_3)$ i $P(s_4)$). In this case it is not possible to find probabilities satisfying the relation $\pi = \pi \times \Pi$ and the source is **not stationary** [1, 3]. It is clear as well from the fact that the state s_3 is an **absorbing** one, and the source cannot go out from this state.

(e) For the **first-order memory sources** the general expression for **entropy** of the source with memory is substantially simplifies

$$H(S) = \sum_{i=1}^{q} P(s_i) \sum_{j=1}^{q} P(s_j/s_i) \, \text{ld} \left(\frac{1}{P(s_j/s_i)} \right).$$

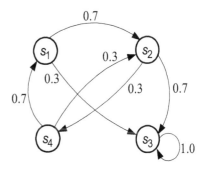

Fig. 2.7 State diagram for (d)

Taking into account the previously calculated state probabilities in d), the additional simplification is achieved comprising only 8 sum elements, instead of $q^2 = 16$ (because some elements of transition matrix are equal to zero)

$$H^{(a)}(S) = P^{(a)}(s_1)\left[0.7 \text{ ld}\left(\frac{1}{0.7}\right) + 0.3 \text{ ld}\left(\frac{1}{0.3}\right)\right]$$

$$+ P^{(a)}(s_2)\left[0.7 \text{ ld}\left(\frac{1}{0.7}\right) + 0.3 \text{ ld}\left(\frac{1}{0.3}\right)\right]$$

$$+ P^{(a)}(s_3)\left[0.1 \text{ ld}\left(\frac{1}{0.1}\right) + 0.9 \text{ ld}\left(\frac{1}{0.9}\right)\right]$$

$$+ P^{(a)}(s_4)\left[0.7 \text{ ld}\left(\frac{1}{0.7}\right) + 0.3 \text{ ld}\left(\frac{1}{0.3}\right)\right]$$

$$= 0.7825 \left[\frac{\text{Sh}}{\text{symb}}\right],$$

the entropy of adjoint source is

$$H^{(a)}(\bar{S}) = \sum_{k=1}^{4} P^{(a)}(s_k) \text{ ld}\left(\frac{1}{P^{(a)}(s_k)}\right) = 1.9937 \left[\frac{\text{Sh}}{\text{symb}}\right].$$

For the source where the transition matrix is doubly stochastic, the state diagram is symmetric, the entropies of the source and the corresponding adjoint source are

$$H^{(b)}(S) = 4 \times 0.25 \times \left[0.7 \text{ ld}\left(\frac{1}{0.7}\right) + 0.3 \text{ ld}\left(\frac{1}{0.3}\right)\right] = 0.8813 \left[\frac{\text{Sh}}{\text{symb}}\right],$$

$$H^{(b)}(\bar{S}) = 4 \times 0.25 \text{ ld}\left(\frac{1}{0.25}\right) = 2 \left[\frac{\text{Sh}}{\text{symb}}\right].$$

For $P(s_1/s_3) = 0$ and $P(s_4/s_3) = 1$, the corresponding values are obtained

$$H^{(c)}(S) = (P^{(c)}(s_1) + P^{(c)}(s_2) + P^{(c)}(s_4))$$

$$\times \left[0.7 \text{ ld}\left(\frac{1}{0.7}\right) + 0.3 \text{ ld}\left(\frac{1}{0.3}\right)\right] = 0.6746 \left[\frac{\text{Sh}}{\text{symb}}\right],$$

$$H^{(c)}(\bar{S}) = \sum_{k=1}^{4} P^{(c)}(s_k) \text{ ld}\left(\frac{1}{P^{(c)}(s_k)}\right) = 1.9867 \left[\frac{\text{Sh}}{\text{symb}}\right],$$

In the case $P(s_1/s_3) = P(s_4/s_3) = 0$ i $P(s_3/s_3) = 1$, the source is not stationary, and the entropy cannot be found.

Maximal entropy would correspond to the case when all the elements of the transition matrix have the values 1/4, the symbols are equally probable as well, yielding

Fig. 2.8 Block-scheme of the equivalent source

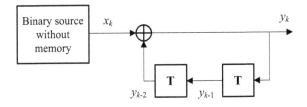

$$H_{\max}(S) = \sum_{i=1}^{4} \frac{1}{4} \sum_{j=1}^{4} \frac{1}{4} \operatorname{ld}(4) = 2 \left[\frac{\text{Sh}}{\text{symb}}\right].$$

In this case, the entropy of the adjoint source has the same value. It can be easily shown that this type of transition matrix corresponds to the zero-memory source.

Problem 2.5 Zero-memory source emits binary sequence x_k at the encoder input as shown in Fig. 2.8, where T denotes the cell delay corresponding to the duration of one symbol and addition is modulo 2.

(a) If encoder and source form an equivalent source emitting sequence y_k, find the type of this sequence. Is it a binary sequence? Whether the equivalent source has memory?
(b) Draw state diagram and trellis for this source. Find the state probabilities and symbol probabilities of the equivalent source.
(c) If the probability of one binary symbol in sequence x_k is $p = 0.2$ find the entropy of equivalent source and entropy of the source adjointed to it.
(d) Draw the entropy of the equivalent source as well as the entropy of its adjoint source as a function of p.

Solution

(a) All operations are in binary field and the sequence y_k is binary sequence as well. Forming of the sequence emitted by the ***equivalent source with memory*** is defined by

$$y_k = x_k \oplus y_{k-2},$$

showing clearly that output depends on the input and on the delayed input, where the delay is equal to the duration of two binary symbols. Therefore, the output symbol depends on two previously emitted symbols and second-order memory source is obtained.

(b) State diagram of the equivalent source can be easily drawn having in view that the source state (second-order memory source—$m = 2$) depends on two previously emitted symbols. These are the symbols at the cell outputs and the current state is $S = (y_{k-2}, y_{k-1})$ while the next state will be $S' = (y_{k-1}, y_k)$. From every state the source can enter only into two other states depending on

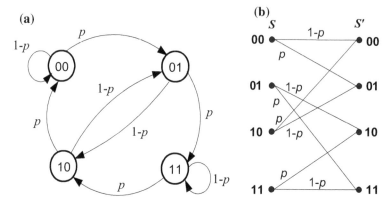

Fig. 2.9 State diagram (**a**) and trellis (**b**) of the second-order memory source

y_k, i.e. on the value of the binary symbol emitted by zero-memory source. Therefore, the transition probabilities are p and $1 - p$. The corresponding state diagram and trellis are shown in Fig. 2.9a, b, where the symbols are denoted by '0' and '1'.

The following set of equations holds

$$P(00) = P(00)(1-p) + P(10)p,$$
$$P(01) = P(00)p + P(10)(1-p),$$
$$P(10) = P(01)(1-p) + P(11)p,$$
$$P(11) = P(01)p + P(11)(1-p).$$

However, this system does not have the unique solution (it is singular!) and a new equation should be added to substitute any one in the set. This equation can be easily written, because the equivalent source must be always in some state

$$P(00) + P(01) + P(10) + P(11) = 1.$$

Steady state probabilities are $P(00) = P(01) = P(10) = P(11) = 1/4$, and all states are equiprobable.

The corresponding probabilities of the binary symbols are

$$P(0) = P(00) + (P(01) + P(10))/2 = 0.5,$$
$$P(1) = P(11) + (P(01) + P(10))/2 = 0.5.$$

It should be noted the perfect symmetry of state diagram as well as of the trellis yielding as a consequence such values of probabilities.

From the state diagram it can be seen that the conditional probability $P(y_k/y_{k-2}y_{k-1})$ equals to the transition probability from the state $S = (y_{k-2}, y_{k-1})$ into the state $S' = (y_{k-1}, y_k)$. Therefore,

$$P(0/00) = P(0/01) = P(1/10) = P(1/11) = 1-p,$$
$$P(1/00) = P(1/01) = P(0/10) = P(0/11) = p.$$

(c) Entropy of the sequence y_k can be found using the general expression for **the entropy of the second-order memory source**

$$H(Y) = \sum_{i_1=1}^{2}\sum_{i_2=1}^{2}\sum_{j=1}^{2} P(y_{i1}, y_{i2})P(y_j/y_{i1}, y_{i2}) \text{ ld}\left(\frac{1}{P(y_j/y_{i1}, y_{i2})}\right),$$

writing equivalently

$$H(Y) = \sum_{i=1}^{4}\sum_{j=1}^{2} P(S_i)P(y_j/S_i) \text{ ld}\left(\frac{1}{P(y_j/S_i)}\right),$$

where $S_i = (y_{i1}, y_{i2})$ denotes the ith source state

$$H(Y) = P(00)P(0/00) \text{ ld}\left(\frac{1}{P(0/00)}\right) + P(00)P(1/00) \text{ ld}\left(\frac{1}{P(1/00)}\right)$$
$$+ P(01)P(0/01) \text{ ld}\left(\frac{1}{P(0/01)}\right) + P(01)P(1/01) \text{ ld}\left(\frac{1}{P(1/01)}\right)$$
$$+ P(10)P(0/10) \text{ ld}\left(\frac{1}{P(0/10)}\right) + P(10)P(1/10) \text{ ld}\left(\frac{1}{P(1/10)}\right)$$
$$+ P(11)P(0/11) \text{ ld}\left(\frac{1}{P(0/11)}\right) + P(11)P(1/11) \text{ ld}\left(\frac{1}{P(1/11)}\right).$$

Taking into account the calculated probabilities

$$H(Y) = \frac{1}{4}(1-p) \text{ ld}\left(\frac{1}{1-p}\right) + \frac{1}{4}p \text{ ld}\left(\frac{1}{p}\right) + \frac{1}{4}(1-p) \text{ ld}\left(\frac{1}{1-p}\right) + \frac{1}{4}p \text{ ld}\left(\frac{1}{p}\right)$$
$$+ \frac{1}{4}p \text{ ld}\left(\frac{1}{p}\right) + \frac{1}{4}(1-p) \text{ ld}\left(\frac{1}{1-p}\right) + \frac{1}{4}p \text{ ld}\left(\frac{1}{p}\right) + \frac{1}{4}(1-p) \text{ ld}\left(\frac{1}{1-p}\right),$$

finally, the following is obtained

$$H(Y) = (1-p) \text{ ld}\left(\frac{1}{1-p}\right) + p \text{ ld}\left(\frac{1}{p}\right).$$

By definition, an adjoint source is a zero-memory source which has the same symbol probabilities as the source with memory to which is it adjoint. In the considered case its entropy is

$$H(\overline{Y}) = \sum_{k=1}^{2} P(y_j) \text{ ld} \left(\frac{1}{P(y_j)}\right) = 0.5 \text{ ld } 2 + 0.5 \text{ ld } 2 = 1 \left[\frac{\text{Sh}}{\text{symb}}\right].$$

The above derived expressions are valid for any value of parameter p. For $p = 0.2$, the corresponding numerical value of entropy for the source with memory is

$$H(S) = -0.8 \text{ ld } (0.8) - 0.2 \text{ ld } (0.2) = 0.7219 \text{ [Sh/symb]},$$

while the entropy of the adjoint source does not depend on p, being always $H(\overline{S}) = 1$ [Sh/symb].

(d) The entropy as a function of p is shown in Fig. 2.10. As one can expect, the entropy of the source with memory is always smaller than the entropy of its adjoint source. It should be noted that for practically all values of p the source having equiprobable symbols '0' i '1' is obtained (it is only not true in the case $p = 0$, but for some initial states only). This means that the condition $P(0) = P(1) = 0.5$ is not sufficient for the entropy to have the maximal value. The same could be concluded from the previous problem, but it should be noted that in this case, the zero-memory source is considered.

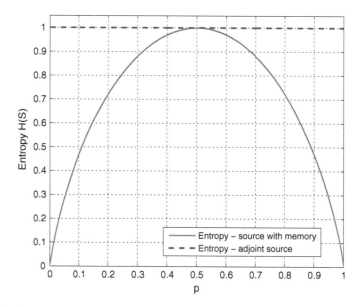

Fig. 2.10 The entropy of the second-order memory source as a function of p

It is interesting to notice that for $p = 0$ and $p = 1$ entropy equals zero. For $p = 0$ the emitted sequence depends on the initial source state. In any case the output sequence will be ***deterministic*** (no information is emitted!) as follows

(1) For the initial state 00 the following sequence is emitted
0000000000000000000...
(2) For the initial state 01 the following sequence is emitted
01010101010101010101...
(3) For the initial state 10 the following sequence is emitted
10101010101010101010...
(4) For the initial state 11 the following sequence is emitted
1111111111111111111...

It is expected from a reader to find the sequence emitted when $p = 1$, depending on the initial state. Do the sequences are deterministic, either they can be random as well?

Problem 2.6 Binary zero-memory source emits sequence x_k entering the encoder input, generating the corresponding encoder output

$$y_k = x_k \oplus a_1 y_{k-1} \oplus a_2 y_{k-2} \oplus \ldots \oplus a_p y_{k-p},$$

where the coefficients a_i are from binary field and addition is modulo 2.

(a) Considering zero-memory source and encoder as an equivalent source (emitting sequence y_k) find entropy of this source.
(b) Draw state diagram and trellis for this source for the case $p = 2$ and $a_1 = a_2 = 1$. Find the corresponding state and symbol probabilities.
(c) If the zero-memory source emits all-zeros sequence and encoder parameters are $p = 5$, $a_2 = a_5 = 1$ and $a_1 = a_3 = a_4 = 1$, draw block-scheme of an equivalent source (with memory), find its entropy and draw autocorrelation function of the sequence y_k.
(d) Find the autocorrelation function of an equivalent source, when its input is generated by zero-memory source emitting binary sequence x_k and (subsequently) encoded by encoder having p cells (p even) according to relation

$$y_k = x_k \oplus y_{k-p}.$$

Solution

(a) All calculations performed by encoder are over binary field and sequence y_k is binary as well, the source memory is of order p. The corresponding entropy is

$$H(Y) = \sum_{i=1}^{2^p} \sum_{j=1}^{2} P(S_i) P(y_j/S_i) \, \text{ld} \left(\frac{1}{P(y_j/S_i)} \right),$$

where $S_i = (y_{i_1}, y_{i_2}, \ldots, y_{i_p})$ denotes the current encoder state. From this state only the transitions into two next states are possible—$(y_{i_1}, y_{i_2}, \ldots, y_{i_p}, 0)$ and $(y_{i_1}, y_{i_2}, \ldots, y_{i_p}, 1)$, depending on the value $y_j = 0$ or $y_j = 1$. Transition probabilities depend on the current state yielding $P(y_j = 1/y_{i_1}, y_{i_2}, \ldots, y_{i_p}) = P_i$ or $P(y_j = 0/y_{i_1}, y_{i_2}, \ldots, y_{i_p}) = 1 - P_i$. The previous expression can be written in the form

$$H(Y) = \sum_{i=1}^{2^p} P(S_i) \left[P_i \, \text{ld} \left(\frac{1}{P_i} \right) + (1 - P_i) \, \text{ld} \left(\frac{1}{1 - P_i} \right) \right].$$

However, taking into account the following

$$P_i = \begin{cases} p, & a_1 y_{k-1} \oplus a_2 y_{k-2} \oplus \ldots \oplus a_p y_{k-p} = 0 \\ 1 - p, & a_1 y_{k-1} \oplus a_2 y_{k-2} \oplus \ldots \oplus a_p y_{k-p} = 0 \end{cases}$$

it is clear that the sum in brackets does not depends on i (one addend will be p, the other $1 - p$), yielding finally

$$H(Y) = \left[p \, \text{ld} \left(\frac{1}{p} \right) + (1 - p) \, \text{ld} \left(\frac{1}{1 - p} \right) \right] \sum_{i=1}^{2^p} P(S_i)$$

$$= p \, \text{ld} \left(\frac{1}{p} \right) + (1 - p) \, \text{ld} \left(\frac{1}{1 - p} \right).$$

Therefore, the encoding cannot change the source memory (but can introduce the memory into emitted sequence).

(b) For $p = 2$ and $a_1 = a_2 = 1$ block-scheme of equivalent source slightly differs from that in Fig. 2.8, because the output of the first cell is introduced into modulo-2 adder as well. The corresponding state diagram and trellis of the equivalent source are constructed in the same way as in the previous problem. They are shown in Fig. 2.11a, b.

Comparing state diagram and trellis to these ones in the previous problem (Fig. 2.9), the complete asymmetry can be noticed here. However, all four states are also equiprobable yielding $P(0) = P(1)$.

(c) The emitting of an all-zeros sequence by zero-memory source is equivalent to the case where it is off, and the corresponding block-scheme of the source is given in Fig. 2.12.

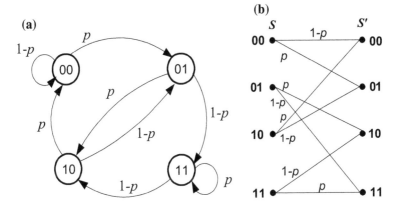

Fig. 2.11 State diagram (a) and trellis (b) of the source having parameters $p = 2$ and $a_1 = a_2 = 1$

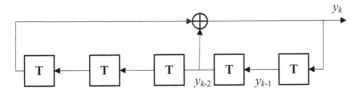

Fig. 2.12 Block-scheme of equivalent source for $p = 5$, $a_2 = a_5 = 1$ and for $x_k = 0$

This block-scheme corresponds to the structure of **pseudorandom binary sequence generator (PN)** [4], PN generator having N delay cells, and where the connections of cells and modulo 2 adder are chosen optimally. During one period it emits $L = 2^N - 1$ bits, from which 2^{N-1} binary "ones" and $2^{N-1} - 1$ binary "zeros". Therefore, the average value is

$$m_y = E\{y_k\} = \frac{1}{L}\sum_{k=1}^{L} y_k = \frac{2^{N-1}}{L} = \frac{(L+1)/2}{L} = \frac{1}{2} + \frac{1}{2L},$$

where $E\{\cdot\}$ denotes mathematical expectation.

The autocorrelation function of this PN sequence can be calculated using the definition of **discrete autocorrelation function** [5]

$$R_y(l) = E\{y_k y_{k+l}\} = \frac{1}{L}\sum_{k=1}^{L} y_n y_{n-k}$$

and by using the relation between two binary numbers

$$ab = (a+b-a\oplus b)/2.$$

the following result is obtained

$$R_y(l) = \frac{1}{2L}\sum_{k=1}^{L}(y_k + y_{k-l} - y_k \oplus y_{k-l}) = m_y - \frac{1}{L}\sum_{k=1}^{L}\frac{y_k \oplus y_{k-l}}{2}.$$

Sum in this expression for $l = 0$ is the sum of the addends being equal to zero, while for $l \neq 0$ it corresponds to one half of the average value yielding finally

$$R_y(l) = \begin{cases} m_y, & l = 0, \pm L, \ldots \\ m_y/2, & l \neq 0, \pm L, \ldots \end{cases}$$

The autocorrelation function is shown in Fig. 2.13. PN sequence is periodical and autocorrelation function will be periodical as well, with the period $L = 2^5 - 1 = 31$. This sequence is periodic and therefore deterministic, without any information carried and the corresponding source entropy equals zero.

(d) In this case sequence y_k can be written as follows

$$y_k = x_k \oplus y_{k-p} = x_k + y_{k-p} - 2x_k y_{k-p}.$$

Average value is

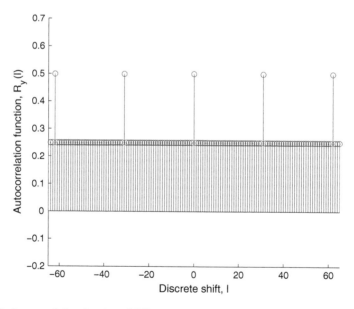

Fig. 2.13 Autocorrelation function of PN sequence generated for $p = 5$ and $a_2 = a_5 = 1$

$$E\{y_k\} = E\{x_k\} + E\{y_{k-p}\} - 2E\{x_k y_{k-p}\},$$

input and output encoder sequences are uncorrelated, yielding

$$m_y = m_x + m_y - 2m_x m_y \Rightarrow m_y = 1/2.$$

PN sequence autocorrelation function can be found using the definition of discrete autocorrelation function

$$\begin{aligned}R_y(l) &= E\{y_k y_{k-l}\} = E\{(x_k + y_{k-p} - 2x_k y_{k-p})(x_{k-l} + y_{k-p-l} - 2x_{k-l} y_{k-p-l})\} \\&= E\{x_k x_{k-l}\} + E\{x_k y_{k-p-l}\} - 2E\{x_k x_{k-l} y_{k-p-l}\} + E\{y_{k-p} x_{k-l}\} \\&\quad + E\{y_{k-p} y_{k-p-l}\} - 2E\{y_{k-p} x_{k-l} y_{k-p-l}\} - 2E\{x_k y_{k-p} x_{k-l}\} \\&\quad - 2E\{x_k y_{k-p} y_{k-p-l}\} + 4E\{x_k y_{k-p} x_{k-l} y_{k-p-l}\} \\&= R_x(l) + m_x m_y - 2R_x(l) m_y + m_x m_y + R_y(l) - 2R_y(l) m_x - 2R_x(l) m_y \\&\quad - 2R_y(l) m_x + 4R_x(l) R_y(l).\end{aligned}$$

For $m_y = 1/2$ a simple relation is further obtained

$$R_y(l) = m_x - R_x(l) + R_y(l)\left[1 - 4R_y(l)m_x + 4R_x(l)\right],$$

and because the input sequence is uncorrelated, it can be written

$$R_x(l) = \begin{cases} m_x, & l = 0, \pm L, \ldots \\ m_x/2, & l \neq 0, \pm L, \ldots \end{cases}$$

In the case $l = 0$ autocorrelation function equals average power of binary signal (and to its average value as well!). It is interesting to consider the following relation

$$R_y(l) = m_x/2 + R_y(l)\left[1 - 2R_y(l)m_x\right], \quad l \neq 0.$$

By using z-transformation the discrete *average power spectrum density* (APSD) can be found

$$\Phi(z) = \frac{m_x/2}{(1-z)(1-(1-2m_x)z^{-p})},$$

and by further using the partial decomposition [6]

$$\Phi(z) = \frac{A}{1-z} + \sum_{j=1}^{p/2} \left(\frac{B_j}{1 - \rho_j z} - \frac{B_j}{1 + \rho_j z}\right),$$

where coefficients A, B_j i ρ_j do not depend on z, the autocorrelation function finally can be written in the form

$$R_y(l) = \sum_{j=1}^{p/2} B_j \rho_j^l, \quad l \geq 0.$$

Autocorrelation functions for $p = 2$ and $p = 8$ are shown in Fig. 2.14a, b. It is easy to notice that autocorrelation function has constant values for all discrete shifts except for integer multiples of p. On the other hand, by considering only the values $R_y(np)$, the exponential decay is observed with the greater values of n. Further, the rate of change of $R_y(np)$ diminishes with the greater values of p. Taking this into account, the memory order (p) can be easily found from the autocorrelation function shape.

Problem 2.7 First-order memory binary source is defined by the conditional probabilities $P(0/1) = P(1/0) = p$.

(a) Draw the corresponding source realization by using zero-memory binary source, one adder and one delay cell.
(b) Draw the corresponding state diagram and trellis. Find the probabilities of binary symbols as well as of stationary states.
(c) For $p = 0.1$ find the corresponding sequence at the output. Find the symbol probabilities for the second extension of the source. Find the order of this extension.
(d) By using definition, find the entropy of source with memory and of its second extension, as well as the entropies of their adjoint sources.
(e) Sketch the source entropy and the entropy of its adjoint source as a function of the extension order for at least three values of p.

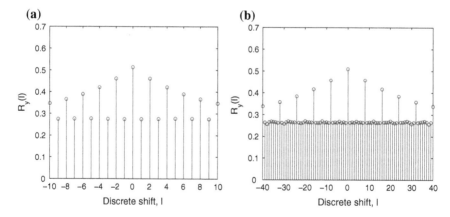

Fig. 2.14 Autocorrelation functions for $m_x = 0, 5$ and $p = 2$ (**a**), $p = 8$ (**b**)

Fig. 2.15 First-order memory binary symmetric source

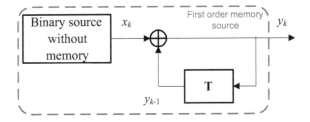

Solution

(a) This is a *first-order memory binary symmetric source*, it can be realized by combining zero-memory binary source and a differential encoder [5], as shown in Fig. 2.15.
(b) State diagram and the corresponding trellis are shown in Fig. 2.16. The source state corresponds to the previously emitted bit and transition probability into the next state is determined only by the bit probability at the zero-order memory binary source output. The source is symmetric (easily concluded from state diagram or the trellis) output binary symbols are equally probable yielding $P(0) = P(1) = 0.5$. The probability for the source to be in the state '0' equals the probability that previously emitted bit is '0' and state probabilities are equal (the source is stationary).
(c) If zero-memory source emits "binary ones" with probability $p = 0.1$, the transition probability into the other state will be small and transition in the sequence at the source output will be rare. *Characteristic sequence* emitted by the source with memory will have the following form

...000000000011111110000001111111111111111111110000000000111111111110000000000...

Source extension of the order $n = 2$ results in grouping two neighboring binary symbols into one compound symbol, the possible compound symbols being A = '00', B = '01', C = '10' and D = '11'. The above binary sequence can now be written using two twice shorter sequence

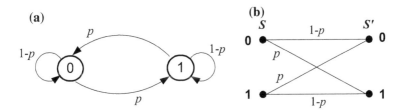

Fig. 2.16 State diagram (**a**) and trellis (**b**) of first-order memory binary symmetric source

...AAAAADDDCAABDDDDDDDDDAAAAABDDDDDAAAAA...

While binary symbols of the original sequence are equally probable, in the new sequence the symbols A and D will have substantially greater probabilities than the symbols B and C, being the consequence of the unequal probabilities in the state diagram. By using the joint probability, it is easy to find

$$P(A) = P(00) = P(0)P(0/0) = 0.5 \times (1-p) = 0.45$$
$$P(B) = P(01) = P(0)P(1/0) = 0.5 \times p = 0.05$$
$$P(C) = P(10) = P(1)P(0/1) = 0.5 \times p = 0.05$$
$$P(D) = P(11) = P(1)P(1/1) = 0.5 \times (1-p) = 0.45$$

To find the entropy of the nth extension of the m-order memory source, the memory order of the extended source firstly has to be found. It is calculated by using formula [7]

$$\mu = \left\lceil \frac{m}{n} \right\rceil,$$

where the operator $\lceil . \rceil$ denotes the taking of upper integer (if the quotient is not an integer). In this case $m = 1$ and $n = 2$ yielding

$$\mu = \lceil 1/2 \rceil = 1,$$

obtaining the first-order memory source by extension. By any extension of source with memory, zero-memory source cannot be obtained because there will be always some statistical dependence at the borders of extension source symbols.

(d) The entropy of the original first-order memory source is

$$H(S) = \sum_{i=1}^{2} \sum_{j=1}^{2} P(s_i) P(s_j/s_i) \text{ ld} \left(\frac{1}{P(s_j/s_i)} \right)$$

or

$$H(S) = 0.5(1-p) \text{ ld} \left(\frac{1}{1-p} \right) + 0.5p \text{ ld} \left(\frac{1}{p} \right) + 0.5p \text{ ld} \left(\frac{1}{p} \right) + 0.5(1-p) \text{ ld} \left(\frac{1}{1-p} \right),$$

yielding finally the well known result

$$H(S) = (1-p) \operatorname{ld}\left(\frac{1}{1-p}\right) + p \operatorname{ld}\left(\frac{1}{p}\right),$$

showing that it is equal to entropy of the zero-memory source at the differential coder input, because the entropy cannot be changed by differential encoding. Entropy of the corresponding adjoint source is, as in the previous example,

$$H(\bar{S}) = \sum_{i=1}^{2} P(s_i) \operatorname{ld}\left(\frac{1}{P(s_i)}\right) = 0.5 \operatorname{ld} 2 + 0.5 \operatorname{ld} 2 = 1 \left[\frac{\text{Sh}}{\text{symb}}\right].$$

The extended source is the first-order memory source as well, and the following expression can be used

$$H(S^2) = \sum_{i=1}^{2}\sum_{j=1}^{4} P(s_i)P(\sigma_j/s_i) \operatorname{ld}\left(\frac{1}{P(\sigma_j/s_i)}\right),$$

where $\sigma_1 = A$, $\sigma_2 = B$, $\sigma_3 = C$ and $\sigma_4 = D$. Stationary compound symbol probabilities are found earlier as $P(\sigma_1) = P(\sigma_4) = (1-p)/2$ and $P(\sigma_2) = P(\sigma_3) = p/2$.

If the symbol σ_j is obtained by combining $s_{j,1}$ and $s_{j,2}$, then sequence $s_i \sigma_j$ corresponds to the sequence s_i, $s_{j,1}$, $s_{j,2}$ emitted by the original source, and the corresponding conditional probabilities can be found by using

$$P(\sigma_j/s_i) = P(s_{j,2}/s_{j,1})P(s_{j,1}/s_i),$$

and after the final ordering of the entropy expression (tedious, but straight procedure) the following is obtained

$$H(S^2) = 2(1-p) \operatorname{ld}\left(\frac{1}{1-p}\right) + 2p \operatorname{ld}\left(\frac{1}{p}\right).$$

By definition, the entropy of the adjoint source is

$$H(\bar{S^2}) = \sum_{i=1}^{2}\sum_{j=1}^{2} P(\sigma_i)\operatorname{ld}\left(\frac{1}{P(\sigma_i)}\right) = -2\frac{(1-p)}{2}\operatorname{ld}\left(\frac{1-p}{2}\right) - 2\frac{p}{2}\operatorname{ld}\left(\frac{p}{2}\right)$$
$$= -(1-p)[\operatorname{ld}(1-p) - 1] - p[\operatorname{ld}(p) - 1] = 1 - (1-p)\operatorname{ld}(1-p) - p\operatorname{ld}(p).$$

One can easily verify that two previously given expressions can be obtained on the basis of the general relations

$$H(S^n) = nH(S),$$
$$H(\bar{S^n}) = (n-1)H(S) + H(\bar{S}),$$

where the last equality is obtained using the fact that the difference between the entropy of the extended source and its adjoint source does not change with the extension order.

For $p = 0.1$ the corresponding numerical values are

$$H(S) = 0.469 \text{ [Sh/s]}, \quad H(S^2) = 0.938 \text{ [Sh/s]}, \quad H(\overline{S^2}) = 1.469 \text{ [Sh/s]},$$

and the values of $H(S^n)$ and $H(\overline{S^n})$ are shown in Fig. 2.17 for three characteristic values of p.

When $p = 0.5$, the transitions into the next possible states are equiprobable and the source loses the memory (generated sequence becomes uncorrelated) yielding $H(S) = H(\overline{S})$ and $H(S^n) = H(\overline{S^n}) = nH(\overline{S})$. On the other hand, if $p = 0.01$ there is a great difference between the entropies of the original source and its adjoint source $(H(S) \ll H(\overline{S}))$ and it even after the extension $H(\overline{S^n})$ has substantially greater value than $H(S^n)$. If p is fixed, the difference between entropy of the extended source and the entropy of its adjoint source does depend of the extension order, being in concordance with relation

$$H(\overline{S^n}) - H(S^n) = H(\overline{S}) - H(S).$$

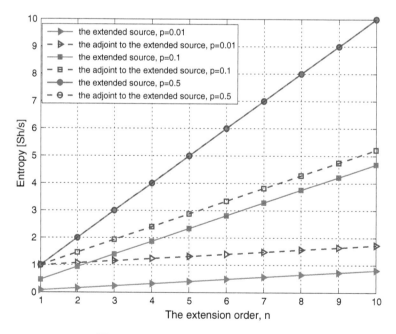

Fig. 2.17 $H(S^n)$ and $H(\overline{S^n})$ as functions of the extension-order n for $p = 0.01$, $p = 0.2$ and $p = 0.5$

Problem 2.8 Zero-memory continuous source generates Gaussian random process x_k ($m_x = 0$, $\sigma_x^2 = 1$) as the input of transversal filter yielding at the output

$$y_k = x_k - \sum_{i=1}^{p} a_i y_{k-i},$$

where coefficients a_i are real numbers and the addition is in decimal system.

(a) Draw the system block-scheme and explain its functioning. Find the average value and autocorrelation function of the output process.
(b) Is it possible to generate Gaussian random process having predefined autocorrelation function?
(c) Find the entropy $H(X)$ at the filter input. Does the entropy of the output random process is smaller or greater than $H(X)$?

Solution

(a) Block-scheme of the system, shown in Fig. 2.18, can be found directly from the corresponding equation. The average value of the random process at the output is

$$E\{y_k\} = E\{x_k\} - \sum_{i=1}^{p} a_i E\{y_{k-i}\},$$

and this process is the stationary Gaussian process as well (autoregressive IIR filter is a linear system!). As $m_x = E\{x_k\} = 0$, it follows

$$m_y \left(1 + \sum_{i=1}^{p} a_i\right) = 0 \Rightarrow m_y = 0.$$

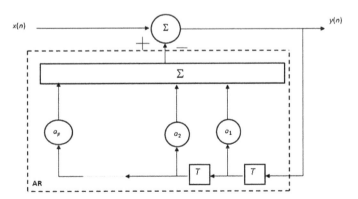

Fig. 2.18 Structure of order p AR filter

By performing the z-transform of y_k, the corresponding **transfer function** [4] is obtained

$$H(z) = \frac{1}{1 - \sum_{i=1}^{p} a_i z^{-i}},$$

and the discrete average power spectrum density of the output Gaussian process is

$$\Phi_y(z) = \frac{N_0}{\left|1 + \sum_{i=1}^{p} a_i z^{-i}\right|^2}.$$

Using Wiener-Khinchin theorem, the corresponding output autocorrelation function is easily obtained

$$R(l) = \frac{1}{2\pi} \int_{-\pi}^{\pi} \frac{N_0 e^{zl}}{\left|1 + \sum_{i=1}^{p} a_i z^{-i}\right|^2} dz \, (l = 0, \pm 1, \cdots, \pm(p-1)).$$

(b) **Autoregressive (AR) model** makes possible to generate process having the predefined autocorrelation function. If the values of $R(l)$ are given in advance, it is possible to define correlation matrix $\mathbf{R}$, coefficients vector of AR filter $\mathbf{a}$ and correlation vector $\mathbf{r}$, as follows

$$\mathbf{R} = \begin{bmatrix} R_y(0) & R_y(1) & \cdots & R_y(p-1) \\ R_y(1) & R_y(0) & \cdots & R_y(p-2) \\ \vdots & \vdots & \ddots & \vdots \\ R_y(p-1) & R_y(p-2) & \cdots & R_y(0) \end{bmatrix}, \quad \mathbf{a} = \begin{bmatrix} a_1 \\ a_2 \\ \vdots \\ a_p \end{bmatrix}, \quad \mathbf{r} = \begin{bmatrix} R_y(1) \\ R_y(2) \\ \vdots \\ R_y(p) \end{bmatrix}.$$

On the basis of $\mathbf{R}$ and $\mathbf{r}$, the coefficients of AR filter can be calculated to obtain output process which has the predefined autocorrelation function. The coefficients are obtained by using the **Yule-Walker equations** [8], which in matrix form is given by [9]

$$\mathbf{a} = -\mathbf{R}^{-1} \mathbf{r}.$$

In Fig. 2.19 it is shown how from an uncorrelated Gaussian random process x_k, a correlated random process y_k can be generated. It is obtained by filtering process x_k using filter with $p = 50$ delay cells, where the delay of one cell corresponds to the sampling period of the process T = 1 [µs]. Estimation was made using the sample

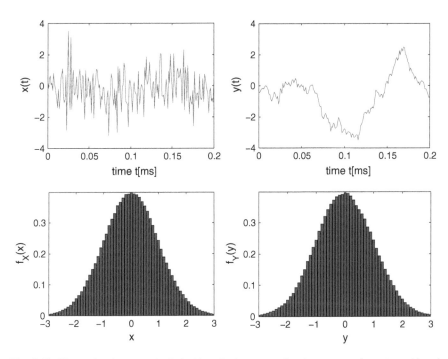

Fig. 2.19 Uncorrelated process (noise) $x(t)$ at the input, correlated process at the output $y(t)$ and the corresponding probability density functions for $l_{max} = 50$

size $N = 10^6$ and the uncorrelated and correlated signals are shown in Fig. 2.19. In the same figure the probability density functions of both processes are shown. It is clear that the first order statistics are not changed when AR filter was used.

On the other hand, second order statistics of input and output processes differ substantially, as shown in Fig. 2.20. The input process is uncorrelated, while, the output process has the autocorrelation function corresponding very good to the predefined linearly decreasing function.

$$R(l) = \begin{cases} (l_{max} - l)/l_{max}, & l \leq l_{max} \\ 0, & l > l_{max} \end{cases}$$

where parameter l_{max} determines the autocorrelation function slope.

(c) **Entropy of the continuous source without memory** is obtained by using the expression

$$H(X) = \int_{-\infty}^{\infty} w(x) \operatorname{ld}\left(\frac{1}{w(x)}\right) dx.$$

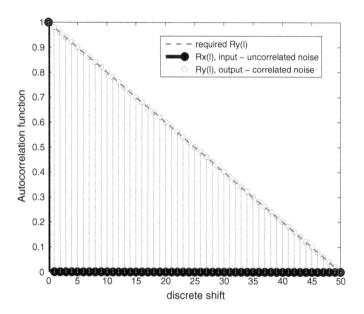

Fig. 2.20 Input and output autocorrelation function of the process

For Gaussian process with parameters $m_x = 0$, $\sigma_x^2 = 1$, probability density function has the form

$$w(x) = \frac{1}{\sqrt{2\pi}} e^{-\frac{x^2}{2}}, \quad -\infty < x < \infty.$$

Taking into account that process is uncorrelated, the above written expression for entropy can be used

$$H(X) = \int_{-\infty}^{\infty} w(x)\,\mathrm{ld}\!\left(\frac{1}{w(x)}\right) dx = \frac{1}{\sqrt{2\pi}} \int_{-\infty}^{\infty} e^{-\frac{x^2}{2}} \mathrm{ld}\!\left((2\pi)^{\frac{1}{2}} e^{\frac{x^2}{2}}\right) ds$$

$$= \frac{\mathrm{ld}\,(2\pi)}{2}\left[\frac{1}{\sqrt{2\pi}}\int_{-\infty}^{\infty} e^{-\frac{x^2}{2}}dx\right] + \frac{\mathrm{ld}\,(e)}{2}\left[\frac{1}{\sqrt{2\pi}}\int_{-\infty}^{\infty} x^2 e^{-\frac{x^2}{2}}dx\right].$$

Both expressions in brackets have 1 as a value, and the entropy of uncorrelated Gaussian process at the AR filter input is

$$H(X) = \frac{\mathrm{ld}\,(2\pi e)}{2}.$$

The output process is correlated, and can be considered as generated by a continuous source with memory. For such sources, the entropy cannot be calculated using the same formula as above. However, it can be noticed that the source generates signal y_k which has the same probability density function as signal x_k and that their first order statistics are identical. Therefore, one can consider the source generating x_k as an adjoint source to that one generating signal y_k, from which follows

$$H(Y) \leq H(\overline{Y}) = H(X).$$

The equality is valid only for $l_{\max} = 1$, while for its higher values, the entropy of the output sequence of AR filter is smaller than entropy at the input, and the difference is greater if autocorrelation function decreases slowly.

Chapter 3
Data Compression (Source Encoding)

Brief Theoretical Overview

In this chapter only the discrete sources are considered. Generally, the **encoding** is a mapping of sequences of source alphabet symbols (S) into sequences of code alphabet symbols (X) (Fig. 3.1).

Number of code alphabet symbols (r) is called **code base**. If the code alphabet consists of two symbols only, **binary code** is obtained. **Block code** maps each of the symbols (or their sequences) of source alphabet S into a fixed sequence of code alphabet X symbols. The obtained sequences are called **code words** (X_i). A number of symbols in a code word is called **code word length**.

Consider the following binary code ($r = 2$) for a source with $q = 4$ symbols. Code alphabet symbols are denoted simply as 0 and 1.

S	X_i
s_1	0
s_2	01
s_3	11
s_4	01

This code is defined by the table. A code can be also defined in some other way, e.g. using mathematical operations. The above code is practically useless because two source symbols have the same code words. The decoding is impossible. It is **singular** code. Therefore, a natural restriction is put—all code words should be distinct. A code is **nonsingular** if all code words are distinct. Consider the following nonsingular code

Fig. 3.1 Encoder block-scheme

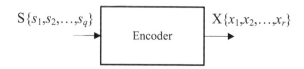

S	X_i
s_1	0
s_2	01
s_3	11
s_4	00

But this code is also useless. The sources are emitting sequences of symbols and the corresponding code words will be received by decoder in sequences. E.g., if the received sequence is 0011, there are two possibilities for decoding—$s_1 s_1 s_3$ and $s_4 s_3$. Therefore, a further restriction should be put. All possible sequences of code words (corresponding to all possible sequences of source symbols) must be different to allow a *unique decodability*. In other words, a nonambiguous decoding must be allowed. A (block) code is uniquely decodable if the nth extension of the code (i.e. the sequence corresponding to the nth source extension) is nonsingular for finite n.

Consider the following uniquely decodable codes

S	(a)	(b)	(c)
s_1	00	0	0
s_2	01	10	01
s_3	10	110	011
s_4	11	1110	0111

From the practical point of view, one more condition should be imposed. All three codes are uniquely decodable. Code words of code (a) have the same code length. Code (b) is a *comma code* (Problems 3.1 and 3.2), where at the end of every word is the symbol not appearing anywhere else. In binary code this symbol can be 0 (and 1 as well). In the above example the code words consist of a series of ones, with a zero at their end. However, code (c) is a little bit different. Here all code words start with 0. It means that decoder has to wait for the beginning of the next code word, to decode the current one. Therefore, an additional condition is put. A uniquely decodable code is *instantaneous* (Problems 3.2–3.4) if it is possible to decode each word in a sequence without reference to neighboring code symbols. A necessary and sufficient condition for a code to be instantaneous is that no complete code word is a prefix of some other code word. Code (c) does not satisfy this condition. In Fig. 3.2 the mentioned subclasses of codes are shown.

Brief Theoretical Overview 47

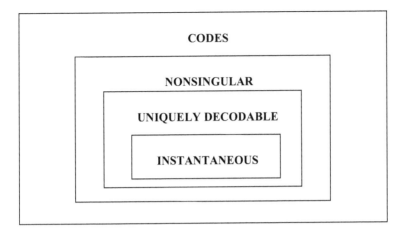

Fig. 3.2 Subclasses of codes

As a suitable way to visualize the code, a *code tree* can be constructed. The tree consists of the root, the nodes and the branches. From the root and from every node, for *r*-base code, there are *r* branches, each one corresponding to one code symbol. The alphabet symbols are at the ends of outer branches. Decoder starts from the root and follows the branch determined by a code symbol. At the end the corresponding source symbol should be found. In Fig. 3.3 the code trees corresponding to the last example are shown.

In the figure is arbitrary chosen that going along left branch corresponds to 0 and along right branch corresponds to 1. Of course, bits in the code words can be

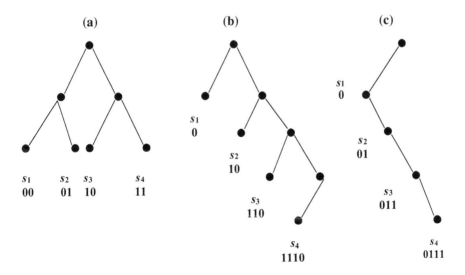

Fig. 3.3 Code trees corresponding to the above example

complemented. It corresponds to the rotations of the branches starting from nodes at corresponding level. Codes (a) and (b) are instantaneous, while the tree corresponding to (c) is degenerated, and the code is not instantaneous. It also can be noticed that the last bit in the code (b) for s_4 is needless and the decoder can make the correct decision without it. It can be concluded also that the use of a non-degenerated code tree guarantees an instantaneous code.

For the binary code ($r = 2$), for every integer q, code tree can be constructed having exactly q final nodes. Generally, for $r \geq 3$, it is not always possible to have exactly q final nodes. It can be easily seen from the example in Fig. 3.4.

It is obvious that ternary coding cannot provide a useful code tree for source with $q = 4$ symbols. Code (b) offers 5 code words and one code word will be unused.

If the source has q symbols, and the code base is r, the **Kraft inequality** (Problems 3.2, 3.4 and 3.5) gives necessary and sufficient condition for the existence of an instantaneous code with word lengths $l_1, l_2, \ldots, l_q$

$$\sum_{i=1}^{q} r^{-l_i} \leq 1.$$

Of course, for a binary code (code base $r = 2$), the inequality is

$$\sum_{i=1}^{q} 2^{-l_i} \leq 1.$$

If this relation is satisfied with equality, the code is **complete** (Problem 3.4). In fact, it is merely a condition on the word lengths of a code and not on the construction of words themselves (in the extreme case, one can put only symbol 0 in all code words). The next example is interesting.

Consider the following binary ($r = 2$) codes for a source with $q = 4$ symbols [3]

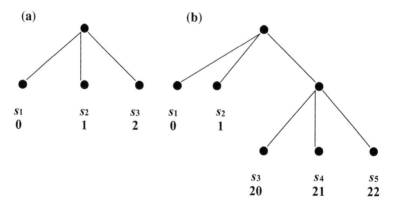

Fig. 3.4 Code trees of ternary code ($r = 3$) for the source with $q = 3$ (a) and $q = 5$ symbols (b)

Brief Theoretical Overview

S	(a)	(b)	(c)	(d)	(e)
s_1	00	0	0	0	0
s_2	01	100	10	100	10
s_3	10	110	110	110	110
s_4	11	111	111	11	11
$\sum_{i=1}^{4} 2^{-l_i}$	1	7/8	1	1	9/8

Codes (a), (b), (c) and (d) satisfy *Kraft inequality*. However, code (d) is not instantaneous because s_4 is a prefix of s_3. But, the code words with such lengths can be used to obtain an instantaneous code—(c). On the other hand, using code words whose lengths correspond to code (e), it is not possible to construct an instantaneous code.

The goal of data compression is to transmit or record the information as economically as possible. The useful measure is the **average code word length** (Problem 3.1)

$$\bar{l} = L = \sum_{i=1}^{q} P_i l_i.$$

where P_i ($i = 1, 2, \ldots, q$) are the corresponding source symbol probabilities. A code is **compact** if its average code word length is less than or equal to the average code word length of all other uniquely decodable codes for the same source and the same code alphabet. One can ask is there the lower limit for the average code word length. The answer is given by **First Shannon theorem** (Problem 3.7). Consider a zero-memory source. Entropy of the nth extension (Problem 2.7) of zero-memory source is $H(S^n) = nH(S)$, where $H(S)$ is the entropy of the original source. Shannon proved the following

$$\lim_{n \to \infty} \frac{L_n}{n} = \frac{H(S)}{\mathrm{ld}\, r}.$$

i.e. by unlimited source extension ($n \to \infty$) the equivalent average code word length per original symbol can be made as close to the entropy as one wish. In praxis, with the finite n, the following is valid

$$L \geq \frac{H(S)}{\mathrm{ld}\, r}.$$

For a binary code

$$L \geq H(S).$$

This result is obvious, because one bit cannot carry more than one shannon.

This theorem can be proved as well for sources with memory. For the nth extension of such source it is obtained (where the *adjoint source* to the source S, is denoted by $\overline{S}$)

$$H(S) + \frac{H(\overline{S}) - H(S)}{n} \leq \frac{L_n}{n} < H(S) + \frac{H(\overline{S}) - H(S) + 1}{n}.$$

It means that with sufficient source extension, the equivalent average code word length can be as near as we wish to the source entropy.

To compare the various codes, some parameters are defined. A *code efficiency* (Problems 3.1, 3.7 and 3.8). is

$$\eta = \frac{H(S)}{L} \times 100\% = 100\%$$

For 100% efficiency a code is *perfect* (Problems 3.1 and 3.3). *Compression ratio* is (Problems 3.1, 3.7 and 3.8).

$$\rho = \frac{\lceil \operatorname{ld} q \rceil}{L}.$$

where operator $\lceil . \rceil$ denotes the first greater (or equal) integer than the argument.

The *maximum possible compression ratio value* (Problem 3.1) is

$$\rho_{max} = \frac{\lceil \operatorname{ld}(q) \rceil}{H(S)} = \frac{\lceil \operatorname{ld}(q) \rceil}{\operatorname{ld}(q)}$$

The interesting features concerning the transmission errors are **maximum code word length** (Problem 3.5). and **sum of code word lengths** as well as **code words length variance** (Problem 3.5). Further, **effective efficiency of code for sources with memory**, can be defined as well as **maximum possible efficiency** and **maximum possible compression ratio** (Problem 3.7) for a given extension order.

The first known procedure for the source encoding was the **Shannon-Fano** one [10,11] (Problems 3.1 and 3.4). It was proposed by Shannon, while Fano made it easier to follow by using the code tree (Problems 3.3, 3.4, 3.6, 3.7 and 3.8). By applying this procedure sometimes more code trees have to be considered. The main reason is that the most probable symbols should have the shortest words. For binary codes it could be achieved by dividing symbols into two groups having equal (or as equal as possible) probabilities. Code words for one group will have the first bit 0, and for the second the first bit will be 1. It is continued into the subgroups until all subgroups contain only one symbol. For $r > 2$, the number of subgroups is r. Consider the next examples:

Brief Theoretical Overview

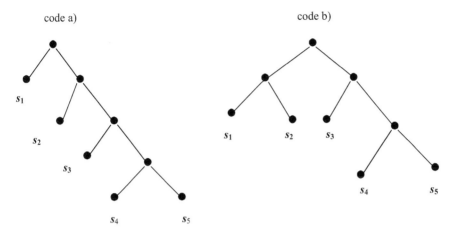

Fig. 3.5 Code trees corresponding to the codes (**a**) and (**b**)

(a)	S	P_i	I division	II division	III division	IV division
	s_1	0.6	0	0	0	0
	s_2	0.2	1	10	10	10
	s_3	0.1	1	11	110	110
	s_4	0.07	1	11	111	1110
	s_5	0.03	1	11	111	1111

(b)	S	P_i	I division	II division	III division
	s_1	0.3	0	00	00
	s_2	0.2	0	01	01
	s_3	0.2	1	10	10
	s_4	0.2	1	11	110
	s_5	0.1	1	11	111

The corresponding code trees are shown in Fig. 3.5 (left branch corresponds to 0, right one to 1).

It is easy to conclude that this procedure is not an algorithm. Generally, for every q and r, each corresponding code trees should be examined to obtain a compact code. In Fig. 3.6, two code trees (from possible 5) are shown for a source with 6 symbols.

The corresponding codes, entropy and average word code lengths are

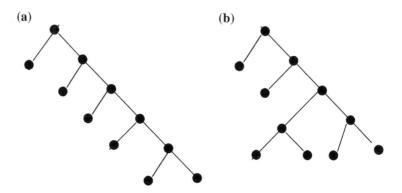

Fig. 3.6 Two possible code trees for a source with 6 symbols

S	P_i	(a)	(b)
s_1	0.65	0	0
s_2	0.15	10	10
s_3	0.08	110	1100
s_4	0.05	1110	1101
s_5	0.04	11110	1110
s_6	0.03	11111	1111
$H(S) = 1.6597$ Sh/symb		L_a=1.74 b/symb	L_b=1.75 b/symb

The difference in code word average length is very small and it is difficult to choose the right one code tree without calculation.

However, **Huffman** [12] (Problems 3.3, 3.4, 3.5, 3.6, 3.7 and 3.8) gave an algorithm resulting in obtaining a compact code. For illustration binary coding will be used. Consider the source without memory with the symbols $s_1, s_2, \ldots, s_q$ and the corresponding symbol probabilities P_i ($i = 1, 2, \ldots, q$). Symbols should be ordered according to nonincreasing probabilities $P_1 \geq P_1 \geq \cdots \geq P_q$. Now, the reductions start. The last two of ordered symbol are combined into one symbol (its probability is $P_{q-1} + P_q$) and a new source is obtained containing $q - 1$ symbols (first reduction). It is continued until only two symbols are obtained (after $q - 2$ reductions). Their code words start with 0 and 1. After that one goes back dividing the compound symbols into ones it was obtained from, adding 0 and 1 into the corresponding code words. At the end, the compact code is obtained.

Consider the previous example. Probabilities of symbols to be combined are denoted by (i.e. ♦) after the probability value, and probability of resulting symbol are denoted with the same sign before the numerical value of the combined symbol probability, $S_1, \ldots, S_4$ are the sources obtained after reduction.

Brief Theoretical Overview

S	P_i	X_i	S_1		S_2		S_3		S_4	
s_1	0.65	0	0.65	0	0.65	0	0.65	0	0.65	0
s_2	0.15	11	0.15	11	0.15	11	•0.20♦	11	♦0.35	1
s_3	0.08	101	0.08	101	♦0.12•	110	0.15♦	110		
s_4	0.05	1001	•0.07♦	1000	0.08•	101		101		
s_5	0.04•	10000	0.05♦	1001						
s_6	0.03•	10001								

Average code word length is $L = 1.74$ b/symb corresponding to the code tree (b) from the Fig. 3.6. The obtained code words are not identical to the previous solution, but the code word lengths are the same. Of course, by complementing the corresponding bits, the identical code words could be obtained.

One more question is of interest. In some cases there will be more than two compound code words with the same probabilities. What will happen if the compound code word is not put at the last place? Consider the following example.

(a) Symbol obtained by reduction is put always at the end of symbols having the same probabilities

S	P_i	X_i	S_1		S_2		S_3		S_4	
s_1	0.5	0	0.5	0	0.5	0	0.5	0	0.5	0
s_2	0.2	11	0.2	11	0.2	11	•0.3♦	10	♦0.5	1
s_3	0.1	101	0.1	101	♦0.2•	100	0.2♦	11		
s_4	0.1	1000	0.1♦	1000	0.1•	101				
s_5	0.07•	10010	•0.1♦	1001						
s_6	0.03•	10011								

Average code word length is $L = 2.1$ b/symb (entropy is $H(S) = 2.0502$ Sh/symb)

(b) Symbol obtained by the first reduction is put always at the first place among of symbols having the same probabilities, and in latter reduction at the last place.

S	P_i	X_i	S_1		S_2		S_3		S_4	
s_1	0.5	0	0.5	0	0.5	0	0.5	0	0.5	0
s_2	0.2	11	0.2	11	0.2	11	•0.3♦	10	♦0.5	1
s_3	0.1	1000	•0.1	101	♦0.2•	100	0.2♦	11		
s_4	0.1	1001	0.1♦	1000	0.1•	101				
s_5	0.07•	1010	0.1♦	1001						
s_6	0.03•	1011								

Average code word length is the same as above $L = 2.1$ b/symb.

The structure of obtained code is different, but it is not important, because both codes are compact (Problem 3.3). It is a nontrivial example of the existence of more compact codes for the same source and the same code alphabet. This fact can be used to impose some additional conditions for the obtained code. For example, the condition can be put that the lengths of the obtained code words have smaller variance. It will lessen the demands for encoder memory. It can be achieved by putting the compound symbol always at the first place among of symbols having the same probabilities.

One drawback of a "classic" Huffman procedure is the need to know the symbol probabilities. However, these probabilities are not known at the beginning (except mainly for the languages). But often there is no time to wait the end of a long message to obtain a corresponding statistics. The messages must be transmitted continuously as they are generated. The problem is solved by using the *adaptive (dynamic) Huffman algorithm*. The code tree is changed continuously to take into account the current symbol probabilities, it is dynamically reordered. One possible solution was proposed independently by **Faller** and **Gallager**, later improved by **Knuth** and the algorithm is known as **FGK algorithm** (Problem 3.9). It is based on the so called *sibling property* (Problems 3.2, 3.3, 3.4 and 3.9). **Vitter** (Problem 3.9) gave the modification that is suitable for very large trees. These long procedures are exposed in details in the corresponding problems.

The Huffman coding does not take into account the source memory. LZ (*Lempel-Ziv*) (Problem 3.10) procedure is some kind of universal coding without any explicit model. It can be thought out as a try to write a program for a decoder to generate the sequence compressed by the encoder at the transmitting part.

For ternary encoding, the last three symbols are combined, and correspondingly for r-ary encoding. If after the last reduction the number of compounded symbols differs from 3 (r), dummy symbols (having the probability equal to zero) should be added before the first reduction to avoid the loose of short code words.

Problems

Problem 3.1 By applying Shannon-Fano encoding procedure find the binary code corresponding to zero-memory source emitting $q = 5$ symbols whose probabilities

(a) are defined by $P(s_i) = \begin{cases} 2^{-i}, i = 1, 2, \ldots, q-1 \\ 2^{-i+1}, i = q \end{cases}$

(b) are mutually equal, $P(s_i) = 1/q$, $i = 1, 2, \ldots, q$.

For both cases find the code efficiency and the compression ratio. For values $2 \leq q \leq 20$ draw efficiency and compression ratio as the functions of q for both sources.

Solution

(a) Procedure for obtaining the code using **Shannon-Fano encoding** is shown in Table 3.1. The encoding steps are:

Table 3.1 Shannon-Fano procedure for the first set (a) of symbol probabilities

s_i	$P(s_i)$	Code words forming				Code word	l_i code word length
s_1	1/2	0				0	1
s_2	1/4	1	0			10	2
s_3	1/8		1	0		110	3
s_4	1/16			1	0	1110	4
s_5	1/16				1	1111	4

1. Symbols are ordered according to nonraising probabilities.
2. Set of symbols is divided into two groups with equal or approximately equal probabilities, but the groups can not include non adjacent symbols.
3. To every obtained group binary symbols "0" or "1" are joined, they will be the first symbol of the code words for symbols in this group.
4. If any group obtained in the previous step contains more than one symbol, its elements are further grouped into two subgroups with approximately equal probabilities. To every obtained subgroup new binary symbol is joined, being the next symbol in the corresponding code words.
5. Described procedure is continued until all the symbols obtain the code words.

In this case a set of symbols in the first step is divided into two equally probable subgroups (s_1 is in the first, all other symbols in the second). In the next steps the subgroups are always further divided into two equally probable sets. Therefore, in this case, **code word length** corresponding to the ith symbol (denoted by l_i) satisfies the relation

$$l_i = \mathrm{ld}\, \frac{1}{P(s_i)},$$

and so called **comma code** is obtained, where the binary zero occurs only at the last position in the code word, except for one of the words that has the maximum length, having binary ones only.

To find the code efficiency, the entropy and the **average code word length** should be found. In this case

$$H(S) = \sum_{i=1}^{5} P(s_i)\mathrm{ld}\frac{1}{P(s_i)} = \frac{1}{2}\mathrm{ld}\,2 + \frac{1}{4}\mathrm{ld}\,4 + \frac{1}{8}\mathrm{ld}\,8 + 2\times\frac{1}{16}\mathrm{ld}\,16 = 1.875[\mathrm{Sh/symb}]$$

$$L = \sum_{i=1}^{5} P(s_i)l_i = \frac{1}{2}\times 1 + \frac{1}{4}\times 2 + \frac{1}{8}\times 3 + 2\times\frac{1}{16}\times 4 = 1.875[\mathrm{b/symb}]$$

and the **perfect code** is obtained, because its **efficiency** equals

$$\eta = \frac{H(S)}{L} \times 100\% = 100\%.$$

Compression ratio shows the reducing of the number of bits at the encoder output comparing to the case when binary (non compression) code is used, it is defined as

$$\rho = \frac{\lceil \operatorname{ld} q \rceil}{L} = \frac{3}{1.875} = 1.6,$$

where operator $\lceil . \rceil$ denotes the first greater (or equal) integer than the argument. In this case, the number of bits at the encoder output is 1.6 times smaller.

For the given probabilities, it is easy to verify that for any number of emitted symbols, a perfect comma code always can be found, and the following relation holds

$$\rho = \frac{\lceil \operatorname{ld}(q) \rceil}{H(S)},$$

where the general expression for entropy, derived in Problem 2.2, yields

$$H(S) = 2 - 2^{2-q}.$$

(b) For the second set of probabilities, the procedure for obtaining the code using Shannon-Fano encoding is shown in Table 3.2.

Entropy and average code word length are now

$$H(S) = \sum_{i=1}^{5} P(s_i) \operatorname{ld} \frac{1}{P(s_i)} = 5 \times \frac{1}{5} \operatorname{ld} 5 = 2.3219 \, [\text{Sh/symb}],$$

$$L = \sum_{i=1}^{5} P(s_i) l_i = 3 \times \frac{1}{5} \times 2 + 2 \times \frac{1}{5} \times 3 = 2.4 \, [\text{b/symb}],$$

Table 3.2 Shannon-Fano procedure for the second set (b) of symbol probabilities

s_i	$P(s_i)$	Obtaining of code word			Code word	l_i code word length
s_1	1/5	0	0		00	2
s_2	1/5	0	1		01	2
s_3	1/5	1	0		10	2
s_4	1/5	1	1	0	110	3
s_5	1/5	1	1	1	111	3

and efficiency and compression ratio

$$\eta = \frac{H(S)}{L} \times 100\% = 96.75\%,$$

$$\rho = \frac{\lceil \mathrm{ld}\,(q) \rceil}{L} = \frac{3}{2.4} = 1.25.$$

For the probabilities in the form $P(s_i) = 1/q$, $i = 1, 2, \ldots, q$, general expression for entropy is

$$H(S) = \sum_{i=1}^{q} P(s_i)\, \mathrm{ld}\, \frac{1}{P(s_i)} = \sum_{i=1}^{q} \frac{1}{q}\, \mathrm{ld}\,(q) = \mathrm{ld}\,(q),$$

while the obtained compression ratio is

$$\rho = \frac{\lceil \mathrm{ld}\,(q) \rceil}{L} = \rho_{\max} \eta,$$

where $\rho_{\max}$ denotes **maximum possible compression ratio value**, not depending on the code used, but on the source and shows the "source compression potential", given by

$$\rho_{\max} = \frac{\lceil \mathrm{ld}\,(q) \rceil}{H(S)} = \frac{\lceil \mathrm{ld}\,(q) \rceil}{\mathrm{ld}\,(q)}.$$

When the number of source symbols corresponds to the power of 2 ($q = 2^n$, n—integer), the code obtained by Shannon-Fano procedure is equivalent to the binary code, and this code is perfect. In this case the efficiency is maximum possible and the obtained compression ratio is equal to the maximum possible compression ratio. However, here it equals 1, meaning that there is no any compression.

Therefore, for the source that has the equiprobable symbols, the compression can be achieved only if the number of symbols is not $q = 2^n$. This code is not perfect, but the compression ratio is greater than 1. The conclusion is that only non perfect codes can be compressed, while perfect codes (for these values of q for which they can be constructed) are useless from compression point of view. The reader should respond to the question is that fact in contradiction with the definition that for some source the perfect code achieves 100% efficiency and the maximum possible compression ratio. The maximum possible compression ratio vs. the number of source symbols is shown in Fig. 3.7. For equiprobable symbols $\rho_{\max}$ has small values. However, in the second analyzed case source compression potential is greater and grows with the number of emitted symbols.

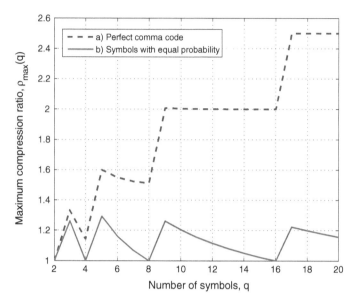

Fig. 3.7 The maximum possible compression ratio vs. the number of source symbols for two cases analyzed

From Fig. 3.7 the following conclusions can be drawn:

1. Compression ratio does not depend on the obtained code efficiency only, but on the source entropy as well. Sources with smaller entropy have greater compression potential, given by ρ_{max}.
2. For the first analyzed case, for any number of source symbols, the obtained comma code is perfect and the achieved compression ratio is equal to ρ_{max}.
3. For the second analyzed source, when number of symbols is increased, the code efficiency approaches to $\eta = 100\%$, but that does not mean that the compression ratio grows monotonously with the number of emitted symbols.
4. For the second analyzed source, the obtained code is perfect only when $q = 2^n$ yielding the compression ratio $\rho = \rho_{max} = 1$. For all other cases, the curve in Fig. 3.7 is only the upper bound for compression ratio. However, it is interesting that by using a non perfect code (if e.g. when combined with the source extensions) for the source having equiprobable symbols for $q \neq 2^n$, the compression ratio $\rho > 1$ can be achieved.

Problem 3.2 Find the binary code for zero-memory source defined by

s_i	s_1	s_2	s_3	s_4
$P(s_i)$	0.6	0.25	0.125	0.025

Problems

(a) Find the efficiency and the code compression ratio.
(b) If the probability of s_1 is smaller for 0.1 and $P(s_4)$ is greater for the same value, find the corresponding efficiency and the compression ratio. Comment the results.
(c) Whether the obtained code is instantaneous? Elaborate the answer!

Solution

(a) It is obvious that comma code is the best solution. The first symbol is encoded by one bit only ('0'), second by two ('10') and third and fourth by three bits ('110' and '111'). The entropy and the average code word length are

$$H(S) = -0.6\,\text{ld}(0.6) - 0.25\,\text{ld}(0.25) - 0.125\,\text{ld}(0.125) - 0.025\,\text{ld}(0.025)$$
$$= 1.4502\,[\text{Sh/symb}]$$
$$L = 0.6 \times 1 + 0.25 \times 2 + 0.125 \times 3 + 0.025 \times 3 = 1.55\,[\text{b/symb}],$$

the efficiency and compression ratio are $\eta = 93.56\%$ and $\rho = \lceil \text{ld}\,(q) \rceil / L = 2/1.55 = 1.2903$.

(b) In the case when the probabilities of the first and the fourth symbol are changed, comma code is the best solution as well. The same code as in (a) will be used. The probabilities are of the type 2^{-i}, and the code is perfect yielding the same values for entropy and the average code word length

$$H(S) = L = 1.75\,[\text{Sh/symb}].$$

The efficiency is $\eta = 100\%$, and the compression ratio is

$$\rho = \lceil \text{ld}\,(q) \rceil / L = 2/1.75 = 1.1429.$$

In this case by changing the probabilities of two symbols (the number of symbols is the same!) the efficiency becomes greater, but the compression ratio becomes smaller. In both cases obtained codes are compact, and in the second case code is even perfect, but the compression ratio is small, because the source is not "suitable for compression". In fact, the source entropy in the second case is greater than in the first one and the maximal values of the compression ratio (defined in the previous problem) decreased from $\rho_{\text{max}} = 1.3791$ to $\rho_{\text{max}} = 1.1429$. Because of that, the compression ratio is smaller.

(c) In both cases the same code was used, with the same code word lengths. For an instantaneous code, the Kraft inequality should be satisfied, as in both cases

$$\sum_{i=1}^{4} 2^{-l_i} = 2^{-1} + 2^{-2} + 2 \times 2^{-3} = 1 \leq 1.$$

Kraft inequality is always satisfied for comma code (in fact with equality), it is obvious that the code can be constructed with the same code word lengths, but it is not instantaneous. For example, the code that has the code words '0', '10' '100' and '111' is not even uniquely decodable. Therefore, it should be verified whether some code word is the prefix of some other code word.

For the codes (a) and (b) it is not the case, and the codes are instantaneous.

Problem 3.3 Find the binary code for zero-memory source defined by:

s_i	s_1	s_2	s_3	s_4	s_5	s_6
$P(s_i)$	0.65	0.05	0.08	0.15	0.04	0.03

(a) Apply the Huffman procedure, find the efficiency. Whether the obtained code is compact?
(b) Draw the corresponding code tree. Whether the code tree satisfies the sibling property?

Solution

(a) *Huffman procedure* consists of the following steps:

1. The symbols are ordered according to the non increasing probabilities.
2. The reduction is carried out by combining two symbols having the smallest probabilities into the one symbol.
3. The obtained symbol (denoted in this example by an asterisk) has the probability equal to sum of the combined symbols probabilities.
4. The procedure is repeated (including the possible reordering of new symbols) until only two symbols remain. One is denoted e.g. by "0", and the second by "1".
5. Code word for a combined symbol is a prefix of the code words for symbols from which it was obtained. After this prefix, "0" and "1" are added.
6. The procedure is repeated until all original symbols obtain corresponding code words (third column in the Table 3.3).

Entropy, average code word length and efficiency are:

$$H(S) = \sum_{i=1}^{6} P(s_i) \, \text{ld} \, \frac{1}{P(s_i)} = 1.66 \, [\text{Sh/symb}],$$

$$L = \sum_{i=1}^{6} P(s_i) l(s_i)$$
$$= 0.65 \times 1 + 0.05 \times 4 + 0.08 \times 2 + 0.15 \times 3 + 0.04 \times 5 + 0.03 \times 5$$
$$= 1.74 \, [\text{b/symb}],$$

Table 3.3 Huffman procedure

s_i	$P(s_i)$	x_i	s_i	$P(s_i)$	x_i	s_i	$P(s_i)$	x_i	s_i	$P(s_i)$	x_i	s_i	$P(s_i)$	x_i
s_1	0.65	0	s_1	0.65	0	s_1	0.65	0	s_1	0.65	0	s_1	0.65	0
s_4	0.15	11	s_4	0.15	11	s_4	0.15	11	$s_2s_3s_5s_6$	**0.20***	10	$s_2 s_3s_4s_5 s_6$	**0.35***	1
s_3	0.08	101	s_3	0.08	101	$s_2s_5s_6$	**0.12***	100	s_4	0.15	11			
s_2	0.05	1001	s_5s_6	**0.07***	1000	s_3	0.08	101						
s_5	**0.04**	10000	s_2	0.05	1001									
s_6	**0.03**	10001												

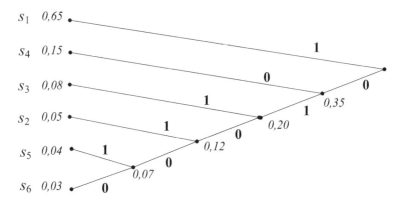

Fig. 3.8 Huffman procedure illustrated by tree

$$\eta = (H(S)/L) \times 100\% = 95.4\%,$$

while the obtained compression ratio is

$$\rho = \lceil \mathrm{ld}\ (q) \rceil / L = 1.8072.$$

Huffman procedure guarantees the obtaining of the *compact* code. The obtained code here is not perfect. However, for this source it is not possible to find an instantaneous code achieving smaller average code word length (higher efficiency).

(b) It is easier to follow the Huffman procedure using the code tree, as shown in Fig. 3.8. The tree is formed from the root (on the right side), two branches starting from it are denoted by different bits and the procedure is repeated going to the lower hierarchical levels. It is not important which branch is denoted by binary zero and which by binary one.

Here the result of Huffman procedure gives slightly different code words than in the previous case—1, 0101, 011, 00, 01001 and 01000. In Fig. 3.9 the ordered code tree is shown where the hierarchical levels can be easily noticed. It is obvious that this tree satisfies the sibling property, because the symbol probabilities, started from the left to the right and from the lowest hierarchical level up, do not decrease. One can verify easily that the same tree topology corresponds to the code from Table 3.3, although the code words differ. There are five nodes at the tree, from each one two branches are going out. Therefore, 32 equivalent compression codes can be formed having the same tree topology, and each one is instantaneous and compact.

Problem 3.4 Zero-memory source is defined by:

s_i	s_1	s_2	s_3	s_4	s_5	s_6	s_7
$P(s_i)$	0.39	0.21	0.13	0.08	0.07	0.07	0.05

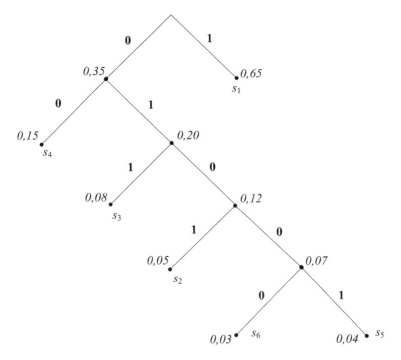

Fig. 3.9 Ordered code tree from Fig. 3.8

(a) Apply Shannon-Fano encoding procedure and find the average code word length.
(b) Whether the obtained code is compact? Whether the code is perfect?
(c) Whether the Kraft inequality is satisfied?

Solution

(a) It is chosen that in the first division symbols s_1 and s_2 form one group, the other symbols forming the other group (probability ratio is 0.6:0.4). The alternatives are s_1 in the first group (0.39:0.61) or s_1, s_2 and s_3 in the first group (0.73:0.27), but according to the Shannon-Fano procedure they are suboptimal solutions. The obtained groups are further subdivided into the subgroups by trying that they have as equal as possible probabilities. The procedure is given in Table 3.4.

Entropy and the average code word length are

Table 3.4 Shannon-Fano procedure

s_i	P (s_i)	Obtaining of the code word					Code word	l_i the code word length
s_1	0.39	0	0				00	2
s_2	0.21		1				01	2
s_3	0.13	1	0	0			100	3
s_4	0.08			1			101	3
s_5	0.07			1	0		110	3
s_6	0.07				1	0	1110	4
s_7	0.05					1	1111	4

$$H(S) = \sum_{i=1}^{7} P(s_i) \operatorname{ld} \frac{1}{P(s_i)} = 2.43 [\text{Sh/symb}],$$

$$L = \sum_{i=1}^{7} P(s_i) l_i = 2.52 [\text{b/symb}],$$

(b) Compactness of the code obtained by Shannon-Fano procedure will be verified by comparison to the code which is compact, obtained by Huffman procedure—Fig. 3.10.
The corresponding average code word length is

$$L = \sum_{i=1}^{7} P(s_i) l_i$$
$$= 0.39 \times 1 + 0.21 \times 3 + 0.13 \times 3 + 0.08 \times 4 + 0.07 \times 4 + 0.07 \times 4 + 0.05 \times 4$$
$$= 2.49 [\text{b/symb}]$$

and the obtained efficiency is greater—η = 97.59%.
The existence of code which has shorter average code length (higher efficiency), confirms that the above code obtained by Shannon-Fano procedure is not compact. It confirms as well that the application of Shannon-Fano procedure does not guarantee the obtaining of a compact code. Of course, non compact code surely is not perfect. It is recommended to the reader to verify whether by using the different subdivisions during Shannon-Fano procedure (when the probabilities in the subgroups are not balanced) the compact code can be obtained.
Compactness can be verified as well by inspection whether the corresponding code tree has the sibling property. The code tree is shown in Fig. 3.11 where the hierarchical levels can be easily seen. The sibling property is not fulfilled because the tree cannot be altered in such a way that the probabilities at the third hierarchical level (denoted by the dashed line) are increasingly ordered without the change of the tree topology. This tree does not correspond to the code obtained by Huffman procedure and the code cannot be compact.

Problems

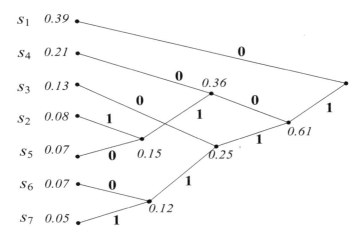

Fig. 3.10 The code tree obtained by Huffman procedure

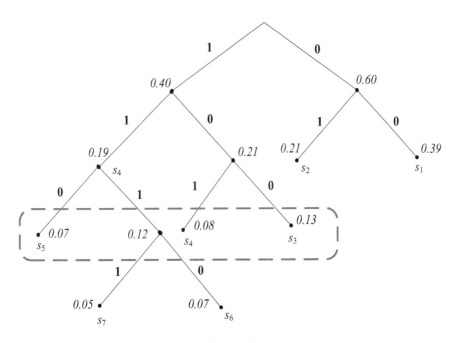

Fig. 3.11 Code tree corresponding to the Shannon-Fano procedure

(c) Kraft inequality is satisfied

$$\sum_{i=1}^{7} 2^{-l_i} = 2 \times 2^{-2} + 3 \times 2^{-3} + 2 \times 2^{-4} = 1 \leq 1,$$

what should be expected, because the Shannon-Fano procedure guarantees the instantaneous code. This relation is here satisfied with equality, and the code is complete.

Problem 3.5 Apply the Huffman procedure for a source defined by the table given below if the symbols obtained by reduction are always put at the last place in the group of equally probable symbols, either they are put arbitrary.

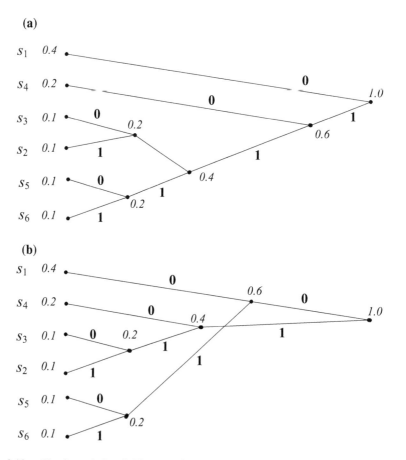

Fig. 3.12 a The first solution. **b** The second solution

Problems

Table 3.5 Huffman codes corresponding to code trees shown in Fig. 3.12a, b

s_i	$P(s_i)$	$x_i^{(1)}$	$l_i^{(1)}$	$x_i^{(2)}$	$l_i^{(2)}$
s_1	0.4	0	1	00	2
s_2	0.2	10	2	10	2
s_3	0.15	1100	4	110	3
s_4	0.1	1101	4	111	3
s_5	0.1	1110	4	010	3
s_6	0.05	1111	4	011	3

s_i	s_1	s_2	s_3	s_4	s_5	s_6
$P(s_i)$	0.4	0.2	0.1	0.1	0.1	0.1

(a) Draw at least two code trees corresponding to obtained codes that have different topologies. Whether the codes satisfy Kraft inequality? Whether these codes have the same average code word length? What are the differences?

(b) If the source emits the sequence $s_2, s_1, s_4, s_1, s_5, s_3, s_6$ and the channel errors occur at the first and eighth bit, find the decoded sequences in both cases.

Solution

In Fig. 3.12a, b two possible results of Huffman procedure use are shown.

The obtained code words are given in Table 3.5. For both cases the average code word length is the same—$L = 2.4$ [b/symb], but the **maximum code word length** is different ($max(l_i) = 5$ for the first code and $max(l_i) = 4$ for the second one) as well as the **sum of code word lengths** ($sum(l_i) = 20$ and $sum(l_i) = 18$).

Also, differ the **code words length variances** found according the formula

$$Var^{(n)} = \frac{1}{q}\sum_{i=1}^{q}(l_i^{(n)} - L)^2,$$

where n denotes the ordinal number of the solution (code). Corresponding numerical values are

$$Var^{(1)} = (1 - 2.4)^2 + (2 - 2.4)^2 + 4 \times (4 - 2.4)^2 = 12.36,$$

$$Var^{(2)} = 2 \times (2 - 2.4)^2 + 4 \times (3 - 2.4)^2 = 1.76.$$

For both topologies it is possible to form a number of different code words sets. There are five nodes in the tree, from each one two branches going out. Therefore 32 different compact codes can be formed for each tree topology. Codes corresponding to one topology have different code words from the second one, but their lengths are the same and their variances are the same.

(b) For the first code when seven symbols were sent, eight symbols were received, one bit error usually cause the error propagation to the next symbol as well.

s_2	s_1	s_4	s_1	s_5	s_3	s_6
1**0**	0	1101	**0**	1**1**10	1100	1111
0 0 0		1101		1101 0		1100 1111
$s_1\ s_1\ s_1$		s_4		$s_4\ \ \ s_1$	s_3	s_6

For the second code when seven symbols were sent, seven symbols were received, and error propagation is not noticeable.

s_2	s_1	s_4	s_1	s_5	s_3	s_6
1**0**	00	111	**0**0	**0**10	110	011
00	00	111	00	110	110	011
s_1	s_1	s_4	s_1	s_3	s_3	s_6

Although in the second case code words length variance is smaller (code words lengths are more uniform) it not resulted in reducing the problems due the bad encoder and decoder synchronization, usually caused by transmission errors. It is obvious that this effect arouses always when the code words have different lengths. In this case the number of symbols in error as well as the error structure depend on code words structure and the consequences generally cannot be easily predicted.

Problem 3.6 Zero-memory source is defined by

s_i	s_1	s_2	s_3	s_4	s_5	s_6	s_7	s_8	s_9	s_{10}
$P(s_i)$	0.18	0.2	0.14	0.15	0.1	0.11	0.05	0.04	0.02	0.01

Apply the Huffman procedure to obtain binary, ternary and quaternary code. Find the corresponding efficiencies.

Solution

The procedure of based m Huffman code obtaining consists of the following steps [1, 7]:

1. In the first step q_0 symbols are grouped where $2 \leq q_0 \leq m$ and q is the smallest integer satisfying

$$rem\left\{\frac{q - q_0}{m - 1}\right\} = 0.$$

where $rem\{\}$ is the reminder after the division
2. In every following step m symbols are grouped, until only one symbol remains.
3. When the tree is formed to every branch starting from the root one of the m-ary symbols is adjoined. All symbols that can be reached by the path using one branch will have the code word starting with the corresponding adjoined symbol.

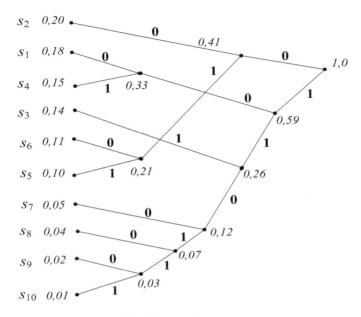

Fig. 3.13 Procedure to obtain Huffman binary code

4. The procedure is repeated by adding a new symbols corresponding to the branches starting from the nodes at the lower hierarchical levels, until the lowest level is reached (where the source symbols are put).

For the considered source the obtaining of binary code is shown in Fig. 3.13. In this case $q_0 = 2$, regardless of the number of source symbols.

For a binary code, the following is obtained

$$H(S) = \sum_{i=1}^{10} P(s_i) \operatorname{ld} \frac{1}{P(s_i)} = 2.981 \,[\mathrm{Sh/symb}], L^{(2)} = 3.02\,[\mathrm{b/symb}], \eta^{(2)} = 98.72\%.$$

The procedure for ternary code forming is shown in Fig. 3.14a. In this case $q = 10$ and as $q - q_0$ should be even, it is easy to find $q_0 = 2$. The corresponding code words are given in Table 3.6. The average code word length and efficiency are

$$L^{(3)} = \sum_{i=1}^{10} P(s_i) l_i^{(3)} = 1.95 \,[\text{ter. symb/symb}], \eta^{(3)} = \frac{H(S)}{\operatorname{ld}(3) L^{(3)}} \times 100\%$$
$$= 96.42\%.$$

The procedure for quaternary code forming is shown in Fig. 3.14b. In this case $m = 4$, $q_0 = 4$ yielding

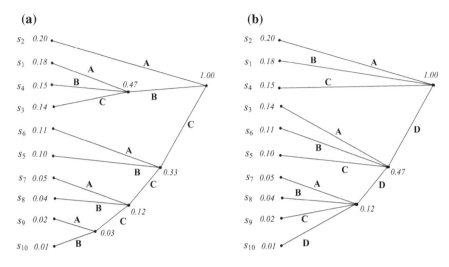

Fig. 3.14 Huffman procedure for three (**a**) and four (**b**) symbols of the code alphabet

$$L^{(4)} = \sum_{i=1}^{10} P(s_i) l_i^{(4)} = 1.59 [\text{quot. symb/symb}], \quad \eta^{(4)} = \frac{H(S)}{\text{ld}(4) L^{(4)}} \times 100\%$$
$$= 90.88\%.$$

Problem 3.7 The source emits two symbols according to the table:

s_i	s_1	s_2
$P(s_i)$	0.7	0.3

(a) By applying the Huffman procedure find the code for the source and for its second and third extension as well. Find the efficiency and the compression ratio for all codes and compare the results to the bounds given by the First Shannon theorem.

Table 3.6 Code words of binary, ternary and quaternary code and their lengths

s_i	s_1	s_2	s_3	s_4	s_5	s_6	s_7	s_8	s_9	s_{10}
$P(s_i)$	0.18	0.2	0.14	0.15	0.1	0.11	0.05	0.04	0.02	0.01
$x_i^{(2)}$	100	00	111	101	011	010	1100	11010	110110	110111
$l_i^{(2)}$	3	2	3	3	3	3	4	5	6	6
$x_i^{(3)}$	BA	A	BC	BB	CB	CA	CCA	CCB	CCCA	CCCB
$l_i^{(3)}$	2	1	2	2	2	2	3	3	4	4
$x_i^{(4)}$	B	A	DA	C	DC	DB	DDA	DDB	DDC	DDD
$l_i^{(4)}$	1	1	2	1	2	2	3	3	3	3

Problems

(b) Draw the efficiency as a function of the extension order (n) for $P(s_i) = 0.05$, $P(s_1) = 0.1$, $P(s_1) = 0.2$, $P(s_1) = 0.3$ and $P(s_1) = 0.4$ for the extension order $1 \leq n \leq 8$.

(c) Find the dependence for the maximal compression ratio on the probability $P(s_1)$.

Solution

(a) Encoding of the original source is trivial—symbol s_1 is encoded by binary zero and symbol s_2 by binary one. Entropy and average code word length are

$$H(S) = -0.7 \text{ ld } (0.7) - 0.3 \text{ ld } (0.3) = 0.8813 [\text{Sh/symb}],$$
$$L_1 = 0.7 \times 1 + 0.3 \times 1 = 1 [\text{b/symb}],$$

the efficiency is $\eta = 88.13\%$ and the compression ratio $\rho = 1/L_1 = 1$.

The encoding of the *second source extension* is given in Table 3.7. Firstly, the combined symbols are formed having the probabilities $P(s_i, s_j) = P(s_i)P(s_j)$, because it is zero-memory source, and after that, the binary Huffman procedure is used

From $H(S^2) = 2H(S) = 1.7626 [\text{Sh/symb}]$, after finding $L_2 = 1.81 [\text{b/symb}]$ the efficiency is increased at $\eta = 97.38\%$ and compression ratio is $\rho = 2/L_2 = 1.105$. For the third extension code words are shown in Table 3.8. Further, it is obtained $H(S^3) = 3H(S) = 2.6439 [\text{Sh/symb}]$, $L_3 = 2.726 [\text{b/symb}]$, $\eta = 96.99\%$ and $\rho = 3/L_3 = 1.1005$.

(b) It is obvious that the efficiency increases with the extension order. The described procedure is repeated for various values of $P(s_1)$, results are given in Table 3.9. The efficiency does not increase always monotonically with the extension order (n)! However, with further increase of n, efficiency will converge to 100%, satisfying the *First Shannon theorem*

$$\lim_{n \to \infty} \frac{H(S^n)}{L_n} = 1.$$

The dependence on the extension order is noticeable easier in Fig. 3.15. Of course, maximum compression ratio is obtained for $P(s_1) = P(s_2)$, it is equal to 100% for any extension order. However, in this case the maximum

Table 3.7 Huffman encoding for the second source extension

s_i	$P(\sigma_i)$	x_i	s_i	$P(s_i)$	x_i	s_i	$P(s_i)$	x_i
$\sigma_1 = s_1 s_1$	0.49	1	σ_1	0.49	1	$\sigma_2\sigma_3\sigma_4$	0.51*	0
$\sigma_2 = s_1 s_2$	0.21	01	$\sigma_3\sigma_4$	0.30*	00	σ_1	0.49	1
$\sigma_3 = s_2 s_1$	0.21	000	σ_2	0.21	01			
$\sigma_4 = s_2 s_2$	0.09	001						

Table 3.8 Huffman encoding for the third source extension

σ_i	$s_1 s_1 s_1$	$s_1 s_1 s_2$	$s_1 s_2 s_1$	$s_1 s_2 s_2$	$s_2 s_1 s_1$	$s_2 s_1 s_2$	$s_2 s_2 s_1$	$s_2 s_2 s_2$
$P(\sigma_i)$	0.343	0.147	0.147	0.063	0.147	0.063	0.063	0.027
x_i	00	010	011	1100	10	1101	1110	1111

Table 3.9 The efficiency as a function of extension order, for various probabilities $P(s_1)$

n	1	2	3	4	5	6	7	8
$P(s_1) = 0.05$	28.64	49.92	66.10	78.00	85.19	90.56	94.29	95.86
$P(s_1) = 0.1$	46.90	72.71	88.05	95.22	97.67	99.75	98.87	98.57
$P(s_1) = 0.2$	72.19	92.55	99.17	97.45	97.83	99.54	98.66	98.59
$P(s_1) = 0.3$	88.13	97.38	96.99	98.82	99.13	99.22	99.64	99.48
$P(s_1) = 0.4$	97.10	97.10	98.94	98.96	99.31	99.45	99.55	99.64

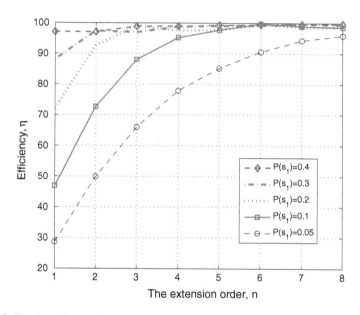

Fig. 3.15 The dependence of efficiency on the extension order

compression ratio does not mean the highest compression ratio. It will be additionally explained in the next part of solution.

(c) The extensions are obtained from the original source and the maximal compression ratio does not depend on the extension because the compression potential does not depend on the extension order

$$\rho_{max}^{(n)} = \frac{\lceil \mathrm{ld}\,(2^n) \rceil}{H(S^n)} = \frac{n \lceil \mathrm{ld}\,(2) \rceil}{nH(S)} = \frac{1}{H(S)} = \rho_{max}^{(1)}.$$

while the obtained compression ratio of binary source ($\mathrm{ld}(q) = 1$) is $\rho^{(n)} = n/L_n$, and on the basis of First Shannon theorem with the increasing of extension order it attains the maximum use of the source compression potential as well

$$\lim_{n \to \infty} \frac{1}{\rho^{(n)}} = \lim_{n \to \infty} \frac{L}{n} = H(S).$$

The dependence of maximum compression ratio on the binary source symbol probability is shown in Fig. 3.16. It is obvious that for the equiprobable symbols the compression ratio is $\rho = 1$, regardless the extension order, while in case when $P(s_1) \gg P(s_2)$ or $P(s_2) \gg P(s_1)$, the obtained compression ratio can be significantly high.

Problem 3.8 First-order memory binary source is defined by the transition matrix

$$\Pi = \begin{bmatrix} 1-p & p \\ p & 1-p \end{bmatrix}.$$

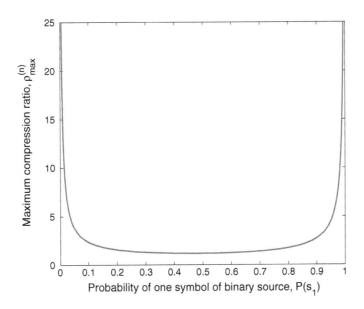

Fig. 3.16 The dependence of maximum compression ratio on the binary source symbol probability

(a) Find the corresponding binary Huffman codes for a source as well as for its second and third extension if $p = 0.01$.
(b) Find the minimum value of the average code word length for the nth extension. Draw the efficiency and compression ratio dependence on the extension order for $p = 0.1$, $p = 0.01$ and $p = 0.001$.

Solution

(a) In solution of Problem 2.6 it was shown that the corresponding stationary binary symbols probabilities are $P(0) = P(1) = 0.5$. Huffman code for such a source is trivial—symbol '0' is coded by '0' and '1' by '1'. Obviously, the average code word length is $L_1 = H(\bar{S}) = 1$[b/symb]. The entropy is

$$H(S) = -0.01 \, \text{ld}(0.01) - 0.99 \, \text{ld}(0.99) = 0.0808 [\text{Sh/symb}],$$

efficiency and compression ratio are

$$\eta = 8.08\%, \quad \rho = 1.$$

Stationary probabilities of combined symbols corresponding to the second extension and the corresponding code words are shown in Table 3.10.

It is easy to verify the following

$$H(S^2) = 2H(S) = 0.1616 [\text{Sh/symb}],$$

$$H(\bar{S^2}) = \sum_{i=1}^{4} P(\sigma_i) \, \text{ld} \, (1/P(\sigma_i)) = 1.0808 \, [\text{Sh/symb}],$$

while the average code word length is

$$L_2 = \sum_{i=1}^{4} P(\sigma_i) l_i = 1.515 [\text{b/symb}],$$

the efficiency is $\eta = 10.67\%$, and the compression ratio $\rho = 2/L_2 = 1.32$. The relation $H(\bar{S^2}) < L$ is satisfied as well.

Table 3.10 Stationary probabilities of the second extension symbols and the corresponding code words

Combined symbols	Probabilities	Code words	l_i
$\sigma_1 = $ '00'	$P(00) = P(0)P(0/0) = 0.495$	0	1
$\sigma_2 = $ '01'	$P(01) = P(0)P(1/0) = 0.005$	110	3
$\sigma_3 = $ '10'	$P(10) = P(1)P(0/1) = 0.005$	111	3
$\sigma_4 = $ '11'	$P(11) = P(1)P(1/1) = 0.495$	10	2

Problems

The probabilities of the third extension combined symbols and the corresponding code words are given in Table 3.11. In this case, it is easy to verify

$$H(S^3) = 3H(S) = 0.2424 \text{ [Sh/symb]},$$

$$H(\overline{S^2}) = \sum_{i=1}^{4} P(\sigma_i)\text{ld}(1/P(\sigma_i)) = 1.1616\text{[Sh/symb]},$$

and the average code word length is

$$L_3 = \sum_{i=1}^{8} P(\sigma_i)l_i = 1.549\text{[b/symb]},$$

yielding the efficiency $\eta = 15.65\%$, and the compression ratio $\rho = 1.94$. It is interesting to note that the relation $H(\overline{S^2}) < L$ in this case is satisfied as well.

(b) The Huffman procedure takes into account only the stationary probabilities of symbols emitted by the extended source and the average code word length cannot be smaller than the entropy of the source adjoined to that extension, i.e. the following must hold

$$L_n \geq H(\overline{S^n}).$$

Table 3.11 Stationary probabilities of the third extension symbols and the corresponding code words

Combined symbols	Conditional probabilities	Probabilities	Code words	l_i
$\sigma_1 = $ '000'	$P(00/0) = P(0/0)P(0/0) = 0.9801$	$P(000) = P(0)P(00/0) = 4.9005 \times 10^{-1}$	0	1
$\sigma_2 = $ '001'	$P(01/0) = P(1/0)P(0/0) = 0.0099$	$P(001) = P(0)P(01/0) = 4.95 \times 10^{-3}$	1111	4
$\sigma_3 = $ '010'	$P(10/0) = P(0/1)P(1/0) = 0.0001$	$P(010) = P(0)P(10/0) = 5 \times 10^{-5}$	111001	6
$\sigma_4 = $ '011'	$P(11/0) = P(1/1)P(1/0) = 0.0099$	$P(011) = P(0)P(11/0) = 4.95 \times 10^{-3}$	1100	4
$\sigma_5 = $ '100'	$P(00/1) = P(0/0)P(0/1) = 0.0099$	$P(100) = P(1)P(00/1) = 4.95 \times 10^{-3}$	1101	4
$\sigma_6 = $ '101'	$P(01/1) = P(1/0)P(0/1) = 0.0001$	$P(101) = P(1)P(01/1) = 5 \times 10^{-5}$	111000	6
$\sigma_7 = $ '110'	$P(10/1) = P(0/1)P(1/1) = 0.0099$	$P(110) = P(1)P(10/1) = 4.95 \times 10^{-3}$	11101	5
$\sigma_8 = $ '111'	$P(11/1) = P(1/1)P(1/1) = 0.9801$	$P(111) = P(1)P(11/1) = 4.9005 \times 10^{-1}$	10	2

Further, the *effective efficiency of Huffman code for sources with memory* can be defined

$$\eta_{ef}^{(n)} = \frac{H(\overline{S^n})}{L_n} \times 100\%,$$

and for considered case it is

$$\eta_{ef}^{(1)} = 100\%, \quad \eta_{ef}^{(2)} = 71.34\%, \quad \eta_{ef}^{(2)} = 74.99\%.$$

It is clear that with the increasing of the extension order, the effective efficiency rapidly converge to the its maximum value even when there is a significant difference between the symbol probabilities, what is here the case. More critically is how the *maximum possible efficiency* and the *maximum possible compression degree* change. The correspond formulas are

$$\eta_{\max}^{(n)} = \frac{H(S^n)}{H(\overline{S^n})} \times 100\%, \quad \rho_{\max}^{(n)} = \frac{\lceil \log_2(q^n) \rceil}{H(\overline{S^n})} \times 100\%.$$

Using the known relations for entropy of nth extension and the source adjoined to it, as well as the fact that the original source is binary, the following expressions are obtained (and the dependences shown in Figs. 3.17 and 3.18)

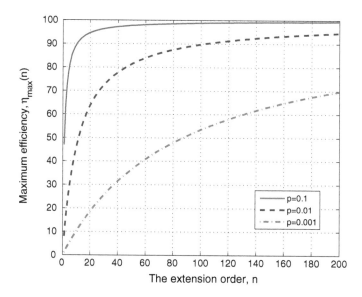

Fig. 3.17 The maximum efficiency that can be achieved by Huffman encoding, first-order memory source, p parameter

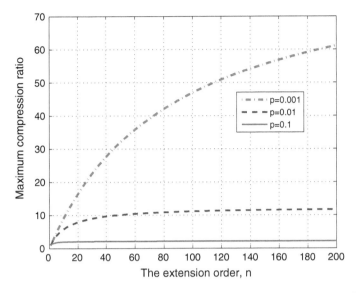

Fig. 3.18 The maximum compression ratio that can be achieved by Huffman encoding, first-order memory source, p parameter

$$\eta_{max}^{(n)} = \frac{nH(S)}{(n-1)H(S)+H(\overline{S})} \times 100\%, \quad \rho_{max}^{(n)} = \frac{n}{(n-1)H(S)+H(\overline{S})}.$$

For the higher p values maximum code efficiency can be achieved with a small extension order, but the corresponding maximum value of compression ratio is significantly smaller. Sources with memory where parameter p has small values have a great compression potential. For $p = 0.001$, the 50th extension provides compression ratio of the order 30. This means that 30 equiprobable bits emitted by the source can be, in average, represented by one bit only at the encoder output. However, in this case the Huffman procedure should be applied for 2^{50} symbols of source alphabet!

Problem 3.9 Discrete source for which the number of symbols and its probabilities are not known in advance, emits the sequence

$$aabcccbd\ldots$$

(a) Apply binary adaptive Huffman encoding by using FGK algorithm.
(b) Explain in details the decoding procedure if its input is generated by the encoder previously considered.
(c) Repeat the procedure from (a) if instead of FGK algorithm, Vitter algorithm is applied.

Solution

In *adaptive Huffman algorithm* encoder and decoder use dynamic change of code tree structure. The procedure of tree rearranging after the reception of every new symbol at the encoder input depends on the applied variant of the adaptive Huffman algorithm, but the code tree in every moment must have the sibling property. The algorithm is based on the fact that the binary code is a Huffman code if and only if the code tree has a sibling property. It practically means that in any moment (for any length of the incoming sequence) the code tree obtained by applying the adaptive Huffman algorithm could be obtained and by applying the static Huffman algorithm as well. Therefore, the code is compact always, although the code tree structure changes in time.

(a) The basis of *FGK algorithm* was independently found by *Newton Faller* (1973) [13] and *Robert Gallager* (1978) [14], while *Donald Knuth* (1985) [15] added improvements into the original algorithm, and the name of the algorithm is derived from the family initials of the authors.

The procedure of the code forming for the input sequence *aabcccbd* is exposed in what follows

1. At the beginning the Huffman tree is degenerated and consists of the zero node only (knot **0**) as shown in Fig. 3.19a. At the encoder input is symbol *a* and it is easy to verify that this symbol is not on the tree. Therefore, the encoder emits code word corresponding to the knot *0* (being 0 as well), emitting simultaneously the corresponding ASCII code for the symbol—ASCII(*a*). As shown in Fig. 3.19b, now a new tree is formed by division of the node *0* into two nodes—*0* and *a*. To node *0* the code word 0 is adjoined and the number of previous appearances of this symbol corresponds to its weight being always w(*0*) = 0 (this symbol does not appear at the input sequence, but it is added artificially, for the easier algorithm realization). To the node corresponding to symbol *a* code word 1 is adjoined and the corresponding node weight is w(*a*) = 1 (in the figures, the weight is always in the rectangle, near the node).

2. Once more into encoder enters *a*, already existing on the tree. The encoder emits code corresponding to the knot *a*—it is now the code word 1. Tree structure does not change because the entering element already exists on the tree. Because two symbols *a* have been entered, the node weight

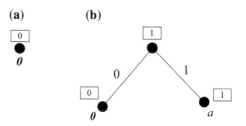

Fig. 3.19 Tree structure at the beginning (**a**) and after the entering of the symbol *a* (**b**)

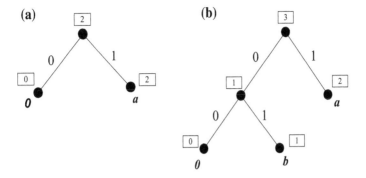

Fig. 3.20 Tree structure after entering of the second (**a**) and of the third (**b**) symbol

increments, now it is $w(a) = 2$, and to this node at the new tree code 1 is adjoined, as shown in Fig. 3.20a.

3. Into encoder enters **b**, not existing yet at the tree. The encoder emits the code corresponding to the node **0** (the code being 0) and after that ASCII (**b**). New tree is formed by separating the node **0** into two nodes—**0** and **b**, as shown in Fig. 3.20b. To the node **0** the code 00 is adjoined with the weight $w(\mathbf{0}) = 0$ while the code 01 is adjoined to the node **b** with the weight $w(\mathbf{b}) = 1$.

4. Now, into encoder enters **c**, not existing yet at the tree. The encoder emits the code corresponding to the node **0**, now being 00 and after that ASCII (**c**). New tree is formed by separating the node **0** into two nodes—**0** and **c**. To the node **0** code 000 is adjoined (it is obvious that the corresponding code for this node will be all-zeros sequence, always with the weight $w(\mathbf{0}) = 0$!), and to the node **b** the code 001 is adjoined, with the weight $w(\mathbf{b}) = 1$, as shown in Fig. 3.21a.

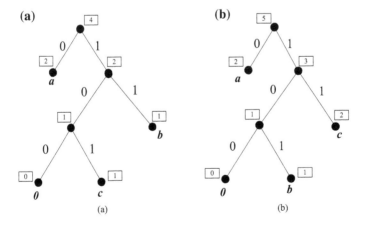

Fig. 3.21 Tree structure after entering of the fourth (**a**) and of the fifth (**b**) symbol

5. Into encoder now enters *c*, already existing on the tree. The encoder emits code corresponding to the node *c* being now 001. The weight of node corresponding to symbol *c* is incremented and a new tree does not have the sibling property. Because of that, symbols *b* and *c* change the places and a new tree is obtained (the tree topology does not change). The rule for this reordering is that the node whose weight increased by incrementation ($w(c) = 2$) change the place with the node which has a smaller weight than it, and from all the nodes that have this property, the one is chosen which is at the lowest hierarchical level. Just because of that, the node *c* goes up for one level towards the root. It should be noted that this reordering is logical, because the same result would be obtained for static Huffman algorithm with the same symbol probabilities (i.e. $P(a) = P(c) = 0.4$ and $P(b) = 0.2$), with the exception that here formally exists zero node always being at the lowest hierarchical level. New tree is shown in Fig. 3.21b (the first level is drawn in a slightly different way to emphasize that the sibling property is satisfied).
6. Now once more enters the symbol *c*. At the encoder output the code corresponding to the node *c* is emitted being now 01. Because the weight incrementation yields now the weight $w(c) = 3$, the node *c* change the place with the node *a* going up one level. New tree is shown in Fig. 3.22a.
7. Into encoder now enters once more the symbol *b*. At the encoder output the code corresponding to the knot *b* is emitted which is now 001. Because totally two symbols *b* have been entered so far, the weight increments is now $w(b) = 2$, and to this node at a new tree code 001 is adjoined, as shown in Fig. 3.22b. Tree structure in this step does not change, because the tree with a new weights satisfies the sibling property.
8. Into encoder enters *d*, not existing yet at the tree. The encoder emits the code corresponding to the knot *0*, now being 000 and after that ASCII(*d*). New tree is formed by separating the node *0* into two nodes—*0* and *d*. It is obvious that the weights observed from left to right starting at the lowest

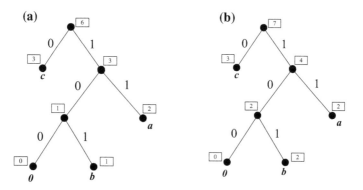

Fig. 3.22 Tree structure after entering of the sixth (**a**) and of the seventh (**b**) symbol

Fig. 3.23 Tree structure after entering of the eighth symbol

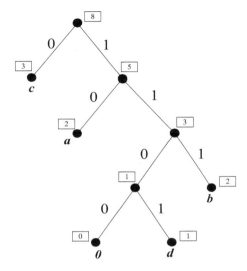

level to the upper levels do not decrease. The tree satisfies the sibling property and further reordering is not needed. Final tree structure for a given sequence is shown in Fig. 3.23.

The emitted bit sequence for the input sequence ***abcccbd*** is

0, ASCII(*a*); 1; 0, ASCII(*b*); 00, ASCII(*c*); 001; 01; 001; 000, ASCII(*d*).

where with the sign ",''the symbols emitted for one symbol at the input are separated, while ";" corresponds to the adjacent input symbols. It should be noted that the same entering sequence was encoded using the Huffman static code corresponding to the tree from Fig. 3.18 (when there is no zero node, and the symbol *d* is represented by 000) is

01; 001; 1; 1; 1; 001; 000,

while for the correct decoding, the decoder must receive in advance ASCII(*a*), ASCII(*b*), ASCII(*c*), ASCII(*d*), enabling it to "know" the correspondence between the symbols and the code words.

Therefore, for FGK algorithm 16 bits and four ASCII codes for letters are transmitted in real time, as the symbols enter into the encoder. For a static Huffman algorithm, 14 bits are transmitted, but four ASCII codes for letters had to be transmitted in advance. The reader should try to explain why, in this case, the static Huffman algorithm yields better results (whether only due to the absence of the zero knot?). Of course, great advantage of adaptive Huffman algorithm is that for the encoder functioning, the input symbol probabilities should not be known in advance.

(b) If the encoder output is connected directly to the decoder input, error-free transmission us supposed. In this case the decoder output should be found, if its input is

0, ASCII(*a*); 1; 0, ASCII(*b*); 00, ASCII(*c*); 001; 01; 001; 000, ASCII(*d*).

The decoding procedure will be described step by step as well:

1. At the beginning, in the decoder is only the degenerated Huffman tree consisting only of the zero node (node *0*), as shown in Fig. 3.19a. Into decoder enters the code word 0 accompanied by ASCII(*a*) and the decoder finds it—it corresponds to the node *0*. Every time when decoder comes to this node, it separate it into two new nodes adjoining to one code word adding the suffix 0 to the existing code for zero node (it is a new zero node, the corresponding code word is 00) and to the other node adjoins the symbol whose ASCII code was received, adding at the existing code word suffix 1 (the symbol is *a*, new code word is 01). The node weight corresponding to the new symbol is incremented and the tree is shown in Fig. 3.19b. The sibling property is satisfied and the next step begins.
2. In the second step the received sequence is 1. At the previously formed tree code 1 corresponds to the symbol *a*, corresponding weight is incremented becoming $w(a) = 2$. The tree is shown in Fig. 3.20a and satisfies the sibling property.
3. In the next step the sequence 0, ASCII(*b*) is received. At the previously formed tree code 0 corresponds to the symbol *0*, the procedure of adding a new symbol starts (node *0* is separated into *0* and *b*) having the weight $w(b) = 1$ and a code word 01. The tree is shown in Fig. 3.20b has the sibling property and the further reordering is not needed.
4. In the fourth step the received sequence is 00, ASCII(*c*). At the previous tree to the bit 0 corresponds the node without adjoined final symbol (more code words start with that bit). For this reason one more bit from the input is taken into account and 00 corresponds to symbol *0*, resulting in the further adding of a new symbol (node *0* is separated into *0* and *c*), with the weight $w(c) = 1$ with the corresponding code word 001. Now, the tree (shown in Fig. 3.21a) has a sibling property and the next step can begin.
5. Sequence at the decoder input in the fifth step is 001. At the previously formed tree 0 and 00 do not define any symbol node while 001 corresponds to the symbol *c* and the corresponding weight is incremented—$w(c) = 2$. The obtained tree does not have a sibling property and symbols *b* and *c* change the places. Due to the using the same procedure for tree reordering, the same result is obtained, shown in Fig. 3.21b.
6. The input sequence in the sixth step is 01. To bit 0 no one symbol node corresponds, while 01 corresponds to symbol *c* and the corresponding weight is incremented—$w(c) = 3$. The obtained tree does not satisfy the

sibling property and symbols **a** and **c** change the places giving the tree shown in Fig. 3.22a.

7. In this step the received sequence is 001. The first and the second bit are not sufficient for symbol location and the third one (the complete sequence) points to the symbol **b** whose weight is incremented to $w(b) = 2$. The tree satisfies the sibling property and does not change the structure, as shown in Fig. 3.22b.

8. In this step the received sequence is 000, ASCII(**d**). The code 000 points to symbol **0**, resulting in the adding of a new symbol (node **0** is separated into **0** and **d**), having the weight $w(d) = 1$ with the corresponding code word 0001. The tree is shown in Fig. 3.23, satisfies the sibling property and further reordering is not needed.

(c) The second variant of adaptive Huffman algorithm was proposed by **Jeffrey Vitter** (1987) [16]. The basic modification is that during the tree reordering symbols nodes are put at as low as possible hierarchical levels. If the two nodes have the same weight, being at the different levels, then during reordering the goal is that a node corresponding to the end symbol has to be put on the lower level. In this procedure the numeration of nodes is introduced from the left to the right and from the lowest level up (underlined numbers in Fig. 3.24). It is not allowed that from the two notes having the same weight, one corresponding to the one symbol and the other being combined, the symbol node has the upper ordinal number. If it happens, these two nodes should change the positions at the tree (the combined node "carries" its successors as well).

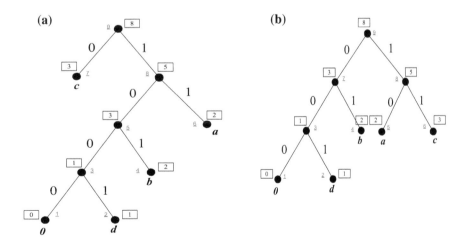

Fig. 3.24 Code trees obtained by FGK procedure (**a**) and by Vitter procedure (**b**)

For the input sequence in this problem (***aabcccbd***) the obtained trees are identical for the first seven steps as for FGK procedure, because the above condition is satisfied. However, the tree for the eighth step differs, as shown in Fig. 3.24. It is obvious that for tree obtained by FGK procedure (Fig. 3.24a) symbol node (No. 6) and the combined node (No. 7) have the same weight, but, the symbol node is at the lower level. If these nodes change the places (symbol No. 7 successors), the tree shown in Fig. 3.24b is obtained as a result of ***Vitter procedure***.

In the previous problems it was mentioned that the number of levels in a code tree corresponds to the maximal code word length, denoted by $max(l_i)$, having here the value 4 for FGK and 3 for Vitter procedure. Sum of the lengths of all code words $sum(l_i)$ is here 14 for FGK and 12 for Vitter procedure. Therefore, the Vitter procedure provides smaller values for both numerical characteristics comparing to FGK. Vitter procedure can easily be expanded for the case when both nodes to be changed are combined and to the higher level (nearer to the root) the node which has a greater value for $max(l_i)$ either $sum(l_i)$ should be put. In such way, the code words lengths are equalized (the variance defined in Problem 3.5 is smaller) and the time for which code word lengths attain the maximum value defined by the corresponding format and by the implementation is longer. If 16 bits are used for the word lengths, for the node weights greater than 4095 the exceeding arises. This problem for the weights is easily solved (when the root weight becomes 4095, the symbol node weights are divided by two) but the problem of maximum code word length stands for very great trees and the Vitter procedure has an important application.

Problem 3.10 The sequence from the binary symmetric source with memory enters into Lempel-Ziv encoder which forms dictionary of length $N = 8$.

(a) For the following relation of source parameters $P(0/1) \gg P(0/0)$ and for the emitted sequence

$$0101010101010101001011010010101010101011010\ldots$$

explain the principle of dictionary forming and find the achieved compression ratio. What are minimum and maximum values for the compression ratio in this case?

(b) If the errors occur at the first and at the tenth bit of encoded sequence, find the decoded sequence.

(c) Repeat the previous for the case $P(0/1) \ll P(0/0)$ if the sequence emitted by source is

$$000000000000000000000000000000011111111\ldots$$

(d) Find the compression ratio for the cases $P(0/1) = 1$ and $P(0/1) = 0$, when dictionary length is N and the sequence is very long. How to estimate the dependence of compression ratio on the dictionary length for single-order memory binary source?

Problems

Table 3.12 Dictionary forming for the sequence from (a)

Program in pseudo language	Dictionary		w	k	wk	?	out
	Address	Contents					
w=nil;	0	0					
loop	1	1	Nil	0	0	+	/
read k;			0	1	01	−	0
if wk in vocabulary	2	01	1	0	10	−	1
w=wk;	3	10	0	1	01	+	/
else			01	0	010	−	2
code of w→out	4	010	0	1	01	+	/
wk→ table of strings			01	0	010	+	/
w=k;			010	1	0101	−	4
end;	5	0101	1	0	10	+	/
end loop;			10	1	101	−	3
	6	101	1	0	10	+	/
			10	1	101	+	/
			101	0	1010	−	6
	7	1010					

Solution

(a) Dictionary forming for **Lempel-Ziv code** [17, 18] including **Welch modification** [19] can be described by program in pseudo language, given in left column in the Table 3.12. The encoder functioning can be described by the entering and the outgoing sequences

$$\underline{010101010101}\ 0101\ 0101\ 1010\ 0101\ 0101\ 0101\ 1010$$
$$\rightarrow \underline{012436}\ 5\ 5\ 7\ 5\ 5\ 5\ 7$$

where the underlined part corresponds to the bits at the input, used for dictionary forming, i.e. to the addresses emitted at the encoder output during the dictionary forming.

Every address is denoted by three bits and the obtained *compression ratio* equals to the quotient of the entering and outgoing bits, i.e.

$$\rho = \frac{N_{ul}}{N_{izl}} = \frac{13+28}{18+21} = 1.0513,$$

where the first addends in numerator and the denominator correspond to the number of bits at the input and the output during the dictionary forming (the transient regime), while the second addends describe the coder functioning in the stationary regime.

Compression ratio is by rule minimum after the dictionary forming

$$\rho_{min} = \frac{13}{18} = 0.7222,$$

while the maximum value is achieved if the sequence for encoding is very long and the transient regime can be neglected, and the sequence has a property that a maximum numbers of symbols (four in this case) from the input, can be represented by one address (three binary symbols) yielding

$$\rho_{max} = \frac{4}{3} = 1.3333.$$

In this case the decoder, on the base of known dictionary and the decoder input (denoted by *in*) estimate the sequence entering its input, denoted by k' (as given in Table 3.12). From decoded sequence, the following is obtained

$$out = 0\ 1\ 2\ 4\ 3\ 6\ 5\ 5\ 7\ 5\ 5\ 5\ 7$$
$$= 000\ 001\ 010\ 100\ 011\ 110\ 101\ 101\ 111\ 101\ 101\ 101\ 111.$$

If the errors occurred at the first and the tenth bit (Fig. 3.25), the received sequence is

$$in = \underline{1}00\ 001\ 010\ \underline{0}00\ 011\ 110\ 101\ 101\ 111\ 101\ 101\ 101\ 111$$
$$= \underline{4}\ 1\ 2\ \underline{0}\ 3\ 6\ 5\ 5\ 7\ 5\ 5\ 5\ 7.$$

Although the error propagation in the decoded sequence does not exist, it is possible that in the decoded sequence occurs smaller or greater number of bits because to the various addresses correspond to the code words having the various length, and the estimation of the encoded sequence (from encoder input) is

010 1 01 **0** 10 101 0101 0101 1010 0101 0101 0101 0101

of course, differing from the original sequence

0 1 01 **010** 10 101 0101 0101 1010 0101 0101 0101 1010

and in this case the total number of bits remained unchanged.

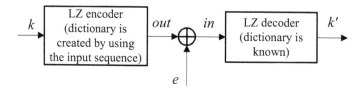

Fig. 3.25 LZ encoder and decoder

Table 3.13 Dictionary forming for the sequence from b

Dictionary Address	Contents	W	k	wk	?	out
0	0					
1	1	Nil	0	0	+	/
		0	0	00	−	0
2	00	0	0	00	+	/
		00	0	000	−	2
3	000	0	0	00	+	/
		00	0	000	+	/
		000	0	0000	−	3
4	0000	0	0	00	+	/
		00	0	000	+	/
		000	0	0000	+	/
		0000	0	00000	−	4
5	00000	0	0	00	+	/
		00	0	000	+	/
		000	0	0000	+	/
		0000	0	00000	+	/
		00000	0	000000	−	5
6	000000	0	0	00		/
		00	0	000		/
		000	0	0000		/
		0000	0	00000		/
		00000	0	000000		/
		000000	0	0000000		6
7	0000000					

(b) For a new sequence, the dictionary forming is given in Table 3.13 and the input and outgoing sequences are

$$\underline{0\ 00\ 000\ 0000\ 00000\ 000000\ 0000000}\ 00000\ 11111111\ldots \rightarrow \underline{023456}\ 5\ ????$$

All symbol groups inscribed in dictionary are the combinations of binary zeros only. The problem arouses when in the stationary regime appears the first binary one—coder in this case cannot find the corresponding word and the encoding becomes impossible.

This example illustrates the fact that the sequence used for dictionary forming has to be representative. It is desirable that during the transient regime the sequence at the coder input has the same statistical properties as the rest of the sequence. For stationary signals it is always possible although the duration of the transient regime can be relatively long, and this duration determines the needed dictionary dimension. For practical purpose is (e.g. for compression of the text corresponding to the speaking language) usually the dictionary dimension of 1024 or 2048 addresses is chosen (every symbol is represented by 10 or 11 bits).

(c) If $P(0/1) = 1$ and if the source is symmetric then $P(1/0) = 1$ and combinations 00 and 01 will never appear in the input encoder sequence. It is obvious that in this case only two sequences of length three (010 and 101), two sequences of length four (1010 and 0101) and two sequences of any length n will appear. Therefore, in the dictionary with $N = 2^n$ positions only the pairs of sequences having the lengths $m = 1, 2, \ldots, N/2$ will be kept.

In the stationary regime only the sequences of the length $N/2$ will be separated and each one is encoded using $n = \text{ld}(N)$ bits, sufficiently for a binary record of every decimal address. For this optimal case the compression ratio is

$$\rho_{opt}^{(1)}(N) = \frac{N}{2 \text{ ld } (N)}.$$

For the case of binary symmetric source with memory where $P(0/1)=P(1/0)=0$, the sequence consists of zeroes only either of ones only (depending on the starting state) and the compression ratio is

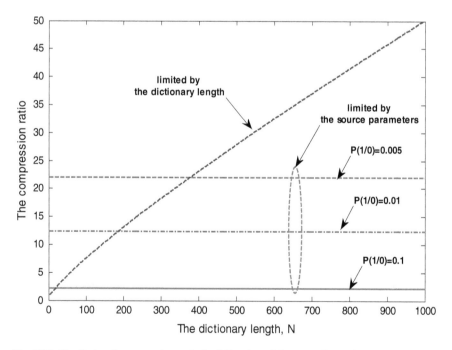

Fig. 3.26 The limits of compression ratio for LZ code and for some first-order memory sources

$$\rho_{opt}^{(2)}(N) = \frac{N}{\mathrm{ld}\,(N)}.$$

Of course, the expressions are valid if the length of the sequence is sufficiently long and the duration of the transition period can be neglected. The dependence of compression ratio on the number of bits representing the address in the dictionary is shown in Fig. 3.26.

It is clear the with the increase of dictionary length, the compression ratio tends to infinity. It is understandable, because the entropy in this case equals zero, and for such binary source

$$\rho_{max} = \frac{1}{H(S)} \to \infty.$$

Of course, if the sequence is not a deterministic one, but has a long memory, the compression ratio must satisfy the condition

$$\lim_{N \to \infty} (\rho(N)) \leq \frac{\lceil \mathrm{ld}\,(q) \rceil}{H(S)}.$$

For $P(0/1) = P(1/0)$, the entropy of first-order memory symmetric source is

$$H(S) = -P(1/0)\mathrm{ld}(P(1/0)) - (1 - P(1/0))\mathrm{ld}(1 - P(1/0)),$$

yielding

$$\lim_{N \to \infty} (\rho(N)) = \rho_{max} = \frac{-1}{P(1/0)\mathrm{ld}(P(1/0)) + (1 - P(1/0))\mathrm{ld}(1 - P(1/0))}$$

and the corresponding upper bound of compression ratio is shown in Fig. 3.26 by dashed line, for various values of $P(1/0)$.

It should be as well noticed that the lines are the limiting cases only, and that the real compression ratio can substantially differ from these values, but it must be below them

$$\rho(N) < \min\left\{\frac{N}{2\mathrm{ld}(N)}, \frac{-1}{P(1/0)ld(P(1/0)) + (1 - P(1/0))ld(1 - P(1/0))}\right\}.$$

The deviation will be greater as the memory is shorter, but some useful conclusion can be drawn as well from the diagram. The compression ratio for a first-order memory binary source where $P(1/0) = 0.01$ in any case cannot be greater than 12.38, but using LZ coder having dictionary length 128 bits (seven-bits addresses) even in the ideal case (with an infinite source extension) cannot achieve the compression ratio greater than 9.14, because the address length is a characteristic of the encoder itself.

One more example can be considered. It is known that the printed English text can be approximately described as a source with memory of the 30th order, emitting 26 letters. The entropy of this source is estimated as $H(S) = 1.30$ [Sh/symb], while $\mathrm{ld}(q) = 4.70$ yielding $\lceil \mathrm{ld}(q) \rceil = 5$ and a maximum compression ratio is approximately $\rho_{max} = 3.85$. From the figure one could conclude that for the corresponding compression a dictionary should have a few tens of bits. It is not the case and the obtained compression ratio for LZ coder increases substantially slower than it is shown in Fig. 3.26, and the needed dictionary length is at least $N = 1024$.

Chapter 4
Information Channels

Brief Theoretical Overview

The information channels, as well as the sources, generally are ***discrete*** or ***continuous***. Physical communication channel is, by its nature, a *continuous* one. Using such an approach the noise and other interferences (other signals, fading etc.) can be directly taken into account. However, it can be simplified introducing the notion of a *discrete* channel (Fig. 1.1), incorporating the signal generation (modulation), continuous channel and signal detection (demodulation). Some channel can be of a *mixed* type. The input can be discrete, as for the case of digital transmission, while at the receiver, the decision is made on the basis of continuous amplitude range, i.e. the output is continuous.

The discrete channel is described by the ***input alphabet*** $(x_i \in X\{x_1, x_2, \ldots, x_i, \ldots, x_r\})$, the ***output alphabet*** $(y_j \in Y\{y_1, y_2, \ldots, y_j, \ldots, y_s\})$ and a set of conditional probabilities $P(y_j/x_i)$, according to Fig. 4.1. These sets are finite (generally, they may be countable as well). This channel is a ***zero-memory*** (***memoryless***) channel, because the conditional probabilities depend on the current symbol only. A good example for such a channel is that one where there is no intersymbol interference and where only the white Gaussian noise is present. In the channel with intersymbol interference, the probability of the received symbols depends as well and on the adjacent symbols and in such way a *memory* is introduced into the channel. Therefore, here $P(y_j/x_i)$ is a conditional probability that at the channel output symbol y_j will appear, if the symbol x_i is sent (emitted). These probabilities can be arranged as the ***transition (channel) matrix*** completely describing the channel (Problems 4.1–4.3).

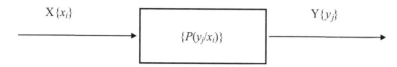

Fig. 4.1 Discrete memoryless channel

$$P \equiv [P_{ij}] = \begin{bmatrix} P_{11} & P_{12} & \cdots & P_{1s} \\ P_{21} & P_{22} & \cdots & P_{2s} \\ \vdots & \vdots & \cdots & \vdots \\ P_{r1} & P_{r2} & \cdots & P_{rs} \end{bmatrix},$$

where $P_{ij} \equiv P(y_j/x_i)$. Index i corresponds to ith row (i.e. input symbol) and index j corresponds to jth column (i.e. output symbol).

Obviously,

$$\sum_{j=1}^{s} P_{ij} = 1 \quad (i = 1, 2, \ldots, r),$$

i.e. matrix P must be stochastic, because at the receiving end the decision must be made—which symbol y_j is received, after symbol x_i was sent. Some authors use P_{ij} to denote the conditional probability that symbol y_i is received, after symbol x_j was sent. In this case, the sum of elements in every column must be equal 1.

A very simple discrete channel is **binary channel** (BC) (Problems 4.1–4.3) where the transition matrix is

$$P_{BC} = \begin{bmatrix} v_1 & p_1 \\ p_2 & v_2 \end{bmatrix},$$

where $v_1 + p_1 = v_2 + p_2 = 1$. Instead of using a matrix, the corresponding graph is very useful (Fig. 4.2a). This channel can be as well completely described by adding

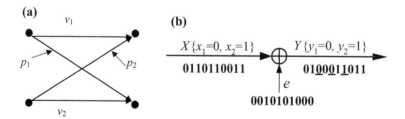

Fig. 4.2 Graph corresponding to a binary channel (**a**) and an equivalent description using error sequence (**b**)

the input sequence using an XOR gate and the *error sequence* (*e*) (Fig. 4.2b). In this sequence binary ones are at the positions where the errors occurred, complementing the transmitted bit.

A special case of binary channel is *binary symmetric channel* (BSC) (Problems 4.4–4.6) where the transition matrix is

$$\mathbf{P}_{\text{BSC}} = \begin{bmatrix} v & p \\ p & v \end{bmatrix} (p+v=1).$$

The notion "symmetry" here means that the probabilities of transition from 0 to 1, and from 1 to 0 are the same. It is the simplest channel described by only one parameter $p(v)$ corresponding in the same time to the probability of error (P_e) (*crossover probability*). In this case P_e does not depend on the probabilities of input symbols. For the simple binary channel probability of error is

$$P_e = P(x_1)p_1 + P(x_2)p_2.$$

As said earlier, a good example for BC is such channel where there is no intersymbol interference and where only the white Gaussian noise is present. It is in fact BSC. Such channels are not often encountered in praxis, but BSC can be used for modeling even for channels with memory as the basic building block.

If BC or BSC is accepted as an idealized model for a digital line, then, for a line with more regenerative sections a model can be used having more BC (BSC) in cascade (Problems 4.4, 4.5, 4.7). The next problem is to form an equivalent channel. In Fig. 4.3 two identical BSC are shown in *cascade*.

The corresponding transition matrix can be found easily

$$\mathbf{P}_{\text{ekv}} = \begin{bmatrix} v^2 + p^2 & 2pv \\ 2pv & v^2 + p^2 \end{bmatrix},$$

i.e., the equivalent channel is BSC as well. Its matrix is obtained by multiplication of the corresponding transition matrices. Using induction method, it can be easily proved that in a general case the matrix of the equivalent channel is obtained by multiplication of matrices of channels is cascade. For identical channels the

Fig. 4.3 Two identical BSC in cascade

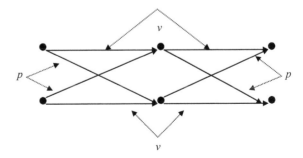

equivalent transition matrix is obtained by exponentiation. Further, it can be proved as well that when the number of cascaded channels increases, the elements of equivalent transition matrix converge to 0.5 regardless on the value of p, meaning that the information transmission is not possible.

The channels can be **extended** as well as the sources by considering sequences of n channel symbols as the symbols of n-th extension. For example, the equivalent matrix of the second extension of BSC—(BSC)^2, is a Kronecker square of BSC matrix [2]

$$P_{(\text{BSC})^2} = \begin{bmatrix} v P_{\text{BSC}} & p P_{\text{BSC}} \\ p P_{\text{BSC}} & v P_{\text{BSC}} \end{bmatrix} = \begin{bmatrix} v^2 & pv & pv & p^2 \\ pv & v^2 & p^2 & pv \\ pv & p^2 & v^2 & pv \\ p^2 & pv & pv & v^2 \end{bmatrix}.$$

One may ask why $r \neq s$? The answer is that in general case such channels are encountered in praxis. For example, BSC can be modified, if it is decided that the transmission is unsuccessful, using the "erasure" of the bit instead of the decision that the other symbol (bit) is received (error!). Such channel is called **binary erasure channel** (BEC) (Problem 3.8). The corresponding transition matrix is

$$P_{\text{BEC}} = \begin{bmatrix} v & p & 0 \\ 0 & p & v \end{bmatrix} (p + v = 1).$$

In Fig. 4.4 a graph of (symmetric) BEC is shown.

This model corresponds to a communication system with a feedback, and the receiver can send to the transmitter the information whether the received symbol is received correctly or not. If the erased symbol is received (usually denoted with E), the transmitter will repeat it until it is correctly received. Probability of this event is $1 - p$, and the corresponding signaling rate should be multiplied with the same factor.

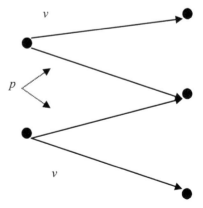

Fig. 4.4 Graph of (symmetric) BEC

Brief Theoretical Overview

For further explanations it is important to consider the probability relations in a channel. Consider discrete channel without memory described by transition matrix **P** with (dimension $r \times s$). Its elements are probabilities $P_{ij}(P(y_j/x_i))$. They can be called as well *transition probabilities*. If the probabilities of the source symbols $P(x_i)$ ($i = 1, 2, ..., r$) are known (*input probabilities*), then the probabilities at the channel output symbols $P(y_j)$ ($j = 1, 2, ..., s$) (*output probabilities*) can be calculated

$$P(y_j) = \sum_{i=1}^{r} P(x_i) P_{ij} \quad (j = 1, 2, ..., s).$$

Therefore, the output probabilities can be found if the input and transition probabilities are known. Of course, the following must hold

$$\sum_{i=1}^{r} P(x_i) = 1, \quad \sum_{j=1}^{s} P(y_j) = 1.$$

Of course, $P(x_i)$ are *a priori input probabilities*. $P(y_j/x_i)$ can be called *a posteriori output probabilities* and $P(y_j)$—a priori *output probabilities*. Generally, on the basis of transition and output probabilities sometimes it is not possible to calculate input probabilities.

Consider the next (trivial) example. BSC is described by $p = v = 0.5$. Let $P(x_1) = a$ and $P(x_2) = b$ ($a + b = 1$). It is easy to calculate

$$P(y_1) = av + bp = 0.5(a + b) = 0.5$$
$$P(y_2) = ap + bv = 0.5(a + b) = 0.5,$$

because $a + b = 1$. Therefore, $P(y_1) = P(y_2) = 0.5$ regardless of input probabilities. Of course, it is an extreme case.

There are two sets of input probabilities. Besides the above mentioned a priori probabilities ($P(x_i)$), the set of *a posteriori input probabilities* is important—$P(x_i/y_j)$ ($i = 1, 2, ..., r$; $j = 1, 2, ..., s$)—when it is known which symbol (y_j) is received. Of course, it is known at the receiving end. These probabilities are calculated as

$$P(x_i/y_j) = \frac{P(x_i) P(y_j/x_i)}{P(y_j)}.$$

This formula (Bayes' rule) is obtained from the *joint probability* of events x_i and y_j

$$P(x_i, y_j) = P(x_i) P(y_j/x_i) = P(y_j) P(x_i/y_j).$$

The a posteriori input probabilities can be calculated as a function of a priori input probabilities and the transition probabilities, i.e. on the basis of known data

$$P(x_i/y_j) = \frac{P(x_i)P(y_j/x_i)}{\sum_{i=1}^{r} P(x_i)P(y_j/x_i)}.$$

Of course, the following must hold as well

$$\sum_{i=1}^{r} P(x_i/y_j) = 1, \quad \sum_{i=1}^{r}\sum_{j=1}^{s} P(x_i, y_j) = 1.$$

Consider the following example [3].
BC is described by matrix

$$P = \begin{bmatrix} \frac{2}{3} & \frac{1}{3} \\ \frac{1}{10} & \frac{9}{10} \end{bmatrix}.$$

Find all unknown probabilities if the input (a priori) probabilities are $P(x_1) = 3/4$ and $P(x_2) = 1/4$.

Output (a posteriori) probabilities are

$$P(y_1) = P(x_1)P(y_1/x_1) + P(x_2)P(y_1/x_2) = \frac{3}{4} \times \frac{2}{3} + \frac{1}{4} \times \frac{1}{10} = \frac{21}{40}$$

$$P(y_2) = P(x_1)P(y_2/x_1) + P(x_2)P(y_2/x_2) = \frac{3}{4} \times \frac{1}{3} + \frac{1}{4} \times \frac{9}{10} = \frac{19}{40}$$

Of course $P(y_1) + P(y_2) = 1$ must hold.
A posteriori input probabilities are

$$P(x_1/y_1) = \frac{P(x_1)P(y_1/x_1)}{P(y_1)} = \frac{20}{21}$$

$$P(x_2/y_2) = \frac{P(x_2)P(y_2/x_2)}{P(y_2)} = \frac{9}{19}.$$

Similarly, the other probabilities can be found. However, the calculation is simpler taking into account the following

$$P(x_1/y_1) + P(x_2/y_1) = 1, \quad P(x_2/y_1) = 1 - P(x_1/y_1) = 1 - \frac{20}{21} = \frac{1}{21}.$$

Analogously

$$P(x_1/y_2) = 1 - P(x_2/y_2) = 1 - \frac{9}{19} = \frac{10}{19}.$$

Joint probabilities can be easily found

$$P(x_1, y_1) = P(y_1)P(x_1/y_1) = \frac{20}{21} \times \frac{21}{40} = \frac{1}{2}.$$

The same result can be obtained in the other way

$$P(x_1, y_1) = P(x_1)P(y_1/x_1) = \frac{3}{4} \times \frac{2}{3} = \frac{1}{2}.$$

Similarly, it is easy to calculate the rest

$$P(x_1, y_2) = 1/4; \quad P(x_2, y_1) = 1/40; \quad P(x_2, y_2) = 9/40.$$

Sum of the last four probabilities equals one.

Entropy, defined in Chap. 2, according to the above consideration, can be called *a priori entropy*.

$$H(X) = \sum_{i=1}^{r} P(x_i) \operatorname{ld}\left(\frac{1}{P(x_i)}\right)$$

because it is calculated using a priori probabilities. However, after the reception of specific symbol y_j, the *a posteriori entropy* of the set X is

$$H(X/y_j) = \sum_{i=1}^{r} P(x_i/y_j) \operatorname{ld}\left(\frac{1}{P(x_i/y_j)}\right).$$

In fact, $H(X)$ is an average measure (per symbol) of the receiver uncertainty about the emitted source symbols, before any symbol reception, while $H(X/y_j)$ is a measure of the remaining part of (a posteriori) uncertainty about the emitted symbol, when specific symbol y_j is received. It can be called *partial a posteriori entropy*.

For the above example

$$H(X) = \frac{3}{4}\operatorname{ld}\frac{4}{3} + \frac{1}{4}\operatorname{ld}4 = 0.811 \ \frac{\text{Sh}}{\text{symb}}$$

$$H(X/y_1) = \frac{20}{21}\operatorname{ld}\frac{21}{20} + \frac{1}{21}\operatorname{ld}21 = 0.276 \ \frac{\text{Sh}}{\text{symb}}$$

$$H(X/y_2) = \frac{9}{19}\operatorname{ld}\frac{19}{9} + \frac{10}{19}\operatorname{ld}\frac{19}{10} = 0.998 \ \frac{\text{Sh}}{\text{symb}}.$$

Therefore, if y_1 is received the uncertainty is smaller, because it is more probable that symbol x_1 was sent then x_2. However, if y_2 is received, the uncertainty is greater, because an a posteriori probabilities of both symbols are very close—9/19 and 10/19. The conclusion is that this uncertainty can increase after the reception of some specific symbol.

However, information theory considers the average values and it is natural to average partial a posteriori entropy over all output symbols to obtain *a posteriori entropy*

$$H(X/Y) = \sum_{j=1}^{s} P(y_j)H(X/y_j)$$

$$= \sum_{j=1}^{s} P(y_j) \sum_{i=1}^{r} P(x_i/y_j) \mathrm{ld}\left(\frac{1}{P(x_i/y_j)}\right)$$

$$= \sum_{i=1}^{r} \sum_{j=1}^{s} P(y_j) P(x_i/y_j) \mathrm{ld}\left(\frac{1}{P(x_i/y_j)}\right)$$

$$= \sum_{i=1}^{r} \sum_{j=1}^{s} P(x_i, y_j) \mathrm{ld}\left(\frac{1}{P(x_i/y_j)}\right).$$

For the above example

$$H(X/Y) = P(y_1)H(X/y_1) + P(y_2)H(X/y_2) = 0.617 \frac{\mathrm{Sh}}{\mathrm{symb}},$$

and the uncertainty is smaller in the average.

Now, it can be mathematically defined how the information is transmitted through the channel. Let the channel output is considered. The uncertainty about specific symbol x_i is $\mathrm{ld}\ (1/P(x_i))$. If the received symbol is y_j, the measure of uncertainty about the symbol x_i is $\mathrm{ld}\ (1/P(x_i/y_j))$. Therefore, the quantity of information transmitted in this case is equal to the difference of a priori uncertainty and a posteriori uncertainty

$$I(x_i, y_j) = \mathrm{ld}\left(\frac{1}{p(x_i)}\right) - \mathrm{ld}\left(\frac{1}{p(x_i/y_j)}\right).$$

Of course, it should be averaged over all pairs of symbols—(x_i, y_j). These average values are in fact a priori and a posteriori entropies, and the average quantity of transmitted information is

$$I(X,Y) = H(X) - H(X/Y) \left[\frac{\mathrm{Sh}}{\mathrm{symb}}\right].$$

Brief Theoretical Overview

It is usually called **mutual information** (Problems 4.1–4.6, 4.8).
The last expression can be written as

$$I(X,Y) = H(X) - H(X/Y)$$

$$= 1 \cdot \sum_{i=1}^{r} P(x_i) \operatorname{ld}\left(\frac{1}{P(x_i)}\right) - \sum_{i=1}^{r}\sum_{j=1}^{s} P(x_i, y_j) \operatorname{ld}\left(\frac{1}{P(x_i/y_j)}\right).$$

Instead of multiplying the first sum by "1", it can be written $1 = \sum_{j=1}^{s} P(y_j/x_i)$, yielding

$$I(X;Y) = \sum_{i=1}^{r}\sum_{j=1}^{s} P(x_i)P(y_j/x_i) \operatorname{ld}\left(\frac{1}{P(x_i)}\right)$$

$$- \sum_{i=1}^{r}\sum_{j=1}^{s} P(x_i, y_j) \operatorname{ld}\left(\frac{1}{P(x_i/y_j)}\right)$$

$$= \sum_{i=1}^{r}\sum_{j=1}^{s} P(x_i, y_j) \operatorname{ld}\left(\frac{P(x_i/y_j)}{P(x_i)}\right)$$

$$= \sum_{i=1}^{r}\sum_{j=1}^{s} P(x_i, y_j) \operatorname{ld}\left(\frac{P(x_i, y_j)}{P(x_i)P(y_j)}\right),$$

where the last row is obtained using Bayes' rule. It should be noted that $I(X, Y)$ is symmetric with regard to X and Y ($I(X, Y) = I(Y, X)$—mutual information!).
For the above example

$$I(X, Y) = H(X) - H(X/Y) = 0.811 - 0.617 = 0.194 \, \frac{\text{Sh}}{\text{symb}}.$$

Therefore, on the average, some quantity information per symbol is transmitted through the channel.

Considering the last row in the above expression, it can be easily concluded that the mutual information will be equal to zero if the input and output symbols are statistically independent in pairs, i.e.

$$P(x_i, y_j) = P(x_i)P(y_j) \quad (i = 1, 2, \ldots, r; j = 1, 2, \ldots, s).$$

This conclusion verifies once more that the approach to definition the quantity of information is intuitively correct. When input and output symbols are statistically independent in pairs, i.e. when any received symbol does not depend on the emitted, there is no any transmission of information.

As said earlier, the entropy of the finite discrete source is limited. It can be easily shown the following

$$I(X,Y) \geq 0,$$

where the equality sign corresponds only to the case of independent symbols. On the other hand, the mutual information is maximal, if the a posteriori entropy equals zero, because then all emitted information is transmitted. Therefore,

$$0 \leq I(X,Y) \leq H(X).$$

Besides the above defined entropies, three more can be defined. The *joint entropy* of X and Y

$$H(X,Y) = \sum_{i=1}^{r} \sum_{j=1}^{s} P(x_i, y_j) \operatorname{ld}\left(\frac{1}{P(x_i, y_j)}\right),$$

the *entropy of output symbols*

$$H(Y) = \sum_{j=1}^{s} P(y_j) \operatorname{ld}\left(\frac{1}{P(y_j)}\right).$$

and *conditional entropy of output symbols*

$$H(Y/X) = \sum_{i=1}^{r} \sum_{j=1}^{s} P(x_i, y_j) \operatorname{ld}\left(\frac{1}{P(y_j/x_i)}\right),$$

Using the above mentioned symmetry and some mathematical manipulation the following expression can be obtained

$$H(X,Y) = H(X) + H(Y) - H(X/Y) = H(X) + H(Y/X) = H(Y) + H(X/Y).$$

Consider BSC defined by $\boldsymbol{P}_{\text{BSC}} = \begin{bmatrix} v & p \\ p & v \end{bmatrix}$ ($v + p = 1$). Let input (a priori) probabilities are $P(x_1) = a$, $P(x_2) = b$ ($a + b = 1$). Probabilities of output symbols are

$$P(y_1) = av + bp \text{ i } P(y_2) = ap + bv.$$

Mutual information is

Brief Theoretical Overview

$$I(X,Y) = H(Y) - H(Y/X) = H(Y) - \sum_{i=1}^{2} P(x_i) \sum_{j=1}^{2} P(y_j/x_i) \operatorname{ld}\left(\frac{1}{P(y_j/x_i)}\right)$$

$$= H(Y) - \sum_{i=1}^{2} P(x_i) \cdot \left(p \cdot \operatorname{ld}\frac{1}{p} + v \cdot \operatorname{ld}\frac{1}{v}\right)$$

$$= H(Y) - (a+b) \cdot \left(p \cdot \operatorname{ld}\frac{1}{p} + v \cdot \operatorname{ld}\frac{1}{v}\right)$$

$$= H(Y) - \left(p \cdot \operatorname{ld}\frac{1}{p} + v \cdot \operatorname{ld}\frac{1}{v}\right).$$

Introducing the corresponding values into $H(Y)$ one obtains

$$I(X,Y) = (av+bp) \cdot \operatorname{ld}\left(\frac{1}{av+bp}\right) + (ap+bv) \cdot \operatorname{ld}\left(\frac{1}{ap+bv}\right)$$
$$- \left(p \cdot \operatorname{ld}\frac{1}{p} + v \cdot \operatorname{ld}\frac{1}{v}\right),$$

or,

$$I(X,Y) = H(ap+bv) - H(p),$$

where the **entropy function** is used

$$H(x) \equiv (1-x) \cdot \operatorname{ld}\left(\frac{1}{1-x}\right) + x \cdot \operatorname{ld}\frac{1}{x}.$$

From

$$p \leq ap + bv \leq v, \quad \text{za} \quad p \leq 0.5,$$

i.e.

$$p \geq ap + bv \geq v \quad \text{za} \quad p \geq 0.5 \; (v \leq 0.5),$$

it follows

$$H(ap+bv) \geq H(p),$$

therefore, $I(X, Y) \geq 0$.

Now, the last step to define the **channel capacity** follows.

Mutual information can be written using input (a priori) probabilities, characterizing the source, and transition probabilities, characterizing the channel.

$$I(X,Y) = \sum_{i=1}^{r}\sum_{j=1}^{s} P(x_i)P(y_j/x_i) \operatorname{ld}\left(\frac{P(y_j/x_i)}{\sum_{i=1}^{r} P(x_i)P(y_j/x_i)}\right).$$

It is easy to conclude that the mutual information depends on transition probabilities, characterizing the channel, as well on input probabilities, characterizing the source. This means that mutual information depends on the way the channel is used as well—which source is at the channel output. However, the quantity should be found, depending only of the channel. The number of input symbols is fixed (r) and such sources can only differ by symbol probabilities. It is natural to chose the input probabilities in such a way to obtain **maximal mutual information**

$$I_{\max} = \max_{P(x_i)} I(X,Y),$$

where the maximization is carried out over the set of input probabilities. Dimension of $I_{\max}$ is Sh/symb corresponding to the maximal quantity of information that can be transmitted in the average by one symbol.

The engineers prefer to know the maximal quantity of information that can be transmitted over the channel in one second. Let the source has symbol rate $v_m(X, Y)$ [symb/s], being the maximal rate for the channel. Then, the **maximal information rate** is

$$C = v_m(X,Y) \cdot I_{\max} \quad \left[\frac{\text{symb}}{\text{s}} \frac{\text{Sh}}{\text{symb}} = \frac{\text{Sh}}{\text{s}}\right].$$

It will be called in this book the **channel capacity** (Problems 4.4, 4.5, 4.10, 4.11). Some authors (especially in the courses on probability) call $I_{\max}$ channel capacity. However, the symbol rates are in a range from a few thousands symbols per second to more terasymbols per second. It is usually taken $v_m(X, Y) = 2f_g$, to avoid the intersymbol interference (First Nyquist criterion).

In reality, the **information rate** (flux) (Problems 4.1, 4.5) is

$$\Phi(X,Y) \equiv v(X,Y) \cdot I(X,Y) \quad \left[\frac{\text{Sh}}{\text{s}}\right],$$

where $v(X, Y)$ and $I(X, Y)$ are the real values for the system. The corresponding **efficiency** can be defined

$$\eta_C \equiv \frac{\Phi(X,Y)}{C},$$

Generally, the calculation of channel capacity is very difficult, usually using the variation calculus or some similar mathematical tool. Here only two interesting case will be given.

Consider a discrete channel where the number of symbols is r. However, further hypothesis is that there is no noise in the channel (***noiseless channel***). Let the transmission (signaling) rate is $v_m(X, Y) = 2f_c$, providing the transmission without intersymbol interference. According to definition

$$I(X, Y) = H(X) - H(X/Y).$$

However, after the receiving of any symbol there would not be any ambiguity, because there is no noise nor intersymbol interference. Hence, $H(X/Y) = 0$, and

$$I_{max} = H(X).$$

According to previous discussion, here the entropy is maximum when the symbols are equiprobable, i.e. when $H_{max}(X) = \mathrm{ld}\, r$. Finally,

$$C = 2f_c\, \mathrm{ld}\, r$$

Of course, the number of symbols (r) in practice may depend on signal-to-noise ratio.

A very important channel is so called ***continuous channel*** (with additive noise). Here the source is continuous, the channel is continuous and the output (being sum of signal and noise) is continuous as well. Te corresponding entropy is defined in Chap. 2. Introducing the corresponding probability densities $w(x)$, $w(y)$, $w(x/y)$, $w(y/x)$ and $w(x, y)$ (the corresponding indexes are omitted!) one obtains

$$H(X) = \int_{-\infty}^{\infty} w(x)\, \mathrm{ld}\left(\frac{1}{w(x)}\right) dx,$$

$$H(Y) = \int_{-\infty}^{\infty} w(y)\, \mathrm{ld}\left(\frac{1}{w(y)}\right) dy,$$

$$H(X/Y) = \int_{-\infty}^{\infty} \int_{-\infty}^{\infty} w(x, y)\, \mathrm{ld}\left(\frac{1}{w(x/y)}\right) dxdy,$$

Here the mutual information is

$$I(X,Y) = \int_{-\infty}^{\infty} \int_{-\infty}^{\infty} w(x, y)\, \mathrm{ld}\left(\frac{w(x, y)}{w(x)w(y)}\right) dxdy,$$

It is supposed that a frequency band is limited, e.g. $0 \div f_c$, and the corresponding rate is $v_m(X, Y) = 2f_c$, yielding

$$C = 2f_c \cdot I_{max}.$$

Let the continuous channel input (x) has the probability density $w(x)$, where

$$m_x = \bar{x} = \int_{-\infty}^{\infty} xw(x)dx = 0$$

$$\sigma_x^2 = \overline{x^2} = \int_{-\infty}^{\infty} x^2 w(x)dx < \infty.$$

The additive channel noise is n, where

$$m_n = \bar{n} = \int_{-\infty}^{\infty} nw(n)dn = 0$$

$$\sigma_n^2 = \overline{n^2} = \int_{-\infty}^{\infty} n^2 w(n)dn < \infty.$$

Further, it is supposed that signal and noise are statistically independent, yielding the output signal (y)

$$y = x + n.$$

where

$$m_y = \bar{y} = m_x + m_n = 0,$$
$$\sigma_y^2 = \overline{y^2} = \sigma_x^2 + \sigma_n^2.$$

Mutual information is

$$I(X,Y) = H(Y) - H(Y/X).$$

The output ambiguity depends only on additive noise (independent from signal) and the following can be shown ($H(N)$ is noise entropy)

$$I(X,Y) = H(Y) - H(N).$$

One interpretation of this result is that the transmitted quantity of information (per sample) is obtained when from the total output quantity of information, the "false quantity of information", introduced by noise is deducted.

Brief Theoretical Overview

Now, a hypothesis about the noise (its probability density) should be made. The worst case, as explained in Chap. 2, for a fixed variance (σ^2), is when the noise has Gaussian probability density, yielding

$$H_{max} = \frac{1}{2} \operatorname{ld}(2\pi e \sigma^2).$$

Therefore,

$$I(X,Y) = H(Y) - \frac{1}{2} \operatorname{ld}(2\pi e \sigma_n^2),$$

yielding

$$I_{max} = \max_{w(x)} H(Y) - \frac{1}{2} \operatorname{ld}(2\pi e \sigma_n^2).$$

Now, the probability density $w(x)$ of the input signal should be found to maximize $H(Y)$. The output signal, sum of the input signal and noise (statistically independent), will have the maximal entropy if its probability density is Gaussian. To obtain it, the input signal should be Gaussian as well because the sum of the independent Gaussian processes is Gaussian process. Therefore,

$$H(Y)|_{max} = \frac{1}{2} \operatorname{ld}\left(2\pi e \sigma_y^2\right) = \frac{1}{2} \operatorname{ld}\left[2\pi e \left(\sigma_x^2 + \sigma_y^2\right)\right],$$

and a maximal mutual information is

$$I_{max} = \frac{1}{2} \operatorname{ld}\left[2\pi e \left(\sigma_x^2 + \sigma_n^2\right)\right] - \frac{1}{2} \operatorname{ld}\left(2\pi e \sigma_n^2\right) = \frac{1}{2} \operatorname{ld}\left(1 + \frac{\sigma_x^2}{\sigma_n^2}\right).$$

Capacity of such channel is

$$C = v_m(x,y) I_{max} = 2 f_c \cdot \frac{1}{2} \operatorname{ld}\left(1 + \frac{\sigma_x^2}{\sigma_n^2}\right) = f_c \cdot \operatorname{ld}\left(1 + \frac{\sigma_x^2}{\sigma_n^2}\right) \quad \left[\frac{Sh}{s}\right].$$

Some authors instead of f_c use symbol B (frequency band) and instead of variances quotient—symbol S/N (from signal-to-noise, being here a natural quotient of signal and noise powers, not in dB!), yielding finally (Problems 4.4, 4.5, 4.9–4.11)

$$C = B \cdot \operatorname{ld}\left(1 + \frac{S}{N}\right).$$

This expression is obtained using mainly abstract consideration, but the used notions are very well known to the communication engineers—frequency band (B) and signal-to-noise ratio (S/N). Therefore, capacity can be increased by increasing either band either signal-to-noise ratio (i.e. by increasing the transmitter power or by suppressing the noise power). Of course, some compensation can be introduced—decreasing the band can be compensated by increasing the transmitter power (but the increase is slow—logarithmic!) and vice versa.

When applying the above expression for the channel capacity it is very important to keep in mind the conditions supposed. The channel is considered with additive Gaussian noise, a capacity is achieved if the input signal is Gaussian as well. Therefore, it is maximum possible value for capacity. In reality, if the signals are used which are not Gaussian (usually some kind of baseband pulses or some kind of digital modulation), the capacity is smaller. In such cases, a calculation of capacity can be very complicated.

The notion of error probability is very well known. Still, it is useful to connect it with so called **decision rule**. Consider BC described by transition matrix

$$\mathbf{P}_{BC} = \begin{bmatrix} v_1 & p_1 \\ p_2 & v_2 \end{bmatrix}.$$

It is obvious that the error will happen as a result of two exclusive events—either x_1 is emitted and y_2 is received or x_2 is emitted and y_1 is received. Therefore, the **average error probability** is

$$P_e = P(x_1)p_1 + P(x_2)p_2,$$

and it is just the **error probability** (Problems 4.1, 4.2, 4.3, 4.7), as usually this notion is used. For $P(x_1) = P(x_2) = 0.5$

$$P_e = 0.5 \cdot (p_1 + p_2).$$

For BSC, where $p_1 = p_2 = p$ ($v_1 = v_2 = v$),

$$P_e = P(x_1)p + P(x_2)p = [P(x_1) + P(x_2)] \cdot p = p,$$

because $P(x_1) + P(x_2) = 1$ and p is just the parameter equivalent to the error probability.

Consider BSC described by transition matrix

$$\begin{bmatrix} 0.99 & 0.01 \\ 0.01 & 0.99 \end{bmatrix}.$$

Brief Theoretical Overview

It is obvious that the error probability is $P_e = p = 0.01$. But, consider the following transition matrix

$$\begin{bmatrix} 0.01 & 0.99 \\ 0.99 & 0.01 \end{bmatrix}.$$

Do the error probability is 0.99? Of course no! One should only "complement" the logic. In previous cases after receiving symbol $y_1(=0)$ the decision was that symbol $x_1(=0)$ is emitted, and vice versa. If now after the receiving $y_1(=0)$ the decision is $x_1(=1)$, the error probability is again $P_e = 0.01$. Therefore, the error probability does not depend on transition probabilities only, but on the *decision rule* as well—on the way the receiver interprets the received symbol. It should be noted that here (binary channel) the worst case is when the error probability is 0.5.

Now, the notion of decision rule will be more precisely defined, Consider the channel having r input symbols $\{x_i\}$ ($i = 1, 2, \ldots, r$) and s output symbols $\{y_j\}$ ($j = 1, 2, \ldots, s$). Decision rule $D(y_j)$ is a function specifying a unique input symbol x_i for each output symbol y_j, i.e.

$$D(y_j) = x_i$$

In the next to last example it was $D(y_1) = x_1$, $D(y_2) = x_2$ and in the last one $D(y_1) = x_2$, $D(y_2) = x_1$.

Generally, there are r^s different decision rules. For BC there are $2^2 = 4$ different decision rules.

The next question is which decision rule one should choose? The answer is clear—that one yielding minimum error probability. Therefore, the relation between decision rule and error probability should be found. Let symbol y_j is received. The receiver will use the decision rule $D(y_j) = x_i$, But, the receiving of symbol y_j can be as well the consequence of emitting some other symbol from $\{X\}$. It can be written as follows

$$P(x_i/y_j) + P(e/y_j) = 1,$$

where $P(e/y_j)$ is a conditional probability that the wrong decision was made if the received symbol is y_j. Error probability is obtained by averaging this probability over all output symbols.

$$P_e = \sum_{j=1}^{s} P(y_j) P(e/y_j).$$

The considered events are mutually exclusive, the addends are nonnegative, and minimization is achieved by minimizing every addend independently. $P(y_j)$ does not depend on decision rule and $D(y_j) = x_i$ should be chosen minimizing every conditional probability for itself. With the chosen decision rule one obtains

$$P(e/y_j) = 1 - P(x_i/y_j) = 1 - P[D(y_j)/y_j].$$

Therefore, to minimize the error probability, for every received symbol y_j the following decision rule should be chosen

$$D(y_j) = x^*,$$

where x^* is defined as

$$P(x^*/y_j) \geq P(x_i/y_j) \quad (i = 1, 2, \ldots, r).$$

In other words, for every received symbol, the error probability will be minimal if to each y_j the input symbol x^* is joined having the maximum a posteriori probability $P(x^*/y_j)$. Of course, if there are more such symbols, anyone can be chosen.

This rule is denoted as MAP (**Maximum A Posteriori Probability rule**) (Problems 4.1, 4.2). To apply this rule the a posteriori probabilities $P(x_i/y_j)$ should be found. These probabilities can be found only if besides the transition probabilities, a priori probabilities of input symbols are known.

On basis of Bayes' rule, previous inequalities can be written as

$$\frac{P(y_j/x^*)P(x^*)}{P(y_j)} \geq \frac{P(y_j/x_i)P(x_i)}{P(y_j)} \quad (i = 1, 2, \ldots, r).$$

If input probabilities are mutually equal

$$P(x_i) = \frac{1}{r} \quad (i = 1, 2, \ldots, r),$$

previous expression can be written in the form

$$P(y_j/x^*) \geq P(y_j/x_i) \quad (i = 1, 2, \ldots, r),$$

yielding a simple way to find the decision rule on the basis of transition matrix only. Simply, for every received symbol y_j the decision is that symbol x_i is emitted for which in the corresponding column in matrix the conditional probability has a maximum value. This rule is denoted as ML (**Maximum Likelihood rule**) (Problem 4.2). Minimal error probability will be obtained if all input symbols are equiprobable.

For MAP it can be shown that the error probability is

$$P_e = \sum_{j=1}^{s} P(e/y_j)P(y_j)$$

$$= \sum_{j=1}^{s} P(y_j)P\left[1 - P(x^*/y_j)\right]$$

$$= \sum_{j=1}^{s} P(y_j) - \sum_{j=1}^{s} P(y_j)P(x^*/y_j)$$

$$= 1 - \sum_{j=1}^{s} P(x^*, y_j)$$

$$= 1 - \sum_{j=1}^{s} P(x^*)P(y_j/x^*).$$

For equiprobable input symbols

$$P_e = 1 - \frac{1}{r}\sum_{j=1}^{s} P(y_j/x^*),$$

and error probability is obtained from transition matrix, by adding maximal values in columns.

The above described decision process can be called as well the **hard decision** (Problems 4.2, 4.4, 4.10). This means that the final decision about the received symbol is made in demodulator (detector). However, it is better to quantize the output signal, i.e. instead of symbol value to give some estimation (i.e. instead of the value "1" to generate real (measured) values—0.8; 1.05; 0.98 etc.). In this case, it is said the **soft decision** (Problems 3.10, 3.11) is made. The drawback is that more bits should be used.

In fact, in digital transmission, the channel input is discrete. By noise and other interferences superposition, the channel output is continuous. By using the corresponding thresholds, the output is quantized becoming discrete. The thresholds positions influence the error probability. These positions should be chosen to minimize the error probability.

Up to now, channels **without memory** are considered. As said earlier, in the channel with intersymbol interference, the error probability of the received symbols depends as well and on the adjacent symbols in such a way introducing *memory* into the channel. Models for channels with memory are often made using a combination of memoryless channels, usually by so called **Markov models** (Problem 4.12).

It is obvious that channel without intersymbol interference with additive Gaussian noise can be modeled as BSC. However, in such channels sometimes

Fig. 4.5 State diagram of Gilbert model for burst noise channel

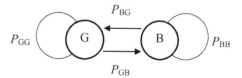

impulsive noise occurs (***burst-noise channel***). It can be considered as a nonstationary channel, because, the error probability changes in time. However, taking into account that during the bursts of noise result in "packets of errors", one can conceive that BSC has two possible states—one without bursts of noise, when one formula for error probability is valid, and the other, during the burst of noise, when the error probability is much higher, e.g. 0.5. Gilbert [20] proposed a simple model. There are two states in the model—"good" (G) and "bad" (B). In every state channel is considered as BSC (where only one parameter is sufficient to describe the channel). In Gilbert model it is supposed the in good state there are no errors ($p_g = 0$), while in bad state error probability can take any value. Usually, it is taken $p_L = 0.5$ (the worst case). Transition matrices are

$$\boldsymbol{P}_G = \begin{bmatrix} 1 & 0 \\ 0 & 1 \end{bmatrix}, \quad \boldsymbol{P}_B = \begin{bmatrix} v_B & p_B \\ p_B & v_B \end{bmatrix} (v_G + p_B = 1).$$

The corresponding transition probabilities are given as well

$$P_{GG} + P_{GB} = 1, P_{BB} + P_{BG} = 1,$$

(where $P_{AB} \equiv P(B/A)$). State diagram is shown in Fig. 4.5.

There are three independent parameters, p_L and two transition from state to state probabilities (other two are then fixed). Transition matrix for the model is

$$\boldsymbol{P}_{GIL} = \begin{bmatrix} P_{GG} & P_{GB} \\ P_{BG} & P_{BB} \end{bmatrix}.$$

Stationary state probabilities are

$$P_G = \frac{P_{BG}}{P_{GB} + P_{GG}}$$

$$P_B = \frac{P_{GB}}{P_{GB} + P_{GG}},$$

and the average error probability

$$P_e = p_B \cdot P_B = \frac{p_B P_{GB}}{P_{GB} + P_{GG}}.$$

Brief Theoretical Overview

The sequence of states can be easily found. The average duration of time intervals (*sojourn time*) in good state corresponds to the intervals without the impulsive noise (without the errors in Gilbert model). The average duration of time intervals in bad state corresponds to the intervals with the impulsive noise. They are

$$N_G = \frac{1}{1 - P_{GG}}, \quad N_B = \frac{1}{1 - P_{BB}}.$$

It is only the first step in modeling real channels from the error statistics point of view. Elliott modified this model by introducing the possibility of errors ($p_G \neq 0$) in the good state (Problem 4.12) [21].

Now, the time came to consider the trough meaning of the channel capacity. It is given by **Second Shannon theorem**. Its complicated proof here will be omitted. According to this theorem, the error probability can be made as small as one wish if the channel information rate is smaller than the capacity ($\Phi(X, Y) < C$). If the signaling rate is $v(X, Y)$, channel information rate is $\Phi(X, Y) = v(X, Y) I(X, Y)$ [Sh/s]. This result is unexpected. The reliable transmission over the unreliable channel is possible! It should be noted that the transmission without any error is not possible. It only means that the error probability can be kept as low as one wish.

This goal can be achieved using **error control coding** (considered in details in the next five chapters). To get some insight, an elementary example will be done here—error control coding using repetitions.

Consider binary memoryless binary source (emitting symbols **0** and **1**). BSC with parameter $p = 10^{-2}$ (crossover probability) is used. Usually it is high error probability. One possible way to lower error probability is to repeat every information bit three times, i.e. **0** is encoded as 000 and **1** as 111. That means that three channel bits are used for every information bit. Of course, the corresponding information rate is three times smaller (i.e. the **code rate** (Problem 4.12) is 1/3). At the receiving end the majority logic is implemented. Three bits are decoded according to the greater number of ones or zeros as one or zero. It is in fact MAP criterion. Probability of one error is this case is $3p(1-p)^2 = 3 \cdot 0.01 \cdot 0.99^2 = 0.0294$, while the probability of two errors is $3 \cdot p^2 \cdot (1 - p) = 3 \cdot 0.0001 \cdot 0.99 = 0.000297$. The probability of undetected error is

$$P_e = \binom{3}{0} p^3 + \binom{3}{1} p^2 (1 - p) = 0.000298.$$

Therefore, single errors in a code word are detected and corrected. If the errors are only detected, then all single and double errors will be detected, but the error probability would be $P_e = p^3 = 10^{-6}$. Considering further fivefold repetition, with error correction, the probability of uncorrected error is

$$P_e = \binom{5}{0} p^5 + \binom{5}{1} p^4 (1 - p) + \binom{5}{2} p^3 (1 - p)^2 \approx 10^{-5}.$$

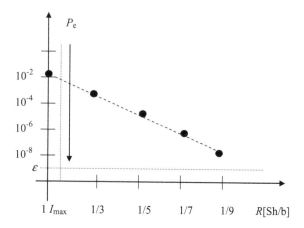

Fig. 4.6 Error probability versus code rate (R) (other explanations in the text)

Of course, the code rate is now $R = 1/5$. The number of repetition can be further increased ($P_e \approx 4 \times 10^{-7}$ for sevenfold repetition and $P_e \approx 10^{-8}$). Corresponding results are shown using fat points in Fig. 4.6. Note the reciprocal values at the abscissa.

By considering this figure, it would be easy to understand the essence of error control coding and the corresponding problems. In this case (n-fold repetition), to obtain small error probability ($<\varepsilon$) a code rate must be drastically decreased. Therefore, the logical question is: does exist some way to decrease the error probability, without such decrease of code rate? Second Shannon theorem just gives a positive answer.

Problems

Problem 4.1 Zero-memory binary source emits symbols with rate $v_s = 100$ [b/s], the probability of one symbol is $P(x_1) = 0.3$. The corresponding channel is described by transition matrix

$$P = \begin{bmatrix} 0.4 & 0.6 \\ 0.75 & 0.25 \end{bmatrix}.$$

(a) Find the entropy and the information rate of the source.
(b) Find the mutual (transmitted) information and the information rate of the channel.
(c) Find the decision rule for the receiver yielding the minimum error probability.
(d) Repeat the same as in (c) for $P(x_1) = 0.2$.

Solution

(a) Entropy and information rate are the source characteristics. Finding $P(x_2) = 1 - P(x_1) = 0.7$, the following is calculated

$$H(X) = \sum_{i=1}^{2} P(x_i) \, \text{ld} \, \frac{1}{P(x_i)} = 0.882 \left[\frac{\text{Sh}}{\text{b}}\right],$$

$$\Phi(X) = H(X)v_s = 88.2 \, [\text{Sh/s}]$$

For the case of equiprobable symbols, the entropy equals one and the information rate has the maximum value—100 [Sh/s].

(b) Transmitted (mutual) information is [1, 2]

$$I(X,Y) = \sum_{i=1}^{2} \sum_{j=1}^{2} P(x_i, y_j) \, \text{ld} \, \frac{P(y_j/x_i)}{P(y_j)},$$

and for calculation the corresponding probabilities should be found:

- Probabilities of channel output symbols are

$$P(y_1) = P(y_1/x_1)P(x_1) + P(y_1/x_2)P(x_2) = 0.645,$$
$$P(y_2) = P(y_2/x_1)P(x_1) + P(y_2/x_2)P(x_2) = 0.355.$$

- Joint probabilities of binary symbols at the input and the output of the channel are

$$P(x_1, y_1) = P(x_1)P(y_1/x_1) = 0.12$$
$$P(x_2, y_1) = P(x_2)P(y_1/x_2) = 0.525$$
$$P(x_1, y_2) = P(x_1)P(y_2/x_1) = 0.18$$
$$P(x_2, y_2) = P(x_2)P(y_2/x_2) = 0.175.$$

Transmitted information is

$$I(X,Y) = 0.12 \, \text{ld} \, \frac{0.4}{0.645} + 0.525 \, \text{ld} \, \frac{0.75}{0.645} + 0.18 \, \text{ld} \, \frac{0.6}{0.355}$$
$$+ 0.175 \, \text{ld} \, \frac{0.25}{0.355} = 0.0793 \left[\frac{\text{Sh}}{\text{b}}\right],$$

and as a source emits symbols (bits) directly into the channel

$$\Phi(X,Y) = v(X,Y)I(X,Y) = v_s I(X,Y) = 7.93 \, [\text{Sh/b}].$$

Transmitted information should always be smaller than the source entropy, and it is obvious that the channel information rate cannot be greater from the source information rate

$$I(X,Y) \leq H(X) \Rightarrow \Phi(X,Y) \leq \Phi(X).$$

(c) Minimum error probability is achieved when the maximum a posteriori probability (*Maximum* A Posteriori *Probability*—MAP) decision rule is used. Simply, this rule leads to the choice of input symbol for which the greatest a posteriori probability of output symbol is obtained. In this case, the a posteriori probabilities are

$$P(x_1/y_1) = P(x_1, y_1)/P(y_1) = 0.186$$
$$P(x_2/y_1) = P(x_2, y_1)/P(y_1) = 0.814$$
$$P(x_1/y_2) = P(x_1, y_2)/P(y_2) = 0.507$$
$$P(x_2/y_2) = P(x_2, y_2)/P(y_2) = 0.493$$

the corresponding decision rule is

$$\left. \begin{array}{l} P(x_1/y_1) = 0.186 \\ P(x_2/y_1) = 0.814 \end{array} \right\} \Rightarrow d_{MAP}(y_1) = x_2,$$

$$\left. \begin{array}{l} P(x_1/y_2) = 0.507 \\ P(x_2/y_2) = 0.493 \end{array} \right\} \Rightarrow d_{MAP}(y_2) = x_1$$

From this follows that the "opposite" decision rule is optimum in this case, and the error probability is

$$P_e = P(x_1)P(y_1/x_1) + P(x_2)P(y_2/x_2) = 0.295,$$

while for a "classic" decision rule the error probability is

$$P'_e = P(x_1)P(y_2/x_1) + P(x_2)P(y_1/x_2) = 0.705.$$

(d) In this case it seems that the optimum decision rule is

$$\left. \begin{array}{l} P(x_1/y_1) = 0.1176, \\ P(x_2/y_1) = 0.8824 \end{array} \right\} \Rightarrow d_{MAP}(y_1) = x_2,$$

$$\left. \begin{array}{l} P(x_1/y_2) = 0.375 \\ P(x_2/y_2) = 0.625 \end{array} \right\} \Rightarrow d_{MAP}(y_2) = x_2,$$

leading to an unexpected result. However, the decision rule must provide that for the different output symbols, the decision must be that different input symbols are emitted. In this case the decision can be based on the a posteriori probabilities, or one have to choose the rule giving a smaller error probability, i.e.

$$P_e = P(x_1)P(y_1/x_1) + P(x_2)P(y_2/x_2) = 0.28,$$
$$P'_e = P(x_1)P(y_2/x_1) + P(x_2)P(y_1/x_2) = 0.72,$$

and the conclusion is that in this case, the optimum decision rule is $d_{MAP}(y_1) = x_2$, $d_{MAP}(y_2) = x_1$.

Problem 4.2 Zero-memory binary source where the probability of one symbol is $P(x_1) = p$ emits the symbols at the input of memoryless channel whose transition matrix is

$$\mathbf{P} = \begin{bmatrix} 0.3 & 0.7 \\ 0.01 & 0.99 \end{bmatrix}.$$

(a) Find the entropies of input and output sequences and conditional entropy of output symbols when the input sequence is known.
(b) Find the transmitted information for $p = 0.1$, $p = 0.5$ and $p = 0.9$. For the same values of p find the decision rule yielding the minimum error probability as well as the probability of error in this case.
(c) Draw the transmitted information as a function of p. Draw the dependence of the error probability as a function of p for the cases when the maximum a posteriori probability (MAP) rule is used and when maximum likelihood (ML) rule is used.

Solution

(a) For zero-memory binary source the entropy is

$$H(X) = p \text{ ld } \frac{1}{p} + (1-p) \text{ ld } \frac{1}{1-p},$$

while the entropy of outgoing sequence is

$$H(Y) = (0.3p + 0.01(1-p)) \text{ ld } \frac{1}{0.3p + 0.01(1-p)} + (0.7p + 0.99(1-p)) \text{ ld } \frac{1}{0.7p + 0.99(1-p)}.$$

Conditional entropy of output symbols when the sequence of the input symbols is known can be expressed as

$$H(Y/X) = 0.3p \text{ ld } \frac{1}{0.3} + 0.01(1-p) \text{ ld } \frac{1}{0.01} + 0.7p \text{ ld } \frac{1}{0.7} + 0.99(1-p) \text{ ld } \frac{1}{0.99}.$$

(b) The transmitted information is

$$I(X,Y) = H(Y) - H(Y/X) = (0.3p + 0.01(1-p)) \text{ ld } \frac{1}{0.3p + 0.01(1-p)}$$
$$+ (0.7p + 0.99(1-p)) \text{ ld } \frac{1}{0.7p + 0.99(1-p)}$$
$$- 0.3p \text{ ld } \frac{1}{0.3} - 0.01(1-p) \text{ ld } \frac{1}{0.01} - 0.7p \text{ ld } \frac{1}{0.7} - 0.99(1-p) \text{ ld } \frac{1}{0.99},$$

after the calculation yielding

$$I(X,Y) \stackrel{p=0.1}{=} 0.0768 \text{ [Sh/b]}, \quad I(X,Y) \stackrel{p=0.5}{=} 0.1412 \text{ [Sh/b]},$$
$$I(X,Y) \stackrel{p=0.9}{=} 0.0417 \text{ [Sh/b]}.$$

To obtain the minimum probability of error MAP decision rule is used as follows

(1) For $p = 0.1$

$$\left. \begin{array}{l} P(x_1/y_1) = 0.7691 \\ P(x_2/y_1) = 0.2308 \end{array} \right\} \Rightarrow d_{MAP}(y_1) = x_1,$$
$$\left. \begin{array}{l} P(x_1/y_2) = 0.0728 \\ P(x_2/y_2) = 0.9272 \end{array} \right\} \Rightarrow d_{MAP}(y_2) = x_2.$$

(2) For $p = 0.5$

$$\left. \begin{array}{l} P(x_1/y_1) = 0.9677 \\ P(x_2/y_1) = 0.0323 \end{array} \right\} \Rightarrow d_{MAP}(y_1) = x_1,$$
$$\left. \begin{array}{l} P(x_1/y_2) = 0.4142 \\ P(x_2/y_2) = 0.5858 \end{array} \right\} \Rightarrow d_{MAP}(y_2) = x_2.$$

(3) For $p = 0.9$

$$\left. \begin{array}{l} P(x_1/y_1) = 0.9963 \\ P(x_2/y_1) = 0.0037 \end{array} \right\} \Rightarrow d_{MAP}(y_1) = x_1,$$
$$\left. \begin{array}{l} P(x_1/y_2) = 0.8642 \\ P(x_2/y_2) = 0.1358 \end{array} \right\} \Rightarrow d_{MAP}(y_2) = x_1.$$

For the first two cases the results are understandable $(d_{MAP}(y_1) = x_1, d_{MAP}(y_2) = x_2)$ while, for the third case the smaller error probability is obtained when "opposite" decision rule is applied. The corresponding error probabilities are

$$P_e^{p=0.1} = 0.0790, \quad P_e^{p=0.5} = 0.355, \quad P_e^{p=0.9} = 0.369.$$

From these results the following can be concluded:

- Transmitted information and the error probability depend not only on the channel, but on the source as well.
- Transmitted information does not depend on decision rule used, while the error probability depends on.
- Maximum transmitted information and minimum error probability are not achieved for the same values of the source parameters (it will be additionally illustrated in the next part of solution).

(c) It is obvious that MAP is not the only possible decision rule. Often the maximum likelihood (ML) rule is used, where the decision depends on channel parameters only

$$\left. \begin{array}{l} P(y_1/x_1) = 0.3 \\ P(y_1/x_2) = 0.01 \end{array} \right\} \Rightarrow d_{ML}(y_1) = x_1, \quad \left. \begin{array}{l} P(y_2/x_1) = 0.7 \\ P(y_2/x_2) = 0.99 \end{array} \right\} \Rightarrow d_{ML}(y_2) = x_2$$

i.e. does not depending in any way on the probabilities of the input binary symbols.

The transmitted information as a function of p is shown in Fig. 4.7 and from the figure it can be found that the maximum corresponds to value $p = 0.409$. Probability of error as a function of p for MAP and ML decision rules are shown in Fig. 4.8. It is obvious that the minimum error probability

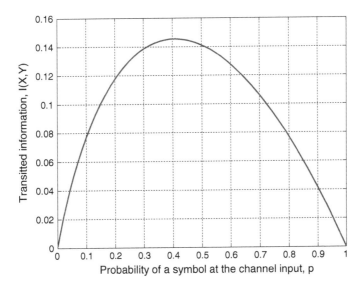

Fig. 4.7 Transmitted information as a function of p

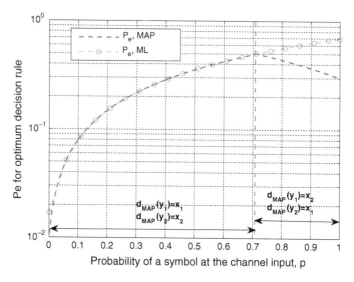

Fig. 4.8 Probability of error as a function of p for MAP and ML

corresponds to $p = 0$, i.e. when the source emits continuously one binary symbol (however, in this case the transmitted information equals to zero, and there is no a symbol probability being optimum according to the both criteria).

Problem 4.3 Zero-memory binary source where the probability of one symbol is $P(x_1)$, emits the symbols at the input of memoryless channel described by the transition matrix

$$P = \begin{bmatrix} 1 - p_0 & p_0 \\ p_1 & 1 - p_1 \end{bmatrix}.$$

(a) For $P(x_1) = 0.1$ find the dependence of $I(X,Y)$ on the channel parameters.
(b) Draw the dependence of $I(X, Y)$ on p_0 for $p_1 = 0.6$ and $P(x_1) = 0.1$; $P(x_1) = 0.5$ and $P(x_1) = 0.9$.
(c) For a fixed value $p_1 = 0.6$, for $p_1 = 0.01$ and $p_1 = 0.99$ find the dependence of $I(X, Y)$ and of the decision rule on the probability of symbol 0 in the incoming sequence.
(d) For $p_0 = p_1 = p$ find the probability $P(x_1)$ yielding the maximum value of $I(X,Y)$.

Solution

(a) The transmitted information for binary asymmetric channel with transition matrix P is

Problems

Fig. 4.9 Transmitted information as a function of binary channel parameters

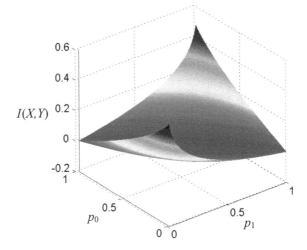

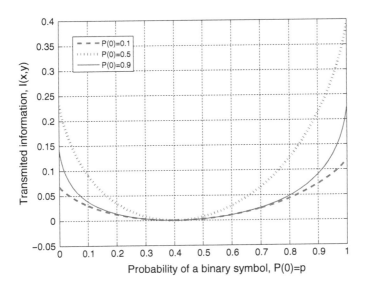

Fig. 4.10 The dependence of transmitted information on p_0 for $p_1 = 0.6$ and some values of $P(x_1)$

$$I(X,Y) = P(x_1)\left[(1-p_0)\ \text{ld}\ \frac{(1-p_0)}{P(y_1)} + p_0\ \text{ld}\ \frac{p_0}{P(y_2)}\right] + P(x_2)\left[p_1\ \text{ld}\ \frac{p_1}{P(y_1)} + (1-p_1)\ \text{ld}\ \frac{(1-p_1)}{P(y_2)}\right].$$

If the parameters p_0 and p_1 are not fixed, the transmitted information can be considered as a three-dimensional function shown in Fig. 4.9. The maximum is

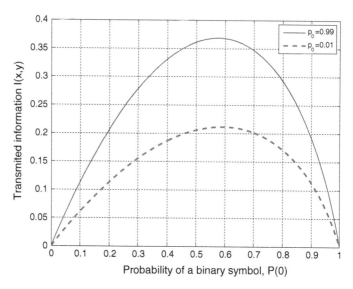

Fig. 4.11 Dependence of transmitted information on $P(x_1)$ for $p_1 = 0.6$ and different values of p_0

achieved for $p_0 = p_1 = 0$ and $p_0 = p_1 = 1$. Minimum is obtained for $p_0 + p_1 = 1$ (diagonal in coordinate system), for any probability of the input symbols.

(b) As shown in Fig. 4.10, for $p_1 = 0.6$, minimum $I(X, Y)$ is obtained for $p_0 = 0.4$, then the transmitted information equals zero for any value of $P(x_1)$. It can be concluded as well that when $p_0 + p_1 \neq 1$ the transmitted information is greater if the probabilities of the source symbols are more equalized.

(c) For a asymmetric channel, maximum $I(X, Y)$ does not corresponds to the case $P(x_1) = P(x_2) = 0.5$, as shown in Fig. 4.11. Putting $P(x_1) = a$, optimal a value is obtained as a solution of the equation

$$\frac{\partial I(X,Y)}{\partial a} = \frac{\partial}{\partial a}\left[a(1-p_0)\, \mathrm{ld}\, \frac{(1-p_0)}{(1-p_0)a + p_1(1-a)} + ap_0\, \mathrm{ld}\, \frac{p_0}{p_0 a + (1-p_1)(1-a)} \right]$$
$$+ \frac{\partial}{\partial a}\left[(1-a)p_1\, \mathrm{ld}\, \frac{p_1}{(1-p_0)a + p_1(1-a)} + (1-a)(1-p_1)\, \mathrm{ld}\, \frac{(1-p_1)}{p_0 a + (1-p_1)(1-a)} \right] = 0.$$

It is easy to verify that it is very difficult to find the optimum value of $P(x_1)$ in a closed form and it is easier to obtain it numerically, by finding the position of transmitted information maximum for a fixed p_0 and p_1. For $p_0 = 0.01$ and $p_1 = 0.6$ maximum value of transmitted information is obtained for $P(x_1)_{max} = 0.5770$ and in the case $p_0 = 0.99$ and $p_1 = 0{,}6$ for $P(x_1)_{max} = 0.5890$.

For $p_0 = p_1 = p$, the transmitted information corresponds to the main diagonal in Fig. 4.9. Then binary channel becomes binary symmetric channel (BSC) and the transmitted information is

$$I(X,Y) = \left((a+p-2ap)\operatorname{ld}\left(\frac{1}{a+p-2ap}\right) + (1-a-p+2ap)\operatorname{ld}\left(\frac{1}{1-a-p+2ap}\right)\right)$$
$$-\left(p\operatorname{ld}\left(\frac{1}{p}\right) + (1-p)\operatorname{ld}\left(\frac{1}{1-p}\right)\right).$$

The optimum value of parameter a is obtained on the basis of

$$\frac{\partial}{\partial a}\left([1-a-p+2ap]\operatorname{ld}\left(\frac{1}{1-a-p+2ap}\right) + (a+p-2ap)\operatorname{ld}\left(\frac{1}{a+p-2ap}\right)\right) = 0,$$

and it is easy to verify that it is satisfied only when

$$1 - a_{opt} - p + 2pa_{opt} = a_{opt} + p - 2pa_{opt},$$

and the optimum value $P(x_1)$ is $a_{opt} = 0.5$ regardless on p value. The maximum transmitted information is

$$I(X,Y)_{max} = 1 - \left(p\operatorname{ld}\left(\frac{1}{p}\right) + (1-p)\operatorname{ld}\left(\frac{1}{1-p}\right)\right).$$

Problem 4.4 The digital transmission system consists of:

1. Binary information source emitting symbols with rate $v_s = 100$ [Mb/s], the probability of one symbol is $P(0) = 0.25$;
2. Line encoder emitting polar pulses of amplitude $U = 1$ [V];
3. Channel consisting of N identical sections. Every section is a linear system with the transfer function corresponding to an ideal low-pass filter with the cutoff frequency $f_c = 1$ [MHz], the noise has a zero mean value and the variance $\sigma^2 = 0.1$ [W];
4. The receiver making decisions on the one sample basis, the threshold is put to zero;
5. The user receiving the reconstructed bit sequence.

   (a) Draw the system block-scheme under assumption that the line encoder, all sections and the receiver (making decisions) can be jointly represented by a discrete channel. Find the source information rate and the channel information rate.
   (b) Consider the case when a greater number of regenerative stations are in cascade, where the line coder, one section and decision block form an equivalent binary symmetric channel. In this case find the channel information rate for $N = 2$, $N = 10$ and $N = 100$.
   (c) Find the capacity of the equivalent discrete channel and draw it as a function of the number of sections. How it can be achieved?

(d) Find the capacity of channel when noise is present, obtained by removing line encoder and decision block (the case N = 1 is considered). How to achieve the capacity in this case?

Solution

(a) System block-scheme is shown in Fig. 4.12. The source information rate is

$$\Phi(S) = v_s \left(\frac{1}{4} \operatorname{ld} 4 + \frac{3}{4} \operatorname{ld} \frac{4}{3} \right) = 811.3 \left[\frac{\text{kSh}}{\text{s}} \right].$$

IT emits symbols through the channel. The channel is equivalently modeled as a binary symmetric channel (the noise influence is the same on the transmission of positive or negative pulses).
The error probability corresponding to this channel is

$$P_e = \frac{1}{2} \operatorname{erfc}\left(\frac{U}{\sqrt{2\sigma}}\right) = \frac{1}{2} \operatorname{erfc}\left(\sqrt{\frac{A_n}{2}}\right) = 7.82 \times 10^{-4},$$

where A_n denotes the signal-to-noise ratio in the channel [here $A_n = 10$ (not dB!)]. The symbols are emitted directly through the binary symmetric channel and transmission rate is

$$\Phi(X,Y) = v_s \left((1/4 + p/2) \operatorname{ld} \left(\frac{1}{1/4 + p/2}\right) + (3/4 - p/2) \operatorname{ld} \left(\frac{1}{3/4 - p/2}\right) \right)$$
$$- v_s \left(p \operatorname{ld} \left(\frac{1}{p}\right) + (1-p) \operatorname{ld} \left(\frac{1}{1-p}\right) \right) = 802.7 \left[\frac{\text{kSh}}{\text{s}} \right].$$

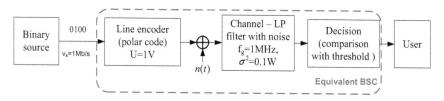

Fig. 4.12 Complete block-scheme of the system

Table 4.1 Parameters of channel having N sections for signal-to-noise ratio $a_n = 10 \log_{10}(A_n) = 10$ dB

N	1	2	10	100
$P_{ekv}^{(N)}$	7.82×10^{-4}	1.56×10^{-3}	7.77×10^{-3}	7.25×10^{-2}
$\Phi_{ekv}^{(N)}(X,Y)$ [Sh/s]	802,700	795,700	751,750	488,630
$C_{ekv}^{(N)}$ [Sh/s]	1,981,600	1,966,300	1,868,700	1,249,600

(b) Cascade of N symmetric binary channels is a binary symmetric channel as well (matrices are multiplied), the equivalent error probability is

$$p_{ekv}^{(N)} = (1 - (1-2p)^N)/2,$$

and the transmission rate is

$$\Phi_{ekv}^{(N)}(X,Y) = v_s \left((1/4 + p_{ekv}^{(N)}/2) \operatorname{ld}\left(\frac{1}{1/4 + p_{ekv}^{(N)}/2}\right) + (3/4 - p_{ekv}^{(N)}/2) \operatorname{ld}\left(\frac{1}{3/4 - p_{ekv}^{(N)}/2}\right) \right)$$

$$- v_s \left(p_{ekv}^{(N)} \operatorname{ld}\left(\frac{1}{p_{ekv}^{(N)}}\right) + (1 - p_{ekv}^{(N)}) \operatorname{ld}\left(\frac{1}{1 - p_{ekv}^{(N)}}\right) \right),$$

The corresponding numerical data are given in Table 4.1.

(c) Channel capacity is a maximum possible value of the transmission rate. The maximum signaling rate was chosen so as to avoid intersymbol interference [22], i.e. $v_{\max}(X,Y) = 2f_c = 2$ [Mb/s], while the maximum transmitting rate is achieved by optimizing the source symbol probabilities [2]

$$C_{ekv}^{(N)} = \max_{P(0)} \left(v_{\max}(X,Y) \Phi_{ekv}^{(N)}(X,Y) \right)$$

$$= v_{\max}(X,Y) \left(1 + p_{ekv}^{(N)} \operatorname{ld}(p_{ekv}^{(N)}) + (1 - p_{ekv}^{(N)}) \operatorname{ld}(1 - p_{ekv}^{(N)}) \right).$$

Capacities of the equivalent channel for 1, 2, 10 and 100 sections is given in Table 4.1, and the corresponding capacity dependence on the channel

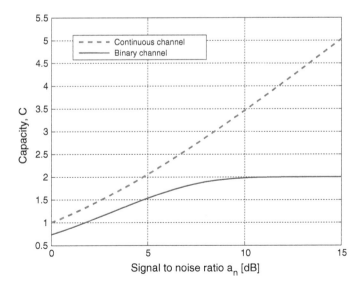

Fig. 4.13 Capacities of discrete and continuous channel corresponding to one section for $A_n = 10$

signal-to-noise ratio is shown in Fig. 4.13. Channel capacity can be used in full if at the channel input the source emits equiprobable symbols with twice greater rate than the earlier source.

If the line encoder and decision block are removed, a continuous channel with noise is obtained. The corresponding capacity is [1]

$$C = f_g \text{ ld } (1 + A_n),$$

and the channel capacity can be achieved only in the case where at his input a continuous information source generating the Gaussian random process is applied. In this case, the process at the channel (linear system) output has the Gaussian distribution as well.

It is obvious that binary channel has a smaller capacity compared to the corresponding continuous channel with noise. This is a consequence of binary signaling used at the transmitting part and of a hard decision at the receiving part, transforming the channel with noise into a binary symmetric channel. In the next problem, it will be shown that by applying a different decision rule at the receiver, this difference could be made smaller.

Problem 4.5 Digital communication system consists of

1. Information source emitting symbols $s_1, \ldots, s_6$, with rate $v_s = 800$ [symb/s]. Symbol probabilities are

$$P(s_1) = 0.29; \quad P(s_2) = 0.24; \quad P(s_3) = 0.21; \quad P(s_4) = 0.14;$$
$$P(s_5) = 0.07.$$

2. Binary encoder performing Huffman coding.
3. Channel consisting of two identical sections. Each one can be considered as a binary symmetric channel. Total error probability is 0.18.

   (a) Draw the system block-scheme, find the entropy and the source information rate.
   (b) Find the average code word length at the encoder output, code efficiency and the compression ratio of the obtained code. Find the probability of ones and zeros at the encoder output.
   (c) Find the transmitted information and the channel information rate if the receiver is positioned after the first channel section.
   (d) Under the condition (c) find the channel capacity if the source emits the symbols with maximum possible (allowed) rate.

Solution

(a) System block-scheme is shown in Fig. 4.14. On the basis of the five given probabilities, the probability of the sixth symbol is

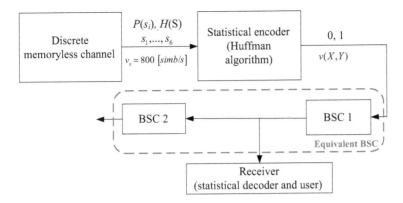

Fig. 4.14 System block-scheme

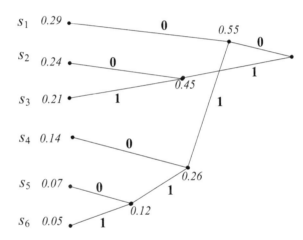

Fig. 4.15 Huffman procedure illustration

$$P(s_6) = 1 - \sum_{i=1}^{5} P(s_i) = 0.05,$$

and the entropy and information rate are

$$H(S) = -\sum_{i=1}^{6} P(s_i) \, \text{ld}\, (P(s_i)) = 2.366 \left[\frac{\text{Sh}}{\text{symb}}\right],$$

$$\Phi(S) = v_s H(S) = 1893.3 \left[\frac{\text{Sh}}{\text{s}}\right].$$

(b) The procedure of the Huffman code obtaining is shown in Fig. 4.15, and it is obvious that to symbols s_1–s_6 correspond the code words '00', '10', '11', '010', '0110' and '0111', respectively. Average code word length and efficiency are

$$L = \sum_{i=1}^{6} P(s_i)l_i = 2.38 \left[\frac{b}{\text{symb}}\right], \quad \eta = \frac{H(S)}{L} \times 100\% = 99.41\%.$$

The probability of zeros and ones can be found by calculating their average number per code words as follows

$$L_0 = 0.29 \times 2 + 0.24 \times 1 + 0.21 \times 0 + 0.14 \times 2 + 0.07 \times 2 + 0.05 \times 1 = 1.29,$$
$$L_1 = 0.29 \times 0 + 0.24 \times 1 + 0.21 \times 2 + 0.14 \times 1 + 0.07 \times 2 + 0,05 \times 3 = 1.09,$$

and

$$P(0) = L_0/L = 0.542 \; P(1) = L_1/L = 0.458,$$

and from the relation $L_0 + L_1 = L$, it follows $P(0) + P(1) = 1$ as well. Under the assumption that the sequence at the encoder output is uncorrelated, the entropy of binary symbols at the channel input can be found

$$H(X) = -0.542 \; \text{ld} \; (0.542) - P(0.458) \; \text{ld} \; (0.458) = 0.9949 \; [\text{Sh/symb}],$$

and as the signaling rate is

$$v(X,Y) = v_s L = 1904 \; [\text{b/s}]$$

the information rate at the encoder output is

$$\Phi(X) = v(X,Y)H(X) = 1894.3 \; [\text{Sh/s}].$$

In this case $\Phi(X) > \Phi(S)$, what is, of course, impossible because the encoder cannot introduce the new information. This illogical result is a consequence of the assumption that the sequence at the encoder output is uncorrelated, what is not true in fact. The introduced imprecision is small and it can be corrected having in mind that the previously found entropy is the entropy of adjoint source, and the true entropy is

$$H(X) = \Phi(S)/v(X,Y) = 0.9944 < 0.9949 = H(\bar{X}).$$

(c) If the probability of binary zero at the channel input is denoted by a, the transmitted information can be calculated using the expression

$$I(X,Y) = \left((a+p_1 - 2ap_1) \operatorname{ld} \left(\frac{1}{a+p_1 - 2ap_1}\right) + (1-a-p_1+2ap_1) \operatorname{ld} \left(\frac{1}{1-a-p_1+2ap_1}\right) \right)$$
$$- \left(p_1 \operatorname{ld} \left(\frac{1}{p_1}\right) + (1-p_1) \operatorname{ld} \left(\frac{1}{1-p_1}\right) \right),$$

where p_1 denotes the error probability corresponding to the one channel section. For two cascaded sections, the error probability in the equivalent channel becomes

$$p = (1 - (1-2p_1)^2)/2 = 2p_1 - 2p_1^2,$$

and

$$p_1 = \frac{1}{2}(1 \pm \sqrt{1-2p}),$$

and the equation has two solutions $p_1^{(1)} = 0.1$ and $p_1^{(2)} = 0.9$. The more logical value of the error probability is $p_1 = 0.1$ (the reader should examine the case $p_1 = 0.9$!), yielding

$$I(X,Y) = 0.5277 \text{ [Sh/simb]}, \quad \Phi(X,Y) = v(X,Y)I(X,Y) = 1004.8 \text{ [Sh/s]}.$$

The source is already using the maximum signaling rate, and the following is valid $v(X,Y) = v_{\max}(X,Y)$. Therefore, the capacity calculation is carried out by maximizing the transmitted information by the choice of the symbol probabilities at the channel input. If these symbols are equiprobable, it is obtained [1]

$$C = v(X,Y)\left(1 - p_1 \operatorname{ld} \left(\frac{1}{p_1}\right) - (1-p_1) \operatorname{ld} \left(\frac{1}{1-p_1}\right) \right) = 1011 \left[\frac{\text{Sh}}{\text{s}}\right].$$

Reader should try to respond to the following questions:

- whether by the different bit position at the Huffman tree the value nearer to the capacity can be obtained?
- whether the insertion of differential encoder between the Huffman encoder and the channel will add to the increase of the information rate and in this case, whether the greater quantity of information in the unit of time would be transmitted?

It is important to note that in this case $C < \Phi(S)$, therefore, it is obvious that for this channel (with a given error probability) it is impossible to transmit all information emitted by the source. However, if the channel state becomes substantially better—the probability of error decreasing to $p_1 = 10^{-4}$, the capacity should increase to $C = 1901.2$ [Sh/s].

It can be noticed that for the information transmission a more unreliable channel can be used if the encoder efficiency is smaller. In the case when the encoder is directly at the channel input, the redundancy introduced by the encoder allows the reliable transmission over the unreliable channel. In other words, without the redundancy, the information without any distortion can be transmitted over the completely reliable channel only. This feature will be used in theory of error control codes where the redundancy will be intentionally introduced to make possible the information transmission over very unreliable channels.

Problem 4.6 Binary source completely defined by the probability $P(0) = a$, sends the bits to the error control encoder which repeats every bit m times. These bits are further sent over the binary symmetric channel where the crossover probability is $p = 0.1$. On the receiving end is the error control decoder using majority logic decision.

Do the following:

(a) Find the entropy and information rate if the source emits the bits with the rate $v_i = 1$ [Mb/s] and if $a = 0.4$.
(b) Find the bit error probability observed by the user. Calculate the transmitted information over the equivalent channel. Draw its dependence as a function of the number of repetitions for $a = 0.1$, $a = 0.2$ and $a = 0.4$. Comment the results.
(c) If the maximum signaling rate over the channel is $v_b = 10$ [Mb/s], find the time needed for transmission of $N = 10^7$ bits under the condition that the error probability observed by the user is smaller than $P_{e,\min} = 10^{-2}$. Whether by the use of some procedure this time can be additionally shortened for $a = 0.2$?

Solution

(a) Entropy and information rate for zero-memory binary source are

$$H(S) = -a \operatorname{ld}(a) - (1-a) \operatorname{ld}(1-a) = 0.9709 \left[\frac{\text{Sh}}{\text{symb}}\right],$$

$$\Phi(S) = v_i[-a \operatorname{ld}(a) - (1-a) \operatorname{ld}(1-a)] = 970.9 \left[\frac{\text{kSh}}{\text{s}}\right].$$

(b) When applying n repetitions error control coding with majority logic and when the channel is symmetric and binary, with the crossover probability p, the error probability observed by user is

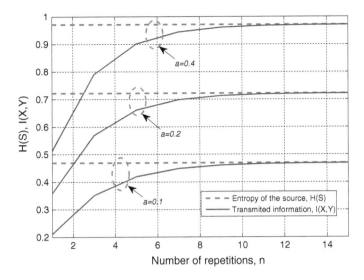

Fig. 4.16 Entropy of input sequence and the transmitted information for $p = 0.1$

$$P_e^{(n)} = \sum_{k=(n-1)/2+1}^{n} \binom{n}{k} p^k (1-p)^{n-k},$$

while the transmitted information through the equivalent channel (consisting of encoder with repetitions, channel and decoder) is

$$I^{(n)}(X,Y) = (a + P_e^{(n)} - 2aP_e^{(n)}) \, \text{ld}\, (\frac{1}{a + P_e^{(n)} - 2aP_e^{(n)}})$$
$$+ (1 - a - P_e^{(n)} + 2aP_e^{(n)}) \, \text{ld}\, (\frac{1}{1 - a - P_e^{(n)} + 2aP_e^{(n)}})$$
$$- P_e^{(n)} \, \text{ld}\, (\frac{1}{P_e^{(n)}}) - (1 - P_e^{(n)}) \, \text{ld}\, (\frac{1}{1 - P_e^{(n)}}).$$

The dependence of transmitted information on the number of repetitions is shown in Fig. 4.16. It is obvious that the error probability depends only on the channel characteristics (for a fixed decision rule), while the transmitted information depends on the source features (parameter a) as well.

The reader should calculate information rate through the equivalent channel (encoder, channel and decoder). Whether it can be increased by the increase of n, if the transmission bit rate through the channel is limited? It should be noticed that the signaling rate in this channel is v_i!

(c) Firstly, the value of n should be found for which the error probability is below the prescribed value, when the information rate is maximum. In this case $p = 10^{-2}$, and it is obtained successively

$$P_e^{(1)} = 10^{-2}, \quad P_e^{(3)} = 2.8 \times 10^{-2}, \quad P_e^{(5)} = 8.56 \times 10^{-3},$$
$$P_e^{(7)} = 2.72 \times 10^{-3}.$$

but the binary rate increases

$$v_b^{(1)} = v_i, \, v_b^{(3)} = 3v_i, \, v_b^{(5)} = 5v_i, \, v_b^{(7)} = 7v_i.$$

It is obvious that the needed transmission quality is achieved for $n \geq 5$, while the minimum binary rate under this condition is obtained just for $n = 5$ and the code with the code rate $R = 1/5$ is the optimum one, yielding the transmission rate

$$v_{i,\max} = Rv_{b,\max} = 2 \, [\text{Mb/s}].$$

For sending $N = 10^7$ bits, the needed time is

$$t = N/v_b = 5 \, [\text{s}].$$

This time can be decreased if a sufficient source extension is done, after that by the use compression encoding and after the repetition encoding as well. The error probability does not change by this procedure and the code with $R = 1/5$ is the optimum one. On the other hand, N bits of the source (S) alphabet can be substituted by $N \times H(S)$ symbols of code alphabet and for $a = 0.2$ it is obtained

$$t_{\min} = N \times H(S)/v_b = 3.61 \, [\text{s}].$$

Problem 4.7 Digital communication system consists of

1. Information source emitting the symbols $s_1, \ldots, s_7$, with the rate $v_s = 1000$ [symb/s]. Symbol probabilities are as follows

s_i	s_1	s_2	s_3	s_4	s_5	s_6	s_7
$P(s_i)$	0.3	0.2	0.15	0.15	0.1	0.05	0.05

2. Binary Huffman encoder and the corresponding decoder at the receiving end.
3. Error control encoder using threefold repetition and the corresponding decoder with majority logic at the receiving end (these blocks are included optionally).
4. Channel consisting of two cascaded sections where one is modeled by binary symmetric channel (crossover probability equals $p = 10^{-2}$) and the other is binary asymmetric channel with parameters $p_1 = 2p_0 = 2p$.

Problems

Find the following:

(a) Source information rate and code efficiency for Huffman encoding
(b) Find the information rate of the equivalent source at the channel input and information rate of the source adjoined to it (draw the system block-scheme)
(c) Channel information rate and the error probability without the error control coding.
(d) Information rate and the error probability when the error control coding is applied and the signaling rate cannot be additionally increased with respect to the case without error control coding. Specially comment how the error control coding influences the information rate!

Solution

(a) The source characteristics are easily obtained

$$H(S) = \sum_{i=1}^{7} P(s_i) \, \mathrm{ld} \, \frac{1}{P(s_i)} = 2.57 \left[\frac{\mathrm{Sh}}{\mathrm{symb}}\right], \quad \Phi(S) = v_s H(S) = 2570 \left[\frac{\mathrm{Sh}}{\mathrm{s}}\right].$$

Code words obtained by Huffman encoding are given in Table 4.2, from which the average code word length, code efficiency and the compression ratio are

$$L = \sum_{i=1}^{7} P(s_i) l_i = 2.6 \left[\frac{\mathrm{b}}{\mathrm{symb}}\right], \quad \eta = 98.88\%, \quad \rho = 1.1538.$$

The complete system block-scheme is shown in Fig. 4.17. It is obvious that the information rate at the channel input (i.e. at the Huffman encoder output) must be equal to the source information rate

$$\Phi(X) = \Phi(S) = 2570 \, [\mathrm{Sh/s}],$$

and while the signaling rate through the channel is

Table 4.2 One possible result of Huffman encoding

Symbols	$P(s_i)$	Code words
s_1	0.30	00
s_2	0.20	10
s_3	0.15	010
s_4	0.15	011
s_5	0.10	110
s_6	0.05	1110
s_7	0.05	1111

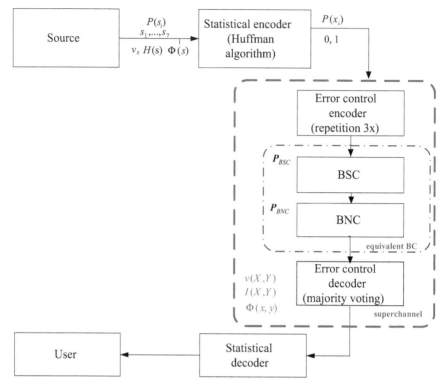

Fig. 4.17 System block-scheme, the notion of superchannel

$$v(x) = v_s L = 2600 \,[\text{b/s}],$$

the entropy of equivalent source consisting of zero-memory source and Huffman encoder is

$$H(X) = \Phi(X)/v(X,Y) = 0.9885 \,[\text{Sh/symb}].$$

To this equivalent source, the source can be joined completely described by symbol probabilities at the channel input

$$P(x_1) = L_0/L = 1.4/2.6 = 0.5385, \quad P(x_2) = L_0/L = 1.2/2.6 = 0.4615,$$

yielding

$$H(\bar{X}) = \sum_{i=1}^{2} P(x_i) \,\text{ld}\, \frac{1}{P(x_i)} = 0.9957 \,[\text{Sh/symb}],$$
$$\Phi(\bar{X}) = v(X,Y)H(\bar{X}) = 2588.82 \,[\text{Sh/symb}].$$

(b) The transmitted information of the equivalent binary source is

$$I(X,Y) = \sum_{i=1}^{2}\sum_{j=1}^{2} P(x_i, y_j) \operatorname{ld} \frac{P(y_j/x_i)}{P(y_j)},$$

where the probabilities correspond to the equivalent channel representing cascade of one binary symmetric channel and one binary (asymmetric) channel described by matrices

$$\mathbf{P}_{BSC} = \begin{bmatrix} v_{BSC,0} & p_{BSC,0} \\ p_{BSC,1} & v_{BSC,1} \end{bmatrix} = \begin{bmatrix} 1-p & p \\ p & 1-p \end{bmatrix},$$

$$\mathbf{P}_{NSC} = \begin{bmatrix} v_{NSC,0} & p_{NSC,0} \\ p_{NSC,1} & v_{NSC,1} \end{bmatrix} = \begin{bmatrix} 1-p & p \\ 2p & 1-2p \end{bmatrix}.$$

The equivalent channel has the transition matrix

$$\mathbf{P}_{ekv} = \begin{bmatrix} 1-p & p \\ p & 1-p \end{bmatrix}\begin{bmatrix} 1-p & p \\ 2p & 1-2p \end{bmatrix}$$
$$= \begin{bmatrix} (1-p)^2 + 2p^2 & p(2-3p) \\ 3p(1-p) & (1-p)(1-2p)+p^2 \end{bmatrix},$$

and for $p = 10^{-2}$ it is obtained

$P(y_1/x_1) = (1-p)^2 + 2p^2 = 0.9803, \quad P(y_2/x_1) = p(2-3p) = 0.0197,$
$P(y_1/x_2) = 3p(1-p) = 0.0297, \quad P(y_2/x_2) = (1-p)(1-2p)+p^2 = 0.9703.$

Probabilities of symbols $y_1 = 0$ and $y_2 = 1$ at the channel output are

$P(y_1) = 0.5279 + 0.0137 = 0.5416, \quad P(y_2) = 0.0106 + 0.4478 = 0.4584,$

and the corresponding joint probabilities are

$P(x_1, y_1) = 0.5385 \times 0.9803 = 0.5279, \quad P(x_1, y_2) = 0.5385 \times 0.0197 = 0.0106,$
$P(x_1, y_1) = 0.4615 \times 0.0297 = 0.0137, \quad P(x_2, y_2) = 0.4615 \times 0.9703 = 0.4478.$

Finally, the numerical value of transmitted information and of information rate can be found

$I(X,Y) = 0.8307 \text{ [Sh/symb]}, \quad \Phi(X,Y) = v(X,Y)I(X,Y) = 2159.82 \text{ [Sh/s]}.$

In this case it is obvious that the optimum decision rule is $d(y_1) = x_1$, $d(y_2) = x_2$ and the error probability without the error control coding is

$$P_{e1} = P(x_1)P(y_2/x_1) + P(x_2)P(y_1/x_2) = 0.0243.$$

(c) The case with three repetitions now will be considered. If at the output of Huffman encoder the binary zero is sent to the error control encoder, at its output the sequence '000' is generated and at the error control decoder output binary zero will be generated in the cases without transmission errors or when only one error occurred at these three bits. Otherwise, the error will occur for binary zero transmission. Therefore, the superchannel (consisting of error control encoder, channel and error control decoder) parameters are

$$P_{SK}(y_1/x_1) = P(y_1/x_1)^3 + 3P(y_1/x_1)^2 P(y_2/x_1) = 0.9989,$$
$$P_{SK}(y_2/x_1) = 3P(y_1/x_1)P(y_2/x_1)^2 + P(y_2/x_1)^3 = 0.0011,$$
$$P_{SK}(y_1/x_2) = 3P(y_2/x_2)P(y_1/x_2)^2 + P(y_1/x_2)^3 = 0.0026,$$
$$P_{SK}(y_2/x_2) = P(y_2/x_2)^3 + 3P(y_2/x_2)^2 P(y_1/x_2) = 0.9974.$$

The probabilities of input symbols are not changed and the output symbol probabilities are

$$P_{SK}(y_1) = 0.5391, \quad P_{SK}(y_2) = 0.4609,$$

while the corresponding joint probabilities are

$$P_{SK}(x_1, y_1) = 0.5279, \quad P_{SK}(x_1, y_2) = 0.0006,$$
$$P_{SK}(x_1, y_1) = 0.0012, \quad P_{SK}(x_2, y_2) = 0.4603.$$

Binary rate in the channel is three times greater than the rate at the coder input – the code rate is $R = 1/3$. If the binary rate in the channel cannot be increased (the channel is the same) information rate at the error control coder must be decreased three times, i.e.

$$v_{SK}(X, Y) = R \times v(X, Y),$$

and the source should emit the symbols three times slower.
Finally, transmitted information and information rate can be calculated

$$I_{SK}(X, Y) = 0.9766 \text{ [Sh/symb]},$$
$$\Phi_{SK}(X, Y) = v_{SK}(X, Y)I_{SK}(X, Y) = 846.39 \text{ [Sh/s]}.$$

Error probability in this case is

$$P_{e1} = P(x_1)P(y_2/x_1) + P(x_2)P(y_1/x_2) = 0.0018,$$

being 13.5 times smaller than in the case without error control coding.
It is important to notice the following

- If the channel is connected directly to the information source, the channel cannot transmit more information (per second) than the source sent, i.e.

$$\Phi(X,Y) \leq \Phi(S)$$

- Channel is capable to transmit the information with success if

$$C \geq \Phi(S),$$

i.e. the channel have to be good enough to transmit all information.
- To transmit the complete information that the source sent, the following must hold

$$\Phi(X,Y) = \Phi(S),$$

where the relation is satisfied only if $\eta R = I(X, Y)$—here for perfect compression code the relation can be satisfied even and for non-ideal channel if $R = I(X, Y)$!
- It is desirable that the ratio $\Phi(X,Y)/C$ should be as great as possible, allowing that for the information transmission the channel having capacity as small as possible can be used. It is also desirable that the signaling rate is as small as possible for a given information rate. By source encoding a non-destructive compression is achieved and the difference between these two rates is smaller (in such a way demands for a frequency band are partially lessened).
- By error control coding the reliability of transmission can be increased (the error probability is decreased) but the information rate cannot be greater, i.e.

$$Rv(X,Y)I_{SK}(X,Y) \leq v(X,Y)I(X,Y).$$

- If the error control code enabling the negligible error probability is applied, the transmitted information through the superchannel is $I_{SK}(X,Y) = 1$, the above relation becoming

$$R \leq I(X,Y),$$

which corresponds to the code rate predicted by the Second Shannon theorem [10].

Problem 4.8 Zero-memory binary source, where the probability of binary zero is $P(x_1)$, is at the input of line encoder which generates polar binary signal with voltage levels ± 1. From line encoder the signal is emitted into the channel with additive white Gaussian noise ($\sigma = 0.5$). The decision block is at the channel output with two thresholds $\pm G$ ($0 \leq G \leq 1$).

(a) Find the transmitted information for $P(x_1) = 0.5$ and $G = 0.5$.
(b) Draw the transmitted information versus G when $P(x_1) = 0.5$.
(c) Draw the transmitted information versus $P(x_1)$ for $G = 0$, $G = 0.5$ and $G = 1$ and compare it with source entropy.

Solution
Signal (amplitude) probability density function at the channel output depends on voltage level emitted by line encoder. It is shown in Fig. 4.18. If the received signal is below the lower threshold ($u < -G$), the decision is that the symbol '0' is at the discrete channel output, while, in the case $u > G$, the decision is '1'. If the received signal is between thresholds, the decision would not be sufficiently reliable. In this case the symbol is considered as *erased*, i.e. to it formally corresponds symbol E. The probabilities for the received signal to be between the thresholds, and above the more distant threshold are

$$p_1 = w_Y(-G < y < G/x = -1) = w_Y(-G < y < G/x = +1) = \frac{1}{2}\text{erfc}\left(\frac{1-G}{\sqrt{2}\sigma}\right) - \frac{1}{2}\text{erfc}\left(\frac{1+G}{\sqrt{2}\sigma}\right),$$

$$p_2 = w_Y(y > G/x = -1) = w_Y(y < -G/x = +1) = \frac{1}{2}\text{erfc}\left(\frac{1+G}{\sqrt{2}\sigma}\right).$$

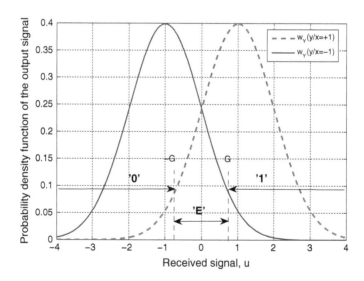

Fig. 4.18 Probability density functions of the signal at channel output, equivalence with BEC (Binary Erasure Channel)

Transition matrix of the binary erasure channel is

$$P = \begin{bmatrix} 1-p_1-p_2 & p_1 & p_2 \\ p_2 & p_1 & 1-p_1-p_2 \end{bmatrix},$$

where for the erasure channel often is supposed $p_2 \approx 0$, not supposed here. Transmitted information is

$$I(X,Y) = P(x_1)\left[(1-p_1-p_2)\operatorname{ld}\frac{(1-p_1-p_2)}{P(y_1)} + p_1\operatorname{ld}\frac{p_1}{P(y_2)} + p_2\operatorname{ld}\frac{p_2}{P(y_3)}\right]$$
$$+ P(x_2)\left[p_2\operatorname{ld}\frac{p_2}{P(y_1)} + p_1\operatorname{ld}\frac{p_1}{P(y_2)} + (1-p_1-p_2)\operatorname{ld}\frac{(1-p_1-p_2)}{P(y_3)}\right],$$

where

$$P(y_1) = (1-p_1-p_2)P(x_1) + p_2 P(x_2),$$
$$P(y_2) = p_1 P(x_1) + p_1 P(x_2),$$
$$P(y_2) = p_2 P(x_1) + (1-p_1-p_2)P(x_2).$$

For $P(x_1) = 0.5$ and $G = \sigma = 0.5$ it is obtained $p_1 = 0.1573$ and $p_2 = 0.0013$ yielding $P(y_1) = 0.4213$, $P(y_1) = 0.1573$, $P(y_1) = 0.4213$ and finally $I(X,Y) = 0.8282$ [Sh/symb].

The transmitted information dependence on the threshold absolute value is shown in Fig. 4.19. It is obvious that for every value of $P(x_1)$, there are the

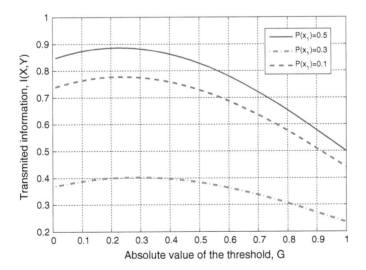

Fig. 4.19 The transmitted information dependence on the threshold G absolute value

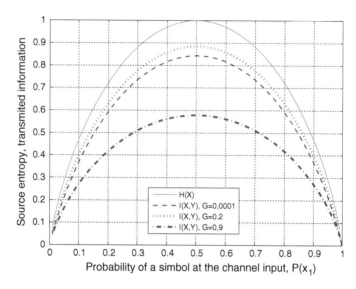

Fig. 4.20 The dependence of the transmitted information on the input signal probabilities

optimum thresholds values $-G$ and G yielding the maximum transmitted information. It is interesting that for $P(x_1) = 0.5$ more information is transmitted when $G = 0.2$ than for $G = 0$ (when BEC is reduced to BSC). The dependence of the transmitted information on the input signal probabilities is shown in Fig. 4.20, leading to the conclusion that the optimal case is when the input symbols are equiprobable.

Problem 4.9 The information source whose rate is $v_s = 10^6$ [symb/s], is defined by the following table

s_i	s_1	s_2	s_3	s_4	s_5	s_6	s_7	s_8	s_9	s_{10}
$P(s_i)$	0.18	0.2	0.14	0.15	0.1	0.11	0.05	0.04	0.02	0.01

Whether it is justified to put the source output at the input of the channel modeled as a pass-band filter with the frequency band $B = 1$ [MHz] if the noise power spectrum density is $N_0 = 10^{-14}$ [W/Hz] and if the signal power is $P_s = 0.1$ [µW]?

Solution
The source information rate is

$$\Phi(S) = -v_s \sum_{i=1}^{8} P(s_i) \operatorname{ld}(P(s_i)) = 2.981 \text{ [MSh/s]},$$

and the continuous channel capacity

$$C = B \operatorname{ld}\left(1 + \frac{P_s}{P_n B}\right) = 10^6 \operatorname{ld}(11) = 3.46 \text{ [MSh/s]}.$$

The information rate is smaller than the channel capacity, and it is justified to use the channel for a given source. According to the first Nyquist criterion for transmission in the transposed frequency band, it should be provided for the signaling rate to be smaller than $v_{max} = 1$ [Msymb/s] (the set of signaling symbols, as a rule, is not the same as emitted by the source!). As the minimum binary rate over the channel is determined by the information rate, it is obvious that BPSK cannot be used, but some M-level modulation scheme could be used, providing $\Phi(S)/\operatorname{ld}(M) < v_{max}$.

Problem 4.10 Zero-memory binary source generating equiprobable symbols is at the line encoder input which generates polar binary signal having the voltage levels $\pm U$. From line encoder the signal is emitted over the channel with additive white Gaussian noise (standard deviation is σ). At the channel output are the quantizer ($s = 2^n$ quantization levels) and a corresponding decision block. The bounds of the quantization intervals are

$$U_{gr} = l \times (4U)/s - 2U, \quad l = 1, \ldots, s-1,$$

and at its output one of the possible M discrete symbols is generated.

(a) Consider equivalent discrete channel consisting of line encoder, channel, quantizer and decision block. Find the transmitted information for equivalent channel for $U = 1$ [V] and $\sigma = 0.5$ [V], when $s = 2$, $s = 4$, $s = 8$, $s = 16$ and $s = 32$.
(b) For a considered discrete channel find the capacity dependence on signal-to-noise ratio for various values of parameter s and compare it to the continuous channel capacity.

Solution

(a) Because of the quantizer, the complete voltage range is divided into s non-overlapping intervals every corresponding to one discrete symbol. When $s = 4$ the signal probability density function at the channel output is shown in

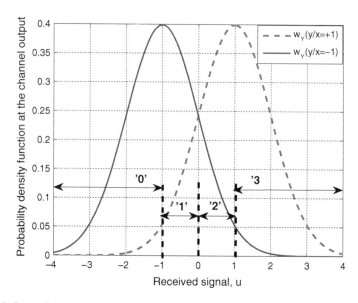

Fig. 4.21 Determining of probabilities of the channel outgoing symbols

Fig. 4.22 Graph corresponding to discrete channel with $r = 2$ inputs and $s = 4$ outputs, to every quantization level corresponds one discrete symbol

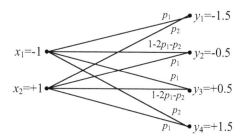

Fig. 4.21, and a graph corresponding to discrete channel with $r = 2$ inputs and $s = 4$ outputs is shown in Fig. 4.22.

The bounds of quantizing intervals are

$$U_{gr}(l) = l \times (4U)/s - 2U, l = 1, \ldots, s-1$$

and by defining $U_x(1) = U$, $U_x(2) = -U$, the channel transition probabilities can be easily found from Fig. 4.21

$$P(y_j/x_i) = p_{|j-i|}$$

$$= \begin{cases} \frac{1}{2}\text{erfc}\left(\min\left\{\frac{U_X(i)-U_{gr}(j-1)}{\sqrt{2}\sigma}, \frac{U_X(i)-U_{gr}(j)}{\sqrt{2}\sigma}\right\}\right) \\ \quad -\frac{1}{2}\text{erfc}\left(\max\left\{\frac{U_X(i)-U_{gr}(j-1)}{\sqrt{2}\sigma}, \frac{U_X(i)-U_{gr}(j)}{\sqrt{2}\sigma}\right\}\right), & i \neq j, 1<j<s \\ \frac{1}{2}\text{erfc}\left(\frac{U_X(i)-U_{gr}(1)}{\sqrt{2}\sigma}\right), & i \neq j = 1 \\ \frac{1}{2}\text{erfc}\left(\frac{U_X(i)-U_{gr}(s-1)}{\sqrt{2}\sigma}\right), & i \neq j = s \\ 1 - \sum_{j \neq i} P(y_j/x_i) & i = j. \end{cases}$$

where $i = 1, 2$ and $j = 1, 2, \ldots, s$.
Transmitted information is

$$I(X, Y) = \sum_{i=1}^{2}\sum_{j=1}^{s} P(x_i, y_j) \, \text{ld} \, \frac{P(y_j/x_i)}{P(y_j)},$$

and having in view the equiprobable binary source symbols ($P(x_1) = P(x_1) = 0.5$) joint probabilities are

$$P(x_i, y_j) = P(y_j/x_i)/2, \quad \forall i, j.$$

The source is symmetric and the maximum transmitted information value is achieved for equiprobable symbols

$$I_{\max}(X, Y) = \frac{1}{2}\sum_{i=1}^{2}\sum_{j=1}^{s} P(y_j/x_i) \, \text{ld} \, \frac{2P(y_j/x_i)}{P(y_j/x_1) + P(y_j/x_2)}.$$

For $U = 1$ [V] and $\sigma = 0.5$ [V], the transmitted information for a given values of s is given in Table 4.3. It is obvious that with the increased number of quantization levels, the transmitted information increases as well, but after some number of levels, the saturation is achieved. In the case when the number of symbols at the channel input is very great—corresponding to the levels being very close each other—the channel with discrete input ($r = 2$) and a continuous output is obtained. It should be noticed as well that the quantization levels here are chosen in a suboptimal way (only the range from $-2U$ to $+2U$ was quantized) and that the additional increase in the transmitted

Table 4.3 Transmitted information for various values of s

s	2	4	8	16	32	64	256	32,768
$I(X,Y)$	0.8434	0.8662	0.8968	0.9087	0.9118	0.9126	0.9128	0.9128

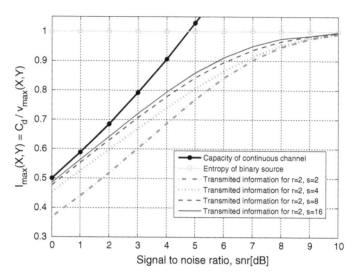

Fig. 4.23 Normalized continuous channel capacity and capacity of the corresponding discrete channel with $r = 2$ inputs and s outputs

information can be achieved, for the same number of quantizing levels, by quantizer optimization.

(b) Capacity of the above defined discrete channel with $r = 2$ inputs and s outputs is

$$C = v_{\max}(X,Y) I_{\max}(X,Y)$$
$$= \frac{1}{2} v_{\max}(X,Y) \sum_{i=1}^{2} \sum_{j=1}^{s} P(y_j/x_i) \operatorname{ld} \frac{2P(y_j/x_i)}{P(y_j/x_1) + P(y_j/x_2)}$$

Having in view that the data for calculating the maximum signaling rate for a channel are not available, the capacity will be normalized to that rate (this approach is very often used in the literature).

The corresponding numerical values are shown in Fig. 4.23. It is obvious that for all signal-to-noise ratio values the increasing of the number of quantization levels increases the transmitted information as well. In such a way it is practically confirmed that *soft* decision is superior to *hard* decision where the received signal is compared to one threshold value only.

It is interesting to find as well the capacity of the continuous channel with noise, the channel corresponding to the system part from line encoder output to the quantizer input.

$$C = B \operatorname{ld}(1 + P_s/P_n).$$

Although the frequency band is not given in this problem, it is obvious that the signaling is in the baseband and the condition $v_{\max}(X,Y) \leq 2B$ must be satisfied, and the maximum channel capacity (corresponding to the system modeled as an ideal low-pass filter) is

$$C/v_{\max}(X,Y) = \text{ld}\left(1+U^2/\sigma^2\right)/2.$$

Of course, it is obvious that the continuous channel capacity is always greater than that of the discrete channel adjoined to it. However, it should be noted that the transmitted information of a discrete source channel cannot be greater than the entropy of the input discrete source.

Problem 4.11 Zero-memory binary source generates equiprobable symbols. Its n-th order extension is led at the line encoder input. Line encoder generates multilevel polar signal having $M = 2^n$ equidistant voltage levels, average power is $P_s = 1$ [W]. Line encoder output is transmitted through the continuous channel modeled by an ideal low-pass filter having the cutoff frequency $f_c = 100$ [kHz], and where the additive Gaussian noise has the average power density spectrum $N_0 = 2.5 \times 10^{-6}$ [W/Hz]. At the channel output are the quantizer (M levels) and the corresponding decision block. Quantization levels bounds are

$$U_{gr}(l) = l \times 2U - MU, l = 1, \ldots, M-1,$$

and at its output one of the possible M discrete symbols is generated.

(a) Draw the block-scheme if the equivalent source consists of binary source and the block for its extension, and the equivalent discrete channel consists of line encoder, channel, quantizer and the decision block. Find the entropy of the equivalent source.
(b) Find the transmitted information for the equivalent channel when $M = 2$, $M = 4$, $M = 8$ and $M = 16$.
(c) For a previously defined discrete source find the dependence of capacity on the signal-to-noise ratio for the various values of parameter s and compare it to the continuous channel capacity.

Solution

(a) System block-scheme is shown in Fig. 4.24. It is obvious that the binary source combined with the extension block emits M equiprobable symbols, and the entropy of the combined source is

$$H(S^n) = \sum_{i=1}^{2^n} \frac{1}{2^n} \text{ld}\left(2^n\right) = nH(S) = n \,[\text{Sh/symb}].$$

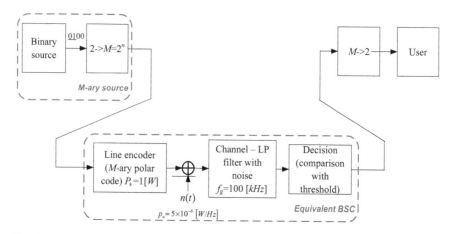

Fig. 4.24 System block-scheme

The receiving part of this system is practically the same as in the previous problem. The transmitting part differs, because n binary symbols are firstly grouped (in the extension block) and every symbol in line coder is represented by one (from M) voltage level

$$U_x(i) = -(M-1)U + 2(i-1)U, \quad i = 1, 2, \ldots, M.$$

Multilevel signal has a unity power and the following relation must hold

$$P_{sr} = \frac{1}{M}\sum_{i=1}^{M} U_x^2(i) = \frac{1}{M}\sum_{i=1}^{M}(2i - M - 1)^2 U^2 = \frac{U^2}{M}\left[4\sum_{i=1}^{M} i^2 - 4(M+1)\sum_{i=1}^{M} i + (M+1)^2\right] = 1,$$

after using the relations

$$\sum_{i=1}^{M} i = \frac{M(M+1)}{2}, \quad \sum_{i=1}^{M} i^2 = \frac{M(M+1)(2M+1)}{6},$$

Fig. 4.25 Graph of symmetric channel with $r = 4$ inputs and $s = 4$ outputs

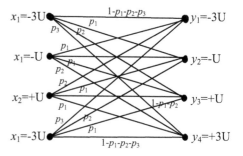

the corresponding value is obtained

$$U = \sqrt{\frac{3}{M^2 - 1}}.$$

(b) Equivalent discrete channel now has M inputs and M outputs and for $M = 4$ is described by the graph shown in Fig. 4.25, and the corresponding transition matrix is

$$\mathbf{P} = \begin{bmatrix} 1 - p_1 - p_2 - p_3 & p_1 & p_2 & p_3 \\ p_1 & 1 - 2p_1 - p_2 & p_1 & p_2 \\ p_2 & p_1 & 1 - 2p_1 - p_2 & p_1 \\ p_3 & p_2 & p_1 & 1 - p_1 - p_2 - p_3 \end{bmatrix}.$$

Transition probabilities are found by using the same equalities as in the previous problem, with only a few changes—parameter s is substituted by M, values $U_{gr}(l)$ are given in the problem text being

$$U_X(i) = (2i - M - 1)U = \frac{9(2i - M - 1)}{\sqrt{M^2 - 1}}, \quad i = 1, 2, \ldots, M.$$

The frequency band is $B = 100$ [kHz], maximum signaling rate is always $v_{s,max} = 2B = 200$ [ksymb/s] for the M-ary symbols. The equivalent binary rate is $v_{b,max} = 2f_c \times \mathrm{ld}(M)$. In this case, for $\sigma^2 = N_0 f_c = 0.25$ [W], transmitted information for some values of M is given in Table 4.4. Normalized capacity of the corresponding continuous channel is $C = 1.1610$ [Sh/symb]. The dependence of transmitted information on the signal-to-noise ratio in the channel is shown in Fig. 4.26. When the number of discrete symbols at the input and at the output of the channel increases, the discrete channel capacity could converge to the continuous channel capacity, seeming not being here completely the case. The basic problem of the above analysis is that is it limited to equiprobable input symbols only.

It should be noticed that in this case the channel is not completely symmetric, which can be easily seen from the graph or from the transition matrix. Although all parameters corresponding to the inputs x_1 and x_4 (combinations '00' and '11' from the binary source) are identical, channel parameters corresponding to the inputs x_2 and x_3 are different. It implicates that it is optimally to choose the probabilities

Table 4.4 Transmitted information for some values of M (equiprobable input symbols)

s	2	4	8	16	32	64	1024
$I(X,Y)$ [Sh/symb]	0.8434	0.9668	1.0691	1.1025	1.1123	1.1153	1.1170

$$P(x_1) = P(x_4),\ P(x_2) = P(x_3) = 1/2 - P(x_1).$$

Each of this probabilities should be in the range from 0 to 0.5, but from the Fig. 4.27 it is obvious that the optimal value of probability $P(x_1) = P(x_4)$ depends on the noise power and has a greater value when the noise power is

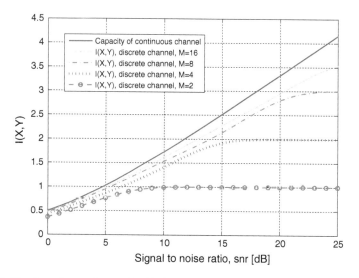

Fig. 4.26 The dependence of transmitted information on the signal-to-noise ratio

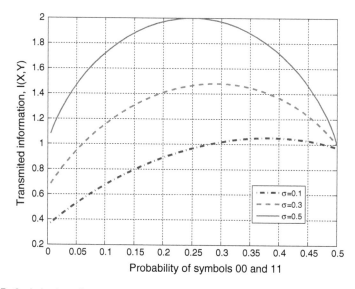

Fig. 4.27 Optimization of parameters $P(00) = P(11)$ for signaling with $M = 4$ levels

greater. It is logical—for a strong noise more distant levels should be used to reduce the noise influence.

Therefore, for $\sigma = 0.5$ [V] ($SNR = 6$ [dB]) by input probabilities optimization the transmitted information can be increased from 0.9668 [Sh/symb] for equiprobable symbols, to 1.0513 [Sh/symb] for the optimum found input probabilities $P(x_1) = P(x_4) = 0.77$, $P(x_2) = P(x_3) = 0.23$.

Of course, an additional gain is possible if for a fixed number of signaling levels at the channel input, an additional quantization is performed at the receiving end in such a way that the number of symbols at the output is greater than at the input, i.e. $r = M$, $s > M$, what will correspond to the multilevel signaling with a soft decision.

Problem 4.12 In one data transmission system zero-memory binary source sends bits with the rate $v_b = 100$ [kb/s] to the BPSK modulator. Modulated signal with the average power $Ps = 5$ [µW] is transmitted over the channel modeled by an ideal pass-band filter. The average power density spectrum of the additive white Gaussian noise in the channel is $N_0 = 2 \times 10^{-11}$ [W/Hz]. The bandwidth is chosen so as to eliminate the intersymbol interference obtaining in the same time the minimum noise power at the channel output.

Beside the Gaussian noise, in the channel there is the impulsive noise which has the average pulse duration $T_L = 50$ [µs] and the average interval between two pulses is $T_D = 500$ [µs]. The impulsive noise has also the Gaussian distribution and the average power spectral density $p_i = 10^{-9}$ [W/Hz], being constant as well in the frequency band of interest. The BPSK signal receiver is using integrate and dump circuit.

(a) Find the signal-to-noise ratio in the channel in intervals without the impulsive noise, as well as when it is active.
(b) Find the error probability in intervals without the impulsive noise as well as when it is active. By which discrete model the corresponding channel error sequence can be modeled?
(c) Find the error probability in the good and in the bad state (impulsive noise), as well as the average number of bits transmitted in every state. Find the channel average error probability.
(d) Draw the autocorrelation function of the corresponding error sequence.

Solution

The channel is modeled as an ideal pass-band filter. The bandwidth is chosen so as to eliminate the intersymbol interference, and for such transmission the rate corresponds to the bandwidth multiplied by the any integer n, being

$$v_{b,\text{no ISI}} = B/n.$$

Minimum noise power at channel output is obtained for $n = 1$ yielding

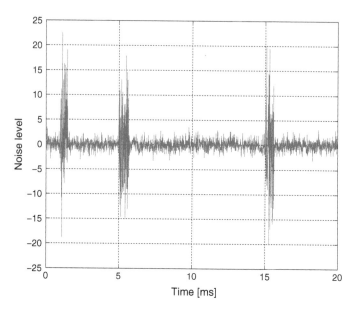

Fig. 4.28 The form of channel noise

$$B_{\min,\text{no ISI}} = v_b = 100\,[\text{kHz}].$$

Let the "good" state corresponds to the intervals without the impulsive noise in the channel. In this case noise power and the average signal-to-noise ratio are

$$\sigma_n^2 = p_n B_{\min,\text{no ISI}} = 2\,[\mu W], \quad SNR_D = \frac{P_s}{\sigma_n^2} = 2.5.$$

The "bad" state corresponds to the intervals when both noises are present, and the impulsive noise power, total noise power and the average signal-to-noise ratio are

$$\sigma_i^2 = p_i B_{\min,\text{no ISI}} = 100\,[\mu W], \sigma_{tot}^2 = \sigma_n^2 + \sigma_i^2 = 102\,[\mu W], SNR_L = \frac{P_s}{\sigma_{tot}^2} = 0.0490.$$

The time form of the channel noise signal is illustrated in Fig. 4.28.

(a) The receiver is realized as a integrate and dump circuit, the probability of error is

$$P_e = \frac{1}{2}\text{erfc}\sqrt{\frac{E_b}{N_0}}.$$

From the relation

Fig. 4.29 Gilbert-Elliott channel model

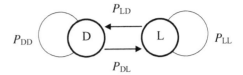

$$\frac{E_b}{N_0} = \frac{P_s/v_b}{P_n/B} = SNR\frac{B}{v_b},$$

and for $v_b = B$, in this case, the error probabilities in good and bad states are

$$p_D = \frac{1}{2}\text{erfc}\sqrt{SNR_D} = 0.0127,$$

$$p_L = \frac{1}{2}\text{erfc}\sqrt{SNR_L} = 0.3771.$$

This channel can be successfully modeled using the Gilbert-Elliott model [20, 21] where it is supposed that every state can be considered as a binary symmetric channel, the corresponding matrices are

$$\boldsymbol{P_G} = \begin{bmatrix} v_G & p_G \\ p_G & v_G \end{bmatrix}, (v_G + p_G = 1) \quad \text{and} \quad \boldsymbol{P_B} = \begin{bmatrix} v_B & p_B \\ p_B & v_B \end{bmatrix}, (v_B + p_B = 1),$$

and the transition from one state into another can be illustrated graphically, as shown in Fig. 4.29.

(b) Stationary probabilities of good and bad state in Gilbert-Elliott model are fully determined by transition probabilities from good into the bad state (P_{GB}) and vice versa (P_{BG}) being

$$\pi_G = \frac{P_{BG}}{P_{GB} + P_{BG}}, \quad \pi_B = \frac{P_{GB}}{P_{GB} + P_{BG}},$$

the number of bits emitted in the states is

$$N_G = 1/P_{GB}, \quad N_B = 1/P_{BG}.$$

On the base of the average duration of time intervals in the good and the bad state ($T_G = 500$ [μs], $T_B = 50$ [μs]) with the signaling interval duration $T_b = 1/v_b = 10$ [μs], a model should be found where in the good state the average number of emitted bits is $N_G = T_G/T_b = 50$, and for every sojourn in the bad state the average number of emitted bits is $N_B = T_B/T_b = 5$. It can be easily found

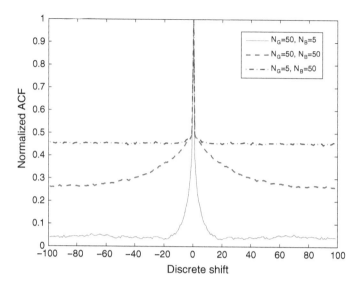

Fig. 4.30 Normalized sequence of error autocorrelation function corresponding to Gilbert-Elliott model

$$P_{BG} = 1/N_B = 0.2; P_{BB} = 1 - P_{BG} = 0.8; P_{GB} = 1/N_G = 0.02;$$
$$P_{GG} = 1 - P_{GB} = 0.98,$$

yielding

$$\pi_G = 0.9091, \quad \pi_B = 0.0909.$$

The channel average error probability is

$$P_e = p_G \pi_G + p_B \pi_B = \frac{p_G P_{BG}}{P_{GB} + P_{BG}} + \frac{p_B P_{GB}}{P_{GB} + P_{BG}} = 0.0458.$$

(d) Typical state sequence is

...GGGGGGGGGGGGBBBGGGGGGGGGGGGGGBBBBBGGGGGGGGGGGGGGGG...

where G denotes that the channel in this signaling interval (corresponding to one bit transmission) is in good state, while B corresponds to the bad state. The corresponding error sequence (it could be obtained by adding modulo 2 sequences of transmitted and received bits) is

...00000100000**0101**000000000000**10101**00000000000000...

where the bold and underlined bits correspond to the bad state.
Normalized sequence of error autocorrelation function obtained on the base

$$R_e(k) = \frac{1}{N}\sum_{n=1}^{N} e(n)e(n+k),$$

it was estimated by Monte Carlo simulation and it is shown in Fig. 4.30.
The autocorrelation function value for zero shift corresponds to the average error probability P_e, it is substantially greater for $N_B > N_G$ than for $N_B < N_G$. For this reason, the normalization is carried out, to allow a more noticeable difference in the function shape. Of course, in the second chapter of this book, it was shown that the discrete sequence autocorrelation function must be discrete, but in this case, to allow a clearer insight, only the envelope of the function is drawn.

It is interesting that in the limiting cases a Gilbert-Elliott model is practically reduced to binary symmetric channel (for $N_B \gg N_G$ the channel is almost always in a bad state, and for $N_G \gg N_B$ the channel is almost always in a good state). In this case the following is valid

$$R_e(k) \approx \begin{cases} p, & k=0 \\ p^2, & k \neq 0 \end{cases}$$

where $p = p_B$ for $N_B \gg N_G$ and $p = p_G$ for $N_G \gg N_B$.
For $N_G = N_B$, the shape of autocorrelation function clearly shows that the error sequence is correlated, i.e. corresponding to the channel with memory.

Chapter 5
Block Codes

Brief Theoretical Overview

In the remaining part of this book error control codes will be considered. As said earlier, these codes enable detection and possible correction of transmission errors. Usually, it is supposed that at the encoder input there is a series of bits, statistically independent and equally probable. It can be a result of previous data compression or scrambling. To detect and correct the errors some redundancy should be added. Depending on the way how the redundancy is added, the error control codes are divided into two families—***block codes*** and ***convolutional codes***. Further subdivision is possible, but it will be postponed and commented in the corresponding chapters. Block codes are invented before convolutional codes. These last will be considered in the next chapter.

From the name it is obvious that block encoding consists of taking a block of input bits (k bits) and representing it with the block of output bits (n bits). This code is denoted as (n, k) code. In this case the ***code rate*** is $R = k/n$ (as used earlier in this book). Generally, block code can be defined by the corresponding table comprising k-tuples of input (information) bits and the corresponding n-tuples of output encoded bits (code words). This table will have 2^k rows. For greater values of k it is not practical and some rule should be used how to obtain the code word from k input bits. Further, the corresponding decoding rule is needed—how to obtain information bits from the received code word. In praxis the ***systematic codes*** are especially interesting, where the block comprising k ***information bits*** is not changed in the code word, and $n - k$ ***control bits*** are added, forming a code word of the length n. In the following, it is primarily supposed that binary codes are considered. But, many of exposed principles and conclusions can be used for forming nonbinary codes as well.

In this chapter will be considered primarily so called linear block codes. At the beginning, some elementary notions will be introduced. ***Repetitions codes*** (Problem 5.1) are considered in the previous chapter as well (Introductory part and

Problems 4.6 and 4.7). At the end of Sect. 4.6 an example where multifold repetition is considered (BSC where crossover probability is $p < 0.5$). Here, only the fivefold repetition will be analyzed—i.e. information bit **0** is encoded as (00000) and information bit **1** is encoded as (11111). Various decoding strategies can be used. There are 32 possible received words. According to one decision rule (all rules are MAP, with majority logic implemented). Five bits are decoded according to the greater number of ones or zeros as one or zero. Therefore, all single errors and double errors in a code word are detected and corrected. The feedback channel is not needed. This procedure is ***Forward Error Control*** (FEC). According to the next decision rule only (00000) and (11111) are decoded as **0** or **1**. In this case all single, double, triple and fourfold bit errors are detected. However, then the repetition of this word is requested over the feedback channel. This procedure is ***Automatic Repeat reQuest*** (ARQ). But, the third decision rule is possible as well. Besides the combination (00000), (no errors!), all combinations having four 0 and one 1 are decoded as **0**, and analogously the combination (11111) and combinations having four 1 are decoded as **1**. For the other combination having 2 or 3 zeros either 2 or 3 ones, the retransmission is requested. It means that single errors are corrected, and all double and triple errors are detected. It is a ***hybrid procedure***.

In fact, these rules are based on ***Hamming distance*** (d) (Problems 5.2 and 5.4). For binary codes Hamming distance between two (binary) sequences of the same length is defined as the number of places (bits) in which they differ. It has the properties of metric. In the above example the Hamming distance between code words is $d = 5$. A discrete five dimensional space can be conceived, having total of 32 points, two of which are code words. Their distance is 5.

According to the first decision rule only the received words identical to code words are decoded, for the other cases, the retransmission is requested (ARQ), It is illustrated in Fig. 5.1a, where the code words are denoted by fat points. According to the second decision the received word is decoded as the nearer word in hamming sense (FEC) as in Fig. 5.1b. For the third decision rule (hybrid procedure), the space is divided into three subspaces. The points in the subspaces around code words are at the distance 1 from code words and single errors are corrected. In the third subspace (in the middle) are the points having Hamming distance 2 or 3 from code words and double and triple errors are detected (Fig. 5.1c). Decoder which every received word decodes as some code word (final decision) is called ***complete decoder***.

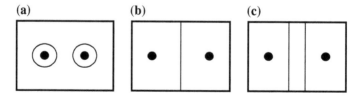

Fig. 5.1 Detection of errors without correction (**a**), with correction of all (up to 4) errors (**b**) and with partial (single error) correction (**c**)

Brief Theoretical Overview

It is obvious that Hamming distance of two sequences equals the number of errors turning one sequence into the other. Therefore, it is easy to find the direct connection of Hamming distance and the numbers of detectable and correctable errors of some code. For $d = 1$, a single error transform one code word into the other and there is even no any error detection. To detect all single errors it must be $d \geq 2$, and to detect q_d errors the following must hold

$$d \geq q_d + 1.$$

Of course, for a code with more (than two) code words, the worst case should be considered, i.e. minimum Hamming distance in the code should be found. To correct q_c errors, the minimum Hamming distance should be

$$q_c < d/2,$$

i.e. (q_c and d are integers)

$$d \geq 2q_c + 1.$$

Step further, to detect q_d errors and of them to correct q_c errors ($q_d \geq q_c$), the following must hold.

$$d \geq q_d + q_c + 1.$$

It is a general formula (previous expressions are obtained for $q_c = 0$, i.e. for $q_d = q_c$). It should be mentioned that if e errors occur ($q_c < e \leq q_e$) these errors will be detected, but not corrected. For $e \leq q_c$ the errors will be detected and corrected. Therefore, the same code can be used for FEC, ARQ or the hybrid procedure.

In the case of an erasure channel (BEC) where the bits "under suspicion" are erased, if q_e bits are erased, the Hamming distance should be

$$d \geq q_e + 1,$$

to allow for distinguishing from the nearest code word. Also, to correct q_c errors—the possibility not explicitly considered in BEC definition—and to erase q_e bits the following must hold

$$d \geq 2q_c + q_e + 1.$$

Codes using single **parity check** are probably the oldest error control codes. Simply, at the end of k-bits information word a single parity check is added generating ($n = k + 1$, k) block code. The total number of ones is even. Hamming distance is $d = 2$ and any odd number of errors can be detected. The code rate is $R = (n - 1)/n$.

However, information bits can be considered in more than one dimension. If they are ordered in two dimensions, as a matrix and parity checks can be calculated for rows and for columns

$$
\begin{matrix}
b_{11} & \cdots & b_{1,k_2} & x \\
\vdots & \vdots & \vdots & \\
b_{k_1,1} & \cdots & b_{k_1,k_2} & x \\
x & \vdots & x & x
\end{matrix}
$$

where the positions of parity bits are denoted with "x". If the block consists of $k_1 k_2$ information bits the obtained block ("codeword") will have $n = (k_1 + 1)(k_2 + 1)$ bits ($k_1 + k_2 + 1$ control bits). Code rate is

$$R = \frac{k_1 k_2}{(k_1 + 1)(k_2 + 1)}.$$

Maximal code rate value is obtained for $k_1 = k_2 = k$, while $n = (k + 1)^2$, and the code rate is

$$R = \frac{k^2}{(k+1)^2}.$$

Hamming in 1950 proposed a class of single-error correcting codes (*Hamming codes*) (Problems 5.3, 5.4, 5.5 and 5.6). The code parameters are $n = 2^b - 1$, $k = n - \mathrm{ld}(n + 1) = n - b$. These codes are described and analyzed in details in the mentioned problems. The result of parity check—*syndrome* ("checking number")— shows, in binary notation, the position of the error. Therefore, this code can correct all single errors, but not any combination of two errors in the code word.

A code that, for some e, can correct all combinations of e errors or less and no others is called a *perfect* code (Problem 5.3). Besides the Hamming codes, it is known that the Golay code (23, 12) is a perfect one, it can correct all single, double or triple errors. Perfect codes satisfy with equality the *Hamming bound* (Problems 5.5 and 5.8) for a (n, k) code having minimum Hamming distance $d_{\min} = 2e_c + 1$

$$q^{n-k} \geq \sum_{t=0}^{e_c} \binom{n}{t} (q - 1)^t$$

where q is the code base ($q = 2$ for binary code). There is also a *Singleton bound* (Problem 5.8) for any code (n, k) having a minimum distance $d_{\min}$

$$d_{\min} \leq 1 + n - k.$$

Any code satisfying that bound with equality is called a *maximum-distance code* (MDS) (Problem 5.8). *Varshamov-Gilbert bound* (Problem 5.11) gives the

Brief Theoretical Overview

lower bound for a maximum possible code words number for a given code base q, code word length n and a minimum code Hamming distance d

$$A_q(n,d) \geq q^n / \sum_{j=0}^{d-1} \binom{n}{j}(q-1)^j.$$

Hamming code can be **expanded** by adding one general parity-check bit (the code word length is now $n = 2^b$). The obtained code can detect practically all even numbers of errors. A code can be **shortened** as well by omitting some information symbols.

It is obvious that for every Hamming (n, k) code, where the syndrome has $n - k$ bits, all single errors can be corrected because

$$n = 2^{n-k} - 1.$$

The corresponding code word lengths are $2^j - 1$ ($j \geq 3$), i.e. codes (7, 4), (15, 11), (31, 26), (63, 57) etc. By adding a general parity check, the codes (8, 4), (16, 11), (32, 26), (64, 57) etc. are obtained, correcting all single errors and detecting all double errors. It is very suitable having in view the byte data structure.

In a previous consideration it was mentioned that generally any block code can be defined by the corresponding table comprising k-tuples of input (information) bits and the corresponding n-tuples of output encoded bits (code words). This table will have 2^k rows. For the corresponding code words there are 2^n possible "candidates". Therefore, there are in total $\binom{2^n}{2^k}$ possible codes (of course, some of them are not good!). For the greater values of n and k it is not practical and some rule should be used how to obtain the code word from k input bits. It means that some simple rule should be defined. These rules (algorithms in fact) can be very different, but there exists a special class of block codes having a simple mathematical description. With the help of discrete mathematics (abstract algebra) apparatus such class was found. In such a way it is possible to construct codes having needed characteristics (Hamming distance, etc.). These codes are **linear block codes**. This class of codes is defined by imposing a strong structural property on the codes. This structure provides guidance in the finding the good codes and helps to practical encoders and decoders realization. At the end of this chapter a short overview of the corresponding part of abstract algebra is included.

The definition of linear block code is very simple. Consider finite field—Galois field $GF(q)$, i.e. field that has q symbols. In the field two operations are defined—addition and multiplication (both are commutative). Sequences of n field elements (vectors) form a vector space V dimension n over the field. In this space vector addition is defined where the corresponding vector elements are added according to the rules from field. The set V is a commutative group under vector addition. Scalar multiplication is defined as well where vectors (i.e. all their elements) are multiplied by field elements (scalars). The product of two n-tuples, $(a_1, a_2, \ldots, a_n)$ and $(b_1, b_2, \ldots, b_n)$, defined as follows

$$(a_1, a_2, \ldots, a_n) \times (b_1, b_2, \ldots, b_n) = a_1 b_1 + a_2 b_2 + \cdots + a_n b_n$$

is called **inner product** or **dot product**. If inner product equals zero (0 from the field), for the corresponding vectors it is said to be **orthogonal**. Definition of linear code is as follows: *A **linear code** is a subspace of vector space over **GF(q)***. In this chapter mainly binary field GF(2) will be considered, but the theory is general, it can be applied for linear codes with q different symbols, if these symbols can be connected 1:1 with the symbols from $GF(q)$. It means that linear block codes exist if the number of elements in the code alphabet is a prime number or exponent of the prime number—2, 3, 4 ($=2^2$), 5, 7, 8 ($=2^3$), 9 ($=3^2$), 11, 13, 16 ($=2^4$) etc. It is very convenient that for any exponent of 2 there exists a linear block code. This approach in describing codes is often called **algebraic coding theory**.

Of course, block codes can be formed for any number of symbols, but then they are not linear. The vector spaces over the finite fields can be used to define code words not as the elements of a subspace. Then, it is usually said that "nonlinear" codes are considered.

Also, at the end of this chapter there are two problems not connected to linear block codes. The **arithmetic block codes** (Problem 5.11) construction is based on the arithmetic operation connecting decimal representations of information and code word. These relations are usually very simple. For **integer codes** (Problem 5.12) information and code words symbols take the values from the set $Z_r = \{0, 1, 2, \ldots, q-1\}$ (integer ring) while all operation during the code words forming are modulo-q.

Previous "mathematical" definition of linear block code can be "translated" by using previously introduced notions. Vector space has 2^n vectors—points—candidates for code words. From these candidates 2^k code words should be chosen. If these words are chosen as a subspace of the considered space, a linear block code is obtained. According to Lagrange's theorem, the number of elements in the group must be divisible by the number of elements in the subgroup. The vectors form a group for addition and for every n and every k there exists linear code (n, k), because 2^n is divisible by 2^k (of course, q^n is divisible by q^k). The identity element is $\mathbf{0}(0, 0, \ldots, 0)$ must be in a subgroup and in a code there must be code word (00 ... 0). The subgroup is closed under addition and the sum of code words will be a code word as well. The Hamming distance between two code words (vectors) equals the **Hamming weight** (i.e. to the number of ones) of their sum. It further means that for any two code vectors there exists code vector obtained by their summation and the Hamming weights of code vectors are in the same time possible Hamming distances in the code. Therefore, to find the minimum Hamming distance in the linear code, the code word having the minimum Hamming weight should be found (of course, not the all zeros code word!). To calculate the error probability, the **weight distribution** should be found called also the **weight spectrum** (Problems 5.2, 5.3, 5.8

and 5.9) of the code. It specifies the number of code words that have the same Hamming weight.

Consider Hamming code (7, 4) (Hamming code is a linear code!). Code words ($2^4 = 16$) are

(0000000), (1101001), (0101010), (1000011), (1001100), (0100101), (1100110), (0001111),
(1110000), (0011001), (1011010), (0110011), (0111100), (1010101), (0010110), (1111111).

Corresponding weight spectrum is: 0(1) 3(7) 4(7) 7(1), shown in Fig. 5.2a. In Fig. 5.2b weight spectrum of extended Hamming code (8, 4) obtained by adding the parity check is shown.

A set of vectors is said to span a vector space if every vector equals at least one linear combination of these vectors. The number of linearly independent vectors is called a vector space **dimension**. Any set of linearly independent vectors spanning the same vector space is called the **basis**. Therefore, there can be more bases of the same vector space. Generally, the group can have the nontrivial subgroups. Similarly, the vector space can have the subspaces. The subspace dimension is smaller than the space dimension. For (n, k) code vector space dimension equals n and code subspace dimension equals k. The subspace can have more bases as well. The subspaces can be orthogonal (dot products of the vectors from bases equal zero).

Consider (n, k) code. Vectors of its base can be ordered in a matrix form, its dimensions are $k \times n$. It is a **generator matrix**) (Problems 5.2, 5.3, 5.4 and 5.9) of the code. Its rows are linearly independent. The matrix rang is k. There are elementary operations over the rows which do not change the matrix rang. Therefore, in this case the code will be the same, only the base would change. Further, by commuting the matrix columns, an **equivalent** code (Problems 5.2 and 5.9) would

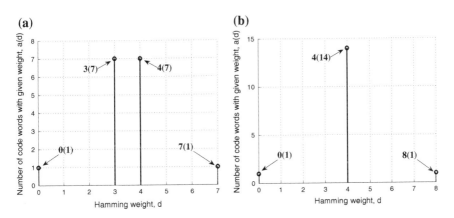

Fig. 5.2 Weight spectrum of Hamming code (7, 4) (**a**) and Hamming code (8, 4) (**b**)

be obtained. This code will have the same weight spectrum as the previous one, i.e. will have the same performances.

Consider linear block code (5, 2). Code words are (00000), (11010), (10101) i (01111), i.e. there are $2^2 = 4$ code words. Combination any two of three nonzero code words forms a generator matrix

$$H_1 = \begin{bmatrix} 1 & 1 & 0 & 1 & 0 \\ 1 & 0 & 1 & 0 & 1 \end{bmatrix}, H_2 = \begin{bmatrix} 1 & 1 & 0 & 1 & 0 \\ 0 & 1 & 1 & 1 & 1 \end{bmatrix}, H_3 = \begin{bmatrix} 1 & 0 & 1 & 0 & 1 \\ 0 & 1 & 1 & 1 & 1 \end{bmatrix}.$$

Of course, any of them can be obtained from the other using elementary operations over the rows.

Such mathematical description provides very simple generation of code words. Consider linear code (n, k) that has generator matrix (dimensions $k \times n$)

$$G = \begin{bmatrix} g_{11} & g_{12} & \cdots & g_{1n} \\ g_{21} & g_{22} & \cdots & g_{2n} \\ \vdots & \vdots & \ddots & \vdots \\ g_{k1} & g_{k2} & \cdots & g_{kn} \end{bmatrix}.$$

To obtain the code word for information bits vector $i(i_1, i_2, \ldots, i_k)$, the following multiplication is used

$$v = i \cdot G,$$

where the resulting code vector (word) $v(v_1, v_2, \ldots, v_n)$ is the sum of G rows corresponding to the places of ones in i. Therefore, a code word must be obtained. Further, the rows of G are linearly independent and to any specific information vector i corresponds code word v, different from other code words. Total number of possible code words is equal to the number of rows of $G - \binom{k}{1}$, plus the number of words obtained by summing two rows of $G - \binom{k}{2}$, plus the number of words obtained by summing three rows of G etc. This number is $2^k - 1$. To this total should be added all zeros code vector, corresponding to all zeros information vector, i.e.

$$\binom{k}{0} + \binom{k}{1} + \cdots + \binom{k}{k} = 2^k.$$

It is just the number of vectors in a code subspace.

Next step is to obtain a systematic code. This problem is solved easily. By elementary operations over the rows (code does not change!) and by permuting the columns, a generator matrix of an equivalent code can be obtained

Brief Theoretical Overview

$$G_s = \begin{bmatrix} 1 & 0 & \cdots & 0 & p_{11} & \cdots & p_{1,n-k} \\ 0 & 1 & \cdots & 0 & p_{21} & \cdots & p_{2,n-k} \\ \vdots & \vdots & \ddots & \vdots & \vdots & \ddots & \vdots \\ 0 & 0 & \cdots & 1 & p_{k1} & \cdots & p_{k,n-k} \end{bmatrix} = [\mathbf{I}_k \ \mathbf{P}],$$

where $\mathbf{I}_k$ is a unity matrix (dimensions $k \times k$) and $\mathbf{P}$ is a matrix dimensions k ($n - k$). Multiplying information vector $\mathbf{i}$ by matrix $\mathbf{G}$, code vector $\mathbf{v}(i_1, i_2, \ldots, i_k, k_1, k_2, \ldots, k_{n-k})$ is obtained. It is a systematic code, information bits are not changed, while the last $n - k$ bits are parity checks

$$k_j = i_1 \cdot p_{1j} \oplus i_2 \cdot p_{2j} \oplus \cdots \oplus i_k \cdot p_{kj},$$

i.e. k_j is parity check taking into account the bits on the positions where $\mathbf{P}$ in jth column has ones. The number of these checks is just $n - k$. In binary encoding these are real *parity checks*, and in general case (nonbinary encoding) these are *generalized parity checks*. It is also possible to construct a *product code* (Problem 5.7). For a systematic product code information bits are ordered in two (or more) dimensions. Then, one error control code is applied for the rows and the other for the columns.

Therefore, the problem of encoding is solved. The encoding is very simple. It is sufficed to know matrix $\mathbf{G}$. The next question is what to do at the receiving end. How to verify do the received vector ($\mathbf{u}$) is a code vector ($\mathbf{v}$)? The abstract algebra gives here an elegant solution as well. As said earlier, if inner product of two vectors equals zero (0 from the field), the vectors are *orthogonal*. Two subspaces are orthogonal if all vectors from one subspace are orthogonal to all vectors of the other subspace. It is shown that for every subspace there is a subspace orthogonal to it. To test the orthogonality, it is sufficient to verify the orthogonality of the corresponding bases. If the bases are orthogonal, then all their corresponding linear combinations are orthogonal. Further, if the dimension of vector space is n, and if the dimension of one subspace is k, the dimension of corresponding orthogonal subspace is $n - k$. Let the base (generator matrix) of one subspace is $\mathbf{G}$ (dimensions $k \times n$) and let $\mathbf{H}$ is the base (generator matrix) of the orthogonal subspace (dimensions $(n - k) \times n$). Taking into account that in matrix multiplication the element a_{ij} is obtained as a dot product of ith row of the first matrix and the jth column of the second matrix, the following is obtained

$$\mathbf{G} \cdot \mathbf{H}^\mathrm{T} = \mathbf{0}.$$

where ($^\mathrm{T}$) denotes transposition and the dimension of matrix $\mathbf{0}$ (all elements equal zero) are $k \times (n - k)$. Of course, by transposing this product one obtains

$$\mathbf{H} \cdot \mathbf{G}^\mathrm{T} = \mathbf{0}^\mathrm{T}.$$

It is obvious, because the subspaces are orthogonal each other. Therefore, instead of using generator matrix $\mathbf{G}$, at the receiving end, from the above matrix equation *parity-check matrix* **H** (rank is $n - k$) (Problems 5.3 and 5.4) should be found and then verify do the received vector is in the subspace orthogonal to the subspace generated by **H**. Solution of the above equation is very simple if matrix **G** corresponds to a systematic code. In this case

$$\mathbf{H} = \begin{bmatrix} -\mathbf{P}^\mathrm{T} & \mathbf{I}_{n-k} \end{bmatrix},$$

where $\mathbf{I}_{n-k}$ is a unity matrix dimensions $(n - k) \times (n - k)$, the sign "−" denotes that for every element of matrix **P** the corresponding inverse element in the field should be taken.

In the case of binary code—$GF(2)$—each element (0 and 1) is inverse to itself and a sign "minus" is not needed. It is obvious that the first $n - k$ elements of every row of matrix **H** are parity-checks and "one" shows the position of the corresponding parity bit. Some authors use a systematic code matrix

$$\mathbf{G}_\mathrm{s} = \begin{bmatrix} \mathbf{P} & \mathbf{I}_k \end{bmatrix}.$$

The corresponding parity-check matrix is

$$\mathbf{H}_\mathrm{s} = \begin{bmatrix} \mathbf{I}_{n-k} & -\mathbf{P}^\mathrm{T} \end{bmatrix},$$

and the orthogonality remains.

One more characteristic is here of interest. Because of mutual orthogonality, if **G** generator matrix of a linear code (n, k) and **H** is a corresponding parity-check matrix, then **H** is a generator matrix of $(n, n - k)$ linear code and **G** is its parity-check matrix. These codes are called **dual codes** (Problem 5.9). The **MacWilliams identities** relate the weight distribution of some code to the weight distribution of its dual code. They are useful if the dual code is small enough for its weight distribution to be found by computer search.

On the base of orthogonality of generator and parity-check matrices it can be shown that a block code can have maximal Hamming distance w if (and only if) every set of $w - 1$ or less columns of matrix **H** is linearly independent. In other words, to obtain the code having minimal Hamming distance w, at least $w - 1$ linearly independent parity checks must exist.

Consider dual code for the above (5, 2) code. It is a code (5, 3). From $\mathbf{H}_1$ the corresponding generator matrix is easily obtained

$$\mathbf{G}_1 = \begin{bmatrix} 1 & 0 & 0 & 1 & 1 \\ 0 & 1 & 0 & 1 & 0 \\ 0 & 0 & 1 & 0 & 1 \end{bmatrix}.$$

From $2^5 = 32$ vectors in the space, set of $2^3 = 8$ vectors form a (code) subspace

(00000), (10011), (01010), (11001), (00101), (10110), (01111), (11100).

Brief Theoretical Overview

The other generator matrices are

$$G_2 = \begin{bmatrix} 1 & 0 & 0 & 1 & 1 \\ 0 & 1 & 0 & 1 & 0 \\ 1 & 0 & 1 & 1 & 0 \end{bmatrix}, G_3 = \begin{bmatrix} 1 & 0 & 0 & 1 & 1 \\ 1 & 1 & 0 & 0 & 1 \\ 1 & 1 & 1 & 0 & 0 \end{bmatrix}.$$

For the earlier mentioned Hamming code (7, 4) generator and parity-check matrices are

$$G = \begin{bmatrix} 1 & 1 & 1 & 0 & 0 & 0 & 0 \\ 1 & 0 & 0 & 1 & 1 & 0 & 0 \\ 0 & 1 & 0 & 1 & 0 & 1 & 0 \\ 1 & 1 & 0 & 1 & 0 & 0 & 1 \end{bmatrix}, H = \begin{bmatrix} 1 & 0 & 1 & 0 & 1 & 0 & 1 \\ 0 & 1 & 1 & 0 & 0 & 1 & 1 \\ 0 & 0 & 0 & 1 & 1 & 1 & 1 \end{bmatrix}.$$

For decoding the received word u, the following vector should be found

$$S = u \cdot H^T.$$

It is called **syndrome**, dimensions are $1 \times (n - k)$ and its elements are parity checks. It is easy to conclude taking into account previous consideration about finding matrix H, when matrix G is given in a systematic form. When the errors are only detected (ARQ procedure), nonzero elements (one or more) of syndrome mean that the error (or errors) occurred. If the errors are corrected (FEC procedure) from the syndrome the corresponding error position(s) should be found. Let code vector v is transmitted and let vector $u = v + e$ is received, where vector e—is an error vector, mentioned earlier having "ones" at the error positions. The following syndrome is obtained

$$S = u \cdot H^T = (v + e) \cdot H^T = v \cdot H^T + e \cdot H^T = 0 + e \cdot H^T = e \cdot H^T.$$

Now, vector e should be found. It is obvious that if the error vectors have the same syndrome, only one of them must be taken into account. Hoping that the smaller number of errors is expected, for the error vector should be taken that one that has a smaller number of ones.

Still further insight into error detection and correction can be obtained using **standard array**. Linear code (n, k) is a subspace of a vector space dimension n. It is in the same time a subgroup of the additive group. Each group can be partitioned with respect to the subgroup. Here the vector space is partitioned with respect to the code subspace. The procedure is described and commented in details in Problem 5.2. Here, a short overview of the procedure will be given. For any linear block code standard array has 2^{n-k} rows and 2^k columns. Every row is called **coset**, code words $c_1, c_2, \ldots, c_{2^k}$ are in the first row (usually starting from the all zeros code word) and the first elements in every row are called **coset leaders**. The element in the ith column and the jth row is obtained by adding c_i and jth coset leader, denoted by e_j. In fact, it is assumed that the code word c_i was transmitted, and that the error vector e_j is added to the code word. *Syndrome* corresponding to this received word is uniquely determined by jth coset leader

$$S = v \otimes H^T = (c_i \oplus e_j) \otimes H^T = e_j \otimes H^T,$$

because the parity-check matrix does not depend on the received word and it is known at the receiving end. The error vectors corresponding to coset leaders are uniquely determined by the syndrome and can be corrected. Therefore, 2^{n-k} different **error patterns** (*vectors*) can be corrected.

As an example consider code (5, 2) with code words (00000), (11010), (10101), (01111). Minimal Hamming distance equals 3 and a code should correct all single errors. It is obvious from standard array

				S
00000	11010	10101	01111	000
00001	11011	10100	01110	101
00010	11000	10111	01101	110
00100	11110	10001	01011	001
01000	10010	11101	00111	010
10000	01010	00101	11111	100
---	---	---	---	
00011	11001	10110	01100	011
00110	11100	10011	01001	111

For coset leaders of two last rows the some error vectors having two errors are chosen (from four possible candidates).

An interesting class of linear codes over GF(2) are **Reed-Muller codes** (Problem 5.10). They can be easy described and by a simple majority logic easy decoded as well. Here the code word at the receiver is found iteratively, without using a syndrome. A notion of orthogonality is introduced (different from orthogonality of vectors). The set of check sums (parity checks) is *orthogonal* to a particular error bit, if this bit is involved in every sum, and no other error bit is checked by more than one sum.

Considered codes are constructed for error control at the channels where single errors are dominant (memoryless channels). However, sometimes the errors occur in *packets* ("bursts"). By using **interleaving** (Problem 5.6) at the transmitter and **deinterleaving** at the receiver the error control code for random errors can successfully combat with packet errors.

Problems

Problem 5.1 Explain a block code construction where bits are repeated n times.

(a) Explain the decoding procedure when the decisions are made using majority logic.
(b) If it is possible to form some other decision rule for this code, explain its advantages and drawbacks with respect to the previously described rule.

(c) Find the residual error probability after decoding and draw it for $n = 5$, $n = 7$, $n = 9$ and $n = 11$, for two different decision rules. Find the code rate for all cases considered.

(d) Comment the quality of this code comparing it to the limiting case given by Second Shannon theorem.

Solution

The construction of a *repetition code* is quite simple and it is based on the fact that a binary zero at the encoder input is represented by a series of n zeros at its output, while to the binary one correspond a series of n successive ones. It is obvious that at the encoder output only two code words can appear, while at the receiver input any bit combination can appear (but not all having the same probability). The decoder, on the basis of the received code word, has to decide which information bit is transmitted.

(a) To attain this goal, usually the *majority decision* rule is used, where it is decided that at the encoder input was binary zero if in the received word there are more zeros than ones, and vice versa. To avoid the case when the received word has an equal number of binary zeros and ones, it is usually taken that a code word length, denoted by n, is an odd number. In this case all received n-bits words are divided into two groups, as illustrated in Fig. 5.3 for $n = 3$. Using this approach the decision error appears if there are errors on at least $(n + 1)/2$ positions. The probability of the residual error for the user just corresponds to the probability of such an event being

$$P_{e,rez1} = \sum_{t=(n+1)/2}^{n} \binom{n}{t} p^t (1-p)^{n-t},$$

where p denotes the crossover probability in the binary symmetric channel.

(b) The *second decision rule* is simpler—when the decoder receives one of two possible code words the decoding procedure is performed. When the received word is not a codeword, correction is not carried out, but the corresponding retransmission is requested. While the first decoding method, based on majority decision, is a typical representative of **FEC** (*Forward Error Correction*) techniques, the second rule is completely based on the code word validity and on the use of **ARQ** (*Automatic Repeat reQuest*) procedure. In this

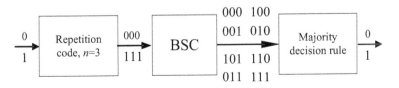

Fig. 5.3 Encoder and decoder procedures when threefold repetition is applied

case the error occurs only if all bits in a code word are hit with the channel errors, and a corresponding probability is

$$P_{e,rez2} = p^n.$$

(c) Using of the previously derived expressions, it is possible to calculate the dependence of the residual error, detected by the user, on the crossover probability in the channel. By direct use of these expressions the dependence of the **residual error after decoding** $P_{e,res}$ on the BSC crossover probability was calculated and shown in Fig. 5.4. The results are shown for both decision rules.

In the figure two effects can be noticed:

(1) For the same value of the crossover probability, the residual error is smaller as the number of repetitions is greater (also, as expected, $P_{e,res}$ is greater if p is greater). On the other hand, the **information rate** for n repetitions is smaller n times than the channel binary rate $v_i = v_b/n$, and generally $v_i = v_b \times R$, where R is the **code rate** (in this case $R = 1/n$). It is obvious that for a repetition code, the increased transmission reliability was paid by the reducing the information rate.

(2) The residual error probability does not depend only on the code construction and the code parameters values. In this case, for an n-repetition code and the constant code rate, it depends on the decoding procedure as well. E.g. for the same $n = 5$ and $R = 1/5$ repetition code the various results are obtained for two different decision rules. On the other hand, the second decision rule

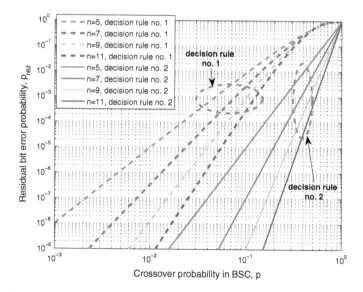

Fig. 5.4 Residual error probability, the repetition code, two decision rules

achieves the better performances, but it implies the feedback channel and ARQ procedure. Because of that, to achieve the small residual error with the moderate *decoder complexity* is a very complicated task.

(d) In Problem 4.7 it was shown that *the transmission reliability can be increased by reducing the channel information rate*. It is interesting to check whether the results obtained here are in concordance with the Second Shannon theorem, formulated in 1948 [10], where it was proved that the reliable transmission can be provided if the code rate (R) is smaller than the transmitted information—$I(X, Y)$—for a given channel.

For BSC, when the input symbol probabilities are equal, the transmitted information depends only on the average transmission error (crossover) probability (p)

$$I_{\max}(p) = 1 - (1-p) \cdot \mathrm{ld}\left(\frac{1}{1-p}\right) - p \cdot \mathrm{ld}\frac{1}{p}.$$

The function $I(p)$ is shown in Fig. 5.5, where for crossover probability the logarithmic scale is used (for $p < 0.5$). In the same figure there are shown the parameter p values for which n-fold repetition provides the residual error probability $P_{e,res} = 10^{-9}$ (for the second decision rule) and at y-axis the corresponding code rate is denoted.

It should be noted that such comparison is not quite correct—$I(p)$ is the upper limit for a code rate for the case of when the residual error is arbitrary small, while,

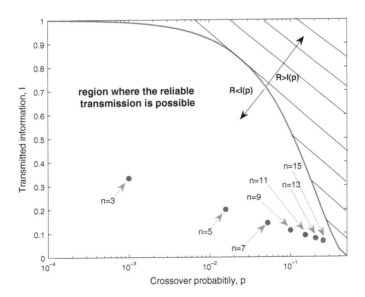

Fig. 5.5 Illustration of the second Shannon theorem for BSC

for the repetition code the values are given for the case when the $P_{e.res}$ is sufficiently small for the practical purposes (for a majority of communication systems $P_{e,res} < 10^{-9}$). However, some interesting effects can be noticed:

(1) For a chosen decision rule, the same value of residual error is achieved for crossover probability values as the code rate is smaller.
(2) In all considered cases $R \ll I(p)$, and it is possible to construct a code having substantially greater code rate, which will provide the same (and even the better) transmission quality that the n-repetition code, for the same crossover probability.
(3) For a very great number of repetitions, the code performances approach to the limits given by the Second Shannon theorem (as $R/I(p) \to 1$) but then R and $I(p)$ approach to zero.

Problem 5.2 Block code is defined by Table 5.1.

(a) Find the code rate. Is it a linear code? If yes, find its generator matrix.
(b) Write all code words and draw the corresponding weight (distribution) spectrum. Is it possible from it to find the code capability to correct the errors?
(c) Find a standard array of a code.
(d) Is the generator matrix of the code unique?
(e) Form some code equivalent to the original code.
(f) By changing one code word form a nonlinear block code.

Solution

(a) The three-bits information words are represented (according to Table 5.1) by five bits code words yielding the code rate $R = 3/5$.

By definition, a ***linear block code*** is a vector subspace of a vector space. For a binary code (n, k) vector space consists of all n-bits combination (2^n), while a vector subspace is formed of 2^k sequences of the length n, satisfying the following

(1) Vector subspace has a neutral (identity) element for addition
(2) Vector subspace is closed for the modulo-2 addition.

Table 5.1 Information words and the corresponding code words of block code

Information word	Code word
0 0 0	0 0 0 0 0
0 0 1	1 1 0 1 1
0 1 0	1 0 1 1 0
0 1 1	0 1 1 0 1
1 0 0	0 1 0 0 1
1 0 1	1 0 0 1 0
1 1 0	1 1 1 1 1
1 1 1	0 0 1 0 0

Problems

To check the code linearity these conditions should be verified. In the code there is a code word consisting of all zeros and the first condition is satisfied. By adding any pair of the code words some code word is obtained (one of the bit combinations in Table 5.1) and code is a linear one.

From Table 5.1 it can be easily found that the **code word** corresponding to **information word** 110 can be obtained by adding the code words corresponding to information words 100 and 010. Now, the general rule can be found, from which follows that ith row of a **generator matrix** should correspond to the code word for an information word having one at the ith position only, and a generator matrix is

$$G = \begin{bmatrix} 1 & 1 & 0 & 1 & 1 \\ 1 & 0 & 1 & 1 & 0 \\ 0 & 1 & 0 & 0 & 1 \end{bmatrix}.$$

All code words can be obtained multiplying all information words by a generator matrix

$$c = i \otimes G$$

where i is an arbitrary information word (generally, any combination of k bits, here $k = 3$) and the operator $\otimes$ denotes modulo-2 matrix multiplication. In this case the above matrix relation is satisfied and eight information words are transformed into code words forming the vector subspace, as shown in Fig. 5.6, and the code is a linear one.

Linear block code can be described by a generator matrix, it completely defines the transformation of 2^k information words into 2^k code words of the length n. These code words form the linear block code. On the other hand, for any linear code, the generation matrix G can be uniquely found.

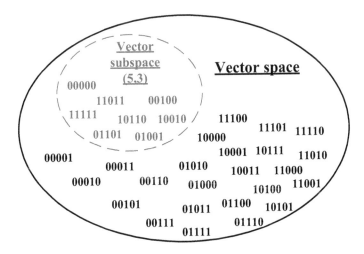

Fig. 5.6 Linear block code (5, 3) as a vector subspace

(b) Code words are given in Table 5.1 and shown in Fig. 5.6 as the elements of vector subspace. Linear block code is a subspace and it is closed under addition modulo-2 (bit-by-bit), and the sum of two code words is a code word as well

$$(11011) \oplus (01101) = (10010).$$

Hamming distance between two code words c_i and c_j, denoted by $d_H(c_i, c_j)$, is determined by a *Hamming weight* of the word $c_i \oplus c_j$ (i.e. by the number of ones in it). It can be written formally as

$$d_H(c_i, c_j) = d_H(c_i \oplus c_j, \mathbf{0}) = w(c_i \oplus c_j),$$

where $w(c_i)$ denotes the Hamming weight of the ith code word. Code words **weight distribution** determines completely **code distances spectrum**. In this code there is one code word having the weight $d = 1$, two words with $d = 2$, two words with $d = 3$, one word with $d = 4$ and one word with $d = 5$. The corresponding weight distribution is shown in Fig. 5.7.

The minimum Hamming distance is $d_{min} = 1$, relation $d_{min} \leq 2e_c + 1$ is satisfied only for $e_c = 0$, and under the previous condition the inequality $d_{min} \leq e_c + e_d + 1$ is satisfied only for $e_d = 0$. It is obvious that this code cannot correct even not to detect all single errors.

(c) **Standard array** of any binary linear block code has 2^{n-k} rows and 2^k columns, as given in Table 5.2. Every row is one **coset**, code words $c_1, c_2, \ldots, c_{2^k}$ are in the first row (usually starting from the all zeros code word) and the first elements in every row are **coset leaders**. The element in the ith column and the jth row is obtained by adding c_i and jth coset leader, denoted by e_j. Under

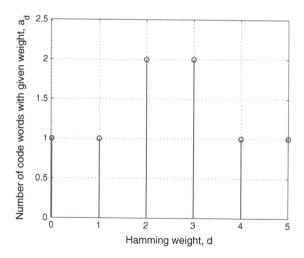

Fig. 5.7 The code (5, 3) weight distribution

Problems

Table 5.2 General clishé for obtaining the standard array of a linear block code

$c_1 = (0, 0, ..., 0)$	c_2	...	c_i	...	c_{2^k}
e_2	$c_2 \oplus e_2$	...	$c_i \oplus e_2$	...	$c_{2^k} \oplus e_2$
...	...	...	...	...	...
$e_{2^{n-k}}$	$c_2 \oplus e_{2^{n-k}}$	...	$c_i \oplus e_{2^{n-k}}$	...	$c_{2^k} \oplus e_{2^{n-k}}$

Table 5.3 Standard array of a code (5, 3)

0 0 0 0 0	**1 1 0 1 1**	**1 0 1 1 0**	**0 1 0 0 1**	**1 0 0 1 0**	**1 1 1 1 1**	**0 1 1 0 1**	**0 0 1 0 0**
1 0 0 0 0	0 1 0 1 1	0 0 1 1 0	1 1 0 0 1	0 0 0 1 0	0 1 1 1 1	1 1 1 0 1	1 0 1 0 0
0 0 0 1 0	1 1 0 0 1	1 0 1 0 0	0 1 0 1 1	1 0 0 0 0	1 1 1 0 1	0 1 1 1 1	0 0 1 1 0
0 0 0 0 1	1 1 0 1 0	1 0 1 1 1	0 1 0 0 0	1 0 0 1 1	1 1 1 1 0	0 1 1 0 0	0 0 1 0 1

assumption that the code word c_i was transmitted, it is obvious that the element of the *j*th row in the *i*th column corresponds to the received word if the *error vector* is equal to the *j*th coset leader. It holds as well as for every column and every row (any *i, j* combination). *Syndrome* corresponding to this code word is uniquely determined by *j*th coset leader

$$S = r \otimes H^T = (c_i \oplus e_j) \otimes H^T = e_j \otimes H^T$$

because the *parity-check matrix* does not depend on the received word and it is known at the receiving end.

The error vectors corresponding to coset leaders are uniquely determined by the syndrome and can be corrected. In general case, it is obvious that 2^{n-k} different *error patterns* can be corrected. To minimize the error probability it is suitable that the coset leaders are the error patterns which occur the most frequently. It is obvious that for BSC (which has a relatively small crossover probability) single errors are more probable than double ones and because of that for coset leaders the combinations with the smallest number of ones should be chosen.

Standard array of the (5, 3) code has 8 columns, 4 rows, as given in Table 5.3. Code words c_i are in bold, while the coset leaders are in italic and underlined. In this example, every element in *i*th column differs from c_i by at most one bit. E.g. second element in the third column equals the sum of $c_3 = (10110)$ and the second coset leader $e_2 = (10000)$. This code can correct the error patterns (10000), (00010) and (00001). Of course, the trivial error pattern (00000) can be "corrected" as well. On the other hand, the error pattern (00100) cannot be corrected and the code cannot correct all single bit error patterns. This pattern is in fact a code word and this error cannot be even detected, confirming the above conclusion that for this code $e_c = e_d = 0$ holds. It is recommended to the reader to verify whether the vector (01000) can be a coset leader as well as whether the code, in this case, could detect and correct the errors.

Let the ith column in the standard array is denoted by D_i. The decoding by using a standard array is simple—if word r was received from the column D_i, the conclusion is that code word c_i is transmitted. It is easy to verify that this way of decoding is based on the maximum likelihood rule, because it is concluded that the code word was transmitted to which the received one has a minimum Hamming distance. The following conclusions can be drawn:

(1) It is possible to decode 2^{n-k} various patterns because there is the same number of different syndromes. For a code having minimum Hamming distance d_{min} for coset leaders the error patterns with weights $e_c \leq (d_{min} - 1)/2$ should be chosen.

(2) If the received word is different from code words, the error can be detected. Therefore, there are in total $2^n - 2^k$ detectable error patterns.

(d) The code generator matrix is not unique. In fact, it can be obtained if as its rows any three independent code words (the third one cannot be the sum of the other two) from the vector subspace shown in Fig. 5.6 are written. It is easy to verify that this condition is satisfied as well as with the matrix

$$G' = \begin{bmatrix} 1 & 0 & 0 & 1 & 0 \\ 1 & 1 & 1 & 1 & 1 \\ 0 & 0 & 1 & 0 & 0 \end{bmatrix}.$$

Of course, the transformation of information words into the code words is not now defined by Table 5.1, but the set of code words will be the same. Therefore, this generator matrix generates the same code. The following should be noticed as well—although for the same code more generator matrices can be found, one generator matrix defines one and only one code.

(e) An **equivalent code** is, by definition, the code having the same code distances spectrum, but at least one its code word differs from the words of the originating code. This code can be formed if two (or more) columns change the places. The following matrix is obtained from the previous generator matrix where the third and the fourth columns have changed the places.

$$G'' = \begin{bmatrix} 1 & 0 & 1 & 0 & 0 \\ 1 & 1 & 1 & 1 & 1 \\ 0 & 0 & 0 & 1 & 0 \end{bmatrix}.$$

The following set of code words corresponds to this generator matrix

(00000), (00010), (11111), (11101), (10100), (10110), (01011), (01001),

Problems

and it is obvious that the code distance spectrum is not changed comparing to the previous case.

(f) The simplest way to form a nonlinear block code is to substitute all zeros code word by some bit combination not belonging to the code. In such a way in the vector subspace (shown in Fig. 5.6) there will be no more the identity element for addition, if the combination (00000) is substituted with (00001). It is easy to verify that, in this case, is not possible to define the generator matrix providing for the transformation of all zeros information word into non zero binary combination (the multiplying of any matrix by a vector consisting of all zeros yields the all zero vector).

Of course, even if a code word "all zeros" exists in the code, it might not be linear. If any code word is changed in such a way that its sum with some other code word does not yield a code word, the closure property regarding the addition is not fulfilled (the sum is a word not belonging to the vector subspace, of course, it belongs to some other vector subspace). In the second variant shown in Table 5.4 the critical word is (01100). In this case, as well, it is not possible to define a generator matrix, because this word cannot be obtained as a linear combination of other three code words.

Problem 5.3 Explain the construction of Hamming code having code word length $n = 2^b - 1$ (b—any integer greater than 1), and after

(a) Explain in details the construction of Hamming (7, 4) code, as well the encoding and decoding process for a code word corresponding to information sequence $i = (1110)$, if during the transmission an error occurred in the third position of the code word;

(b) Decode the received word, if the same information sequence as above was transmitted for the cases without errors and when the errors occurred in the

Table 5.4 Transformation to obtain a nonlinear block code

Nonlinear block code, variant 1		Nonlinear block code, variant 2	
Information word	Code word	Information word	Code word
0 0 0	*0 0 0 0 1*	0 0 0	0 0 0 0 0
0 0 1	1 1 0 1 1	0 0 1	1 1 0 1 1
0 1 0	1 0 1 1 0	0 1 0	1 0 1 1 0
0 1 1	0 1 1 0 1	0 1 1	*0 1 1 0 0*
1 0 0	0 1 0 0 1	1 0 0	0 1 0 0 1
1 0 1	1 0 0 1 0	1 0 1	1 0 0 1 0
1 1 0	1 1 1 1 1	1 1 0	1 1 1 1 1
1 1 1	0 0 1 0 0	1 1 1	0 0 1 0 0

third and the fifth position of the code word, comment the corresponding syndromes;
(c) Find the generator and the parity-check matrices for this code;
(d) Draw the distance spectrum of the code.

Solution
In this problem a class of codes proposed by Richard Hamming in 1950 [23] will be described. Suppose that the aim is to construct a *Hamming code* when $n = 2^b - 1$ bits. If the bit positions are numbered as $l = 1, 2, ..., n$, then the parity-check bits are in the code word c positions $l = 2^m$ ($m = 1, 2, ..., b$), while information bits are in the other places. It is obvious that the total number of parity-check bits in the code word is just $b = \mathrm{ld}(n + 1)$, and the information word length at the encoder input should be $k = n - \mathrm{ld}(n + 1)$. Some possible (n, k) combinations, according to this expression, are given in Table 5.5.

Parity-check bit m, which is in the position $l = 2^m$ of the code word c, is obtained by adding modulo-2 (XOR) of all information bits which are at the positions having 1 at mth position in their binary record.

After the superimposing the channel errors, the code word r is at the decoder input. Here, the syndrome is formed consisting of $b = n - k$ bits. The value of mth syndrome bit is obtained by adding the bits of the received word which are at the positions having 1 at mth position in their binary record. If during the transmission only one of these bits was inverted (error!), the syndrome value, starting from the bits with the higher ordinal number to the position having the smaller ordinal number is a binary record of the error position. General block scheme of the Hamming encoder and decoder are shown in Fig. 5.8.

(a) Hamming code (7, 4) construction can be simply described by a clishé given in Table 5.6. The code word length is $n = 7$, the number of parity-check bits is $b = \mathrm{ld}(7 + 1) = 3$.

Number l, from the first column determines the ordinal bit number in the code word. The position of the first 1 in the mth ($m = 1, 2, 3$) position of a binary record of l, determines the position of mth **parity-check bit** in code word, this bit is denoted by z_m. The value of mth parity check bit is obtained by summing modulo-2 the

Table 5.5 Some possible (n, k) pairs

b	n	k
2	3	1
3	7	4
4	15	11
5	31	26
6	63	57
7	127	120
...	...	...

Problems

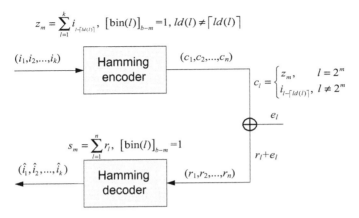

Fig. 5.8 General block scheme of Hamming encoder and decoder

Table 5.6 Clishé for constructing code words for Hamming code (7, 4)

l	bin (l)	c_i
1	00**1**	z_1
2	0**1**0	z_2
3	011	i_1
4	**1**00	z_3
5	101	i_2
6	110	i_3
7	111	i_4

information bits in the positions where the binary record of ordinal number of l in position $3 - m$ has 1. For the obtaining the first parity-check bit the last column in a binary record is used, and for the third one the first column is used, yielding

$$z_1 = c_3 \oplus c_5 \oplus c_7 = i_1 \oplus i_2 \oplus i_4,$$
$$z_2 = c_3 \oplus c_6 \oplus c_7 = i_1 \oplus i_3 \oplus i_4,$$
$$z_3 = c_5 \oplus c_6 \oplus c_7 = i_2 \oplus i_3 \oplus i_4.$$

The code word has a structure $(c_1 c_2 c_3 c_4 c_5 c_6 c_7) = (z_1 z_2 i_1 z_3 i_2 i_3 i_4)$, and a sequence at the receiver input is changed because of channel errors. Therefore, lth code word bit is $r_l = c_l \oplus e_l$ and the decoder, on the basis of received sequence, calculates the parity-check sums, and the mth syndrome component is formed by adding modulo-2 the bits at the positions where the binary record at position $b - m$ has 1.

Similarly to the parity-check bits calculation, for the first syndrome component s_1 the last column in binary record is used. However, in this case all bits in the received word are added, taking into account and position of the first 1 in every column of the clishé in Table 5.6. Denoting by e_l the lth bit in the error vector, **syndrome components** can be written as

$$s_1 = r_1 \oplus r_3 \oplus r_5 \oplus r_7 = (i_1 \oplus i_2 \oplus i_4) \oplus e_1 \oplus i_1 \oplus e_3 \oplus i_2 \oplus e_5 \oplus i_3 \oplus e_7 = e_1 \oplus e_3 \oplus e_5 \oplus e_7,$$
$$s_2 = r_2 \oplus r_3 \oplus r_6 \oplus r_7 = (i_1 \oplus i_3 \oplus i_4) \oplus e_2 \oplus i_1 \oplus e_3 \oplus i_3 \oplus e_6 \oplus i_3 \oplus e_7 = e_2 \oplus e_3 \oplus e_6 \oplus e_7,$$
$$s_3 = r_4 \oplus r_5 \oplus r_6 \oplus r_7 = (i_2 \oplus i_3 \oplus i_4) \oplus e_4 \oplus i_2 \oplus e_5 \oplus i_2 \oplus e_6 \oplus i_3 \oplus e_7 = e_4 \oplus e_5 \oplus e_6 \oplus e_7.$$

On the basis of the above relations, it is easy to prove that the corresponding parity-checks (ordered by descending indexes) form the **syndrome** which determines the position of the code word bit where the error occurred, recorded in binary system $S = (s_3\, s_2\, s_1)$. By complementing this bit, the error can be corrected.

When the transmitted sequence is $\boldsymbol{i} = (1110)$, information bits are $i_1 = 1$, $i_2 = 1$, $i_3 = 1$, $i_4 = 0$ and the corresponding parity-checks are

$$z_1 = i_1 \oplus i_2 \oplus i_4 = 1 \oplus 1 \oplus 0 = 0,$$
$$z_2 = i_1 \oplus i_3 \oplus i_4 = 1 \oplus 1 \oplus 0 = 0,$$
$$z_3 = i_2 \oplus i_3 \oplus i_4 = 1 \oplus 1 \oplus 0 = 0,$$

and at the encoder output is the code word $\boldsymbol{c} = (0010110)$. Due to the transmission error at the fourth position, the received sequence is $\boldsymbol{r} = (0011110)$, the parity-check sums have the values

$$s_1 = r_1 \oplus r_3 \oplus r_5 \oplus r_7 = 0 \oplus 1 \oplus 1 \oplus 0 = 0,$$
$$s_2 = r_2 \oplus r_3 \oplus r_6 \oplus r_7 = 0 \oplus 1 \oplus 1 \oplus 0 = 0,$$
$$s_3 = r_4 \oplus r_5 \oplus r_6 \oplus r_7 = 1 \oplus 1 \oplus 1 \oplus 0 = 1.$$

It is obvious that syndrome is $S = (100) = 4$ and that the error is detected at the fourth position. The decoder inverts the fourth bit, the error is corrected and a correctly reconstructed code word is obtained, from which the information bits are extracted

$$\hat{\boldsymbol{c}} = [0010110] = \boldsymbol{c} \Rightarrow \hat{\boldsymbol{i}} = [1110] = \boldsymbol{i}.$$

(b) Firstly, it should be noted that the code word structure depends on the code construction as well as on the input information sequence. If the same code is applied, to the unchanged information word corresponds the same code word $\boldsymbol{c} = (0010110)$ because the transformation is 1:1.

If during the transmission the error did not occurred, the received word is $\boldsymbol{r} = \boldsymbol{c}$, parity-checks are

$$s_1 = r_1 \oplus r_3 \oplus r_5 \oplus r_7 = 0 \oplus 1 \oplus 1 \oplus 0 = 0,$$
$$s_2 = r_2 \oplus r_3 \oplus r_6 \oplus r_7 = 0 \oplus 1 \oplus 1 \oplus 0 = 0,$$
$$s_3 = r_4 \oplus r_5 \oplus r_6 \oplus r_7 = 0 \oplus 1 \oplus 1 \oplus 0 = 0,$$

and syndrome is $S = (000) = 0$, corresponding to the transmission without errors. In this case decoder set aside the code word bits 3, 4, 6 and 7, and the reconstructed information word is the same as in the previous case.

When the errors occurred in the 3rd and the 5th bit, the received word is $r = (0000010)$

$$s_1 = r_1 \oplus r_3 \oplus r_5 \oplus r_7 = 0 \oplus 0 \oplus 0 \oplus 0 = 0$$
$$s_2 = r_2 \oplus r_3 \oplus r_6 \oplus r_7 = 0 \oplus 0 \oplus 1 \oplus 0 = 1$$
$$s_3 = r_4 \oplus r_5 \oplus r_6 \oplus r_7 = 0 \oplus 0 \oplus 1 \oplus 0 = 1$$

and the syndrome $S = (110) = 6$ indicates that the error occurred at the 6th bit. After the corresponding inversion, the estimations of the code and information words are

$$\hat{c} = (0000000) \Rightarrow \hat{i} = (0000) \neq i.$$

It is obvious that the decoding in this case was unsuccessful because even three information word bits are wrongly decoded. Here the decoding made situation even the worse—channel errors were at two information bits, while the decoder itself introduced an additional error. It does not happen always that the channel (or decoder) introduces the errors on the information bits, but such situation is not desirable.

As a Hamming code syndrome has $n - k = b$ bits, it is easy to calculate that the total number of possible syndrome values is $2^{n-k} = n + 1$. From it, n syndrome values show the single error positions, and the last one (all zeros) corresponds to the transmission without errors. This relation shows that here the **Hamming bound**

$$q^{n-k} \geq \sum_{t=0}^{e_c} \binom{n}{t}(q-1)^t$$

is satisfied with equality for $e_c = 1$ (for a binary code the code word consists only of zeroes and ones, i.e. $q = 2$), the code is perfect. Therefore, this code can correct all single errors, but not any combination of two errors in the code word. Besides the Hamming code, it is known that the Golay code (23, 12) is as well a **perfect** one, it can correct all single, double or triple errors [24].

Now it is interesting to consider how often the single and double errors will occur. Generally, it depends on the code word length, type of the channel errors and the average value of error probability. For BSC the probability that from n transmitted bits, the errors occurred in t positions is given by a binomial distribution

$$P_{e,t} = \binom{n}{t} p^t (1-p)^{n-t}, \quad t \le n.$$

For illustration, let suppose that the channel error probability is $p = 10^{-3}$. In this case the probabilities of single or double channel errors in the code word of length $n = 7$ are

$$P_{e,1} = \binom{7}{1} p(1-p)^6 \approx \frac{7!}{1!6!} \times 10^{-3} = 7 \times 10^{-3},$$

$$P_{e,2} = \binom{7}{2} p^2 (1-p)^5 \approx \frac{7!}{2!5!} \times (10^{-3})^2 = 2.1 \times 10^{-5}$$

Similarly, it is possible to calculate the probabilities for triple or fourfold error sequences, but they will have the substantially smaller values. It is obvious that the uncorrectable error in the code word, for a code (7, 4), for the above channel conditions, is a few hundred times less frequent than the single errors. Therefore, even such simple code can substantially decrease the error probability for the user.

(c) The generator matrix of any linear block code can be found from the relation $c = i \otimes G$, for the considered Hamming code yielding

$$[z_1 z_2 i_1 z_3 i_2 i_3 i_4] = [i_1 i_2 i_3 i_4] \otimes \begin{bmatrix} g_{11} & g_{12} & \cdots & g_{17} \\ g_{21} & g_{22} & \cdots & g_{27} \\ g_{31} & g_{32} & \cdots & g_{37} \\ g_{41} & g_{42} & \cdots & g_{47} \end{bmatrix}.$$

The previous relation, combined with the parity-check bits defining expressions, yields the Hamming code (7, 4) generator matrix

$$\begin{aligned} z_1 &= i_1 \oplus i_2 \oplus i_4, \\ z_2 &= i_1 \oplus i_3 \oplus i_4, \\ z_3 &= i_2 \oplus i_3 \oplus i_4. \end{aligned} \Rightarrow G = \begin{bmatrix} 1 & 1 & 1 & 0 & 0 & 0 & 0 \\ 1 & 0 & 0 & 1 & 1 & 0 & 0 \\ 0 & 1 & 0 & 1 & 0 & 1 & 0 \\ 1 & 1 & 0 & 1 & 0 & 0 & 1 \end{bmatrix}.$$

On the other hand, the Hamming code parity-check matrix can be found generally from the relation

$$S = r \otimes H^T \Rightarrow [s_1 s_2 s_3] = [r_1 r_2 r_3 r_4 r_5 r_6 r_7] \otimes \begin{bmatrix} h_{11} & h_{21} & h_{31} \\ h_{12} & h_{22} & h_{32} \\ \cdots & \cdots & \cdots \\ h_{17} & h_{27} & h_{37} \end{bmatrix},$$

Problems

and in this case, the **parity-check matrix** can be found practically without difficulties

$$s_1 = r_1 \oplus r_3 \oplus r_5 \oplus r_7$$
$$s_2 = r_2 \oplus r_3 \oplus r_6 \oplus r_7 \, s$$
$$s_3 = r_4 \oplus r_5 \oplus r_6 \oplus r_7$$

$$\Rightarrow \quad \mathbf{H}^T = \begin{bmatrix} 1 & 0 & 0 \\ 0 & 1 & 0 \\ 1 & 1 & 0 \\ 0 & 0 & 1 \\ 1 & 0 & 1 \\ 0 & 1 & 1 \\ 1 & 1 & 1 \end{bmatrix} \quad \Rightarrow \quad \mathbf{H} = \begin{bmatrix} 1 & 0 & 1 & 0 & 1 & 0 & 1 \\ 0 & 1 & 1 & 0 & 0 & 1 & 1 \\ 0 & 0 & 0 & 1 & 1 & 1 & 1 \end{bmatrix}.$$

(d) It is obvious that this code has $2^k = 16$ code words. They can be obtained easily if all four-bit information words are multiplied by the generator matrix, resulting in the following combinations

$c_1 = (0000000)$,	$c_2 = (1101001)$,	$c_3 = (0101010)$,	$c_4 = (1000011)$,
$c_5 = (1001100)$,	$c_6 = (0100101)$,	$c_7 = (1100110)$,	$c_8 = (0001111)$,
$c_9 = (1110000)$,	$c_{10} = (0011001)$,	$c_{11} = (1011010)$,	$c_{12} = (0110011)$,
$c_{13} = (0111100)$,	$c_{14} = (1010101)$,	$c_{15} = (0010110)$,	$c_{16} = (1111111)$.

Let two code words from this set are chosen at random, e.g. c_4 and c_{11}. Their sum is the code word as well, because

$$c_4 \oplus c_{11} = (1000011) \oplus (1011010) = (0011001) = c_{10}.$$

As in a vector subspace (and the linear code by definition is it) the closure for addition must be satisfied, the sum of any two code words must be a code word as well. As a consequence is that the Hamming distance between the addends will be equal to the Hamming weight of their sum (i.e. to the number of ones in the obtained code word). Because of that, every code word is a sum of some other two code words, and its weight is the Hamming distance between some other code words and there is no any two words whose Hamming distance does not correspond to the Hamming weight of some third code word.

As a further consequence, it is sufficient to count ones in the code words, because it completely determines the spectrum of all possible distances between the code words. In the above code there is one word consisting of all zeros, seven words having three ones, seven words having four ones and one word having seven ones. The corresponding weight spectrum is shown in Fig. 5.9.

Minimum Hamming distance in this case is $d_{\min} = 3$, the number of **correctable** bits (errors) is given by $d_{\min} \geq 2e_c + 1$ and $e_c = 1$, while the number of errors that can be **detected** obtained from $d_{\min} \geq e_c + e_d + 1$ is $e_d = 1$. This code obviously can detect only one error, and to correct it. If $e_c = 0$, then $e_d = 2$. Of course, it does not mean that all combinations of three errors (in the code word) result in an

Fig. 5.9 Hamming code (7, 4) weight spectrum

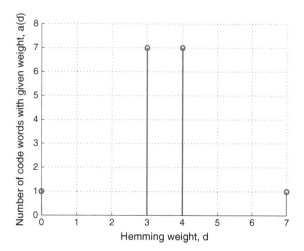

undetectable error—it is the case only for combinations having the same pattern as the code words.

Problem 5.4 Explain the construction of Hamming code having code word length $n = 2^b$, b—any integer greater than 1, and analyze the Hamming code (8, 4) in details.

(a) Illustrate the procedures of encoding and decoding for the information word $i = (0011)$ if the error occurred at the third position of the code word.
(b) Explain in details the decoding of sequences (00000101) and (00101100).
(c) Find the minimum Hamming distance of the code and verify the number of errors the code can correct and detect.
(d) Analyze the code correction capabilities for $p = 10^{-3}$.
(e) Find the generator and the parity-check matrix for systematic (8, 4) Hamming code.

Solution
Hamming code having the code word length $n = 2^b$ bits, is constructed beginning from the previously formed code $(2^b - 1, 2^b - b - 1)$. Then, at the last position one parity-check bit is added, while after the reception one additional syndrome bit is calculated

$$z_{b+1} = \sum_{l=1}^{2^b-1} c_l, \quad s_{b+1} = \sum_{l=1}^{2^b} r_l.$$

In such a way the code which has parameters $(2^b, 2^b - b)$ is obtained, and the additional parity-check bit just makes possible the detection of one more error. The obtained code can correct one and detect two errors in the code word.

Problems

The construction of (8, 4) Hamming code differs from (7, 4) code construction only in **adding one bit obtained by simple parity-check**, i.e. the fourth parity-check bit is calculated as follows

$$z_4 = \sum_{i=1}^{7} c_i = z_1 \oplus z_2 \oplus i_1 \oplus z_3 \oplus i_2 \oplus i_3 \oplus i_4.$$

If the number of ones in code word of the basic code (7, 4) is odd, then $z_4 = 1$, and if it is even, then $z_4 = 0$. In any case, after the inclusion of this bit, the sum (modulo-2) of all bits in the code word must be equal to zero, and the additional bit of the syndrome can be written as

$$s_4 = \sum_{i=1}^{8} r_i = \sum_{i=1}^{8} (c_i \oplus e_i) = \sum_{i=1}^{8} e_i.$$

Therefore, $s_4 = 1$ indicates an odd number of errors during transmission, while $s_4 = 0$ corresponds to the even number of errors. Syndrome indicating the estimated error position is formed using the first three parity-check bits.

(a) The procedure for encoding the information word $i = (1110)$ using the Hamming code (8, 4) can be subdivided into two parts. During the first phase the code word corresponding to (7, 4) code is formed, as explained in details in the previous problem, resulting in code word $c = (0010110)$. During the second phase, the ones in the first seven positions of code word are counted, and because here there are three ones (an odd number), the code word of (8, 4) is obtained by adding the bit $z_4 = 1$.

If during the transmission, the error occurred at the third position, the received word is $r = (00001101)$, and the parity-check bits are

$$s_1 = r_1 \oplus r_3 \oplus r_5 \oplus r_7 = 0 \oplus 0 \oplus 1 \oplus 0 = 1,$$
$$s_2 = r_2 \oplus r_3 \oplus r_6 \oplus r_7 = 0 \oplus 1 \oplus 1 \oplus 0 = 1,$$
$$s_3 = r_4 \oplus r_5 \oplus r_6 \oplus r_7 = 1 \oplus 1 \oplus 1 \oplus 0 = 0,$$
$$s_4 = \sum_{i=1}^{8} r_i = 0 \oplus 0 \oplus 1 \oplus 1 \oplus 1 \oplus 1 \oplus 0 \oplus 1 = 1.$$

From $S = (s_3\, s_2\, s_1) = (0\,1\,1) = 3$ and $s_4 = 1$, it is obvious that during the transmission (in the first seven positions!) occurred an odd number of errors, i.e. one, three, five or seven errors. As it will be shown later, usually it is justifiable to suppose that only one error occurred at the position indicated by syndrome.

The receiving word is corrected and the information bits are "extracted"

$$\hat{c} = (00101101) \Rightarrow \hat{i} = (1110).$$

One can conclude that the error control coding was successful, i.e. the user obtained the information bits without errors in spite of the fact that one error occurred during transmission

(b) If the received word is $r = (00000101)$, the parity-check bits are:

$$s_1 = r_1 \oplus r_3 \oplus r_5 \oplus r_7 = 0 \oplus 1 \oplus 0 \oplus 0 = 0$$
$$s_2 = r_2 \oplus r_3 \oplus r_6 \oplus r_7 = 0 \oplus 1 \oplus 1 \oplus 0 = 1$$
$$s_3 = r_4 \oplus r_5 \oplus r_6 \oplus r_7 = 1 \oplus 0 \oplus 1 \oplus 0 = 1$$
$$s_4 = \sum_{i=1}^{8} r_i = 0 \oplus 0 \oplus 1 \oplus 1 \oplus 0 \oplus 1 \oplus 0 \oplus 1 = 0$$

From $s_4 = 0$ follows that an even number of errors occurred during the transmission corresponds to the case when the errors cannot be corrected. Now the syndrome is $S = (s_3 s_2 s_1) = (110) = 6$, but it is not the error position.

It could be noticed that this case corresponds to the (b) from the previous problem, but here the parity bit $z_4 = 1$ is added. The errors there were at the same positions (third and fifth bit), but the decoder could not correct the errors (even introduced additional error at the sixth position). In this case the decoder only detects the occurrence of an even number of errors and does not try to correct them, but the retransmission is asked for. In such a way the possibility that the decoder introduces new errors is avoided, and if the channel state is better during the retransmission (if not more than one error in the code word occurred) the information word can be successfully transmitted.

For $r = (00101100)$, the parity-check bits are

$$s_1 = r_1 \oplus r_3 \oplus r_5 \oplus r_7 = 0 \oplus 1 \oplus 0 \oplus 0 = 0,$$
$$s_2 = r_2 \oplus r_3 \oplus r_6 \oplus r_7 = 0 \oplus 1 \oplus 1 \oplus 0 = 0,$$
$$s_3 = r_4 \oplus r_5 \oplus r_6 \oplus r_7 = 1 \oplus 0 \oplus 1 \oplus 0 = 0,$$
$$s_4 = \sum_{i=1}^{8} r_i = 0 \oplus 0 \oplus 1 \oplus 1 \oplus 0 \oplus 1 \oplus 0 \oplus 1 = 1.$$

The result $s_4 = 1$ suggests an odd number of channel errors, probably one. However, the syndrome value $S = (s_3 s_2 s_1) = (000) = 0$ shows that there were no errors.

As stressed earlier, the syndrome is calculated on the basis of the first seven code word bits, while, the last parity-check includes and the last—eighth bit. It makes possible to find that the error occurred at the last code word bit, being here the case. In such a case the decoder has only to invert the last code word bit (it is not even needed, why?). The information sequence $\hat{i} = (1110)$ is then extracted. Of course, the decoding is successful in this case.

(c) It is obvious that the inclusion of general parity-check does not affect the all zeros word, while the Hamming weight for all ones word increases from $d = 7$ at $d = 8$. All the words having three ones for (7, 4) code now will have an additional one at the eighth position, while the words having weight $d = 4$ will have zero at the eighth position. It is the reason that in the spectrum there are no more words having the weight $d = 3$, while the number of words having four ones increases from 7 at 14.

It is the reason that Hamming code weight becomes $d_{min} = 4$. The relation $d_{min} \geq 2e_c + 1$ must be satisfied, and the maximum corresponding integer value is $e_c = 1$. The number of errors that could be detected is found from $d_{min} \geq e_c + e_d + 1$, and as the value for e_c is determined ($e_c = 1$), now the value $e_d = 2$ is obtained. Therefore, the Hamming code (8, 4) detects two errors, but only one can be corrected.

(d) Because the code detects an even number of errors in the code word, the probability to occur any number of odd errors should be estimated. Similarly to the previous case, the channel will be considered as memoryless channel and that it can be modeled as a binary symmetric channel. Supposing the average error probability $p = 10^{-3}$, the probabilities of occurrence of one, three, five or seven errors in the code word are, respectively

$$P_{e,1} = \binom{8}{1} p(1-p)^7 \approx 8 \times 10^{-3}, P_{e,2} = \binom{8}{2} p^2(1-p)^6 \approx 2.8 \times 10^{-5},$$

$$P_{e,3} = \binom{8}{3} p^3(1-p)^5 \approx 5.6 \times 10^{-8}, \cdots$$

The drawback of this code is that it is incapable to correct double errors and, if the decoder is designed to correct one error per codeword, it is not capable to detect triple errors. In such a case, **the probability for code word wrong detection** for $p = 10^{-3}$ is mainly determined by the probability of triple error

$$P_{ed} \approx P_{e,3} = \binom{8}{3} p^3(1-p)^5 \approx 5.6 \times 10^{-8}.$$

For such transmission conditions, the probability of an undetected error is about 500 times smaller when compared with the case when additional parity check bit is not added, i.e. in the case when Hamming (7, 4) is applied. It is a substantial improvement, but it is conditioned by the feedback channel use, which allows the request for retransmission of words (packets) when an even number of errors is detected.

In the case when the decoder is designed not to correct any error, it is capable to detect up to three errors in any positions but it is not capable to detect some patterns with larger weights that corresponds to the code words (analysis is related with part c) of this problem and similar to Problem 5.2). For this code, there are 14 critical eight-bit critical error patterns having four ones and one error pattern with weight eight, for which the syndrome equals zero! In this case, the probability for code word wrong detection is

$$P_{ed} = 14 p^4 (1-p)^4 + p^8 \approx 1.4 \cdot 10^{-11}.$$

(e) A general parity-check bit can be written as well as follows

$$z_4 = z_1 \oplus z_2 \oplus i_1 \oplus z_3 \oplus i_2 \oplus i_3 \oplus i_4$$
$$= (i_1 \oplus i_2 \oplus i_4) \oplus (i_1 \oplus i_3 \oplus i_4) \oplus i_1 \oplus (i_2 \oplus i_3 \oplus i_4) \oplus i_2 \oplus i_3 \oplus i_4$$
$$= i_1 \oplus i_2 \oplus i_3$$

and the generator matrix of a Hamming code (7, 4) from the previous problem should be slightly modified

$$G = \begin{bmatrix} 1 & 1 & 1 & 0 & 0 & 0 & 0 & 1 \\ 1 & 0 & 0 & 1 & 1 & 0 & 0 & 1 \\ 0 & 1 & 0 & 1 & 0 & 1 & 0 & 1 \\ 1 & 1 & 0 & 1 & 0 & 0 & 1 & 0 \end{bmatrix}.$$

Generator matrix of systematic Hamming code is formed by the changing the columns places, this operation does not change the code spectrum nor the code capability to correct the errors. If the columns of the previous matrix are reordered as 3-5-6-7-1-2-4-8, the **systematic generator matrix** is obtained

$$G_s = \begin{bmatrix} 1 & 0 & 0 & 0 & | & 1 & 1 & 0 & 1 \\ 0 & 1 & 0 & 0 & | & 1 & 0 & 1 & 1 \\ 0 & 0 & 1 & 0 & | & 0 & 1 & 1 & 1 \\ 0 & 0 & 0 & 1 & | & 1 & 1 & 1 & 0 \end{bmatrix} = [\mathbf{I}_4, \mathbf{P}].$$

The corresponding parity-check matrix can be easily found by noting that to the submatrix $\mathbf{P}$ correspond the last four columns of the generator matrix, yielding as a result

$$H_s = \begin{bmatrix} \mathbf{P}^T, \mathbf{I}_3 \end{bmatrix} = \begin{bmatrix} 1 & 1 & 0 & 1 & | & 1 & 0 & 0 & 0 \\ 1 & 0 & 1 & 1 & | & 0 & 1 & 0 & 0 \\ 0 & 1 & 1 & 1 & | & 0 & 0 & 1 & 0 \\ 1 & 1 & 1 & 0 & | & 0 & 0 & 0 & 1 \end{bmatrix}.$$

Problems

Problem 5.5

(a) Perform the encoding of sequence (10101010101) by (15, 11) code and decode the corresponding code word if the error occurred at the ninth position. How many errors can be corrected and how many detected by this code?

(b) How to modify the code to make possible the detection of two errors? For the previous information word, analyze the case when the errors occurred at the positions 6 and 16.

(c) Decode the sequences (111111000000) and (000000010001) if it is known that the code used can correct one error and detect one error.

(d) Decode the sequences (001001000001) and (101111000011) if it is known that Hamming (12, 7) code is used.

Solution

(a) The code word length is $n = 15$, the first higher power of two is $2^4 = 16$ and the code will be constructed using four columns from the cliché and there are $b = 4$ parity-check bits. Because of $n - k = b$, the code have the possibility to detect and to correct one error. The cliché for forming the code, encoding and decoding procedures is shown below.

l	bin(l)	c_l	-encoding	-decoding
1	0001	z_1	$i = (10101010101)$	$r = (101101000010101)$
2	0010	z_2	$z_1 = c_3 \oplus c_5 \oplus c_7 \oplus c_9 \oplus c_{11} \oplus c_{13} \oplus c_{15}$	$s_1 = r_1 \oplus r_3 \oplus r_5 \oplus r_7 \oplus r_9 \oplus$
3	0011	i_1	$= i_1 \oplus i_2 \oplus i_4 \oplus i_5 \oplus i_7 \oplus i_9 \oplus i_{11} = 1$	$\oplus r_{11} \oplus r_{13} \oplus r_{15} = 1$
4	0100	z_3	$z_2 = c_3 \oplus c_6 \oplus c_7 \oplus c_{10} \oplus c_{11} \oplus c_{14} \oplus c_{15}$	$s_2 = r_2 \oplus r_3 \oplus r_6 \oplus r_7 \oplus r_{10} \oplus$
5	0101	i_2	$= i_1 \oplus i_3 \oplus i_4 \oplus i_6 \oplus i_7 \oplus i_{10} \oplus i_{11} = 0$	$\oplus r_{11} \oplus r_{14} \oplus r_{15} = 0$
6	0110	i_3	$z_3 = c_5 \oplus c_6 \oplus c_7 \oplus c_{12} \oplus c_{13} \oplus c_{14} \oplus c_{15}$	$s_3 = r_4 \oplus r_5 \oplus r_6 \oplus r_7 \oplus r_{12} \oplus$
7	0111	i_4	$= i_2 \oplus i_3 \oplus i_4 \oplus i_8 \oplus i_9 \oplus i_{10} \oplus i_{11} = 1$	$\oplus r_{13} \oplus r_{14} \oplus r_{15} = 0$
8	1000	z_4	$z_4 = c_9 \oplus c_{10} \oplus c_{11} \oplus c_{12} \oplus c_{13} \oplus c_{14} \oplus c_{15}$	$s_4 = r_8 \oplus r_9 \oplus r_{10} \oplus r_{11} \oplus r_{12} \oplus$
9	1001	i_5	$= i_5 \oplus i_6 \oplus i_7 \oplus i_8 \oplus i_9 \oplus i_{10} \oplus i_{11} = 0$	$\oplus r_{13} \oplus r_{14} \oplus r_{15} = 1$
10	1010	i_6	$c = (z_1 \, z_2 \, i_1 \, z_3 \, i_2 \, i_3 \, i_4 \, z_4 \, i_5 \, i_6 \, i_7 \, i_8 \, i_9 \, i_{10} \, i_{11})$	$S = (s_4 s_3 s_2 s_1) = (1001) = 9$
11	1011	i_7	$= (101101001010101)$	$\hat{c} = (101101001010101)$
12	1100	i_8		$\hat{i} = (10101010101)$
13	1101	i_9		
14	1110	i_{10}		
15	1111	i_{11}		

Therefore, the error occurred at the position 9 was detected and corrected, and the user received the correct information sequence.

(b) To made the detection of two errors possible one parity-check bit should be added on the position 16. The **extended Hamming code** has the parameters $(n, k) = (16, 12)$ and the added parity bit is

$$z_4 = \sum_{i=1}^{15} c_i = 0,$$

yielding the code word $c = (1011010010101010)$.

The received code word, after the inversion of bits at the 6th and the 16th position is $r = (1011000010101011)$ while syndrome and additional parity-check bit are

$$S = (s_4 s_3 s_2 s_1) = (0110) = 6, \; s_5 = \sum_{i=1}^{16} r_i = 0.$$

In this case syndrome shows the position of one of the errors, but because of $s_5 = 0$ the receiver detects the double error and does not start the error correction, but asks for retransmission. If the decoder "knew" that one error occurred at the parity-check bit, it could correct both errors.

According to the solution from the previous problem the syndrome can be as well written as

$$s_1 = e_1 \oplus e_3 \oplus e_5 \oplus e_7 \oplus e_9 \oplus e_{11} \oplus e_{13} \oplus e_{15}$$
$$s_2 = e_2 \oplus e_3 \oplus e_6 \oplus e_7 \oplus e_{10} \oplus e_{11} \oplus e_{14} \oplus e_{15}$$
$$s_3 = e_4 \oplus e_5 \oplus e_6 \oplus e_7 \oplus e_{12} \oplus e_{13} \oplus e_{14} \oplus e_{15}$$
$$s_4 = e_8 \oplus e_9 \oplus e_{10} \oplus e_{11} \oplus e_{12} \oplus e_{13} \oplus e_{14} \oplus e_{15}$$

and the combination $S = (0110)$, $s_5 = 0$ appears as well in the case when the errors occurred at the positions 7 and 14. Therefore, the various error combinations can result in the same syndrome value and the errors positions cannot be found exactly.

The same can be clearly seen as well from the Hamming bound in this case

$$2^5 \geq \sum_{t=0}^{e_c} \binom{16}{t} = 1 + \binom{16}{1} = 17,$$

this relation being satisfied only for $e_c = 1$.

(c) From $n = 12$ and the fact that code can detect one error follows that it is not possible to add the general parity-check bit. Because of that 12th position is taken into account when forming the table and parity-check bits for the syndrome. The cliché is given in the table having 4 columns ($12 < 2^4$) and a **shortened Hamming code** (12, 8) is obtained. The decoding procedure is as follows:

Problems

l	bin(l)	c_l	The first received word	The second received word
1	000**1**	z_1	$r = (111111000000)$	$r = (000000010001)$
2	00**1**0	z_2	$s_1 = r_1 \oplus r_3 \oplus r_5 \oplus r_7 \oplus r_9 \oplus r_{11} = 1$	$s_1 = r_1 \oplus r_3 \oplus r_5 \oplus r_7 \oplus r_9 \oplus r_{11} = 0$
3	00**11**	i_1	$s_2 = r_2 \oplus r_3 \oplus r_6 \oplus r_7 \oplus r_{10} \oplus r_{11} = 1$	$s_2 = r_2 \oplus r_3 \oplus r_6 \oplus r_7 \oplus r_{10} \oplus r_{11} = 0$
4	0**1**00	z_3	$s_3 = r_4 \oplus r_5 \oplus r_6 \oplus r_7 \oplus r_{12} = 1$	$s_3 = r_4 \oplus r_5 \oplus r_6 \oplus r_7 \oplus = 1$
5	0**1**0**1**	i_2	$s_4 = r_8 \oplus r_9 \oplus r_{10} \oplus r_{11} \oplus r_{12} = 0$	$s_4 = r_8 \oplus r_9 \oplus r_{10} \oplus r_{11} \oplus r_{12} = 0$
6	0**11**0	i_3	$S = (s_4 s_3 s_2 s_1) = (0111) = 7$	$S = (s_4 s_3 s_2 s_1) = (0100) = 4$
7	0**111**	i_4	$\hat{c} = (111111100000)$	$\hat{c} = (000100010001)$
8	**1**000	z_4	$\hat{i} = (11110000)$	$\hat{i} = (00000001)$
9	**1**00**1**	i_5		
10	**1**0**1**0	i_6		
11	**1**0**11**	i_7		
12	**11**00	i_8		

(d) When a Hamming (12, 7) code is used, it is obvious that it is shortened, because the code word length cannot be written as $n = 2^b - 1$ nor $n = 2^b$. Now the numbers of correctable errors will be checked.

(1) The relation $n - k = \lceil \mathrm{ld}(n) \rceil$, because $n - k = 5$ and $\lceil \mathrm{ld}(12) \rceil = 4$, is not satisfied, the code cannot detect nor correct one error.

(2) The relation $n - k = \lceil \mathrm{ld}(n - 1) \rceil + 1$ is satisfied and the first $n - 1 = 11$ bits form the cliché to obtain the code having $b = \lceil \mathrm{ld}(n - 1) \rceil = 4$ redundant (parity-check) bits. The fifth redundant bit is general parity-check and it is put on the position 12. Therefore, the code can correct one and detect two errors at the code word.

In this case for construction of (12, 7) code, firstly the code (11, 7) is formed and there is not 11th bit according to the cliché, and after a general parity-check bit is added. Procedure for decoding the first received word is as follows.

l	bin(l)	c_l	$r = (001001000001)$
1	000**1**	z_1	$s_1 = r_1 \oplus r_3 \oplus r_5 \oplus r_7 \oplus r_9 \oplus r_{11} = 1$
2	00**1**0	z_2	$s_2 = r_2 \oplus r_3 \oplus r_6 \oplus r_7 \oplus r_{10} \oplus r_{11} = 0$
3	00**11**	i_1	$s_3 = r_4 \oplus r_5 \oplus r_6 \oplus r_7 = 1$
4	0**1**00	z_3	$s_4 = r_8 \oplus r_9 \oplus r_{10} \oplus r_{11} = 0$
5	0**1**0**1**	i_2	$S = (s_4 s_3 s_2 s_1) = (0101) = 5$
6	0**11**0	i_3	$s_5 = 1$
7	0**111**	i_4	$\hat{c} = (111111100000)$
8	**1**000	z_4	$\hat{i} = (1110000)$
9	**1**00**1**	i_5	
10	**1**0**1**0	i_6	
11	**1**0**11**	i_7	

Differently from the previous case, the decoding of the word (101111000011) yields the following syndrome values

$$s_1 = r_1 \oplus r_3 \oplus r_5 \oplus r_7 \oplus r_9 \oplus r_{11} = 0$$
$$s_2 = r_2 \oplus r_3 \oplus r_6 \oplus r_7 \oplus r_{10} \oplus r_{11} = 1$$
$$s_3 = r_4 \oplus r_5 \oplus r_6 \oplus r_7 = 1$$
$$s_4 = r_8 \oplus r_9 \oplus r_{10} \oplus r_{11} = 1$$
$$S = (s_4 s_3 s_2 s_1) = (1110) = 14$$
$$s_5 = 1$$

It is obvious that an even number of errors did not occur, but it is unusual that the syndrome gives the position 14, not existing in the code. The reader should found which error can result in such syndrome value.

Problem 5.6 Uplink of one geostationary satellite system is considered. Due to the frequency limitations maximum signaling rate is $v_b = 200$ [kb/s]. The bad weather conditions in some intervals (not longer than 0.01 [ms]) cause the errors in packets. These intervals have a period $T = 0.14$ [ms]. Due to the long propagation delay, the satellite link is not suitable for ARQ procedure.

(a) Propose the solution enabling transmission without errors under the circumstances. Find the code rate of the proposed error control code.
(b) If the demodulator output sequence is:

    0110100100010011111101000111110010100111101 01

comment the syndrome values, decode the information word and determine the type of channel errors.
(c) If from the terrestrial station a sequence of $N = 5 \times 10^6$ symbols from the set $\{A, B, C, D, E\}$ with the corresponding probabilities $\{0.45; 0.35; 0.1; 0.07; 0.03\}$ is emitted, find a minimum theoretical time for their emitting, if the error control code from (a) is used. If the time for this sequence transmission should not be greater than $t_{max} = 80$ [s], propose a compression code that satisfies this condition.

Solution
The satellite is at the geostationary orbit, the distance from the terrestrial station is at least $d = 33600$ [km]. The electromagnetic waves propagate as a light ($c = 3\times 10^8$ [m/s]), minimum propagation time (delay) on link is $\tau = d/c = 112$ [ms]. The satellite is in fact the relay between two terrestrial stations, and if the errors occur, for a complete retransmission, the total delay for a packet retransmission is $8\tau = 896$ [ms] (the initial transmission earth-satellite-earth, the return path for a negative acknowledgment for a packet reception, once more the complete path for a positive acknowledgement after the packet reception for resetting a transmission buffer). Taking into account that when using ARQ procedure sometimes more

retransmissions are used, it is obvious that the total delay can attain a few seconds. Due to this reason, in satellite systems instead of ARQ procedure, the FEC techniques are more suitable, where the codes having a possibility to correct all the detected errors have the advantage (as a difference from e.g. computer networks).

(a) The signaling rate is $v_b = 200$ [kb/s] and the time for one bit transmission is $T_b = 1/v_b = 5$ [μs]. During one period (0.14 [ms]) 28 bits are transmitted, and during the bad interval (0.01 [ms]) two neighboring bits may be inverted. The interleaving period always equals the number of bits transmitted during one period ($L_I = 28$) and after the deinterleaving it should provide the maximum distance between the errors. If the disturbances appear regularly (strictly periodically and approximately having the same duration for every period), it is suitable to use matrix block-interleaver where one matrix dimension is determined by the number of errors in one disturbance period (denoted by l) and the other by a code word length n, where $nl = L_I$ yielding $n = 28/2 = 14$.

There are two Hamming codes having this code word length. A code (14, 9) can detect two errors, but it is not interesting, because there is no the retransmission. Therefore, code (14, 10) is chosen having the greater code rate as well as the same capability for error correction as the code (14, 9).

(b) The procedure for obtaining the Hamming code (14, 10) code words is given by the following cliché, where the calculating the parity-check bits and syndrome are shown as well

```
 1   0 0 0 1  |z₁|
 2   0 0 1 0  |z₂|
 3   0 0 1 1  |i₁|
 4   0 1 0 0  |z₃|
 5   0 1 0 1  |i₂|
 6   0 1 1 0  |i₃|
 7   0 1 1 1  |i₄|
 8   1 0 0 0  |z₄|
 9   1 0 0 1  |i₅|
10   1 0 1 0  |i₆|
11   1 0 1 1  |i₇|
12   1 1 0 0  |i₈|
13   1 1 0 1  |i₉|
14   1 1 1 0  |i₁₀|
15   1 1 1 1
```

$z_1 = i_1 \oplus i_2 \oplus i_4 \oplus i_5 \oplus i_7 \oplus i_9$,
$z_2 = i_1 \oplus i_3 \oplus i_4 \oplus i_6 \oplus i_7 \oplus i_{10}$
$z_3 = i_2 \oplus i_3 \oplus i_4 \oplus i_8 \oplus i_9 \oplus i_{10}$
$z_4 = i_5 \oplus i_6 \oplus i_7 \oplus i_8 \oplus i_9 \oplus i_{10}$

$s_1 = r_1 \oplus r_3 \oplus r_5 \oplus r_7 \oplus r_9 \oplus r_{11} \oplus r_{13}$
$s_2 = r_2 \oplus r_3 \oplus r_6 \oplus r_7 \oplus r_{10} \oplus r_{11} \oplus r_{14}$
$s_3 = r_4 \oplus r_5 \oplus r_6 \oplus r_7 \oplus r_{12} \oplus r_{13} \oplus r_{14}$
$s_4 = r_8 \oplus r_9 \oplus r_{10} \oplus r_{11} \oplus r_{12} \oplus r_{13} \oplus r_{14}$

The *matrix block-interleaver* is shown in Fig. 5.10. The input sequence obtained from the encoder is firstly entered into the interleaver row-by-row, but is red out column-by-column for emitting through the channel. The packet errors occur in the channel, and in the example shown in Fig. 5.10 it is supposed that they occurred in the three neighboring bits denoted by y_{l1}, y_{12}, y_{22}. The writing in deinterleaver is column-by-column and the reading row-by-row. In such a way, the

$$c \rightarrow \begin{bmatrix} x_{11} & x_{12} & \cdots & x_{1n} \\ x_{21} & x_{22} & \cdots & x_{2n} \\ \cdots & \cdots & \cdots & \cdots \\ x_{l1} & x_{l2} & \cdots & x_{ln} \end{bmatrix}$$
(from encoder)
$\downarrow$

$x = (x_{11}, x_{21}, \ldots, x_{l1}, x_{12}, x_{22}, \ldots, x_{l2}, \ldots, x_{1n}, x_{2n}, \ldots, x_{ln})$
from the interleaver to the channel

from the channel to the deinterleaver
$y = (y_{11}, y_{21}, \ldots, y_{l1}, y_{12}, y_{22}, \ldots, y_{l2}, \ldots, y_{1n}, y_{2n}, \ldots, y_{ln})$
$\downarrow$

$$\begin{bmatrix} y_{11} & y_{12} & \cdots & y_{1n} \\ y_{21} & y_{22} & \cdots & y_{2n} \\ \cdots & \cdots & \cdots & \cdots \\ y_{l1} & y_{l2} & \cdots & y_{ln} \end{bmatrix} \begin{array}{l} \rightarrow r \\ \text{(in decoder)} \end{array}$$

Fig. 5.10 The illustration of the working principle of matrix interleaver and deinterleaver

first row in the deinterleaver matrix corresponds to the first interleaver column and the errors are not more concentrated into a packet, but they are at maximum distances. If not more than l successive errors occurred and the period having the length $L_I = nl$, at the decoder input there would not be any received word having more than one error.

For a given sequence at the receiver input, the forming of code words by interleaver is as follows.

$$y = (00111010111011111100100010001)$$
$\downarrow$

$$\begin{bmatrix} 0 & 1 & 1 & 1 & 1 & 1 & 1 & 0 & 0 & 0 & 0 & 0 & 0 \\ 0 & 1 & 0 & 0 & 1 & 0 & 1 & 1 & 1 & 1 & 0 & 1 & 0 & 1 \end{bmatrix} \begin{array}{l} \rightarrow r^{(I)} \\ \rightarrow r^{(II)} \end{array}$$

1 2 3 4 5 6 7 8 9 10 11 12 13 14

The decoding of the first word is as follows

$s_1 = r_1 \oplus r_3 \oplus r_5 \oplus r_7 \oplus r_9 \oplus r_{11} \oplus r_{13} = 0 \oplus 1 \oplus 1 \oplus 1 \oplus 0 \oplus 0 \oplus 0 = 1$
$s_2 = r_2 \oplus r_3 \oplus r_6 \oplus r_7 \oplus r_{10} \oplus r_{11} \oplus r_{14} = 1 \oplus 1 \oplus 1 \oplus 1 \oplus 0 \oplus 0 \oplus 0 = 0$
$s_3 = r_4 \oplus r_5 \oplus r_6 \oplus r_7 \oplus r_{12} \oplus r_{13} \oplus r_{14} = 1 \oplus 1 \oplus 1 \oplus 1 \oplus 0 \oplus 0 \oplus 0 = 0$
$s_4 = r_8 \oplus r_9 \oplus r_{10} \oplus r_{11} \oplus r_{12} \oplus r_{13} \oplus r_{14} = 1 \oplus 0 \oplus 0 \oplus 0 \oplus 0 \oplus 0 \oplus 0 = 1$

and because the error is at the ninth position (1001), the nearest code word and decoded sequences are

$\hat{c}^{(I)} = (01111111100000) \Rightarrow \hat{i}^{(I)} = (1111100000)$.

The second word decoding is as follows

$$s_1 = r_1 \oplus r_3 \oplus r_5 \oplus r_7 \oplus r_9 \oplus r_{11} \oplus r_{13} = 0 \oplus 0 \oplus 1 \oplus 1 \oplus 1 \oplus 0 \oplus 0 = 1$$
$$s_2 = r_2 \oplus r_3 \oplus r_6 \oplus r_7 \oplus r_{10} \oplus r_{11} \oplus r_{14} = 1 \oplus 0 \oplus 0 \oplus 1 \oplus 1 \oplus 0 \oplus 1 = 0$$
$$s_3 = r_4 \oplus r_5 \oplus r_6 \oplus r_7 \oplus r_{12} \oplus r_{13} \oplus r_{14} = 0 \oplus 1 \oplus 0 \oplus 1 \oplus 1 \oplus 0 \oplus 1 = 0$$
$$s_4 = r_8 \oplus r_9 \oplus r_{10} \oplus r_{11} \oplus r_{12} \oplus r_{13} \oplus r_{14} = 1 \oplus 1 \oplus 1 \oplus 0 \oplus 1 \oplus 0 \oplus 1 = 1$$

and the error is again in the ninth position.

After the error correction, the code word and the decoded sequence are

$$\hat{c}^{(II)} = (01001011010101) \Rightarrow \hat{i}^{(II)} = (0101010101).$$

In both cases, the errors were in the same position in the code words (corresponding to the neighboring positions in the interleaver entering sequence)—it is **packet error** having the length $l = 2$.

(c) The complete block-scheme of the system for data transmission is shown in Fig. 5.11. The symbol probabilities determine the average information per symbol emitted by a source, i.e. the entropy

$$H(S) = -\sum_{i=1}^{q} P(s_i) ld(P(s_i)) = 1.801 \text{ [Sh/symb]}.$$

The average code word length cannot be smaller than the entropy. Therefore, $N_s = 5 \times 10^6$ symbols after the compression cannot be represented by less than $N_b = N \times H(S)$. After adding the parity-check bits in Hamming encoder (14, 10) total number of bits emitted through the channel is

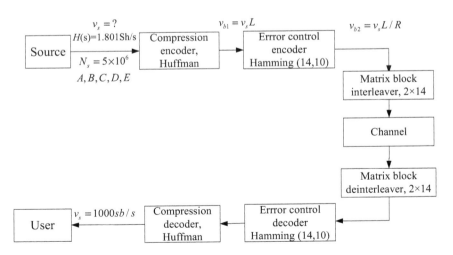

Fig. 5.11 Block-scheme of the complete transmission system

Table 5.7 The result of Huffman procedure

s_i	A	B	C	D	E
$P(s_i)$	0.45	0.35	0.1	0.07	0.03
Code word	0	10	110	1110	1111

$$N_{b,tot} = N_S H(S)/R = 12.607 \times 10^6,$$

and because the transmitting rate is $v_b = 200$ [kb/s], the minimum time for this sequence transmission is

$$t_{min} = N_{b,tot}/v_b = 63.05 \text{ [s]}.$$

The simplest practical solution for a compression encoder is based on a Huffman algorithm. For a memoryless source (it is supposed because in the problem text it is not explicitly written that the source is with memory) this algorithm guarantees to obtain a compact code. For the example from this problem the result of Huffman procedure is shown in Table 5.7.

It is interesting that the obtained code is in the same time so called comma code. In this case, the average code word length is

$$L = \sum_{i=1}^{q} P(s_i) l_i = 0.45 \times 1 + 0.35 \times 2 + 0.1 \times 3 + 0.07 \times 4 + 0.03 \times 4$$
$$= 1.85 \text{ [bit/symb]},$$

and the code efficiency

$$\eta = \frac{H(S)}{L} = 97.35\%.$$

The time needed for transmission, if the Huffman code is used, is

$$t_{Huff} = \frac{N_S L}{R v_b} = 64.75 \text{ [s]},$$

and because it is shorter than that in the problem text, the given condition is satisfied. An additional shortening of the time needed for transmission could be achieved if with the Huffman coding the source extension is combined. However, this time in any case, cannot be shorter than t_{min}.

Problem 5.7 Explain the product code construction using a cliché 2×3 and write the code word structure, i.e. the positions of information and parity-check bits. Find the code rate, write the code word corresponding to information sequence (000111) and explain the decoding procedure if during the transmission the error occurred at the third position.

Solution

In this case the information bits are written in the matrix having the dimensions 2×3 and then one parity-check bit is calculated for every row and for every column. The code word of the ***product code*** is formed by reading the bit sequence starting from the first row of the cliché as follows

$$\begin{array}{ccc|c} i_1 & i_2 & i_3 & z_1 \\ i_4 & i_5 & i_6 & z_2 \\ \hline z_3 & z_4 & z_5 & z_6 \end{array} \Rightarrow c = (i_1, i_2, i_3, z_1, i_4, i_5, i_6, z_2, z_3, z_4, z_5, z_6),$$

where the parity-check bits are

$$z_1 = i_1 \oplus i_2 \oplus i_3 \oplus i_4, z_2 = i_5 \oplus i_6 \oplus i_7 \oplus i_8, z_3 = i_9 \oplus i_{10} \oplus i_{11} \oplus i_{12}, z_4 = i_1 \oplus i_5 \oplus i_9$$
$$z_5 = i_2 \oplus i_6 \oplus i_{10}, z_6 = i_3 \oplus i_7 \oplus i_{11}, z_7 = i_4 \oplus i_8 \oplus i_{12}, z_8 = z_1 \oplus z_2 \oplus z_3 = z_4 \oplus z_5 \oplus z_6 \oplus z_7.$$

The received word is written into the same cliché

$$\begin{array}{ccc|c} r_1 & r_2 & r_3 & r_4 \\ r_5 & r_6 & r_7 & r_8 \\ \hline r_9 & r_{10} & r_{11} & r_{12} \end{array}$$

where the parity-checks are calculated for the rows and for columns. If there is no more than one error, its position can be easily found and the error can be corrected.

If the number of information bits in every row is k_1, and in every column k_2, the code rate is

$$R = \frac{k_1 k_2}{(k_1 + 1)(k_2 + 1)},$$

in this case $R = 6/12 = 1/2$. The encoder from sequence (000111) generates the code word

$$\begin{array}{ccc|c} 0 & 0 & 0 & 0 \\ 1 & 1 & 1 & 1 \\ \hline 1 & 1 & 1 & 1 \end{array} \Rightarrow c = (000011111111).$$

After the channel error occurred at the third position, at the receiving end the cliché is formed

$$r = (001011111111) \Rightarrow \begin{array}{cccc} 0 & 0 & 1 & 0 & 1 \\ 1 & 1 & 1 & 1 & 0 \\ 1 & 1 & 1 & 1 & 0 \\ 0 & 0 & 1 & 0 \end{array}$$

and the error is at the intersection of the first row and the third column, i.e. on the third position in the code word. The information sequence is correctly decoded. The reader should find the generator matrix of the code, to calculate the minimum Hamming distance and to check the number of errors correctable and detectable by the code.

Problem 5.8 Consider a linear block code described by the generator matrix

$$G = \begin{bmatrix} 1 & 1 & 1 & 0 & 0 & 0 \\ 1 & 0 & 0 & 1 & 1 & 0 \\ 0 & 1 & 0 & 1 & 0 & 1 \end{bmatrix}.$$

(a) Find generator and parity-check matrix of the equivalent systematic code
(b) Find the weight spectrum of the code. Comment the code correcting and detecting capabilities.
(c) Find syndromes corresponding to correctable error patterns of the systematic code. Is this code perfect? Is it a MDS code?
(d) For a binary symmetric channel calculate the probability of the unsuccessful error detection. Find the probability that the code does not correct the error.
(e) If the code words (bits) are represented by polar pulses (amplitudes +1 and −1) and are transmitted over the channel with additive white Gaussian noise (AWGN), explain the optimum decision rule for decoding. Illustrate the rule for the case when the received word is (0.1; 0.35; −0.2; −1; 1; 0.15). How the error probability can be estimated for such procedure?

Solution
It is a linear block code (6, 3), code rate $R = 1/2$.

(a) A systematic code is simply obtained by permuting the columns to obtain as the first (or last) three columns a unity submatrix I_3

$$G_s = \begin{bmatrix} 1 & 0 & 0 & 1 & 1 & 0 \\ 0 & 1 & 0 & 0 & 1 & 1 \\ 0 & 0 & 1 & 1 & 0 & 1 \end{bmatrix} = [I_k|P],$$

the parity-check matrix is

$$H_s = [P^T|I_{n-k}] = \begin{bmatrix} 1 & 0 & 1 & 1 & 0 & 0 \\ 1 & 1 & 0 & 0 & 1 & 0 \\ 0 & 1 & 1 & 0 & 0 & 1 \end{bmatrix},$$

it is easy to verify the following

Problems

$$G_s \otimes H_s^T = \begin{bmatrix} 1 & 0 & 0 & 1 & 1 & 0 \\ 0 & 1 & 0 & 0 & 1 & 1 \\ 0 & 0 & 1 & 1 & 0 & 1 \end{bmatrix} \otimes \begin{bmatrix} 1 & 1 & 0 \\ 0 & 1 & 1 \\ 1 & 0 & 1 \\ 1 & 0 & 0 \\ 0 & 1 & 0 \\ 0 & 0 & 1 \end{bmatrix} = \begin{bmatrix} 0 & 0 & 0 \\ 0 & 0 & 0 \\ 0 & 0 & 0 \end{bmatrix}.$$

(b) The list of all codewords of the systematic code is given in Table 5.8, and the corresponding code weight spectrum is shown in Fig. 5.12. It is obvious that a nonsystematic code has the same spectrum, although the code words are different (equivalent codes are considered). The minimum Hamming distance is $d_{\min} = 3$, relation $d_{\min} \leq 2e_c + 1$ is satisfied for $e_c \leq 1$ and the code corrects one error in the code word. When $e_c = 1$ the relation $d_{\min} \leq e_c + e_d + 1$ is satisfied for $e_d \leq 1$ (one error can be detected and corrected). If the decision rule is used where $e_c = 0$, two errors in the code word can be detected.

Table 5.8 Code words list

i_1	i_2	i_3	z_1	z_2	z_3
0	0	0	0	0	0
0	0	1	1	0	1
0	1	0	0	1	1
0	1	1	1	1	0
1	0	0	1	1	0
1	0	1	0	1	1
1	1	0	1	0	1
1	1	1	0	0	0

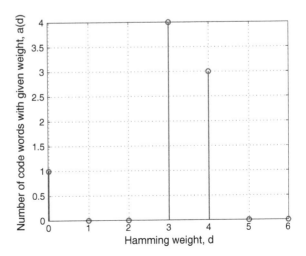

Fig. 5.12 Weight spectrum of code (6, 3)

Table 5.9 Syndromes corresponding to error vectors (optimum coset leaders)

Ordinal pattern number	Pattern e_j	Syndrome S
1	(000000)	(000)
2	(000001)	(001)
3	(000010)	(010)
4	(000100)	(100)
5	(001000)	(101)
6	(010000)	(011)
7	(100000)	(110)
8	(100001)	(111)

(c) Standard array of the code has $2^{n-k} = 8$ rows and seven error patterns can be corrected [besides the trivial one (000000)]. On the basis of relation

$$S = e_j \otimes H_s^T$$

it is easy to verify that to every error vector containing only single one (i.e. one error) corresponds a unique syndrome, as shown in Table 5.9. It is a direct consequence of the fact that parity-check matrix H_s has not two identical columns. It is obvious that every linear code having this feature, and where $n < 2^{n-k}$ holds, can correct the single errors. Unused syndrome value (111) clearly corresponds to double error, and the sum of the corresponding columns of parity-check matrix yields just this value. One of such error vectors is (100001), but it is not a unique one—the same condition is fulfilled as well for the patterns (010100) and (001010).

The Hamming bound generally is

$$q^{n-k} \geq \sum_{t=0}^{e_c} \binom{n}{t}(q-1)^t,$$

where q denotes the code basis (the number of symbols to construct a code words). Because this code is binary ($q = 2$, the code word consists of zeros and ones) and one error at code word can be corrected, the previous relation reduces to

$$2^{n-k} \geq \binom{n}{0} + \binom{n}{1} = 1 + n.$$

In this case $n = 6$ and $n - k = 3$, the Hamming bound is not satisfied with equality. It can be interpreted as well as follows—the code can correct all single errors, but cannot correct two errors in the code word, because it is not clear do the syndrome value (111) corresponds to the error vector (100001) either (010100) or (001010), because all are equiprobable. All syndrome values are not efficiently used and the code is not perfect.

Singleton bound can be written in the form

$$d_{min} \leq n - k + 1,$$

and, as in this case $d_{min} = n-k = 3$, the bound is not satisfied with equality. Because of that the code is not *Maximum Distance Separable* (MDS), meaning that for the same number of parity bits, it may be possible to construct code having the greater minimum Hamming distance ($d_{min} = 4$).

(d) If only the code possibility to detect the errors is considered, it is sufficient to verify does the received word belong to the set of possible code words. If it does not belong to this set, it is obvious that the errors occurred during transmission. However, it is possible to emit one code word and to receive the other—in this case the error will be undetected.

When the transmission system can be modeled as a binary symmetric channel, the probability of occurrence of exactly d errors at n positions is

$$P(n, d) = p^d(1-p)^{n-d},$$

and the **probability of error undetectability** equals the probability that error vector is the same as the code word. It can be found on the weight spectrum basis [25, 26]

$$P_{e,d}(p) = \sum_{d=d_{min}}^{n} a(d) p^d (1-p)^{n-d},$$

where $a(d)$ denotes the number of code words having Hamming weight d.

On the other hand, the **probability of the uncorrectable error** equals the probability that the error vector differs from the coset leaders in the standard array. If the maximum weight of error patterns which correspond to coset leaders is denoted by l, and the number of patterns having weight $i \leq l$ by $L(i)$, this probability is

$$P_{e,c}(p) = 1 - \sum_{i=0}^{l} L(i) p^i (1-p)^{n-i},$$

and for a considered case $l = 2$, $L(0) = 1$, $L(1) = 6$, $L(2) = 1$.

On the basis of previously found weight spectrum, the corresponding numerical values for the probabilities that the error is not detected/corrected are shown in Fig. 5.13. The relations

$$a(d) \leq \binom{n}{d}, \quad L(i) \leq \binom{n}{i},$$

are always valid, and when the weight spectrum is not known, the coefficients $a(d)$ and $L(i)$ can be substituted by a binary coefficients yielding the upper bound of

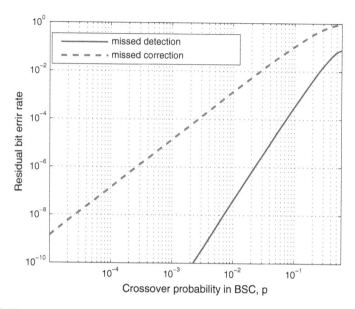

Fig. 5.13 The probability that the error is not detected/corrected, BSC

the corresponding probability. For a perfect code these inequalities become equalities.

(e) Block scheme of a system where the code words are represented by vectors x_i (were the symbols are -1 and 1) and the transmission is through the channel with additive white Gaussian noise (variance σ^2) is shown in Fig. 5.14. If the emitted binary symbols are equiprobable and if at the receiving end, before the decoder, a decision block is inserted having the threshold put at zero, the channel can be modeled as a binary symmetric channel. In this case the *hard decision* is optimal, based on the syndrome and the standard array, where the nearest code word is chosen based on the minimum Hamming distance.

However, if at the receiver input the hard decision block is not inserted, it is possible to find a more efficient decision rule. It is based on the choice of the code word which minimize the conditional probability density function of the received sequence

$$p(y|x) = \prod_{i=1}^{n} p(y_i|x_i) = \prod_{i=1}^{n} \frac{1}{\sqrt{2\pi}\sigma} e^{-\frac{(y_i-x_i)^2}{2\sigma^2}},$$

where it is supposed that the adjacent noise samples are statistically independent.

The previous expression can be minimized if the *Maximum Likelihood Decoding* (ML) is applied, which reduces in this case to choice of the sequence having a minimum **squared Euclid distance** defined by Morelos-Zaragoza [26]

Problems

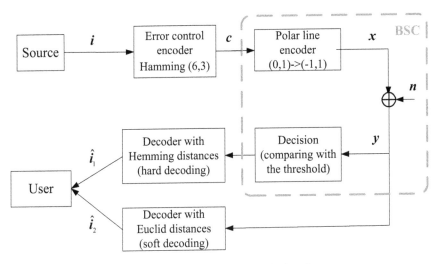

Fig. 5.14 The illustration of two decision rules for linear block codes

$$D^2(\mathbf{x},\mathbf{y}) = \sum_{i=1}^{n} (y_i - x_i)^2.$$

If this metric is used, it can be considered as a *soft decision*, regardless if it was done according to ML or some other decision rule.

The decoding procedure for a received word $\mathbf{y} = (0.1; 0.35; -0.2; -1; 1; 0.15)$ by comparison to the possible emitted words is shown in Table 5.10. The code word $\mathbf{c} = (010011)$ is chosen for which the vector $\mathbf{y}$ has a minimum squared Euclidean distance to the received word, and the decoded information sequence is $\mathbf{i} = (010)$.

It is obvious that even for soft decision, the *decision error* can be made if the received vector, in Euclidian sense, is nearer to some other code word. The probability of such event can be upper bounded [26]

Table 5.10 Illustration of soft decoding of a linear block code

i	c	X	$D^2(x, y)$
0 0 0	0 0 0 0 0 0	−1 −1 −1 −1 −1 −1	8.995
0 0 1	0 0 1 1 0 1	−1 −1 +1 +1 −1 +1	13.195
0 1 0	**0 1 0 0 1 1**	**−1 +1 −1 −1 +1 +1**	**2.995**
0 1 1	0 1 1 1 1 0	−1 +1 +1 +1 +1 −1	8.395
1 0 0	1 0 0 1 1 0	+1 −1 −1 +1 +1 −1	8.595
1 0 1	1 0 1 0 1 1	+1 −1 +1 −1 +1 +1	4.795
1 1 0	1 1 0 1 0 1	+1 +1 −1 +1 −1 +1	10.595
1 1 1	1 1 1 0 0 0	+1 +1 +1 −1 −1 −1	7.995

$$P_{e,c,\text{meko}}(E_b/N_0) \leq \sum_{d=d_{\min}}^{n} \frac{a(d)}{2} \operatorname{erfc}\left(\sqrt{d \times R \frac{E_b}{N_0}}\right),$$

where R denotes a code rate and E_b/N_0 is energy per bit divided by the noise power density spectrum. The crossover probability for the equivalent BSC is

$$p(E_b/N_0) = \frac{1}{2} \operatorname{erfc}\left(\sqrt{R \frac{E_b}{N_0}}\right),$$

and the probability of residual error versus E_b/N_0 for hard decision can be written as

$$P_{e,c,\text{tvrdo}}(E_b/N_0) = 1 - \sum_{i=0}^{l} L(i) \left[\frac{1}{2} \operatorname{erfc}\left(\sqrt{R \frac{E_b}{N_0}}\right)\right]^{i} \left[1 - \frac{1}{2} \operatorname{erfc}\left(\sqrt{R \frac{E_b}{N_0}}\right)\right]^{n-i}.$$

The *coding gain* (G) is usually defined as a saving in E_b/N_0 ratio in comparison to the case when the error control code was not applied, for some fixed error probability

$$P_{e,c,\text{tvrdo}}(E_b^{(1)}/N_0) = p(E_b^{(0)}/N_0) = 10^{-a} \Rightarrow G_{\text{tvrdo}}(10^{-a}) = E_b^{(0)}/N_0 - E_b^{(1)}/N_0,$$
$$P_{e,c,\text{meko}}(E_b^{(2)}/N_0) = p(E_b^{(0)}/N_0) = 10^{-a} \Rightarrow G_{\text{meko}}(10^{-a}) = E_b^{(0)}/N_0 - E_b^{(2)}/N_0.$$

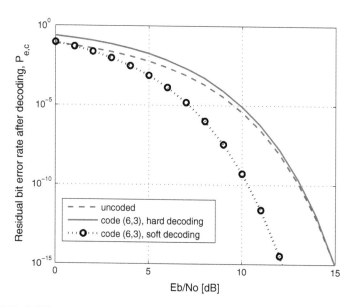

Fig. 5.15 Probability of residual error versus E_b/N_0

Problems

For the considered code (6, 3) the corresponding numerical results are shown in Fig. 5.15.

From the figure the following coding gains can be found:

- for the crossover probability 10^{-4};
  - $G_{hard}(10^{-5}) = -0.3$ dB,
  - $G_{soft}(10^{-5}) = 2.6$ dB,

- for the crossover probability 10^{-10};
  - $G_{hard}(10^{-15}) = 0.1$ dB,
  - $G_{soft}(10^{-15}) = 3$ dB,

On the basis of these results the following can be concluded:

- Hard decision becomes efficient only for very high signal-to-noise ratios and the coding gain becomes positive only for very small values of the error probability registered by the user;
- The difference between asymptotic coding gains is about 3 dB to the advantage of soft decision.

One should have in mind that the residual error probability (y-axis) is the error probability for a code word. To obtain the bit error probability, the corresponding scaling should be done, depending on the probability of occurring of the various error combinations. If the crossover probability is sufficiently small (the ratio E_b/N_0 sufficiently large), the approximate expression is

$$P_{e,b} \approx P_{e,c}/(e_c + 1).$$

Problem 5.9 Consider a linear block code described by generator matrix

$$G = \begin{bmatrix} 1 & 0 & 0 & 1 & 0 & 0 & 0 & 0 & 0 & 0 & 0 & 0 \\ 0 & 1 & 0 & 0 & 0 & 0 & 0 & 0 & 0 & 0 & 1 & 0 \\ 0 & 0 & 1 & 0 & 0 & 0 & 0 & 0 & 0 & 0 & 1 & 1 \\ 0 & 0 & 0 & 1 & 0 & 0 & 0 & 0 & 0 & 0 & 0 & 1 \\ 0 & 0 & 0 & 0 & 1 & 0 & 0 & 0 & 0 & 0 & 1 & 1 \\ 0 & 0 & 0 & 0 & 0 & 1 & 0 & 0 & 0 & 0 & 0 & 0 \\ 0 & 0 & 0 & 0 & 0 & 0 & 1 & 0 & 0 & 0 & 1 & 1 \\ 0 & 0 & 0 & 0 & 0 & 0 & 0 & 1 & 0 & 0 & 0 & 1 \\ 0 & 0 & 0 & 0 & 0 & 0 & 0 & 0 & 1 & 0 & 1 & 0 \\ 0 & 0 & 0 & 0 & 0 & 0 & 0 & 1 & 1 & 0 & 1 & 0 & 0 \end{bmatrix}.$$

(a) Find the generator matrix of an equivalent systematic code. How the corresponding dual code can be described?
(b) Find the weight spectrum of the code defined by the generator matrix G.

(c) Find the probability that this code does not detect the transmission error. Find the probability that dual code does not detect the transmission error

Solution

(a) From the theory of linear block codes it is known that the weight spectrum does not change (the code is equivalent) if the following operations with the generator matrix are performed:

(1) Permutation of any two columns
(2) Adding of one row multiple to the other row
(3) Multiplication of one row or column with a nonzero element

Generator matrix of the corresponding systematic code has the form $G_s = [\mathbf{I}_k, \mathbf{P}]$, its parameters are $n = 12$ and $k = 10$. To achieve that first eight rows of matrix G form a unity matrix, the binary one from the first row, the fourth column, should be removed, as well as from 7th and 8th columns in the last row. The first row of systematic generator matrix is obtained by addition of the first and fourth row of the matrix G. Finally, by adding 7th, 8th and the last row of G the last row of new matrix is obtained

$$G_s = \begin{bmatrix} 1 & 0 & 0 & 0 & 0 & 0 & 0 & 0 & 0 & 0 & 0 & 1 \\ 0 & 1 & 0 & 0 & 0 & 0 & 0 & 0 & 0 & 0 & 1 & 0 \\ 0 & 0 & 1 & 0 & 0 & 0 & 0 & 0 & 0 & 0 & 1 & 1 \\ 0 & 0 & 0 & 1 & 0 & 0 & 0 & 0 & 0 & 0 & 0 & 1 \\ 0 & 0 & 0 & 0 & 1 & 0 & 0 & 0 & 0 & 0 & 1 & 1 \\ 0 & 0 & 0 & 0 & 0 & 1 & 0 & 0 & 0 & 0 & 0 & 0 \\ 0 & 0 & 0 & 0 & 0 & 0 & 1 & 0 & 0 & 0 & 1 & 1 \\ 0 & 0 & 0 & 0 & 0 & 0 & 0 & 1 & 0 & 0 & 0 & 1 \\ 0 & 0 & 0 & 0 & 0 & 0 & 0 & 0 & 1 & 0 & 1 & 0 \\ 0 & 0 & 0 & 0 & 0 & 0 & 0 & 0 & 1 & 1 & 0 \end{bmatrix} = [\mathbf{I}_8, \mathbf{P}].$$

Parity-check matrix of a systematic code is

$$H_s = [\mathbf{P}^T, \mathbf{I}_2] = \begin{bmatrix} 0 & 1 & 1 & 0 & 1 & 0 & 1 & 0 & 1 & 1 & 1 & 0 \\ 1 & 0 & 1 & 1 & 1 & 0 & 1 & 1 & 0 & 0 & 0 & 1 \end{bmatrix},$$

the dual code of this code has a generator matrix

$$G_{d,s} = \begin{bmatrix} 0 & 1 & 1 & 0 & 1 & 0 & 1 & 0 & 1 & 1 & 1 & 0 \\ 1 & 0 & 1 & 1 & 1 & 0 & 1 & 1 & 0 & 0 & 0 & 1 \end{bmatrix}.$$

Code words of this code (12, 2) are all linear combinations of the rows of the matrix, i.e. $c_1 = (000000000000)$, $c_2 = (011010101110)$, $c_3 = (101110110001)$ and $c_4 = (110100011111)$. It is obvious that the minimum Hamming distance of dual code is $d_{min} = 7$.

(b) Weight spectrum of the code defined by **G** can be found directly, by a computer search. In this case all information words ($2^8 = 256$) should be found, everyone should be separately multiplied by generator matrix and find the Hamming weights of the obtained code words. The weight spectrum of the code can be written as a polynomial

$$A(x) = a_0 + a_1 x^1 + \cdots + a_n x^n,$$

where the number of words having weight d is denoted by a_d ($\leq d \geq n$). The codes defined by generator matrix **G** and its systematic version are equivalent, i.e. they have the same weight spectrum. Because of this, the previous relation determines the systematic code weight spectrum as well. But, for large values of n and k, the finding of weight spectrum by computer search is very time consuming.

However, the procedure can be accelerated having in view that for the case $k > (n - k)$ it is simpler to find the weight spectrum of a dual code. If with $B(x)$ the spectrum polynomial of dual code is denoted, where the coefficients B_d determine the spectrum, than the *MacWilliams identities* can be formulated giving the correspondence between the dual codes spectra written as polynomials [27]

$$A(x) = 2^{-(n-k)}(1+x)^n B\left(\frac{1-x}{1+x}\right), \quad B(x) = 2^{-k}(1+x)^n A\left(\frac{1-x}{1+x}\right).$$

In this case the spectrum of dual code (for which $G_{d,s} = H_s$) is described by a polynomial

$$B(x) = 1 + 2x^7 + x^8,$$

and the spectrum of the code defined by **G** is determined by a first MacWilliams identity

$$A(x) = 2^{-2}(1+x)^0 B\left(\frac{1-x}{1+x}\right) = 2^{-2}(1+x)^{10}\left[1 + 2\left(\frac{1-x}{1+x}\right)^7 + \left(\frac{1-x}{1+x}\right)^8\right]$$
$$= 2^{-2}\left[(1+x)^{10} + 2(1-x)^7(1+x)^3 + (1-x)^8(1+x)^2\right]$$
$$= 1 + x + 15x^2 + 63x^3 + 122x^4 + 186x^5 + 238x^6 + 206x^7 + 117x^8 + 53x^9 + 19x^{10} + 3x^{11}.$$

Spectra of a dual code (12, 2) and the original code (12, 10) are shown in Fig. 5.16a and b, respectively. The dual code has a possibility to correct all single,

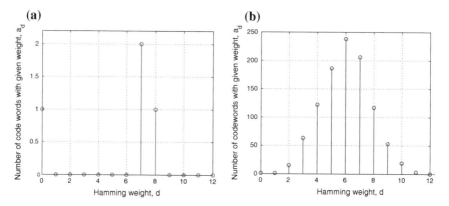

Fig. 5.16 Spectrum of dual code (12, 2) (**a**) and the original code (12, 10) (**b**) defined by **G**

double and triple errors in the code word having the length $n = 12$ bits, while the code defined by **G** cannot even detect all single errors.

(c) As it was explained in the previous problems, the ***probability that this linear code does not detect the error*** is

$$P_{e,d}(p) = \sum_{d=d_{min}}^{n} a_d p^d (1-p)^{n-d} = (1-p)^n \sum_{d=d_{min}}^{n} a_d \left(\frac{p}{1-p}\right)^d,$$

and because $a(0) = 1$ and $a(d) = 0$ for $1 \leq d \leq d_{min}$ always holds, the previous relation becomes

$$P_{e,d}(p) = (1-p)^n \left[\sum_{d=0}^{n} a_d \left(\frac{p}{1-p}\right)^d - 1\right] = (1-p)^n \left[A\left(\frac{p}{1-p}\right) - 1\right].$$

Use of MacWilliams identity yields

$$P_{e,d}(p) = (1-p)^n \left[2^{-(n-k)}\left(1 + \frac{p}{1-p}\right)^n B\left(\frac{1-p/(1-p)}{1+p/(1-p)}\right) - 1\right],$$

and finally

$$P_{e,d}(p) = 2^{-(n-k)} \sum_{d=0}^{n} b_d (1-2p)^d - (1-p)^n.$$

On the other hand, the probability that the dual code (12, 2) does not detect the transmission error is given by

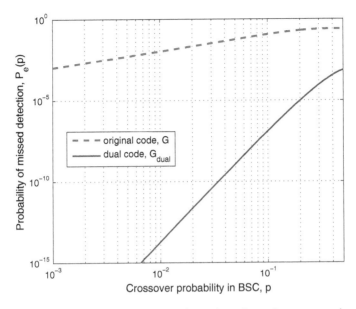

Fig. 5.17 The probability that the codes do not detect the code word error versus the crossover probability

$$P_{e,d,\text{dual}}(p) = \sum_{d=d_{\min,\text{dual}}}^{n} b_d p^d (1-p)^{n-d}.$$

The probability that the codes do not detect the error versus the channel error probability is shown in Fig. 5.17.

Problem 5.10

(a) Explain the construction of the Reed-Muller (RM) codes with parameters (8, 4) and (8, 7). Draw code distances spectra and find their parity-check matrices.

(b) Draw the spectrum of the code obtained by forming 5th, 6th and 7th matrix G of the code RM(8, 7) if instead of vector multiplication the vector addition was used. Comment the result.

(c) Explain the majority decoding logic procedure for the code (8, 4) and then decode the word $r = (00000010)$.

(d) If it is known that in a data transmission system is used Reed-Muller code which corrects one and detects two errors in the code word, decode the received word $r_2 = (0000001000000000)$.

Solution

It is possible to construct for any natural number m and any natural number $r < m$ a Reed-Muller code having code word length $n = 2^m$ with minimum Hamming

Table 5.11 Table for the Boolean function "or"

x_1	0	0	1	1
x_2	0	1	0	1
$f(x_1, x_2)$	0	1	1	1

distance $d_{\min} = 2^{m-r}$ [28, 29]. To Reed-Muller code of rth order, denoted by RM $(2^m, k, 2^{m-r})$, corresponds code word length

$$k = \sum_{i=0}^{r} \binom{m}{i}, \; r \leq m.$$

To describe the Reed-Muller code construction consider Boolean function of two variables [corresponding to the operator logical "or", i.e. $f(x_1, x_2) = \mathrm{OR}(x_1, x_2)$], given in Table 5.11.

If every row of the table is rewritten as a vector, i.e. $x_1 = (0011)$, $x_2 = (0101)$ and $f = (0111)$, the relation $f = x_1 \oplus x_2 \oplus x_1 \otimes x_2$ is valid, where $\oplus$ and $\otimes$ denote the operators of vector addition and multiplication modulo-2 (the operations are carried out bit-by-bit).

It can be easily shown that any Boolean function of m variables can be written as a linear combination of the corresponding vectors, i.e.

$$f = a_0 \oplus \mathbf{1} + a_1 x_1 \oplus a_2 x_2 \cdots \oplus a_m x_m \oplus a_{1,2}(x_1 \otimes x_2) a_{1,3}(x_1 \otimes x_3) \cdots \oplus a_{m-1,m}(x_{m-1} \otimes x_m)$$
$$\oplus a_{1,2,3}(x_1 \otimes x_2 \otimes x_3) \cdots \oplus a_{m-2,m-1,m}(x_{m-2} \otimes x_{m-1} x_m) \oplus \cdots \oplus a_{1,2,\ldots,m}(x_1 \otimes x_2 \otimes \cdots \otimes x_m).$$

where **1** is a vector consisting of 2^m binary ones. Generator matrix of Reed-Muller code is completely determined by the above equality in such a way that to ith row of matrix $\mathbf{G}$ corresponds ith element of this sum, the number of generator matrix rows ($k < n$) is determined by above relation as well.

(a) For a code word length $n = 8$ it is obtained $m = \mathrm{ld}(n) = 3$, and the minimum Hamming distance is always $d_{\min} = 2^{m-r}$, therefore

- for $r = 1$, $d_{\min,1} = 2^{3-1} = 4$, and code corrects $e_c = \lfloor (d_{\min,1} - 1)/2 \rfloor = 1$ error,
- for $r = 2$, $d_{\min,2} = 2^{3-2} = 2$, and code corrects $e_c = \lfloor (d_{\min,2} - 1)/2 \rfloor = 0$ errors,

the corresponding information words lengths are

$$k_1 = 1 + \binom{3}{1} = 4, \; k_2 = 1 + \binom{3}{1} + \binom{3}{2} = 7.$$

For the case $r = 1$, the generator matrix rows are determined by vectors $\mathbf{1}, x_1, x_2, x_3$, where x_1 corresponds to a bit of the maximum weight (MSB) and x_3 corresponds to a bit of minimum weight (LSB) of the corresponding three-bit combinations resulting in

$$G_{(8,4)} = \begin{bmatrix} 1 & 1 & 1 & 1 & 1 & 1 & 1 & 1 \\ 0 & 0 & 0 & 0 & 1 & 1 & 1 & 1 \\ 0 & 0 & 1 & 1 & 0 & 0 & 1 & 1 \\ 0 & 1 & 0 & 1 & 0 & 1 & 0 & 1 \end{bmatrix} = \begin{bmatrix} G_{(8,4)}^{(0)} \\ G_{(8,4)}^{(1)} \end{bmatrix}.$$

Weight spectrum of the code is shown in Fig. 5.18a. It is obvious that the minimum code Hamming distance is $d_{\min,1} = 4$ and that the code is equivalent to Hamming code (8, 4) which has the code rate $R_1 = 4/8 = 0.5$. It is easy to verify that this code is selfdual, because of

$$G_{(8,4)} G_{(8,4)}^T = 0_{4 \times 4} \Rightarrow H_{(8,4)} = G_{(8,4)}.$$

When $r = 2$, generator matrix rows are determined by vectors **1**, $x_1, x_2, x_3, x_1 \otimes x_2, x_1 \otimes x_3, x_2 \otimes x_3$ and the corresponding generator matrix is

$$G_{(8,7)} = \begin{bmatrix} 1 & 1 & 1 & 1 & 1 & 1 & 1 & 1 \\ 0 & 0 & 0 & 0 & 1 & 1 & 1 & 1 \\ 0 & 0 & 1 & 1 & 0 & 0 & 1 & 1 \\ 0 & 1 & 0 & 1 & 0 & 1 & 0 & 1 \\ 0 & 0 & 0 & 0 & 0 & 0 & 1 & 1 \\ 0 & 0 & 0 & 0 & 0 & 1 & 0 & 1 \\ 0 & 0 & 0 & 1 & 0 & 0 & 0 & 1 \end{bmatrix} = \begin{bmatrix} G_{(8,7)}^{(1)} \\ G_{(8,7)}^{(2)} \\ G_{(8,7)}^{(3)} \end{bmatrix}.$$

It is easy to verify that the corresponding parity-check matrix is

$$H_{(8,7)} = \begin{bmatrix} 1 & 1 & 1 & 1 & 1 & 1 & 1 & 1 \end{bmatrix}.$$

and this code corresponds to a simple parity check, its spectrum being shown in Fig. 5.18b.

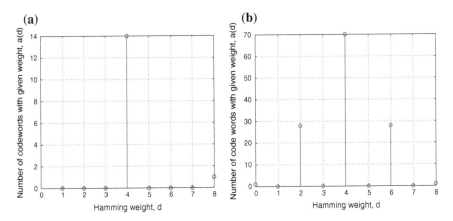

Fig. 5.18 Weight spectrum of RM(8, 4) (**a**) and RM(8, 7) (**b**) code

If the decoding is carried out by using the relation

$$S = rH_{(8,4)}^T,$$

it is clear that the result can be equal to zero or to one, i.e. the syndrome consists of one bit (scalar). If $S = 0$ the conclusion can be drawn that during the transmission there were no errors (or an even number of errors occurred), while if $S = 1$ the conclusion can be drawn that one error occurred (or an odd number of errors occurred). Therefore, single errors can be detected, but not corrected, because their position cannot be found, which corresponds to the value $d_{min,2} = 2$ used for a code construction.

(b) If for obtaining the last three rows of the code (8, 7) generator matrix, instead of multiplying (as for Reed-Muller algorithm), the adding was used, the generator matrix rows (vectors) would have been $\mathbf{1}, x_1, x_2, x_3, x_1, \oplus x_2, x_1 \oplus x_3, x_2, \oplus x_3$ yielding

$$G_{(8,7)}^* = \begin{bmatrix} 1 & 1 & 1 & 1 & 1 & 1 & 1 & 1 \\ 0 & 0 & 0 & 0 & 1 & 1 & 1 & 1 \\ 0 & 0 & 1 & 1 & 0 & 0 & 1 & 1 \\ 0 & 1 & 0 & 1 & 0 & 1 & 0 & 1 \\ 0 & 0 & 1 & 1 & 1 & 1 & 0 & 0 \\ 0 & 1 & 1 & 0 & 0 & 1 & 1 & 0 \\ 0 & 1 & 0 & 1 & 1 & 0 & 1 & 0 \end{bmatrix},$$

and to this code corresponds the parity-check matrix

$$H_{(8,7)}^* = [1\ \ 1\ \ 1\ \ 1\ \ 1\ \ 1\ \ 1\ \ 1].$$

equivalent to the simple parity-check.

However, from code weight spectrum (shown in Fig. 5.17) it is clear that minimum Hamming weight is $d_{min} = 4$ showing that the code has the minimum Hamming distance $d_{min} = 4$ as well, meaning further that it can be used to correct one and to detect two errors in the code word, but with the higher code rate ($R = 7/8 = 0.875$).

The spectrum in Fig. 5.19 is correctly found, but the conclusions drawn from its pattern are wrong. It is clear that there are eight code words consisting of all zeros and 112 code words consisting of four ones. Therefore, to the various information bits combinations corresponds the same code words. The same can be concluded by observing that the number of possible "four ones" combinations at eight positions is

Fig. 5.19 Weight spectrum of the code having the generator matrix G_2^*

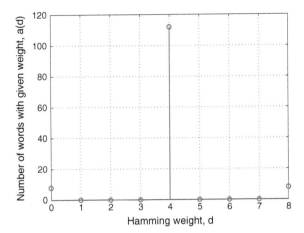

$$\binom{8}{4} = \frac{8!}{2 \times 4!} = 70 < 112.$$

Therefore, the transformation done by the encoder is not of 1:1 type, and the set of words at the encoder output is not even a block code (and, of course, not a linear block code). This is a reason that in this case (and generally as well) on the basis of minimum Hamming weight the minimum Hamming distance cannot be determined, and especially the number of errors detectable and correctable by this code.

(b) From the generator matrix

$$G = \begin{bmatrix} 1 & 1 & 1 & 1 & 1 & 1 & 1 & 1 & 1 & 1 & 1 & 1 & 1 & 1 & 1 \\ 0 & 0 & 0 & 0 & 0 & 0 & 0 & 1 & 1 & 1 & 1 & 1 & 1 & 1 & 1 \\ 0 & 0 & 0 & 0 & 1 & 1 & 1 & 1 & 0 & 0 & 0 & 0 & 1 & 1 & 1 & 1 \\ 0 & 0 & 1 & 1 & 0 & 0 & 1 & 1 & 0 & 0 & 1 & 1 & 0 & 0 & 1 & 1 \\ 0 & 1 & 0 & 1 & 0 & 1 & 0 & 1 & 0 & 1 & 0 & 1 & 0 & 1 \end{bmatrix} = \begin{bmatrix} G_1 \\ G_2 \end{bmatrix}.$$

it is obvious that the code words are formed according to rules

$$c_1 = i_1, c_2 = i_1 \oplus i_4, c_3 = i_1 \oplus i_3, c_4 = i_1 \oplus i_3 \oplus i_4, c_5$$
$$= i_1 \oplus i_2, c_6 = i_1 \oplus i_2 \oplus i_4, c_7 = i_1 \oplus i_2 \oplus i_3, c_4 = i_1 \oplus i_2 \oplus i_3 \oplus i_4.$$

Sums orthogonal to the fourth information bit are

$$S_1^{(4)} = r_1 \oplus r_2 = c_1 \oplus c_2 \oplus e_1 \oplus e_2 = (i_1) \oplus (i_1 \oplus i_4) \oplus e_1 \oplus e_2 = i_4 \oplus e_1 \oplus e_2$$
$$S_2^{(4)} = r_3 \oplus r_4 = c_3 \oplus c_4 \oplus e_3 \oplus e_4 = (i_1 \oplus i_3) \oplus (i_1 \oplus i_3 \oplus i_4) \oplus e_3 \oplus e_4 = i_4 \oplus e_3 \oplus e_4$$
$$S_3^{(4)} = r_5 \oplus r_6 = c_5 \oplus c_6 \oplus e_5 \oplus e_6 = (i_1 \oplus i_2) \oplus (i_1 \oplus i_2 \oplus i_4) \oplus e_5 \oplus e_6 = i_4 \oplus e_5 \oplus e_6$$
$$S_4^{(4)} = r_7 \oplus r_8 = c_7 \oplus c_8 \oplus e_7 \oplus e_8 = (i_1 \oplus i_2 \oplus i_3) \oplus (i_1 \oplus i_2 \oplus i_3 \oplus i_4) \oplus e_7 \oplus e_8 = i_4 \oplus e_7 \oplus e_8$$

each depending only on the fourth information bit and the error vector bit. Further, any error vector bit does not influence to values of two sums (either to more sums).

Sums orthogonal to the third information bit are

$$S_1^{(3)} = r_1 \oplus r_3 = i_4 \oplus e_1 \oplus e_3, S_2^{(3)} = r_2 \oplus r_4 = i_4 \oplus e_2 \oplus e_4,$$
$$S_3^{(3)} = r_5 \oplus r_7 = i_4 \oplus e_5 \oplus e_7, S_4^{(3)} = r_6 \oplus r_8 = i_4 \oplus e_6 \oplus e_8,$$

while sums orthogonal to the second information bit are

$$S_1^{(2)} = r_1 \oplus r_5 = i_2 \oplus e_1 \oplus e_5, S_2^{(2)} = r_2 \oplus r_6 = i_2 \oplus e_2 \oplus e_6,$$
$$S_3^{(2)} = r_3 \oplus r_7 = i_2 \oplus e_3 \oplus e_7, S_4^{(2)} = r_4 \oplus r_8 = i_2 \oplus e_4 \oplus e_8.$$

On the basis of the calculated orthogonal sums, by majority logic decoding the corresponding information bit values are estimated

$$\left(S_1^{(4)}, S_2^{(4)}, S_3^{(4)}, S_4^{(4)}\right) \to i'_4$$
$$\left(S_1^{(3)}, S_2^{(3)}, S_3^{(3)}, S_4^{(3)}\right) \to i'_3$$
$$\left(S_1^{(2)}, S_2^{(2)}, S_3^{(2)}, S_4^{(2)}\right) \to i'_2$$

Using the obtained estimations, the code sequence is formed, with the removed influence of decoded information bits

$$\boldsymbol{r}' = \boldsymbol{r} \oplus \begin{bmatrix} i'_2 & i'_3 & i'_4 \end{bmatrix} \otimes \boldsymbol{G}_{(8,4)}^{(1)},$$

and by majority logic from $\boldsymbol{r}'$ the information bit i_1 is estimated.

For a considered case $\boldsymbol{r} = (00001110)$, and $r_5 = r_6 = r_7 = 1$, while the other vector $\boldsymbol{r}$ values are equal to zero. Majority decision is based on the estimations

$$\left(S_1^{(4)}, S_2^{(4)}, S_3^{(4)}, S_4^{(4)}\right) = (0,0,0,1) \to i'_4 = 0$$
$$\left(S_1^{(3)}, S_2^{(3)}, S_3^{(3)}, S_4^{(3)}\right) = (0,0,0,1) \to i'_3 = 0$$
$$\left(S_1^{(2)}, S_2^{(2)}, S_3^{(2)}, S_4^{(2)}\right) = (1,1,1,0) \to i'_2 = 1$$

Problems

finally yielding

$$r' = r \oplus (001) \otimes G_1 = (00001110) \oplus (00001111)$$
$$= (00000001) \rightarrow i'_1 = 0.$$

The decoded sequence is $i' = (0100)$ corresponding to the code word written in the generator matrix second row $c' = (00001111)$. The difference between the received word and reconstructed code word is in one bit only, and the conclusion can be drawn that the decoder corrected the error on the 8th code word bit.

(c) For a code word length $n = 16$ the value $m = \text{ld}(n) = 4$ is obtained and for minimum Hamming distance $d_{\min} = 4$ the values $e_c = \lfloor (d_{\min} - 1)/2 \rfloor = 1$ and $e_d = d_{\min} - e_c - 1 = 2$ are obtained, while the code word length is

$$k = 1 + \binom{4}{1} = 5,$$

therefore, it is RM(16, 11) code having the generator matrix

$$G = \begin{bmatrix} 1 & 1 & 1 & 1 & 1 & 1 & 1 & 1 & 1 & 1 & 1 & 1 & 1 & 1 & 1 & 1 \\ 0 & 0 & 0 & 0 & 0 & 0 & 0 & 0 & 1 & 1 & 1 & 1 & 1 & 1 & 1 & 1 \\ 0 & 0 & 0 & 0 & 1 & 1 & 1 & 1 & 0 & 0 & 0 & 0 & 1 & 1 & 1 & 1 \\ 0 & 0 & 1 & 1 & 0 & 0 & 1 & 1 & 0 & 0 & 1 & 1 & 0 & 0 & 1 & 1 \\ 0 & 1 & 0 & 1 & 0 & 1 & 0 & 1 & 0 & 1 & 0 & 1 & 0 & 1 & 0 & 1 \end{bmatrix} = \begin{bmatrix} G_1 \\ G_2 \end{bmatrix}.$$

Orthogonal sums for i_5 include bits of the received word having the ordinal number determined by the ordinal numbers of column pairs which should be summed (element by element) to obtain only one "one" in the fifth row, i.e.

i_5: $S_1 = r_1 \oplus r_2, S_2 = r_3 \oplus r_4, S_3 = r_5 \oplus r_6, S_4 = r_7 \oplus r_8, S_5 = r_9 \oplus r_{10}, S_6 = r_{11} \oplus r_{12}, S_7 = r_{13} \oplus r_{14}, S_8 = r_{15} \oplus r_{16}$,

therefore, the adjacent bits are summed and the decision is based on the majority logic.

Orthogonal sums corresponding to bits i_4, i_3 and i_2 are

i_4: $S_1 = r_1 \oplus r_3, S_2 = r_2 \oplus r_4, S_3 = r_5 \oplus r_7, S_4 = r_6 \oplus r_8, S_5 = r_9 \oplus r_{11}, S_6 = r_{10} \oplus r_{12}, S_7 = r_{13} \oplus r_{15}, S_8 = r_{14} \oplus r_{16}$,
i_3: $S_1 = r_1 \oplus r_5, S_2 = r_2 \oplus r_6, S_3 = r_3 \oplus r_7, S_4 = r_4 \oplus r_8, S_5 = r_9 \oplus r_{13}, S_6 = r_{10} \oplus r_{14}, S_7 = r_{11} \oplus r_{15}, S_8 = r_{12} \oplus r_{16}$,
i_2: $S_1 = r_1 \oplus r_9, S_2 = r_2 \oplus r_{10}, S_3 = r_3 \oplus r_{11}, S_4 = r_4 \oplus r_{12}, S_5 = r_5 \oplus r_{13}, S_6 = r_6 \oplus r_{14}, S_7 = r_7 \oplus r_{15}, S_8 = r_8 \oplus r_{16}$,

and the received bits being apart for 2, 4 and 8 positions are summed. As for i_5, the value of every transmitted bit is found based on the majority decision, and the estimations are denoted with i_2', i_3', i_4' i i_5'.

In the considered case $r = (0000001000000000)$ and only $r_7 = 1$, while all others received word bits are equal zero. Majority decision is based on the estimations

$$i_5: \quad (S_1\ S_2\ S_3\ S_4\ S_5\ S_6\ S_7\ S_8) = (000100000) \rightarrow i'_5 = 0$$
$$i_4: \quad (S_1\ S_2\ S_3\ S_4\ S_5\ S_6\ S_7\ S_8) = (001000000) \rightarrow i'_4 = 0$$
$$i_3: \quad (S_1\ S_2\ S_3\ S_4\ S_5\ S_6\ S_7\ S_8) = (001000000) \rightarrow i'_3 = 0$$
$$i_2: \quad (S_1\ S_2\ S_3\ S_4\ S_5\ S_6\ S_7\ S_8) = (000000010) \rightarrow i'_2 = 0$$

finally yielding

$$r' = r \oplus (0000) \otimes G_1 = (0000001000000000) \rightarrow i'_1 = 0.$$

The decoded information word is $i = (00000)$ and the reconstructed code word (all zeros) differs from the received one in one bit.

Problem 5.11 Explain the construction of the following arithmetic codes

(a) binary Brown code defined by relation $c = 19i + 61$, for the base $b = 8$;
(b) binary Varshamov code, code word length $n = 4$;
(c) ternary Varshamov code, code word length $n = 4$;

For every problem part find the code words, weight spectrum, minimum Hamming distance and the number of correctable errors by the code. Explain the notion of asymmetric error and verify whether the codes (b) and (c) satisfy the Varshamov-Gilbert bound.

Solution

The **arithmetic block codes** construction is based on the arithmetic operation connecting decimal representations of information and a code word. These relations are usually very simple and an overview of some typical constructions follows.

(a) Construction method proposed by David Brown in 1960 uses the rule $c = Ai + B$, where c represents a decimal equivalent of the code word and i— a decimal equivalent of the information word [30]. Coefficients A and B are chosen so as that the binary representations of all code words have suitable features and that a decimal equivalent of information word has to fulfill the condition $n \leq b - 1$. The basis b can be any integer greater than two, and in the case $b = 2^k$ binary equivalent of number i can be any combination of k bits, and the set of code words consists of 2^k n-bits combinations.

For $b = 8$ in this case, decimal equivalents of information words are numbers $0 \leq i \leq 7$ and decimal equivalents of code words can be found using the relation $c = 19i + 61$, the corresponding set of code words is given in Table 5.12. The code word length is determined by the number of bits needed to represent the largest decimal number c (here $c_{\max} = 194$) yielding here $n = 8$, the corresponding code rate is $R = k/n = 3/8$.

Problems

Table 5.12 Illustration of Brown code (8, 3) construction

Information words		Code words, $c = 19i + 61$	
Binary representation, i	Decimal representation, i	Decimal representation, c	Binary representation, c
000	0	61	00111101
001	1	80	01010000
010	2	99	01100011
011	3	118	01110110
100	4	137	10001001
101	5	156	10011100
110	6	175	10101111
111	7	194	11000010

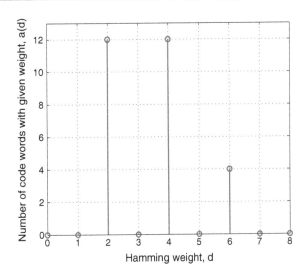

Fig. 5.20 Code distances spectrum of Brown code (8, 3)

It is obvious that the set of code words does not include combination "all zeros" leading to the conclusion that the code is not a linear one. This is the reason that the minimum Hamming distance cannot be found from the code words weights, because the sum of two code words may not be a code word. The distance spectrum of non linear codes is obtained by comparing all pairs of code words. In this case there is total of $2^k(2^k - 1)/2 = 28$ code word pairs and the corresponding spectrum is shown in Fig. 5.20. Minimum Hamming distance is $d = 3$ and the code can correct all single errors, by a search to find the code word which differs from the received word in one bit. It is interesting that these codes besides the single errors can correct as well some specific types of multiple errors.

(b) Rom Varsharmov (Ром Варшамов) in 1965 proposed a method for block code construction where the ordered n-tuples $c = (c_1, c_2, \ldots, c_n)$ are considered, where every symbol can be an integer from the set $\{0, 1, \ldots, q-1\}$ [31]. This ordered n-tuple is a code word only if the following holds

$$\left(\sum_{i=1}^{n} i x_i\right) \bmod (n+1) = 0.$$

Therefore, all the words $c = (c_1, c_2, \ldots, c_n)$ satisfying the above condition form so called integer block code not being necessarily linear in the general case. This procedure can be modified by summation using some other modulo $m > n$ (e.g. $m = 2n + 1$), the remainder being equal to some in advance fixed number from the set $\{0, 1, \ldots, m-1\}$.

For $q = 2$, it is a binary code, and in a case $n = 4$ and $m = n+1 = 5$ only the words (0000), (1001), (0110) and (1111) satisfy the relation

$$\left[\sum_{i=1}^{4} i x_i\right] \bmod 5 = 0,$$

it is easy to verify that the code is linear and that its generator matrix is

$$G = \begin{bmatrix} 1 & 0 & 0 & 1 \\ 0 & 1 & 1 & 0 \end{bmatrix},$$

while the weight spectrum is determined by $a(0) = 1$, $a(2) = 2$, $a(3) = 0$ and $a(4) = 1$ (Fig. 5.21a).

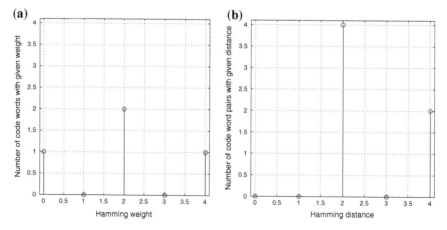

Fig. 5.21 Weight spectrum (**a**) and the code distances spectrum of binary Varshamov code (4, 2) (**b**)

It is interesting to note that the weight spectrum is not fully equivalent to the distances spectrum even when the code is linear one. The code distances spectrum is found by the comparison of all different code words (giving always $a(0) = 0!$), and there is a total of $2^k(2^k - 1)/2 = 6$ combinations, shown in Fig. 5.21b. E.g. there are two code words pairs ((0000) and (1111), (0110) and (1001)) where the Hamming distance equals 4, but there is one word only having the Hamming weight $d = 4$. However, minimum distance and minimum weight are same, and the spectrum shapes are the same for $d \neq 0$.

Minimum Hamming distance is $d_{min} = 2$ and the code cannot correct all single errors. However, if it is supposed that the error probability for binary zero is much higher than the error probability for binary one (the corresponding binary channel has parameters $P(1/0) \gg P(0/1) \approx 0$, what is the case for some magnetic recording systems), i.e. the errors are "*asymmetric*" and an asymmetric error can result in following transitions

$$(0000) \rightarrow (0000), (0001), (0010), (0100), (1000); (1111) \rightarrow (1111);$$
$$(0110) \rightarrow (0110), (1110), (0111); (1001) \rightarrow (1001); (1101), (1101).$$

Every received vector, corresponding to the right sides of denoted transitions, can originate from only one code word, and the code can correct one asymmetric error. Reader should verify whether this code can correct one error when $P(0/1) \gg P(1/0) \approx 0$, and whether this code can correct two asymmetric errors.

Varshamov-Gilbert bound gives the lower bound of a maximum possible code words number for a given code basis q, code word length n and a minimum code Hamming distance d_{min}

$$A_q(n, d) \geq q^n / \sum_{j=0}^{d-1} \binom{n}{j}(q-1)^j.$$

For a considered example $q = 2$, $n = 4$ and $d = 2$ yielding $A_q(n, d) \geq 2^4/(1+4) = 3, 2$, what means that it is possible to construct a code having at least four code words for a given conditions. It is obvious that the constructed linear block code confirms that it is possible to construct at least one code satisfying the bound.

(c) For a *ternary* code, the code words are formed by the numbers from set $\{0, 1, 2\}$. For $n = 4$ and $m = 2n + 1 = 9$ the relation [32]

$$\left(\sum_{i=1}^{4} ic_i \right) \mod 9 = 0$$

Fig. 5.22 Code distance spectrum for the ternary Varshamov code (4, 2)

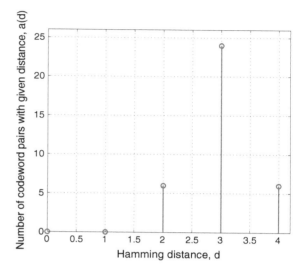

is satisfied by code words (0000), (0111), (0222), (1002), (1120), (1201), (2011), (2122), (2210) ($r^k = 9$ code words). This code is not linear, although includes identity element for addition, because e.g.,

$$(0111) + (1002) = (1110),$$

and it is not a code word meaning that the closure for addition is not satisfied.

Varshamov-Gilbert bound for $q = 3$, $n = 4$ and $d = 2$ yields

$$A_q(n,d) \geq \frac{3^4}{1 + 4 \times 2^1} = 9,$$

in this case satisfied with equality.

Code distance spectrum was found by comparison all $3^k(3^k - 1)/2 = 36$ code words pairs, shown in Fig. 5.22. In this case the minimum Hamming distance is $d_{\min} = 2$ as well, and a code corrects the error at one symbol in the code word, if it is asymmetric. It is important to note that for an integer code, the correcting capability does not depend on the error weight. In this case the error in the symbols can result in transitions $0 \rightarrow 1$ and $0 \rightarrow 2$, but the code is capable to correct the error in one symbol, for every of these weight.

Problem 5.12 An integer linear block code is defined over the integer ring Z_3^4 has the parity-check matrix

$$\mathbf{H} = \begin{bmatrix} 0 & 1 & 1 & 1 \\ 1 & 0 & 1 & 2 \end{bmatrix}.$$

Problems

(a) Explain the notion of an integer linear block code.
(b) Find the generator matrix of the code. Is it unique?
(c) List all code words, draw its spectrum and find a minimum Hamming distance.
(d) Write the generalized Hamming bound for integer linear block codes. Which error weights (and in which positions) this code can correct? Find the corresponding syndromes and comment the result.
(e) Find the correction capability of a code defined over the ring Z_5^6 having the parity-check matrix.

$$H_2 = \begin{bmatrix} 0 & 1 & 2 & 3 & 4 & 1 \\ 1 & 1 & 1 & 1 & 1 & 0 \end{bmatrix}.$$

Solution

(a) For *integer codes* information and code words symbols take the values from the set $Z_r = \{0, 1, 2, \ldots, q-1\}$ while all operation during the code words forming are modulo-q.

For any two elements a and b from ring Z_q two operations are defined—addition and multiplication. Ring properties (axioms) are:

1. The set $\{0, 1, 2, \ldots, q-1\}$ is an additive Abelian group and 0 is an identity element for addition modulo-q (denoted by $a + b$).
2. For the multiplication (denoted by ab), the product is in the ring $ab \in Z_r$ (closure).
3. The multiplication is associative $a(bd) = (ab)d$.
4. The multiplication is distributive in relation to addition, i.e. $a(b + d) = ab + ad$, $(b + d)a = ba + da$.

A ring can be commutative (if $ab = ba$ for any pair of elements), but it is not always the case. Every ring element has an inverse element for addition (their sum equals 0), but it should be noted that an inverse element for multiplication is not needed, and the division cannot be defined in that case. If these inverse elements are in the ring Z_q as well and if the multiplication group is commutative, the ring would become a field. The ring of integers is commutative and the subtraction is defined by

$$a - b = a + q - b,$$

but the inverse elements for multiplication are not defined (the rational numbers are not in a set).

New ring Z_q^n has q^n ordered n-tuples, which have symbols from the originating ring Z_q, and Z_q^n can be regarded as an nth extension of the ring Z_q. Now an integer code with a basis q and parameters (n, k) can be defined as a set of code words, denoted by c, satisfying the conditions [33]

$$\left\{ c \in Z_q^n, cH^T = 0 \right\}.$$

The first condition shows that a code word must be an ordered n-tuple from the ring Z_q^n and matrix H is a parity-check matrix, dimensions $(n - k) \times n$, with the elements from the ring Z_q. Of course, the capability of integer code to correct the errors depends on parity-check matrix.

(b) In this case $n = 4$ and $n - k = 2$, and the generator matrix can be found on the base of relation

$$GH^T = 0 \Rightarrow \begin{bmatrix} g_{11} & g_{12} & g_{13} & g_{14} \\ g_{21} & g_{22} & g_{23} & g_{24} \end{bmatrix} \begin{bmatrix} 0 & 1 \\ 1 & 0 \\ 1 & 1 \\ 1 & 2 \end{bmatrix} = \begin{bmatrix} 0 & 0 \\ 0 & 0 \end{bmatrix},$$

the code is ternary one and all operations are modulo $q = 3$. The previous matrix equation can be written in an expanded form

$$g_{12} + g_{13} + g_{14} = 0, \quad g_{22} + g_{23} + g_{24} = 0$$
$$g_{11} + g_{13} + 2g_{14} = 0, \quad g_{21} + g_{23} + 2g_{24} = 0,$$

and it is obvious that this system has not a unique solution. E.g. one solution is $g_{11} = 2$, $g_{12} = 2$, $g_{13} = 0$, $g_{14} = 1$, $g_{21} = 0$, $g_{22} = 1$, $g_{23} = 1$, $g_{24} = 1$ and the other could be $g_{11} = 0$, $g_{12} = 1$, $g_{13} = 1$, $g_{14} = 1$, $g_{21} = 2$, $g_{22} = 2$, $g_{23} = 0$, $g_{24} = 1$.

It is interesting to note that the code is self-dual, because

$$\begin{bmatrix} 0 & 1 & 1 & 1 \\ 1 & 0 & 1 & 2 \end{bmatrix} \begin{bmatrix} 0 & 1 \\ 1 & 0 \\ 1 & 1 \\ 1 & 2 \end{bmatrix} = \begin{bmatrix} 0 & 0 \\ 0 & 0 \end{bmatrix},$$

and one solution for a generator matrix is

$$G = H = \begin{bmatrix} 0 & 1 & 1 & 1 \\ 1 & 0 & 1 & 2 \end{bmatrix}.$$

(c) All code words of the code can be found multiplying all possible information words by generator matrix yielding

$$c_1 = i_1 G = [0\ 0]\begin{bmatrix} 0 & 1 & 1 & 1 \\ 1 & 0 & 1 & 2 \end{bmatrix} = [0\ 0\ 0\ 0], \quad c_2 = i_2 G = [0\ 1]\begin{bmatrix} 0 & 1 & 1 & 1 \\ 1 & 0 & 1 & 2 \end{bmatrix} = [1\ 0\ 1\ 2],$$

$$c_3 = i_3 G = [0\ 2]\begin{bmatrix} 0 & 1 & 1 & 1 \\ 1 & 0 & 1 & 2 \end{bmatrix} = [2\ 0\ 2\ 1], \quad c_4 = i_4 G = [1\ 0]\begin{bmatrix} 0 & 1 & 1 & 1 \\ 1 & 0 & 1 & 2 \end{bmatrix} = [0\ 1\ 1\ 1],$$

$$c_5 = i_5 G = [1\ 1]\begin{bmatrix} 0 & 1 & 1 & 1 \\ 1 & 0 & 1 & 2 \end{bmatrix} = [1\ 1\ 2\ 0], \quad c_6 = i_6 G = [1\ 2]\begin{bmatrix} 0 & 1 & 1 & 1 \\ 1 & 0 & 1 & 2 \end{bmatrix} = [2\ 1\ 0\ 2],$$

$$c_7 = i_7 G = [2\ 0]\begin{bmatrix} 0 & 1 & 1 & 1 \\ 1 & 0 & 1 & 2 \end{bmatrix} = [0\ 2\ 2\ 2], \quad c_8 = i_8 G = [2\ 1]\begin{bmatrix} 0 & 1 & 1 & 1 \\ 1 & 0 & 1 & 2 \end{bmatrix} = [1\ 2\ 0\ 1],$$

$$c_9 = i_9 G = [2\ 2]\begin{bmatrix} 0 & 1 & 1 & 1 \\ 1 & 0 & 1 & 2 \end{bmatrix} = [2\ 2\ 1\ 0].$$

Because the code is linear, the Hamming distance between any two code words corresponds to the Hamming weight of the word obtaining by their addition. The Hamming weight corresponds to the number of non-zero elements in the code words, and it is obvious that the minimum Hamming distance in this code is $d_{min} = 4$. The corresponding distance spectrum is shown in Fig. 5.23.

An integer code can correct e_c errors of weight t if it is possible to correct all error vectors $e = (e_1, e_2, \ldots, e_n)$ having the Hamming weight $w(e) \leq e_c$, where $e_i \in \{-t, -t+1, \ldots, t-1, t\}$. To achieve this feature, the syndromes corresponding to all correctable error patterns must be different, and a **Hamming bound generalization** can be defined

$$q^{n-k} \geq \sum_{i=0}^{e_c} \binom{n}{i} (2t)^i.$$

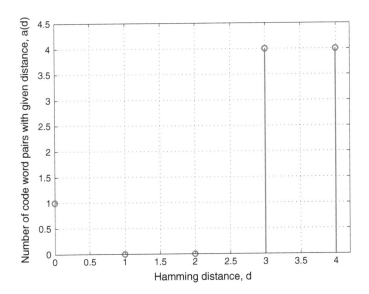

Fig. 5.23 Distance spectrum of the integer code defined by matrix H

As previously concluded $q = 3$ and $e_c = 1$, yielding

$$3^{4-2} \geq 1 + \binom{4}{1} \times 2t \Rightarrow 9 \geq 1 + 8t.$$

Therefore, the errors of the type -1 and $+1$ at any code word position can be corrected. Taking into account that for operations modulo-3, the subtraction of one (-1) corresponds to the adding of two ($+2$), the correctable error patterns are (0001), (0002), (0010), (0020), (0100), (0200), (1000) and (2000).

It is obvious that the code can correct all possible error patterns, and the corresponding syndromes are defined by

$$S = rH^T = eH^T,$$

where e is the corresponding error vector and $r = c + e$ is a received vector corresponding to code word c.

List of syndromes corresponding to correctable error patterns is as follows

$S_1 = [0\ 0\ 0\ 1]H^T = [1\ 2], \quad S_2 = [0\ 0\ 0\ 2]H^T = [2\ 1],$
$S_3 = [0\ 0\ 1\ 0]H^T = [1\ 1], \quad S_4 = [0\ 0\ 2\ 0]H^T = [2\ 2],$
$S_5 = [0\ 1\ 0\ 0]H^T = [1\ 0], \quad S_6 = [0\ 2\ 0\ 0]H^T = [2\ 0],$
$S_7 = [1\ 0\ 0\ 0]H^T = [0\ 1], \quad S_8 = [2\ 0\ 0\ 0]H^T = [0\ 2].$

Obviously, when there are no errors, the same syndrome is always obtained

$$S_0 = [0\ 0\ 0\ 0]H^T = [0\ 0].$$

(d) In this case $n = 6$, and $n - k = 2$, the generator matrix has the dimensions 4×6 and can be written as

$$\begin{bmatrix} g_{11} & g_{12} & g_{13} & g_{14} & g_{15} & g_{16} \\ g_{21} & g_{22} & g_{23} & g_{24} & g_{25} & g_{26} \\ g_{31} & g_{32} & g_{33} & g_{34} & g_{35} & g_{36} \\ g_{41} & g_{42} & g_{43} & g_{44} & g_{45} & g_{46} \end{bmatrix} \begin{bmatrix} 0 & 1 & 2 & 3 & 4 & 1 \\ 1 & 1 & 1 & 1 & 1 & 0 \end{bmatrix}^T = \begin{bmatrix} 0 & 0 \\ 0 & 0 \\ 0 & 0 \\ 0 & 0 \end{bmatrix},$$

the code basis is $q = 5$ and the equation system corresponding to the first generator matrix column is

$$g_{12} + 2g_{13} + 3g_{14} + 4g_{15} + g_{16} = 0$$
$$g_{11} + g_{12} + g_{13} + g_{14} + g_{15} = 0$$

whose one solution is $g_{11} = g_{12} = g_{13} = g_{14} = g_{15} = 1$, $g_{16} = 0$, and the other $g_{11} = 1$, $g_{12} = g_{15} = g_{16} = 0$, $g_{13} = g_{14} = 2$.

The equation of a similar form can be written as well for the other rows of generator matrix, but the solutions for the various rows of the matrix G have to be mutually different. One possible form of generator matrix is

$$G = \begin{bmatrix} 1 & 1 & 1 & 1 & 1 & 0 \\ 1 & 2 & 0 & 2 & 0 & 2 \\ 2 & 2 & 1 & 0 & 0 & 1 \\ 0 & 0 & 2 & 1 & 2 & 0 \end{bmatrix},$$

where one should be careful that the third and the fourth row are not the linear combinations of the upper two (the operations are modulo-5).

The corresponding distance spectrum is shown in Fig. 5.24. Total number of code words is $5^4 = 625$ and a minimum Hamming distance is $d_{min} = 3$, the code can correct $e_c = 1$ error in the code word.

In this case $q = 5$ and $e_c = 1$, and a generalized Hamming bound is

$$5^{6-4} \geq 1 + \binom{6}{1} \times 2t \Rightarrow 25 \geq 1 + 12t,$$

here $t = 2$ and the errors of the type $-2, -1, 0, +1, +2$ can be corrected at any position in the code word. Correctable error patterns are

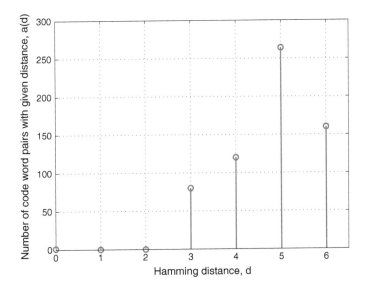

Fig. 5.24 Distance spectrum for integer code defined by matrix H_2

(000001), (000002), (000003), (000004), (000010), (000020), (000030), (000040), (000100), (000200), (000300), (000400), (001000), (002000), (003000), (004000), (010000), (020000), (030000), (040000), (100000), (200000), (300000), (400000),

therefore, the code can correct the error of any weight at any position in the code word.

Brief Introduction to Algebra I

The aim of this subsection is to facilitate the study of block codes. Practically all proofs are omitted. The corresponding mathematical rigor can be found in many excellent textbooks [25, 27, 34].

Algebraic systems satisfy rules which are very often the same as applied to an "ordinary number system".

Groups

Let G be a set of elements $(a, b, c, \ldots)$. An operation ($\otimes$) is defined for which certain axioms hold. The operation is a procedure applied to two elements to obtain the uniquely defined third one (it is in fact a *binary operation*). It can be denoted

$$f(a, b) = c,$$

but it is customary to write

$$a \otimes b = c.$$

In the following, we it will be written very often

$$a \cdot b = c \quad \text{or} \quad a + b = c,$$

calling the operation "multiplication" or "addition" although the operation is not necessarily the multiplication or addition of ordinary numbers.

Axioms for the group:

G1. (*Closure*) An application of operation to any two group elements results always in an element of the same group.

$$\forall a, b \in G: \quad a \otimes b = c \in G.$$

G2. (*Associativity*) The order of performing the operation is immaterial.

$$\forall a, b, c \in G: \quad a \otimes (b \otimes c) = (a \otimes b) \otimes c = a \otimes b \otimes c.$$

G3. (*Identity element*) There is a unique identity (neutral) element $e \in G$ satisfying (even the operation is not commutative).

$$\forall a \in G: \quad e \otimes a = a \otimes e = a.$$

If operation is denoted by "+" (i.e. if it is called addition), the neutral element is written as "0", and if operation is denoted by "·" (i.e. if it is called multiplication), it is written as "1".

G4. (*Inverse elements*) For every element of a set a there is always an unique inverse (symmetric) element of a set b (even the operation is not commutative)

$$\forall a \in G, \exists b \in G: \quad a \otimes b = b \otimes a = e.$$

If operation is denoted by "+" (i.e. if it is called addition), the inverse element is written as "$-a$", and if operation is denoted by "·" (i.e. if it is called multiplication), it is written as "a^{-1}".

Additionally, if a group satisfies the *commutative law*, i.e.

$$\forall a, b \in G: \quad a \otimes b = b \otimes a,$$

the group is called *Abbelian* or *commutative*.

The uniqueness of the identity element as well of the inverse elements can be easily proved.

In every group (was it commutative or not) the following is satisfied

$$(a \otimes b)^{-1} = b^{-1} \otimes a^{-1}.$$

If the number of elements in a group is finite, the group is called a *finite group* (the number of elements is called *order* of G.

Examples:

1. The set of integers (positive, negative and zero) is a commutative group under addition.
2. The set of positive rationals is a commutative group under multiplication.
3. The set of real numbers (excluding zero) is a commutative group under multiplication.
4. The set of $n \times n$ real-valued matrices is a commutative group under matrix addition.
5. The set of $n \times n$ nonsingular ($\det \neq 0$) matrices is a non-commutative group under the matrix multiplication.
6. The functions $f_1(x) = x$ and $f_2(x) = 1/x$ under the operation

$$f_1(x) \otimes f_2(x) = f_1(f_2(x))$$

are the only elements of a binary group (i.e. the group consisting of only two elements). The corresponding table (for every finite group this table can be constructed) for the operation is

$$\begin{array}{cc} f_1(x) & f_2(x) \\ f_1(x) \quad f_1(x) & f_2(x) \\ f_2(x) \quad f_2(x) & f_1(x) \end{array}$$

The identity element is $f_1(x)$ and every element is its own inverse. Is a finite group of the order 2.

7. The linear transformations (rotation or reflections) of an equilateral triangle into itself are the elements of an algebraic group of the order 6.

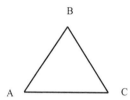

By denoting the transformation corresponding to the 120° counterclockwise rotation as

$$\begin{pmatrix} ABC \\ CAB \end{pmatrix},$$

and introducing the following notation:

$1 = \begin{pmatrix} ABC \\ ABC \end{pmatrix}$, (no rotation nor reflection)

$a = \begin{pmatrix} ABC \\ CAB \end{pmatrix}$, (counterclockwise rotation 120°)

$b = \begin{pmatrix} ABC \\ BCA \end{pmatrix}$, (counterclockwise rotation 240°)

$c = \begin{pmatrix} ABC \\ ACB \end{pmatrix}$, (reflection about bisector of angle A)

$d = \begin{pmatrix} ABC \\ CBA \end{pmatrix}$, (reflection about bisector of angle B)

$e = \begin{pmatrix} ABC \\ BAC \end{pmatrix}$, (reflection about bisector of angle A)

defining the operation

Brief Introduction to Algebra I

$$a \otimes b$$

as a transformation a followed by a transformation b, the following table is obtained:

$\otimes$	1	a	b	c	d	e
1	1	a	b	c	d	e
a	a	b	1	d	e	c
b	b	1	a	e	c	d
c	c	e	d	1	b	a
d	d	c	e	a	1	b
e	e	d	c	b	a	1

The obtained group is not commutative (the table is not symmetric with respect to the main diagonal). The corresponding inverses can be easily found from the table

8. The set of integers $G = \{0, 1, ..., m - 1\}$ is a group under modulo-m addition (i.e. the result is obtained as the remainder from dividing the sum by m). For $m = 3$, the table is the following

$\oplus$	0	1	2
0	0	1	2
1	1	2	0
2	2	0	1

The group is commutative.

9. The set of integers $G = \{0, 1, ..., p\}$ is not generally a group under modulo-p multiplication (i.e. the result is obtained as the remainder from dividing the product by p). However, if p is a *prime*, G is the group under modulo-p multiplication.

Comments:

1. There is a group consisting of only one element. It must be the identity element. All other axioms also hold.
2. There is a group with two elements. One must be the identity element (for example 0). Denote the other with a. According to G.3: $a + 0 = 0 + a = a$. Therefore, a must be its own inverse, i.e. $a + a = 0 \Rightarrow -a = a$. The table is as follows

+	0	a
0	0	a
a	a	0

The table is practically the same as for Example 6. In fact, there is a unique table for all groups consisting of two elements (all groups of order 2 are isomorphic). The same table is obtained for modulo-2 addition (by putting $a = 1$).

3. All groups of order 3 are also isomorphic (corresponding to modulo-3 addition).
4. There are 2 different groups of order 4 and both are commutative. One is obtained on the basis of modulo-4 addition.

Subgroup

Let G be a group. Let H be a subset of G. H is a subgroup of G if its elements satisfy all the axioms for the group itself with the same operation. The identity element must be in the subgroup.

Examples:

1. In the group of 6 transformations of the equilateral triangle (Ex. 7), the following sets are subgroups: (1, *a*, *b*), (1, *c*), (1, *d*) and (1, *e*).
2. In the group of all integers (Example 1), the subset of integers that are multiples of any integer is a subgroup.

Rings

The introducing of another "operation" results in an algebraic structure called *ring*. One operation is called *addition* (denoted as $a + b$), the other is called *multiplication* (denoted as ab). In order for a set R to be a ring, some axioms must hold.

Axioms for the ring:

R1. The set R is a commutative group under addition (the identity element is denoted by 0).

R2. (*Closure*) The operation "multiplication" is defined for any two elements of R, the product being also always element of R, i.e.

$$\forall a, b \in R,\ ab \in R.$$

R3. (*Associativity*) The multiplication is associative, i.e.

$$a(bc) = (ab)c.$$

R4. (*Distributivity*) The multiplication is distributive over addition, i.e.

$$a(b+c) = ab + ac, \quad (b+c)a = ba + ca.$$

The ring is commutative if its multiplication operation is commutative. It can be easily proved that in any ring

Brief Introduction to Algebra I

$$0a = a0 = 0, \quad a(-b) = (-a)b = -(ab).$$

Note that the existence of identity element for multiplication (denoted e.g. by 1) is not supposed. A ring having this identity element is called a *ring with identity*. In such a ring

$$a1 = 1a = a.$$

It can be shown that identity element 1 is unique.

Note also that in the ring there need not be any inverse elements. In ring with identity the inverse elements may exist. In the noncommutative ring left and right inverse elements may exist. If they exist both, then they are mutually equal, i.e. there is a unique inverse element. An element of the ring having an inverse element is called a *unit*. Further, under ring multiplication the set of units form a group.

Examples:

1. The set of integers (positive, negative and zero) is a commutative ring under (ordinary) addition and multiplication.
2. The set of real numbers is a commutative ring under (ordinary) addition and multiplication.
3. The set of all polynomials in one variable (e.g. x) with integer coefficients is a commutative ring. This means that polynomials may be added and multiplied giving as a result polynomial with integer coefficients. Inverse elements for addition are polynomials of the same degree having as coefficients the inverse elements of the corresponding integers. On the other hand, there are not inverse elements for multiplication because rational functions are not included in the ring. Still, the identity element exists, it is zero-degree polynomial $p(x) = 1$.

Comments:

1. There is a ring consisting of only one element. It must be the addition identity element (0). The corresponding rules are: $0 + 0 = 0$, $(0)(0) = 0$.
2. There are two possible (different!) rings with two elements. One element is 0, denote the other with a. The corresponding table for additive commutative group is (as it must be!)

+	0	a
0	0	a
a	a	0

Of course, $0a = a0 = 0$. The question is what is the value of aa? As either $aa = a$ or $aa = 0$ both satisfy associative and distributive laws, two different tables for multiplication are possible

·	0	a		·	0	a
0	0	0	or	0	0	0
a	0	a		a	0	0

yielding two rings with different structure.

Fields

The group can be considered as a set in which additions and subtraction are possible, the ring—as a structure where besides the addition and subtraction, the multiplication is also possible. To include division, a more powerful algebraic structure is needed. This is a *field*.

Axioms for the field:

F1. The set F is a commutative group under addition.
F2. The set F is closed under multiplication. All set nonzero ($\neq 0$) elements form a commutative group under multiplication.
F3. The distributive law holds for all field elements, i.e.

$$\forall a, b, c \in \text{F}: \quad a(b+c) = ab + ac, \quad (a+b)c = ac + bc.$$

Therefore, the field can be considered as a commutative ring with a multiplicative identity element ("ring with identity") in which every nonzero element has a multiplicative inverse (i.e. all nonzero elements are "units").

Examples:

1. The set of rational numbers is a field.
2. The set of real numbers is a field.
3. The set of complex numbers is a field.
4. The set of the numbers of the type $a + 2\sqrt{b}$, where a and b are rational numbers, is a field.
5. The set of integers is not a field (the inverse elements for multiplication—rational numbers—are not included).

The field can have an infinite number of elements (as in Examples 1–4). If a field with finite number of (q) elements exists, it is called a *finite field*, or a *Galois field*, denoted by GF(q). The number of elements is called *order* of a field. In the following only the finite fields will be considered.

Comments:

1. The smallest field must have two elements—identity element for addition (0) and identity element for multiplication (1). It is GF(2). The corresponding tables are

+	0	1		·	0	1
0	0	1		0	0	0
1	1	0		1	0	1.

2. It can be shown that the field with q elements exists only if q is a prime (p) or a power of a prime (p^n). All fields of the same order have the same tables (they are isomorphic).

Brief Introduction to Algebra I

3. The tables for the field of order p (a prime) are obtained by modulo-p addition and modulo-p multiplication. The field is also called a *prime field*.

Examples:
1. The tables for GF(3) (instead of "2" as a third field element a can be written):

+	0	1	2		·	0	1	2
0	0	1	2		0	0	0	0
1	1	2	0		1	0	1	2
2	2	0	1		2	0	2	1.

2. The tables for GF(4) = GF(2^2). The other two field elements are denoted by a and b.

+	0	1	a	b		·	0	1	a	b
0	0	1	a	b		0	0	0	0	0
1	1	0	b	a		1	0	1	a	b
a	a	b	0	1		a	0	a	b	1
b	b	a	1	0		b	0	b	1	a.

Note that GF(2) is a *subfield* of GF(2^2)—the corresponding tables of GF(2) are contained in tables of GF(2^2) (but not in tables of GF(3)!). GF(2^2) is an *extension field* of GF(2).

3. The tables for modulo-4 addition and multiplication are:

+	0	1	2	3		·	0	1	2	3
0	0	1	2	3		0	0	0	0	0
1	1	2	3	0		1	0	1	2	3
2	2	3	0	1		2	0	2	0(!)	2(!)
3	3	0	1	2		3	0	3	2	1

The table for modulo-4 multiplication even does not correspond to the table of the group (because $2 \cdot 2 = 4$, the element 2 does not have the inverse element—as commented earlier the set of integers 0, 1, ..., q is a group under modulo-q multiplication only if q is a prime number). By comparing these tables to tables for GF(4) it can be noted that even the table for addition in GF(4) does not correspond to modulo-4 addition.

The further analysis of the finite fields as well as of the way to obtain the tables for GF(p^n) will be postponed for the next chapter.

Coset Decomposition and Factor Groups

Suppose that the elements of a finite group G are $g_1, g_2, g_3, \ldots$ and the elements of a subgroup H are $h_1, h_2, h_3, \ldots$ (h_1 is the identity element). Construct the array as follows—the first row is the subgroup itself with the identity element at the left (i.e. $h_1, h_2, \ldots$). The first (leading) element in the second row (g_1) is any element not appearing in the first row (i.e. not being the element of the subgroup), the other elements are obtained by the corresponding (group) operation $g_1 \otimes h_i$ (in fact, the first element of the row can be envisaged as $g_1 \otimes h_1 = g_1$). The next row is formed by choosing any previously unused group element in the first column (e.g. g_2). The construction is continued until all the elements of a (finite) group appear somewhere in the array. The row is called a *left coset* (if the commutativity is supposed, it is simply a *coset*), and the first element the *coset leader*. The array itself is a *coset decomposition* of the group. If the commutativity is not supposed there exists also a *right coset*.

The array is (the multiplication is supposed):

$$\begin{array}{cccccc}
h_1=1 & h_2 & h_3 & . & . & h_n \\
g_1h_1=g_1 & g_1h_2 & g_1h_3 & . & . & g_1h_n \\
g_2h_1=g_2 & g_2h_2 & g_2h_3 & . & . & g_2h_n \\
. & . & . & & & . \\
. & . & . & & & . \\
g_mh_1=g_m & g_mh_2 & g_mh_3 & . & . & g_mh_n
\end{array}$$

Further, it can be proved that every element of the group will appear only once in a coset decomposition. It can be proved also that the group elements g' and g'' are in the same (left) coset if (and only if) $(g')^{-1}g''$ is an element of the subgroup.

There is a well known Lagrange's theorem stating that the order of a subgroup must divide the order of a group. It is clearly seen (but not directly proved!) from this rectangular decomposition. It means further, that in a group whose order is a prime number there are no subgroups, except the trivial ones—the whole group and the identity element.

Example:

Consider noncommutative group corresponding to the linear transformations of an equilateral triangle (Example 7 for groups).

(a) Decomposition using subgroup $H_1(1, a, b)$

$$\begin{array}{ccc} 1 & a & b \\ c & e & d \end{array} \quad \text{or} \quad \begin{array}{ccc} 1 & a & b \\ e & d & c \end{array} \quad \text{or} \quad \begin{array}{ccc} 1 & a & b \\ d & c & e \end{array}$$

(b) Decomposition using subgroup $H_2(1, c)$

$$\begin{array}{cc} 1 & c \\ a & d \\ b & e \end{array} \quad \text{or} \quad \begin{array}{cc} 1 & c \\ b & e \\ d & a. \end{array}$$

A subgroup H is called *normal* (or *invariant*) if

$$\forall h \in H \land \forall g \in G: \quad g^{-1}hg \in H.$$

For a normal subgroup every left coset is also a right coset. In the commutative groups every subgroup is normal. For the preceding example, H_1 is a normal subgroup.

Further, for a normal subgroup it is possible to define an operation on the cosets. In fact a new group is formed, the cosets being its elements. The new group is called the *factor group*. The corresponding coset containing the element g is denoted $\{g\}$. The operation definition is

$$\{g_1\}\{g_2\} = \{g_1 g_2\}$$

valid only if no matter which element is chosen from the cosets, the resulting coset is the same. The identity element is the subgroup itself—H = {1}

$$\{1\}\{g\} = \{1g\} = \{g\}$$

and the inverse is the coset containing the corresponding inverse element $\{g^{-1}\}$, i.e.

$$\{g\}\{g^{-1}\} = \{gg^{-1}\} = \{1\}.$$

If the group is commutative, the factor group will be also commutative.

Examples:

1. The linear transformations of an equilateral triangle ABC. The subgroup H_1 is normal. Starting from the following decomposition

$$\begin{array}{ccc} 1 & a & b \\ c & e & d \end{array}$$

denoting the cosets by I = {1} and C = {c}, the following factor group is obtained

$$\begin{array}{ccc} & I & C \\ I & I & C \\ C & C & I \end{array}$$

Of course, the group is isomorphic with the other groups of two elements.

2. Let G be the set of integers (positive, negative and zero)—a commutative group under addition. Let H be the subgroup consisting of multiples of an integer—n. All the numbers from 0 to $n - 1$ are in different cosets. They can be taken as the coset leaders. For $n = 3$ the cosets are

0	3	-3	6	-6	9	-9	.	.	.
1	4	-2	7	-5	10	-8	.	.	.
2	5	-1	8	-4	11	-7	.	.	.

Denoting the cosets by {0}, {1} and {2} the corresponding table is obtained
In fact, it is the addition modulo-3.

+	{0}	{1}	{2}
{0}	{0}	{1}	{2}
{1}	{1}	{2}	{0}
{2}	{2}	{0}	{1}

Vector Spaces

The notion of a vector space is a well known. However, in the following a specific kind of vector space will be considered whose vectors have the elements from a finite field.

Axioms for the vector space:

A set of elements (*vectors*) V is called a *vector space* over a field F (whose elements are called *scalars*) if it satisfies the following axioms:

V1. (*Closure under addition*) The set V is a commutative group under addition.
V2. (*Scalar multiplication*) The product $c\mathbf{v} \in V$ is defined ($c \in F, \mathbf{v} \in V$). It can be called an *outer operation*.
V3. (*Distributivity*) Scalar multiplication is distributive over vector addition, i.e.

$$c(\mathbf{u}+\mathbf{v}) = c\mathbf{u} + c\mathbf{v}.$$

V4. (*Distributivity*) $(c + d)\mathbf{v} = c\mathbf{v} + d\mathbf{v}$
V5. (*Associativity*) $(cd)\mathbf{v} = c(d\mathbf{v})$

By introducing now the vector multiplication a *linear associative algebra* A over a field F can be obtained. It will be needed later (for the next chapter).

Axioms for the linear associative algebra:

A1. (*Vector space*) The set A is a vector space over F.
A2. (*Closure under multiplication*) For any two elements $\mathbf{u}, \mathbf{v} \in A$, a product is defined $\mathbf{uv} \in A$.
A3. (*Associativity*) For any three elements $\mathbf{u}, \mathbf{v}, \mathbf{w} \in A$, $(\mathbf{uv})\mathbf{w} = \mathbf{u}(\mathbf{vw})$.
A4. (*Bilinearity*) $\mathbf{u}(c\mathbf{v} + d\mathbf{w}) = c\mathbf{uw} + d\mathbf{uw}$, $(c\mathbf{v} + d\mathbf{w})\mathbf{u} = c\mathbf{vu} + d\mathbf{wu}$. ($c, d \in F, \mathbf{u}, \mathbf{v}, \mathbf{w} \in V$)

Let denote an ordered set of n field elements (an *n-tuple*) with $(a_1, a_2, \ldots, a_n)$. If the addition and multiplication of n-tuples are defined as follows

$$(a_1, a_2, \ldots, a_n) + (b_1, b_2, \ldots, b_n) = (a_1 + b_1, a_2 + b_2, \ldots, a_n + b_n),$$
$$(a_1, a_2, \ldots, a_n)(b_1, b_2, \ldots, b_n) = (a_1 b_1, a_2 b_2, \ldots, a_n b_n),$$

as well as a multiplication of an n-tuple by a field element

$$c(a_1, a_2, \ldots, a_n) = (ca_1, ca_2, \ldots, ca_n),$$

it can be verified easily that a set of n-tuples forms a vector space over a field.

The identity element is $\mathbf{0}(0, 0, \ldots, 0)$. 0 is the identity element for addition in the field. The following relations hold:

Brief Introduction to Algebra I

$$0v = \mathbf{0}, a\mathbf{0} = \mathbf{0}, (-v) = (-1)v$$

(–1 is the inverse element for addition of the multiplication identity element in the field).

A *subspace* is a subset of a vector space satisfying the axioms for a vector space.

In a vector space a sum of the form (vector)

$$u = a_1v_1 + a_2v_2 + \cdots + a_kv_k$$

is called a *linear combination* of vectors.

A set of vectors $v_1, v_2, \ldots, v_k$ is *linearly dependent* if (and only if) there is a set of scalars $c_1, c_2, \ldots, c_k$, not all being equal to zero, such that

$$c_1v_1 + c_2v_2 + \cdots + c_kv_k = \mathbf{0}.$$

A set of vectors is *linearly independent* if it is not linearly dependent.

A set of vectors *spans* (*generates*) a vector space if every vector in a space equals their linear combination. By eliminating the linearly dependent vectors from the set, a new set of linearly independent vectors will be obtained spanning the same vector space.

The number of linearly independent vectors in the set that span a vector space is called the *dimension* of a space, while the set is called a *basis* of the space. It can be shown that in a k-dimensional vector space every set of more than k vectors must be linearly dependent. On the other hand, in a k-dimensional vector space every set of k linearly independent vectors is a basis. Therefore, a vector space has more bases. In fact, every suitable transformation (will be defined later) of the basis leads to a new basis.

In fact, any n-tuple (vector) can be always written as follows

$$u = a_1(1,0,\ldots,0) + a_2(0,1,0,\ldots,0) + \cdots + a_n(0,0,\ldots,0,1).$$

The vectors (1, 0, ..., 0), (0, 1, 0, ..., 0) ... can be considered as *orts*. They are linearly independent. Therefore, the dimension of the vector space consisting of the vectors with n elements equals n.

The product of two n-tuples defined as follows

$$(a_1, a_2, \ldots, a_n)(b_1, b_2, \ldots, b_n) = a_1b_1 + a_2b_2 + \cdots + a_nb_n$$

is called *inner product* or *dot product*. If inner product equals zero (0 from the field), for the vectors is said to be *orthogonal*. For inner product the commutativity and distributivity can be easily verified on the basis of the field axioms.

Matrices

In this subsection only the corresponding elements needed for the study of the codes are exposed.

The *matrix* **M** ($m \times n$) is an ordered set (a rectangular array) of elements of m rows and n columns

$$\mathbf{M} = \begin{bmatrix} a_{11} & a_{12} & \cdots & a_{1n} \\ a_{21} & a_{22} & \cdots & a_{2n} \\ \vdots & \vdots & \ddots & \vdots \\ a_{m1} & a_{m2} & \cdots & a_{mn} \end{bmatrix} = [a_{ij}].$$

In the following exposition matrix elements are also the elements of a finite field. Therefore, the matrix rows can be considered as a n-tuples of field elements—vectors (the same for the columns—vectors of m elements).

The number of linearly independent rows is called the *row rank* (the same for the number of linearly independent columns—the *column rank*). The row rank equals column rank being the *rank* of the matrix.

The following set of *elementary row operations* does not change the matrix rank:

1. Interchange of any two rows.
2. Multiplication of any row by a (nonzero) field element
3. Multiplication of any row by a (nonzero) field element and addition of the result to another row.

Now, the *row-space* of a matrix can be defined—a space whose basis consists of the rows of the matrix. Its dimension equals the rank of the matrix. By performing the elementary row operations (and obtaining another matrix from the first), the basis is transformed (the space does not change). These matrices have the same row-space. From the row-space point of view, linearly dependent rows can be omitted from the matrix.

A suitable simplification of a matrix can be obtained using elementary row operations. An *echelon canonical form* has the following properties:

1. Every leading term of a nonzero row is 1 (multiplication identity element from the field).
2. Every column containing such a leading term has all other elements 0 (addition identity element from the field).
3. The rows are ordered in such way that the leading term of the next row is to the right of the leading term in the preceding row. All zero rows (if any) are below the other rows.

This procedure corresponds to that one when solving the system of linear equations by the successive elimination of the variables.

If the rows of $n \times n$ matrix are linearly independent, matrix is *non-singular* (det $\neq 0$), its rank is n and at the end of procedure an *identity matrix* is obtained (corresponding the complete solution of the system of equations).

When multiplying an $m \times n$ matrix $[a_{ij}]$ by an $n \times p$ matrix $[b_{jk}]$ the $m \times p$ matrix $[c_{ik}]$ is obtained where c_{ik} is the inner product of the ith row of $[a_{ij}]$ by the kth column of $[b_{jk}]$.

Brief Introduction to Algebra I

The number of vectors (of n elements) in an n-dimensional vector space over GF (q) equals q^n. It can be shown (because the subspace is a subgroup under addition) that the number of vectors in a subspace must be $q^k (k \leq n)$. The dimension of the corresponding subspace is k.

Let V be the vector space of the dimension n. Let V_1 be the subspace of a dimension k. The subspace V_2 is called the *null space* of V_1 if every vector of V_2 is orthogonal to every vector of V_1. To assert the orthogonality of the subspaces it will suffice to assert the orthogonality of the corresponding bases (vectors of the basis), because all the other vectors are their linear combinations. In other words, if the bases of the orthogonal subspaces V_1 and V_2 (they are mutually orthogonal—null spaces for each other) are considered as a rows of the corresponding matrices ($\mathbf{M}_1$ and $\mathbf{M}_2$), their product is

$$\mathbf{M}_1 \mathbf{M}_2^T = \mathbf{0} \quad \text{or} \quad \mathbf{M}_2 \mathbf{M}_1^T = \mathbf{0}^T,$$

where the corresponding matrix types are: $\mathbf{M}_1(k \times n)$, $\mathbf{M}_2((n-k) \times n)$ and $\mathbf{0}$ ($k \times (n-k)$). By using the transposing (^{T}) the multiplication of rows of matrix $\mathbf{M}_1$ by the rows of matrix $\mathbf{M}_2$ is achieved. Therefore, null space V_2 for the subspace V_1 of dimension k must have the dimension $n - k$, if the dimension of the whole space V equals n.

Chapter 6
Cyclic Codes

Brief Theoretical Overview

Cyclic codes (*cyclic subspaces*) are one of the subclasses of linear codes. They are obtained by imposing on an additional strong structure requirement. A brief overview of the corresponding notions of abstract algebra comprising some examples and comments is at the end of this chapter. Here only the basic definitions are given. Thus imposed structure allows a successful search for good error control codes. Further, their underlying Galois field description leads to efficient encoding and decoding procedures. These procedures are algorithmic and computationally efficient. In the previous chapter a notion of ring was introduced. For any positive integer q, there is the *ring of integers* obtained by modulo-q operations. If q is a prime integer (p), Galois field GF(p) is obtained. The number of field elements is just p—all remainders obtained by division modulo-p. These fields are called *prime fields*. A subset of field elements is called a *subfield* if it is a field under the inherited operations. The original field can be considered as an *extension* field. Prime numbers cannot be factored and they have not nontrivial subfields. Consider now a polynomial over a field GF(q), i.e. the polynomial having the highest power $q - 1$, its coefficients being from GF(q). If its leading coefficient equals one, it is called a *monic* polynomial. For any monic polynomial $p(x)$ a *ring of polynomials* modulo $p(x)$ exists with polynomial addition and multiplication. For a binary **ground** field every polynomial is monic. If $p(x)$ is a *prime polynomial*, i.e. if it is irreducible (it cannot be further factored) and monic, then a *ring of polynomials modulo $p(x)$ is a field* (Galois field). In any Galois field the number of elements is a power of a prime. In fact, polynomial coefficients are taken from a prime field. For any prime (p) and any integer (m) there is a Galois field with p^m elements. These fields can have a nontrivial subfields. A *primitive element* of GF(q) is an element (usually denoted by α) such that every field element can be expressed as its power. Every Galois field has a primitive element. Therefore, a multiplicative group in Galois field is *cyclic*—all elements are obtained by exponentiation of the primitive element

(0 is not in the multiplicative group!). A ***primitive polynomial*** is a prime polynomial over GF(q) if it has a primitive element of the extension field as a root. The smallest positive integer e such that for any element g, $g^e = 1$ is called ***order*** of the element g. The order of a primitive element of GF(q) is $q - 1$. The prime polynomial of smallest degree over GF(q) such that it has an element of the extension field as a root is called the ***minimal polynomial*** of that element. It is unique. The elements of the extension field sharing the same minimal polynomial are called ***conjugates***. All roots of irreducible polynomial have the same order. If α is a primitive element, then all its conjugates are primitive elements as well. All non-zero elements of the extension field of GF(q) are the roots of $x^{q^m-1} - 1$. In the following exposition ground filed is a binary field ($x^n + 1$ is the corresponding polynomial). However, the same approach is valid for a nonbinary ground field, taking into account the corresponding relations in this case. Therefore, this polynomial can be uniquely factored using the minimal polynomials. Euclidean algorithm can be used for polynomials as well as for integers.

A vector is an ordered n-tuple of elements from some field. An equivalent is a polynomial where the elements are ordered as the polynomial coefficients. Therefore, vectors and polynomials can be considered as a different way to denote the same extension field elements.

For vector

$$v(a_0, a_1, \ldots, a_{n-1})$$

the corresponding polynomial is

$$v(x) = a_{n-1}x^{n-1} + \ldots + a_1x + a_0.$$

E.g. vector (110010) can be written as polynomial $x^5 + x^4 + x + 1$. It should be noted that this polynomial can be written as well as $1 + x + x^4 + x^5$ or $x^5 + x + x^4 + 1$, in fact in arbitrary order of powers, because the power of x denotes the position of a coefficient.

To multiplying a polynomial by x, corresponds a cyclic shift of vector coefficients to the right for one place.

$$x(a_0 + a_1x + \ldots + a_{n-1}x^{n-1}) = a_{n-1} + a_0x + \ldots + a_{n-2}x^{n-1}$$

because $x^n = 1$.

An ***ideal*** is a subset of elements of a ring if it is a subgroup of the additive group and a result of multiplication of an ideal element by any ring element is in the ideal. In an integer ring a set of integer is an ideal if and only if it consists of all multiples of some integer. Similarly, a set of polynomials is an ideal (in a polynomial ring) if and only if it consists of all multiples of some polynomial. It was shown that a subspace is the cyclic subspace if and only if it is an ideal.

Let the power of $g(x)$ is $n - k$, then linearly independent residue classes (polynomials) are

$$0, g(x), xg(x), \ldots, x^{k-1}g(x).$$

The power of polynomial $x^k g(x)$ is n, and after division by $x^n - 1$ (in binary field $-1 = 1!$) it will be represented by the corresponding residue.

Now, cyclic code can be defined. Consider vector space dimension n (length of code vectors). **Subspace of this space is called cyclic (code) if for any vector**

$$v(a_0, a_1, \ldots, a_{n-1})$$

vector

$$v_1(a_{n-1}, a_0, a_1, \ldots, a_{n-2})$$

obtained by cyclic permutation of its elements is as well in a cyclic subspace. Therefore, cyclic codes are a subclass of linear block codes as shown in Fig. 6.1.

Furthermore, from a mathematical point of view it can be said that a *cyclic code is an ideal in polynomial algebra modulo polynomial $x^n - 1$ over the coefficients field*. It is specified completely by the corresponding polynomial that divides $x^n + 1$ (for a binary ground field $-1 = 1$ and n—a code word length).

In fact, $x^n + 1$ can be factored comprising all *irreducible* polynomials. Every such polynomial as well as all their products can be *generator polynomials* $g(x)$ of a cyclic code. The *parity check polynomial* $h(x)$ is obtained from $x^n + 1 = g(x)h(x)$. In Problem 6.1 it is analyzed in details for $n = 7$. As a consequence, cyclic codes can be encoded and decoded by multiplying and dividing the corresponding polynomials. It provides for efficient encoding and decoding procedures. Cyclic codes can be shortened as well as obtained in the systematic form.

Now, the vector orthogonality and "polynomial orthogonality" will be compared. If $a(x)b(x) = 0$, then the vector corresponding to $a(x)$ is orthogonal to the vector corresponding to $b(x)$ with the order of its components reversed as well as to every cyclic shift of this vector. Of course, the multiplication is commutative, therefore, $a(x)b(x) = 0 \Rightarrow b(x)a(x) = 0$ and the same rule can be applied keeping

Fig. 6.1 Cyclic codes as a subclass of linear block codes

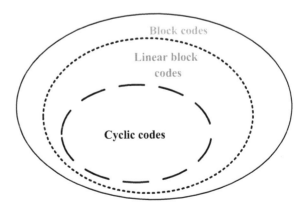

components of $b(x)$ in normal order and reverting and shifting the components of $a(x)$.

The following short analysis shows the extreme simplicity in description of cyclic codes:

- For a complete description of (n, k) block code all code words (2^k) should be known.
- For a complete description of linear block code with the same parameters (n, k), its basis should be known—k code words of length n (generator matrix).
- For a complete description of the corresponding cyclic code it is sufficient to know only one code word of length n. All others code words are obtained by its cyclic shifts and as the linear combinations of these shifts.

There are some ***trivial cyclic codes***. Repetition code $(n, 1)$ is trivial (only two code words), as well as the code (n, n) (full vector space, but no error control!). The binary code where the number of ones is even is trivial as well. E.g. code (3, 2) which has code words (000), (011), (101) i (110), and the code (4, 3) which has code words (0000), (0011), (0101), (0110), (1001), (1010), (1100) i (1111). It is interesting that for some codewords length, e.g. for $n = 19$, binary trivial codes are in the same time unique existing cycling codes.

Consider polynomial $x^7 + 1$ over GF(2). Here

$$x^7 + 1 = (1+x)(1+x+x^3)(1+x^2+x^3).$$

Let the generator polynomial is

$$g(x) = 1 + x^2 + x^3$$

The cyclic code (7, 4) is obtained ($n = 7$, $n - k = 3$).

One code word can be obtained by using coefficients of generator polynomial, and the other three linearly independent by its cyclical shifting

$$g(x) = (1011000)$$
$$xg(x) = (0101100)$$
$$x^2g(x) = (0010110)$$
$$x^3g(x) = (0001011)$$

and the corresponding generator matrix is

$$G = \begin{bmatrix} 1 & 0 & 1 & 1 & 0 & 0 & 0 \\ 0 & 1 & 0 & 1 & 1 & 0 & 0 \\ 0 & 0 & 1 & 0 & 1 & 1 & 0 \\ 0 & 0 & 0 & 1 & 0 & 1 & 1 \end{bmatrix}.$$

Brief Theoretical Overview

The other code words (vectors, i.e. polynomials) are their linear combinations. In Fig. 6.2 all code words are shown as well as the way to obtain them using polynomials corresponding to the rows of generator matrix. There are total seven code words forming a "cyclic set" (all possible cyclic shifts of the word corresponding to $g(x)$). Hamming weight is $d = 3$. These words are linearly independent and by summing two words from this set, the code word having Hamming weight $d = 4$ is obtained. By its cyclic shifts the other "cyclic set" is obtained having seven code words as well. Including combinations (0000000) ans (1111111) a complete code (16 code words) is obtained. It is equivalent to Hamming (7, 4) code which has the same weight spectrum.

Parity check polynomial is

$$h(x) = (1+x)(1+x+x^3) = 1+x^2+x^3+x^4.$$

It is here obtained by multiplication of the other factors, but it could have been obtained by division as well, i.e. $h(x) = (x^7 + 1) : g(x)$.

Polynomial $h(x)$ forms its code (ideal)

$$h(x) = (1011100)$$
$$xh(x) = (0101110)$$
$$x^2h(x) = (0010111).$$

If the product of two polynomilas equals zero, i.e. if

$$g(x) \cdot h(x) = x^n - 1 = 0(\mod(x^n - 1)),$$

then the corresponding coefficients vectors are orthogonal (the coefficients of one of them should be taken in inverse order). Parity-check matrix is here

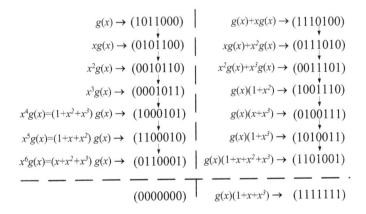

Fig. 6.2 Coded words of cyclic code (7, 4)

$$H = \begin{bmatrix} 0 & 0 & 1 & 1 & 1 & 0 & 1 \\ 0 & 1 & 1 & 1 & 0 & 1 & 0 \\ 1 & 1 & 1 & 0 & 1 & 0 & 0 \end{bmatrix}.$$

It is easy to verify

$$GH^T = 0.$$

It should be noticed that the rows in this matrix are as well the cyclic permutation of one row. Taking H ($h(x)$) as a generator matrix (polynomial) a **dual code** is obtained.

The same field can be obtained modulo any primitive polynomial. Consider FG (2^3). Two primitive polynomials are $x^3 + x + 1$ and $x^3 + x^2 + 1$:

	x^3+x+1				x^3+x^2+1			
	(primitive element α)				(primitive element $\beta=\alpha^3$)			
(100)	$\alpha^0=$	1			$\beta^0=$	1		
(010)	$\alpha^1=$		α		$\beta^1=$		β	
(001)	$\alpha^2=$			α^2	$\beta^2=$			β^2
(110)	$\alpha^3=$	1	$+\ \alpha$		$\beta^3=$	1		$+\ \beta^2$
(011)	$\alpha^4=$		α	$+\ \alpha^2$	$\beta^4=$	1	$+\ \beta$	$+\ \beta^2$
(111)	$\alpha^5=$	1	$+\ \alpha$	$+\ \alpha^2$	$\beta^5=$	1	$+\ \beta$	
(101)	$\alpha^6=$	1		$+\ \alpha^2$	$\beta^6=$		β	$+\ \beta^2$
(100)	$\alpha^7(\alpha^0)=1$				$\beta^7(\beta^0)=1$			

The finite fields with the same number of elements are isomorphic—they have the unique operation tables. The difference is only in the way of naming their elements.

The fact that in both matrix the next rows are only the cyclic shifts of preceding rows contributes significantly to the elegant and economic encoder and decoder construction.

From

$$g(x) \cdot h(x) = x^n - 1 = 0 \,(\mathrm{mod}(x^n-1)),$$

it is obvious that codes generated by $g(x)$ and $h(x)$ are dual codes. Therefore, to generate cyclic code word from information polynomial $i(x)$ (power $k-1$, because the are k information bits) one should obtain code polynomial (power $n-1$), divisible by generator polynomial $g(x)$. At the receiver, the received word $u(x)$ should be divided by $g(x)$ and if the remainder equals zero, the conclusion can be drawn that there were no errors. Also, $u(x)$ could b multiplied by $h(x)$ and verify does the result (not remainder!) equals zero (modulo $x^n - 1$!).

Brief Theoretical Overview

Hardware implementation of cyclic codes is based on using linear switched circuits to multiply and divide polynomials. Corresponding shift registers have the number of cells equal to the polynomial power. Therefore, for using $g(x)$ $n - k$ cells are needed, and for $h(x)$ k cells are needed. It is not important for short codes ($n - k$ and k do not differ significantly), but for long codes more economical is to divide. It is just the reason to implement division for CRC **Cyclic Redundancy Check**) procedure, to be considered later.

The next question is how to generate a *systematic cyclic code*. Let the information (bit) polynomial is

$$i(x) = i_{k-1}x^{k-1} + \ldots + i_1 x + i_0$$

and the generator polynomial

$$g(x) = g_{n-k}x^{n-k} + \ldots + g_1 x + g_0.$$

By multiplying, the following is obtained

$$v(x) = i(x)g(x) = v_{n-1}x^{n-1} + \ldots + v_1 x + v_0.$$

It must be code polynomial, because one factor is $g(x)$. To obtain systematic code, information polynomial should be in advance multiplied by x^{n-k}, yielding new information polynomial

$$i*(x) = x^{n-k}i(x) = i_{k-1}x^{n-1} + \ldots + i_1 x^{n-k-1} + i_0 x^{n-k}.$$

This result should be divided by $g(x)$ (power $n - k$). It is obvious that the power of remainder will be $n - k - 1$, having in total $n - k$ coefficients

$$r(x) = r_{n-k-1}x^{n-k-1} + \ldots + r_1 x + r_0.$$

Code word is obtained by the following subtraction

$$v(x) = i^*(x) - r(x) = x^{n-k}i(x) - r(x).$$

The corresponding code vector is

$$(i_{k-1}, \ldots, i_1, i_0, r_{n-k-1}, \ldots, r_1, r_0),$$

and information bits are not changed, while the rest of bits are parity-checks. Therefore, a systematic code is obtained. Here, parity-checks are obtained in a more elegant way comparing to classic way to introduce parity-checks. From the "hardware point of view" it means that the information bits are put in the upper part of shift register and parity-checks in the rest of cells.

Golay code (23, 12) (Problem 6.5) is a very known cyclic code. It was invented before the invention of cyclic codes and later it was recognized as a cyclic code. It satisfies the Hamming bound with equality, it is the **perfect** code.

In communications the **Cyclic Redundancy Check** (CRC) (Problems 6.6, 6.7, 6.8 and 6.9) is often used, where the cyclic code is applied to detect the errors. In its systematic version, the number of information bits is not fixed, but it can vary in some broad range. A main advantage of code shortening is a possibility to adjust the code word length to a length of the message entering the encoder. In such a way, the same generator polynomial can be used for encoding the information words of various lengths. CRC codes are generally used for a channels which have a low noise power which (except for rare packet errors) can be modeled as a BSC where the channel error probability is sufficiently small ($p \ll 10^{-2}$), and the codes are optimized for this channel type.

CRC procedure is based on forming systematic cyclic code. The information bits transmitted in one frame are encoded using cyclic codes as follows

$$c(x) = i(x)x^{n-k} + rem\left\{\frac{i(x)x^{n-k}}{g(x)}\right\} = i(x)x^{n-k} + r(x),$$

where $rem\{\cdot\}$ denotes the corresponding remainder. The information and the parity-check bits are separated. It can be conceived that, to transmitted information bits, CRC appendix is "attached". Its dimension is usually substantially shorter than the rest of he packet ($n - k \ll k$). During the transmission an error polynomial $e(x)$ is superimposed. At the receiver, the received sequence is divided by $g(x)$. The corresponding remainder is

$$r'(x) = rem\left\{\frac{c(x) + e(x)}{g(x)}\right\}.$$

If the remainder equals zero, it is concluded that there were no errors. In the opposite case, the retransmission can be requested. Problems 6.6, 6.7, 6.8 and 6.9 deal in details with CRC procedure showing as well encoder and decoder hardware realization.

Cyclic codes are a subclass of linear codes and can be described as well by matrices. Of course, as said earlier, the polynomial description is more elegant because the generator polynomial coefficients just form one row of generator matrix, other rows are its cyclic shifts. Up to now it seemed that one can firstly chose a generator polynomial and after that has to determine the number of correctable errors. However, a special class of cyclic codes guarantees the required number of correctable errors in advance. These are **BCH codes** (discovered separately by Bose and Ray-Chaudhuri at one side, and by Hocquenghem at the other side) (Problems 6.10, 6.11, 6.12 and 6.13). They are obtained using a constructive way, by defining generator polynomial roots. Simply, to correct e_c errors, the generator polynomial should have $2e_c$ roots expressed as the consecutive powers of the primitive element. If it starts from α, the code is called **primitive**, otherwise, it is

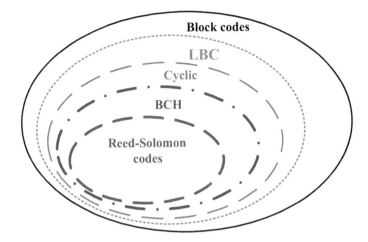

Fig. 6.3 BCH and RS codes as a subset of cyclic codes

not a primitive one. Of course, to find the generator polynomial, the corresponding conjugate elements should be taken into account and sometimes a number of correctable errors is augmented (the conjugates continue a series of the consecutive powers). These codes are very known cyclic codes. In the Fig. 6.3 their position in the family of linear block codes is shown. **Reed-Solomon (RS) codes** (considered later) are a special subset of nonbinary BCH codes.

Any cyclic code can be specified using generator polynomial $g(x)$ roots. Let these roots from the field extension are

$$\alpha_1, \alpha_2, \ldots, \alpha_r.$$

The received polynomial (vector) $u(x)$ is a code word if and only if $\alpha_1, \alpha_2, \ldots, \alpha_r$ are roots of $u(x)$—what is equivalent to the fact that $u(x)$ is divisible by. $g(x)$. However, the coefficients are from ground field and $u(x)$ should have all others conjugate elements as a roots—meaning that $u(x)$ must be divisible by the other minimal polynomials

$$m_1(x), m_2(x), \ldots, m_r(x).$$

In fact, generator polynomial is obtained as

$$g(x) = \text{LCM}\{m_1(x), m_2(x), \ldots, m_r(x)\}$$

because some roots can have the same minimal polynomial. Of course, $x^n - 1$ is divisible by $g(x)$.

It can be shown that for every positive integer m_0 and $d_{\min}$ (minimal Hamming distance) exists BCH code [generator polynomial $g(x)$], only if it is the minimum power polynomial with the coefficients from GF(r), which has the roots from the extension field

$$\alpha_0^{m_0}, \alpha_0^{m_0+1}, \ldots, \alpha_0^{m_0+d_{min}-2}$$

where α_0 is some element from extension field. Code word length is LCM of the roots order.

Generally, it is not necessary for r to be a prime. It can be as well power of the prime. It is only important that $GF(r)$ exists. Used symbols are from this field, but the generator polynomial roots are from the extension field. Therefore, coefficients can be from $GF(p^n)$ (p—prime) and the roots from $GF((p^n)^k)$. Of course, some roots can be from the ground field, but it is the subfield of the extension field.

Consider the following example. Let $p = 2$ (binary ground field) and let GF(2^3) is generated by primitive polynomial $p(x) = x^3 + x + 1$. Find cyclic code with the minimal number of parity-check bits, over this field, if one root of generator polynomial is $\alpha_1 = \alpha$.

Complete GF(2^3) is shown in Table 6.1 (it should be noticed that $\alpha^n = \alpha^{n-7}$ and that identity addition element is omitted (it belongs to the field and sometimes is denoted as $\alpha^{-\infty}$). Groups of conjugate elements (roots) are denoted with symbols A, B and C.

Conjugate elements for α are α^2 and α^4, yielding

$$g(x) = (x+\alpha)(x+\alpha^2)(x+\alpha^4) = x^3 + x + 1.$$

The same generator polynomial is obtained for $\alpha_1 = \alpha^2$ and for $\alpha_2 = \alpha^4$.

Parity-check polynomial is obtained from $g(x)h(x) = 1$, i.e. $h(x) = (x+1)(x^3 + x^2 + 1)$.

If for roots of generator polynomial $\alpha_1 = \alpha$ and $\alpha_2 = \alpha^3$ are chosen, two different minimal polynomials are obtained

$$m_1(x) = (x+\alpha)(x+\alpha^2)(x+\alpha^4) = x^3 + x + 1$$
$$m_2(x) = (x+\alpha^3)(x+\alpha^6)(x+\alpha^5) = x^3 + x^2 + 1 (\alpha^{12} = \alpha^5)$$

Table 6.1 Galois field GF(2^3)

Exponential equivalent	Polynomial equivalent			Binary equivalent	Group of conjugate roots
α^0			1	001	A
α^1		α		010	B
α^2	α^2			100	B
α^3		$\alpha+$	1	011	C
α^4	α^2+	α		110	B
α^5	α^2+	$\alpha+$	1	111	C
α^6	α^2+		1	101	C

and

$$g(x) = m_1(x)m_2(x) = (x^3+x+1)(x^3+x^2+1) = x^6+x^5+x^4+x^3+x^2+x+1,$$

yielding

$$h(x) = x+1.$$

Further, for $\alpha_1 = 1$ and $\alpha_2 = \alpha$, the corresponding minimal polynomials are:

$$m_1(x) = x+1$$
$$m_2(x) = (x+\alpha)(x+\alpha^2)(x+\alpha^4) = x^3+x+1$$

yielding

$$g(x) = m_1(x)m_2(x) = (x+1)(x^3+x+1) = x^4+x^3+x^2+1$$

and

$$h(x) = x^3+x^2+1.$$

Generally, consider $GF(2^m)$. It can be shown that for any two positive integers m ($m \geq 3$) and e_c ($e_c < 2^{m-1}$) exists binary BCH code having the following characteristic:

- Code word length $\quad n = 2m - 1$
- Number of control bits $\quad n - k \leq me_c$
- Minimal Hamming distance: $\quad d_{min} \geq 2e_c + 1.$

This code can correct all combinations of e_c of smaller number of errors. Generator polynomial is a minimal power polynomial which has for roots successive powers of α:

$$\alpha, \alpha^2, \alpha^3, \ldots, \alpha^{2e_c}.$$

Some of conjugate elements can continue a series of successive powers yielding the **obtained** Hamming distance greater than the **projected** one (Problem 6.11).

The above considered codes ($m_0 = 1$) are called **primitive** BCH codes. However, the series of successive powers can start from any nonprimitive element (β). It is only important that $2e_c$ successive powers of this element ($\beta, \beta^2, \ldots$) are the roots of generator polynomial. In such way **nonprimitive** BCH codes are obtained.

Generally, error correction using cyclic code is not an easy problem. Of course, the syndrome components can be found as the received word values for code roots. Syndromes can be obtained as well from **error locators** (locator $X_i = \alpha^k$ corresponds to ith error which occurred at the kth position in a code word) (Problem 6.12). The corresponding equations are called **power-sum symmetric functions**

(Problem 6.12), they form a system of nonlinear algebraic equations of more variables and they are not suitable for solution by a direct procedure. **Peterson** (Problems 6.12 and 6.13) showed that, by introducing a notion of **error locator polynomial**, these equations can be transformed into a series of linear equations. It is a polynomial which has as the roots the error locators. In the special case, when all operations are in the binary field and when there are exactly $v = e_c$ errors, a system is additionally simplified. The next algorithm was proposed by **Berlekamp** (Problems 6.13). To apply his algorithm, firstly a **syndrome polynomial** is defined. Further, an **error magnitude polynomial** is formed. In a binary case all errors have the same **magnitude** (1). It is interesting that for nonbinary (RS) codes. Berlekamp algorithm can now be defined as a procedure for finding coefficients of special polynomial, by solving the series of smaller problems. Berlekamp algorithm is more difficult to understand than Peterson approach, but it provides for the substantially efficacy implementation. A complexity of Peterson technique increases with the square of the number of corrected errors, having efficient application for binary BCH decoders correcting a small number of errors. On the other hand, the Berlekamp algorithm complexity increases linearly, allowing forming efficacy decoders correcting more tens of errors.

Reed-Solomon codes are an important class of nonbinary cyclic codes (Problems 6.14, 6.15, 6.16, 6.17, 6.18 and 6.19). They do not use symbols from GF(2), but from some field extension. They are discovered in 1960 but it was soon recognized that they are a special case of nonbinary BCH codes. Here code symbols and generator polynomial roots are from the same field. Generally, code symbols are from $GF(q)$ (q primitive or primitive power) and the roots are from $GF(q^m)$. Therefore, here $m = 1$. If α is a primitive element, code word length is

$$n = q^m - 1 = q - 1$$

and RS codes are relatively "short". Minimal polynomial for any element β is just

$$m_\beta(x) = x - \beta.$$

Generator polynomial of RS code correcting e_c (symbols!), if one starts from α, is

$$g(x) = (x-\alpha)(x-\alpha^2)\ldots(x-\alpha^{2e_c}),$$

its power is always $2e_c$ because conjugate elements do not exist. Number of control symbols is $n - k = 2e_c$ and a code word is obtained by multiplication of information polynomial and generator polynomial.

In original paper the case $q = 2^b$ was considered (binary code for alphanumerical characters), i.e. field $GF(2^b)$. Minimum Hamming distance for RS (n, k) code is

$$d = n - k + 1,$$

satisfying Singleton bound with equality. Therefore, it is a MDS code. In other words, for a specified (n, k), any other code can not have greater minimum Hamming distance than RS code. For $q = 2^b$, RS code parameters are:

- code word length: $n = b(2^b - 1)$;
- number of control bites: $n - k = 2be_c$ (e_c—number of bit errors);
- code words number: $|C| = 2^{mk}$ [there are 2^{mk} information polynomials, each of k coefficients can be any field element (2^m)].

Let power of $g(x)$ is $n - k$, (k—number of information bits). If $g(x)$ generates linear cyclic code over $GF(2^r)$, the generator matrix is

$$G = \begin{bmatrix} g(x) \\ xg(x) \\ \vdots \\ x^{k-1}g(x) \end{bmatrix}.$$

RS code can be shortened as well.

Consider $GF(2)$ and the extension field $GF(2^b = 4)$, obtained modulo polynomial $x^2 + x + 1$ where the symbols are 0, 1, α and α^2. Find RS codes for correcting one and two errors.

Code (3, 2) has the generator polynomial

$$g(x) = x - a.$$

Corresponding $16 (= 2^4)$ code words are

000   101 $\alpha 0 \alpha$ $\alpha^2 0 \alpha^2$ 011   110 $\alpha 1 \alpha^2$ $\alpha^2 1 \alpha$
$0\alpha\alpha$ $1\alpha\alpha^2$ $\alpha\alpha 0$ $\alpha^2 \alpha 1$ $0\alpha^2\alpha^2$ $1\alpha^2\alpha$ $\alpha\alpha^2 1$ $\alpha^2\alpha^2 0$

yielding code with one "parity-check" ($\alpha^2 = \alpha + 1$).

For (3, 1) code, generator polynomial is ($\alpha^3 = 1$)

$$g(x) = (x - \alpha)(x - \alpha^2) = x^2 - (\alpha + \alpha^2)x + \alpha^3 = x^2 + x + 1,$$

and, repetition code is obtained

000   111 $\alpha\alpha\alpha$ $\alpha^2\alpha^2\alpha^2$.

For a different (3, 1) code, generator polynomial is ($\alpha^0 = 1$)

$$g(x) = (x - 1)(x - \alpha) = x^2 - (\alpha + 1)x + \alpha = x^2 + \alpha^2 x + \alpha,$$

and the code words are

$$000 \quad 1\alpha^2\alpha \quad \alpha 1\alpha^2 \quad \alpha^2\alpha 1.$$

Consider now field $GF(2^b = 8)$ (Table 6.1). Starting from the primitive element α, find RS code for correcting one error. There is no need to find the conjugate elements. The generator polynomial roots are α and α^2

$$g(x) = (\alpha + x)(\alpha^2 + x) = \alpha^3 + \alpha^4 x + x^2,$$

and $n = 7$, $n - k = 2$ yielding RS code (7, 5), the number of code words is $|C| = 2^{3(7-2)} = 32768$. Generator matrix (it can be also given in a binary form (15 rows, 7 columns)) is

$$G = \begin{bmatrix} \alpha^3 & \alpha^4 & 1 & 0 & 0 & 0 & 0 \\ 0 & \alpha^3 & \alpha^4 & 1 & 0 & 0 & 0 \\ 0 & 0 & \alpha^3 & \alpha^4 & 1 & 0 & 0 \\ 0 & 0 & 0 & \alpha^3 & \alpha^4 & 1 & 0 \\ 0 & 0 & 0 & 0 & \alpha^3 & \alpha^4 & 1 \end{bmatrix}.$$

Gorenstein and *Zierler* (Problem 6.17) modified Peterson algorithm to decode the nonbinary codes. **Berlekamp-Massey** algorithm for RS codes decoding starts from a fact that the error position polynomial does not depend on error magnitudes in the code word. It makes possible to use Berlekamp algorithm to find the error position, and later on to determine the error magnitudes and to do their correction. *Forney* algorithm (Problem 6.18) for roots finding allows to determine the error magnitude if its location is known, as well as error locator polynomial and magnitude error polynomial, but to find a magnitude of error, the position of other errors are not explicitly used. The *erasure* of symbols is a simplest way of soft decision. For nonbinary codes the location of erased symbol is known, but not the error magnitude. Here *erasure locator polynomial* is calculated by using known erasure locators. Generally, at the beginning the error locator polynomial and be found [e.g. by using Berlekamp-Massey (Problem 6.18) or *Euclidean algorithm* (Problem 6.19)], and then a main equation for errors and erasures decoding is used to obtain a final solution.

Problems

Problem 6.1 Consider the linear block codes, word code length $n = 7$.

(a) How many cyclic codes can be constructed which have this word length? Find their generator polynomials.
(b) For one of the possible generator polynomials find all code words and code weight spectrum. Comment the code possibilities concerning the error correction.

(c) Whether $g(x) = 1 + x^3 + x^4$ can be the generator polynomial of cyclic (7, 3) code?
(d) Form a systematic generator matrix for the code with the generator polynomial $g(x) = 1 + x$.

Solution
Consider a vector space whose elements are all vectors of the length n over GF(q). A subspace of this space is called **cyclic** if for any vector $(c_0, c_1, \ldots, c_{n-1})$, a vector $(c_{n-1}, c_0, c_1, \ldots, c_{n-2})$, obtained by cyclic permutation of its coefficients also belongs to this subspace. For an easier analysis a code vector $(c_0, c_1, \ldots, c_{n-1})$ can be represented as well as a polynomial, which has the coefficients equal to the vector elements [34].

$$c(x) = c_{n-1}x^{n-1} + \ldots + c_1 x + c_0.$$

A vector subspace is cyclic if it consists of all polynomials divisible by a **generator polynomial** $g(x)$, which is in the same time a divisor of polynomial $x^n - 1$, where n is a code word length. In a binary field ($q = 2$) the subtraction is equivalent to the addition, and for binary cyclic codes it can be written

$$x^n + 1 = g(x)h(x),$$

where $h(x)$ is a **parity-check polynomial**. If a degree of polynomial $g(x)$ is $n - k$, a subspace dimension is k, the power of polynomial $h(x)$ is k and a cyclic code (n, k) is obtained.

Because of this the first step to find the generator polynomial of binary code is a factorization of polynomial $x^n + 1$ according to

$$x^n + 1 = \prod_{i=1}^{l} m_i(x),$$

where $m_1(x), m_2(x), \ldots$ are **irreducible (minimal) polynomials** over field GF(2).

(a) For $n = 7$ the factorization is as follows

$$x^7 + 1 = (1+x)\left(1 + x + x^3\right)\left(1 + x^2 + x^3\right),$$

the procedure to find the minimal polynomials will be described in details later.

According to previously given rules, a generator polynomial can be any combination of minimal polynomials. There are three irreducible factors and it is possible to form $2^{n-k} - 1 = 7$ generator polynomials (polynomial $g(x) = x^7 + 1$ would generate a code word without information bits), each one defining one cyclic code. Taking into account that the generator polynomial highest power is always $n - k$, all possible generator polynomials and the corresponding code parameters are:

(1) $g(x) = 1 + x \Rightarrow n - k = 1$, (7, 6) code
(2) $g(x) = 1 + x + x^3 \Rightarrow n - k = 3$, (7, 4) code
(3) $g(x) = 1 + x^2 + x^3 \Rightarrow n - k = 3$, (7, 4) code
(4) $g(x) = (1 + x)(1 + x + x^3) \Rightarrow n - k = 4$, (7, 3) code
(5) $g(x) = (1 + x)(1 + x^2 + x^3) \Rightarrow n - k = 4$, (7, 3) code
(6) $g(x) = (1 + x + x^3)(1 + x^2 + x^3) \Rightarrow n - k = 6$, (7, 1) code
(7) $g(x) = 1 \Rightarrow n - k = 0$, (7, 7) code

In the first case only one parity-check bit is added corresponding to a simple parity-check, and a parity-check polynomial is $h(x) = (1 + x + x^3)(1 + x^2 + x^3)$. In the sixth case, the obtained code is a sevenfold repetition (n-fold repetition is always a cyclic code!). In the last case the parity-check bits are not added, the information sequence is unchanged and a code word is the same as the information word.

(b) Previous analysis shows that one possible cyclic code generator polynomial is

$$g_1(x) = (1+x)\left(1+x+x^3\right) = 1+x^2+x^3+x^4$$
$$= g_0 + g_1 x + g_2 x^2 + g_3 x^3 + g_4 x^4.$$

For arbitrary generator polynomial the elements of the first row of generator matrix are determined by its coefficients (starting from the lowest power), while every next row is obtained by shifting the row above for one place at the right as follows

$$G_1 = \begin{bmatrix} g_0 & g_1 & g_2 & g_3 & g_4 & 0 & 0 \\ 0 & g_0 & g_1 & g_2 & g_3 & g_4 & 0 \\ 0 & 0 & g_0 & g_1 & g_2 & g_3 & g_4 \end{bmatrix} = \begin{bmatrix} 1 & 0 & 1 & 1 & 1 & 0 & 0 \\ 0 & 1 & 0 & 1 & 1 & 1 & 0 \\ 0 & 0 & 1 & 0 & 1 & 1 & 1 \end{bmatrix}.$$

In a linear block code there is always all zeros word. The second code word can be red from generator polynomial coefficients (it corresponds to the combination (100) at the encoder input) and other six can be obtained as its cyclic shifts.

$$(1011100) \rightarrow (0101110) \rightarrow (0010111) \rightarrow (1001011) \rightarrow (1100101)$$
$$\rightarrow (1110010) \rightarrow (0111001).$$

It is obvious that a code have one word with weight zero and seven words with weight four ($a(0) = 1$, $a(4) = 7$), and it can correct one error and detect two errors in the code word. Reader should verify to which information words correspond written code words (it can be easily found by using a generator matrix).

Problems

(c) Suppose at a moment that

$$g_2(x) = 1 + x^3 + x^4$$

is a generator polynomial of a cyclic code (7, 3). The polynomial $g(x)$ highest power is $n - k = 4$ and it seems possible that a cyclic code is obtained. If $g(x)$ is cyclic code generator polynomial, then $xg(x)$ i $x^2g(x)$ must be also the code polynomials and a generator matrix is

$$G_2 = \begin{bmatrix} 1 & 0 & 0 & 1 & 1 & 0 & 0 \\ 0 & 1 & 0 & 0 & 1 & 1 & 0 \\ 0 & 0 & 1 & 0 & 0 & 1 & 1 \end{bmatrix}.$$

Code words of this linear block code are obtained multiplying all possible three-bit information vectors by a generator matrix, given in Table 6.2.

It is clear that by cyclic shifts of the code word $c^{(1)} = (000000)$ no other code word can be obtained. Cyclic shifts of a word (1001100) yield $c^{(5)} = (1001100) \rightarrow c^{(3)} = (0100110) \rightarrow c^{(2)} = (0010011) \rightarrow (\underline{1001001}) \rightarrow (\underline{1100100}) \rightarrow (\underline{0110010}) \rightarrow (\underline{0011001})$, and the last four combinations are not code words. Because of the existence of at least one code word whose all cyclic shifts are not code words, this code obviously is not cyclic.

It can be noted that the same conclusion could be drawn from the fact that the supposed generator polynomial $g(x) = 1 + x^3 + x^4$ does not divide a polynomial $x^7 + 1$ without remainder, because it cannot be written as a product of its minimal polynomials. It should be noted as well that the spectrum of a code defined by generator matrix G_2 is

$$a(d) = \begin{cases} 1, d = 0, \\ 3, d = 3, \\ 2, d = 4, \\ 1, d = 5, \\ 1, d = 6, \end{cases}$$

Table 6.2 Transformations defining one block code

Word ordinal number	Information word	Code word
1	0 0 0	0 0 0 0 0 0 0
2	0 0 1	0 0 1 0 0 1 1
3	0 1 0	0 1 0 0 1 1 0
4	0 1 1	0 1 1 0 1 0 1
5	1 0 0	1 0 0 1 1 0 0
6	1 0 1	1 0 1 1 1 1 1
7	1 1 0	1 1 0 1 0 1 0
8	1 1 1	1 1 1 1 0 0 1

and this code has a minimum Hamming distance $d_{min} = 3$, it detects one error and can correct it as well. However, the code is not cyclic and cannot be defined by a generator polynomial $g_2(x)$, i.e. the transition from $g_2(x)$ to $\mathbf{G}_2$, was not correct.

(d) Starting from a generator polynomial $g(x) = 1 + x = g_0 + g_1 x$ by previously described procedure a starting nonsystematic generator matrix of this cyclic code is obtained

$$\mathbf{G} = \begin{bmatrix} g_0 & g_1 & 0 & 0 & 0 & 0 & 0 \\ 0 & g_0 & g_1 & 0 & 0 & 0 & 0 \\ 0 & 0 & g_0 & g_1 & 0 & 0 & 0 \\ 0 & 0 & 0 & g_0 & g_1 & 0 & 0 \\ 0 & 0 & 0 & 0 & g_0 & g_1 & 0 \\ 0 & 0 & 0 & 0 & 0 & g_0 & g_1 \end{bmatrix} = \begin{bmatrix} 1 & 1 & 0 & 0 & 0 & 0 & 0 \\ 0 & 1 & 1 & 0 & 0 & 0 & 0 \\ 0 & 0 & 1 & 1 & 0 & 0 & 0 \\ 0 & 0 & 0 & 1 & 1 & 0 & 0 \\ 0 & 0 & 0 & 0 & 1 & 1 & 0 \\ 0 & 0 & 0 & 0 & 0 & 1 & 1 \end{bmatrix},$$

the addition of all six rows and putting the result in the first row yields

$$\mathbf{G}^{(I)} = \begin{bmatrix} 1 & 1 & 0 & 0 & 0 & 0 & 0 \\ 0 & 1 & 1 & 0 & 0 & 0 & 0 \\ 0 & 0 & 1 & 1 & 0 & 0 & 0 \\ 0 & 0 & 0 & 1 & 1 & 0 & 0 \\ 0 & 0 & 0 & 0 & 1 & 1 & 0 \\ 0 & 0 & 0 & 0 & 0 & 1 & 1 \end{bmatrix} \rightarrow \mathbf{G}^{(II)} = \begin{bmatrix} 1 & 0 & 0 & 0 & 0 & 0 & 1 \\ 0 & 1 & 1 & 0 & 0 & 0 & 0 \\ 0 & 0 & 1 & 1 & 0 & 0 & 0 \\ 0 & 0 & 0 & 1 & 1 & 0 & 0 \\ 0 & 0 & 0 & 0 & 1 & 1 & 0 \\ 0 & 0 & 0 & 0 & 0 & 1 & 1 \end{bmatrix}.$$

Further, in matrix $\mathbf{G}^{(II)}$ the rows 2–6 are added and the result is put into the second row of a new matrix $\mathbf{G}^{(III)}$, where the rows 3–6 are added and put in the next one

$$\mathbf{G}^{(II)} = \begin{bmatrix} 1 & 0 & 0 & 0 & 0 & 0 & 1 \\ 0 & 1 & 1 & 0 & 0 & 0 & 0 \\ 0 & 0 & 1 & 1 & 0 & 0 & 0 \\ 0 & 0 & 0 & 1 & 1 & 0 & 0 \\ 0 & 0 & 0 & 0 & 1 & 1 & 0 \\ 0 & 0 & 0 & 0 & 0 & 1 & 1 \end{bmatrix} \rightarrow \mathbf{G}^{(III)} = \begin{bmatrix} 1 & 0 & 0 & 0 & 0 & 0 & 1 \\ 0 & 1 & 0 & 0 & 0 & 0 & 1 \\ 0 & 0 & 1 & 1 & 0 & 0 & 0 \\ 0 & 0 & 0 & 1 & 1 & 0 & 0 \\ 0 & 0 & 0 & 0 & 1 & 1 & 0 \\ 0 & 0 & 0 & 0 & 0 & 1 & 1 \end{bmatrix}$$

$$\rightarrow \mathbf{G}^{(IV)} = \begin{bmatrix} 1 & 0 & 0 & 0 & 0 & 0 & 1 \\ 0 & 1 & 0 & 0 & 0 & 0 & 1 \\ 0 & 0 & 1 & 0 & 0 & 0 & 1 \\ 0 & 0 & 0 & 1 & 1 & 0 & 0 \\ 0 & 0 & 0 & 0 & 1 & 1 & 0 \\ 0 & 0 & 0 & 0 & 0 & 1 & 1 \end{bmatrix}.$$

The procedure, which is in fact the Gaussian elimination method, is repeated two times more to form the fourth and the fifth row, and a generator matrix finally has a form

$$G^{(IV)} = \begin{bmatrix} 1 & 0 & 0 & 0 & 0 & 0 & 1 \\ 0 & 1 & 0 & 0 & 0 & 0 & 1 \\ 0 & 0 & 1 & 0 & 0 & 0 & 1 \\ 0 & 0 & 0 & 1 & 1 & 0 & 0 \\ 0 & 0 & 0 & 0 & 1 & 1 & 0 \\ 0 & 0 & 0 & 0 & 0 & 1 & 1 \end{bmatrix} \rightarrow G^{(V)} = \begin{bmatrix} 1 & 0 & 0 & 0 & 0 & 0 & 1 \\ 0 & 1 & 0 & 0 & 0 & 0 & 1 \\ 0 & 0 & 1 & 0 & 0 & 0 & 1 \\ 0 & 0 & 0 & 1 & 0 & 0 & 1 \\ 0 & 0 & 0 & 0 & 1 & 1 & 0 \\ 0 & 0 & 0 & 0 & 0 & 1 & 1 \end{bmatrix}$$

$$\rightarrow G^{(VI)} = \begin{bmatrix} 1 & 0 & 0 & 0 & 0 & 0 & 1 \\ 0 & 1 & 0 & 0 & 0 & 0 & 1 \\ 0 & 0 & 1 & 0 & 0 & 0 & 1 \\ 0 & 0 & 0 & 1 & 0 & 0 & 1 \\ 0 & 0 & 0 & 0 & 1 & 0 & 1 \\ 0 & 0 & 0 & 0 & 0 & 1 & 1 \end{bmatrix}.$$

It is obvious that $G^{(VI)}$ is a systematic form of code generator matrix corresponding to the general parity-check. The parity-check bit of this code is a sum of all information bits modulo-2, what can be clearly seen from the last matrix. Obviously, the same procedure can be used for code $(n, n-1)$ construction, for an arbitrary code word length.

Problem 6.2 A cyclic code, code word length $n = 7$, is defined by the generator polynomial $g(x) = 1 + x + x^3$.

(a) Verify whether the code obtained by this generator polynomial is cyclic. Find the corresponding parity-check polynomial, generator matrix and parity-check matrix.
(b) Describe in details the procedure to find the code word corresponding to information word (1010). List all code words of the code and comment its capability to correct errors.
(c) Describe in details the decoding procedure when in transmitted code word (corresponding to (0111) information word) if the error occurred at the third position.
(d) Whether the code obtained by shortening of this code for one bit is cyclic as well?

Solution

(a) A polynomial is generator one if it divides the polynomial $x^n + 1$ without remainder. To verify it, the polynomial division is carried out:

$$
\begin{array}{r}
(x^7+1) : (x^3+x+1) = x^4+x^2+x+1 \\
\underline{x^7+x^5+x^4} \\
x^5+x^4+1 \\
\underline{x^5+x^3+x^2} \\
x^4+x^3+x^2+1 \\
\underline{x^4+x^2+x} \\
x^3+x+1 \\
\underline{x^3+x+1} \\
0
\end{array}
$$

The remainder equals zero and the code is cyclic. Parity-check polynomial is obtained as the result of a division

$$h(x) = 1 + x + x^2 + x^4.$$

For a cyclic code the polynomials corresponding to the code words must be divisible by $g(x)$. Therefore, the generator matrix rows are determined by coefficients of polynomials $g(x)$, $xg(x)$, $x^2g(x)$, $x^3g(x)$

$$G = \begin{bmatrix} 1 & 1 & 0 & 1 & 0 & 0 & 0 \\ 0 & 1 & 1 & 0 & 1 & 0 & 0 \\ 0 & 0 & 1 & 1 & 0 & 1 & 0 \\ 0 & 0 & 0 & 1 & 1 & 0 & 1 \end{bmatrix}.$$

Parity-check polynomial $h(x)$ defines the corresponding code as well, and the corresponding parity-check matrix rows are determined by coefficients of polynomials $h(x)$, $xh(x)$, $x^2h(x)$ red out in a reverse order

$$H = \begin{bmatrix} 1 & 1 & 1 & 0 & 1 & 0 & 0 \\ 0 & 1 & 1 & 1 & 0 & 1 & 0 \\ 0 & 0 & 1 & 1 & 1 & 0 & 1 \end{bmatrix}.$$

The following relation holds

$$g(x)h(x) = x^7 + 1 = 0 \left((x^7 + 1) \right),$$

and every code vector is orthogonal to the every row of H yielding

$$GH^\mathrm{T} = 0.$$

(b) Information word (1010) is represented by polynomial $i(x) = 1 + x^2$ and a corresponding **code polynomial** is obtained as follows

Table 6.3 Cyclic code generated by a polynomial $g(x) = 1 + x + x^3$

Information word	Code word	Polynomial
(0 0 0 0)	(0 0 0 0 0 0 0)	$0 = 0 \times g(x)$
(1 0 0 0)	(1 1 0 1 0 0 0)	$1 + x + x^3 = g(x)$
(0 1 0 0)	(0 1 1 0 1 0 0)	$x + x^2 + x^3 + x^4 = xg(x)$
(1 1 0 0)	(1 0 1 1 1 0 0)	$1 + x^2 + x^3 + x^4 = (1+x)g(x)$
(0 0 1 0)	(0 0 1 1 0 1 0)	$x^2 + x^3 + x^5 = x^2 g(x)$
(1 0 1 0)	(1 1 1 0 0 1 0)	$1 + x + x^2 + x^5 = (1 + x^2)g(x)$
(0 1 1 0)	(0 1 0 1 1 1 0)	$x + x^3 + x^4 + x^5 = (x + x^2)g(x)$
(1 1 1 0)	(1 0 0 0 1 1 0)	$1 + x^4 + x^5 = (1 + x + x^2)g(x)$
(0 0 0 1)	(0 0 0 1 1 0 1)	$x^3 + x^4 + x^6 = x^3 g(x)$
(1 0 0 1)	(1 1 0 0 1 0 1)	$1 + x + x^4 + x^6 = (1 + x^3)g(x)$
(0 1 0 1)	(0 1 1 1 0 0 1)	$x + x^2 + x^3 + x^6 = (x + x^3)g(x)$
(1 1 0 1)	(1 0 1 0 0 0 1)	$1 + x^2 + x^6 = (1 + x + x^3)g(x)$
(0 0 1 1)	(0 0 1 0 1 1 1)	$x^2 + x^4 + x^5 + x^6 = (x^2 + x^3)g(x)$
(1 0 1 1)	(1 1 1 1 1 1 1)	$1 + x + x^2 + x^3 + x^4 + x^5 + x^6 = (1 + x^2 + x^3)g(x)$
(0 1 1 1)	(0 1 0 0 0 1 1)	$x + x^5 + x^6 = (x + x^2 + x^3)g(x)$
(1 1 1 1)	(1 0 0 1 0 0 1 1)	$1 + x^3 + x^5 + x^6 = (1 + x + x^2 + x^3)g(x)$

$$c(x) = i(x)g(x) = \left(1+x^2\right)\left(1+x+x^3\right) = 1+x+x^3+x^2+x^3+x^5$$
$$= 1+x+x^2+x^5.$$

The highest power of generator polynomial is $n - k = 3$, and it is obvious that the code (7, 4) is obtained, the corresponding code word in this case is (1110010). Code words corresponding to all possible information words are given in Table 6.3, as well as their polynomial representations.

This code has seven words with the weights $d = 3$ and $d = 4$, one with the weight $d = 0$ and one with the weight $d = 7$. It is equivalent to Hamming (7, 4) code. However, these codes are not identical because they do not have the same code words. Both codes are equivalent to the cyclic code whose generator polynomial is $g(x) = 1 + x^2 + x^3$ (given in the Brief Introductory Remarks). It is obvious that a minimum Hamming distance is $d_{\min} = 3$ and the code can detect and correct no more than one error in the code word.

(c) From Table 6.3 it is easy to find that to information word $i = (0111)$ corresponds the code word $c = (0100011)$. If during the transmission an error occurred at the third position, the received sequence is $r = (0110011)$ and the corresponding polynomial is $r(x) = x + x^2 + x^5 + x^6$. The decoding procedure is based on finding the remainder after the division of received word polynomial by a generator polynomial

$$\begin{array}{r} (x^6+x^5+x^2+x) : (x^3+x+1) = x^3+x^2+x \\ \underline{x^6+x^4+x^3} \\ x^5+x^4+x^3+x^2+x \\ \underline{x^5+x^3+x^2} \\ x^4+x \\ \underline{x^4+x^2+x} \\ x^2 \end{array}$$

The obtained remainder is (001), it is not equal to zero meaning that an error occurred. To find its position, one should find to what error type corresponds this remainder. This code can correct all single errors, the syndrome value is found from $S = rH^T = (c \oplus e)H^T = eH^T$ and to the error at jth position the syndrome corresponds being equal to the jth column of a parity-check matrix, as shown in Table 6.4.

(d) Generator matrix of the code obtained from the originating one by *shortening* it for one bit is obtained by eliminating the last row and the last column from the basic code matrix. There are six cyclic shifts of the code word and all words (given in Table 6.5) have not the same Hamming weight, it is obvious that the *shortening disturbed the cyclical property*. However, the minimum

Table 6.4 Syndromes corresponding to correctable error patterns

Error pattern	Syndrome
1 0 0 0 0 0 0	1 0 0
0 1 0 0 0 0 0	0 1 0
0 0 1 0 0 0 0	0 0 1
0 0 0 1 0 0 0	1 1 0
0 0 0 0 1 0 0	0 1 1
0 0 0 0 0 1 0	1 1 1
0 0 0 0 0 0 1	1 0 1
0 0 0 0 0 0 0	0 0 0

Table 6.5 Code words of the code (6, 3) obtained by cyclic code (7, 4) shortening

Information word	Code word
0 0 0	0 0 0 0 0 0
0 0 1	0 0 1 1 0 1
0 1 0	0 1 1 0 1 0
0 1 1	0 1 0 1 1 1
1 0 0	1 1 0 1 0 0
1 0 1	1 1 1 0 0 1
1 1 0	1 0 1 1 1 0
1 1 1	1 0 0 0 1 1

Hamming distance was not changed and the code still can detect and correct one error in the code word (but having a smaller code rate $R = 1/2$). It is easy to verify that this code is equivalent to shortened Hamming code (6, 3) from Problem 4.8.

$$G(1) = \begin{bmatrix} 1 & 1 & 0 & 1 & 0 & 0 & 0 \\ 0 & 1 & 1 & 0 & 1 & 0 & 0 \\ 0 & 0 & 1 & 1 & 0 & 1 & 0 \\ 0 & 0 & 0 & 1 & 1 & 0 & 1 \end{bmatrix}$$

$$G(1) = \begin{bmatrix} 1 & 1 & 0 & 1 & 0 & 0 \\ 0 & 1 & 1 & 0 & 1 & 0 \\ 0 & 0 & 1 & 1 & 0 & 1 \end{bmatrix}.$$

Problem 6.3 Generator polynomial of a cyclic code (code word length $n = 7$) is

$$g(x) = (1+x)\left(1+x^2+x^3\right).$$

(a) Find the code generator matrix and list all code words.

(b) Explain the procedure to obtain one code word of the systematic version of the code. Write all code words and a corresponding generator matrix.
(c) Do the generator matrix of a systematic code can be obtained from the corresponding matrix of the nonsystematic code and how?

Solution

(a) By applying the procedure from the previous problem, it is easy to find the generator polynomial

$$g(x) = 1 + x + x^2 + x^4,$$

and, as $g_0 = g_1 = g_2 = 1$, $g_3 = 0$ and $g_4 = 1$, the generator matrix is

$$G = \begin{bmatrix} g_0 & g_1 & g_2 & g_3 & g_4 & 0 & 0 \\ 0 & g_0 & g_1 & g_2 & g_3 & g_4 & 0 \\ 0 & 0 & g_0 & g_1 & g_2 & g_3 & g_4 \end{bmatrix} = \begin{bmatrix} 1 & 1 & 1 & 0 & 1 & 0 & 0 \\ 0 & 1 & 1 & 1 & 0 & 1 & 0 \\ 0 & 0 & 1 & 1 & 1 & 0 & 1 \end{bmatrix}.$$

To obtain a nonsystematic code, the code polynomial is found from $c(x) = i(x) g(x)$, the list of all code words is given in Table 6.6.

(b) Code polynomial of a **systematic cyclic code** is found using the relation

$$c(x) = i(x)x^{n-k} + rem\left\{\frac{i(x)x^{n-k}}{g(x)}\right\}$$

where $rem\{.\}$ denotes the remainder (polynomial) after division. In this case $n - k = 4$ and to information word $i = (111)$ corresponds the polynomial $i(x) = 1 + x + x^2$, while the code polynomial is

$$c(x) = \left(1 + x + x^2\right)x^4 + rem\left\{\frac{(1+x+x^2)x^4}{1+x+x^2+x^4}\right\}.$$

The remainder after division is obtained as follows

Table 6.6 Code words of cyclic code (7, 3), nonsystematic and systematic version

Information word	Code word of the nonsystematic code	Code word of the systematic code
0 0 0	0 0 0 0 0 0 0	0 0 0 0 0 0 0
1 0 0	1 1 1 0 1 0 0	1 1 1 0 1 0 0
0 1 0	0 1 1 1 0 1 0	0 1 1 1 0 1 0
0 0 1	0 0 1 1 1 0 1	1 1 0 1 0 0 1
1 1 0	1 0 0 1 1 1 0	1 0 0 1 1 1 0
0 1 1	0 1 0 0 1 1 1	1 0 1 0 0 1 1
1 0 1	1 1 0 1 0 0 1	0 0 1 1 1 0 1
1 1 1	1 0 1 0 0 1 1	0 1 0 0 1 1 1

$$\begin{array}{r}x^6+x^5+x^4 : x^4+x^2+x+1=x^2+x \\ \underline{x^6+x^4+x^3+x^2} \\ x^5+x^3+x^2 \\ \underline{x^5+x^3+x^2+x} \\ x\end{array}$$

as $rem\{(i(x) \cdot x^{n-k})/g(x)\} = x$, a code word is obtained from the code polynomial coefficients

$$c(x) = x^4 + x^5 + x^6 + x = x + x^4 + x^5 + x^6 \rightarrow c = (0100111).$$

To information words $i_1 = (100)$, $i_2 = (010)$ and $i_3 = (001)$ correspond the code polynomials

$$c_1(x) = x^4 + rem\left\{\frac{x^4}{1+x+x^2+x^4}\right\} = 1+x+x^2+x^4$$

$$c_2(x) = x^5 + rem\left\{\frac{x^5}{1+x+x^2+x^4}\right\} = x+x^2+x^3+x^5$$

$$c_3(x) = x^6 + rem\left\{\frac{x^6}{1+x+x^2+x^4}\right\} = 1+x+x^3+x^6$$

and a generator matrix of systematic code can be easily found

$$G_s = \begin{bmatrix} 1 & 1 & 1 & 0 & 1 & 0 & 0 \\ 0 & 1 & 1 & 1 & 0 & 1 & 0 \\ 1 & 1 & 0 & 1 & 0 & 0 & 1 \end{bmatrix}.$$

Using the linear combinations of the generator matrix rows, the code words of this systematic code can be easily found (given in the third column of the Table 6.6). It is obvious that both systematic and nonsystematic codes are cyclic, because by a cyclic shift of any code word, the code word is obtained as well. In this case a set of code words is identical, the code being systematic or not.

(c) Generator matrix of the systematic code can be easily found, its third row will be obtained by summing the first and the third row of the nonsystematic code generator matrix.

$$G = \begin{bmatrix} 1 & 1 & 1 & 0 & 1 & 0 & 0 \\ 0 & 1 & 1 & 1 & 0 & 1 & 0 \\ 0 & 0 & 1 & 1 & 1 & 0 & 1 \end{bmatrix} \longrightarrow \begin{bmatrix} 1 & 1 & 1 & 0 & 1 & 0 & 0 \\ 0 & 1 & 1 & 1 & 0 & 1 & 0 \\ 1 & 1 & 0 & 1 & 0 & 0 & 1 \end{bmatrix} = G_s.$$

Problem 6.4 The cyclic code generator polynomial is:

$$g(x) = (x+1)^2(x^3+x+1)(x^3+x^2+1)^2$$

(a) Verify do the code is a cyclic one! Find all code words and draw its weight spectrum,
(b) How many errors can be corrected and how many can be detected? Verify whether the Hamming and Singleton bounds are satisfied.
(c) Find the probability that the emitted code word is not corrected if the channel can be modeled as BSC (crossover probability $p = 10^{-2}$).

Solution
A generator polynomial has to satisfy the condition to be a divisor of polynomial $x^n + 1$ (n—code word length). It means that for given $g(x)$ a minimum n should be found so as that

$$rem\left\{\frac{x^n+1}{g(x)}\right\} = 0$$

In this case, for $g(x) = (x+1)^2(x^3+x+1)(x^3+x^2+1)^2$, minimum n value satisfying the above equality is $n = 14$, because the factoring can be carried out in field GF(2):

$$x^{14} + 1 = (x^7+1)^2 = (x+1)^2(x^3+x+1)^2(x^3+x^2+1)^2 = g(x)(x^3+x+1),$$

where the second element at the right side is the parity-check polynomial, while the generator polynomial is

$$g(x) = (x+1)^2(x^3+x+1)(x^3+x^2+1)^2 = 1+x+x^2+x^4+x^7+x^8+x^9+x^{11}.$$

(a) To information polynomial $i(x) = 1 + x^2$ corresponds the code polynomial

$$c(x) = i(x)g(x) = 1+x+x^3+x^6+x^7+x^8+x^{10}+x^{13},$$

and, procedure for a direct obtaining of a code word from the information word $i(x) = 1 + x^2$ and the generator polynomial coefficients is as follows

```
   1 1 1 0 1 0 0 1 1 1 0 1 × 1 0 1
   1 1 1 0 1 0 0 1 1 1 0 1
         1 1 1 0 1 0 0 1 1 1 0 1
c=(1 1 0 1 0 0 1 1 1 0 1 0 0 1).
```

Table 6.7 Code words of a cyclic code (14, 3)

Information word	Code word
0 0 0	00000000000000
0 0 1	00111010011101
0 1 0	01110100111010
0 1 1	01001110100111
1 0 0	11101001110100
1 0 1	11010011101001
1 1 0	10011101001110
1 1 1	10100111010011

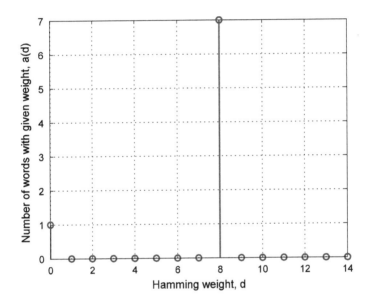

Fig. 6.4 Weight spectrum of the code (14, 3)

It is easy to show that six code words more can be obtained by cyclic shifts of the obtained code word. As to the information word $i = (000)$ always corresponds all zero code word, list of all code words is given in Table 6.7 and the corresponding weight spectrum is shown in Fig. 6.4.

(b) Minimum Hamming distance is $d_{min} = 8$, it is obvious that the code corrects up to $e_c = 3$ errors and detects $e_d = 4$ errors in the code word. If the code capability to correct the errors ($e_c = 0$) is not used, the code can detect up to $e_d \leq d_{min} - 1 = 7$ errors in the code word.

The code parameters are $n = 14$, $k = 3$, a Hamming bound is satisfied because

$$2^{14-3} \geq \binom{14}{0} + \binom{14}{1} + \binom{14}{2} + \binom{14}{3} \Rightarrow 2048 > 1 + 14 + 91 + 364$$
$$= 470,$$

and a *Singleton bound* is

$$d_{\min} \leq n - k + 1 \Rightarrow 8 \leq 12.$$

It is obvious that both bounds are satisfied with a very large margin, meaning that it is possible to construct a code having better performances with the same code parameters. It can also be noticed that for the same code word length and the information word length, the following relation is satisfied as well

$$2^{14-3} \geq \sum_{t=1}^{4} \binom{14}{t},$$

meaning that a Hamming bound predicts the existence of a linear block code having the parameters (14, 3) correcting $e_c = 4$ words at the code word. Singleton bound in this case is less severe and it even predicts the code existence with $d_{\min} \leq 12$, correcting up to $e_c = 5$ error at the code word.
A code (14, 3) correcting five errors at a code word does not fulfill a Hamming bound, it does not exist, and there is no sense to look for it. However, it even does not mean that a code (14, 3) correcting four errors exists. The fulfilling of both bounds only means that this code may exist (and one has to find it!). The Hamming and Singleton bounds consider the possibility of existence the linear block codes which has given parameters, and even if such a code exists, it does not mean that it is a cyclic code.

(c) In Problem 5.8 it was shown that the probability not to detect errors corresponds to the probability that an error vector corresponds to the code word, and it can be easily calculated using the weight spectrum

$$P_{e,d}(p) = \sum_{d=d_{\min}}^{n} a(d) p^d (1-p)^{n-d},$$

for the crossover probability $p = 10^{-2}$ it is

$$P_{e,d}(10^{-2}) = 7 \times (10^{-2})^8 \times (1 - 10^{-2})^{14-8} = 6{,}59 \times 10^{-16}.$$

On the other hand, the probability that the error is not corrected corresponds to the probability that an error vector does not corresponds to the coset leader in a standard array. If the maximum weight of the pattern corresponding to the coset

Fig. 6.5 Golay code (23, 12) weight spectrum

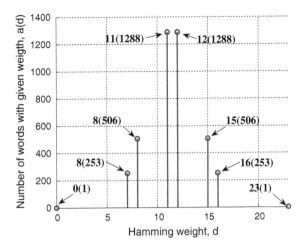

leaders is denoted by l, and the number of patterns of the weight $i \leq l$ is denoted by $L(i)$, this probability is

$$P_{e,c}(p) = 1 - \sum_{i=0}^{l} L(i) p^i (1-p)^{n-i}.$$

The code corrects single, double and triple errors and it is obvious that all error vectors having one, two or three ones (total 470) can be coset leaders, and $L(0) = 1$, $L(1) = 14$, $L(2) = 91$ and $L(3) = 364$. Now, the coset leaders corresponding to the syndrome values left over should be found. The reader should verify that, for this code, all error vectors having four ones result in the various syndrome values and to find the probability not to correct the errors. It is the most easier to use a computer search.

Problem 6.5 The Golay cyclic code (code word length $n = 23$) generator polynomial is:

$$g(x) = 1 + x^2 + x^4 + x^5 + x^6 + x^{10} + x^{11}.$$

(a) Explain the forming procedure of one code word.
(b) Draw weight spectrum and find the number of errors correctable by this code. Whether the code is perfect? Is it a MDS code?
(c) Show how to form a Golay code (24, 12) and explain its possibilities.
(d) Explain how to form a systematic variant of Golay (24, 12) code,

Solution

The code word length is $n = 23$ and generator polynomial power is $n - k = 11$, therefore the information word length is $k = 12$. It is the *Golay code* (23, 12) described in [24].

(a) The forming of a code word corresponding to information word $i(x) = 1 + x^5 + x^{11}$ is carried out by finding the code polynomial coefficients

$$c(x) = i(x)g(x) = (1+x^5+x^{12})(1+x^2+x^4+x^5+x^6+x^{10}+x^{11}) = 1+x^2+x^4+x^5+x^6+x^{10}$$
$$+x^{11}+x^5+x^7+x^9+x^{10}+x^{11}+x^{15}+x^{16}+x^{12}+x^{14}+x^{16}+x^{17}+x^{18}+x^{22}+x^{23}$$
$$= 1+x^2+x^4+x^6+x^7+x^9+x^{12}+x^{14}+x^{15}+x^{17}+x^{18}+x^{22}+x^{23},$$

and the corresponding code word is c = (10101011010010110110011).

(b) Golay code (23, 12) weight spectrum was found by a computer search by repeating the above exposed procedure to find all 2^{12} code words. The result is shown in Fig. 6.5. Minimum Hamming distance is $d_{min} = 7$ and the code can correct up to three errors in the code word.

In this case the Hamming bound is satisfied with equality because

$$2^{23-12} = \sum_{t=1}^{3} \binom{23}{t} = 2048,$$

and the Golay (23, 12) code is perfect. Singleton bound is satisfied as well

$$d_{min} \leq n - k + 1 \Rightarrow 7 \leq 12$$

but not with the equality and it is not an MDS code.

(c) The extended Golay code which has the parameters (24, 12) is obtained by adding a general parity-check to the Golay (23, 12) code. Its generator polynomial is

$$g(x) = (1+x)(1+x^2+x^4+x^5+x^6+x^{10}+x^{11})$$
$$= 1+x+x^2+x^3+x^4+x^7+x^{10}+x^{12},$$

a weight spectrum of the code is determined by coefficients $a(0) = a(24) = 1$, $a(8) = a(16) = 759$, $a(12) = 2576$.

(d) The systematic Golay code (24, 12) has the code words determined by the coefficients of the polynomial

$$c(x) = i(x)x^{n-k} + rem\left\{\frac{i(x)x^{n-k}}{g(x)}\right\},$$

and to information word $i = (100000000000)$ corresponds a code polynomial

$$c_1(x) = x^{12} + rem\left\{\frac{x^{12}}{1+x+x^2+x^3+x^4+x^7+x^{10}+x^{12}}\right\}$$
$$= 1+x+x^2+x^3+x^4+x^7+x^{10}+x^{12}.$$

$$c_2(x) = x^{13} + rem\left\{\frac{x^{13}}{1+x+x^2+x^3+x^4+x^7+x^{10}+x^{12}}\right\}$$
$$= 1+x+x^2+x^3+x^4+x^7+x^{10}+x^{13}.$$

This procedure is repeated for all information words having only one "one" resulting in the corresponding code words. By using these words as the rows of a matrix which has 12 rows and 24 columns, a systematic generator matrix is obtained

$$G = [\mathbf{P}, \mathbf{I}_{12}].$$

This matrix is usually additionally modified so as that the unity matrix is at the first 12 columns, while the positions of other columns are changed so as their rows and columns become identical:

$$G = [\mathbf{I}_{12}, \mathbf{A}] = \begin{bmatrix} 1 & 0 & 0 & 0 & 0 & 0 & 0 & 0 & 0 & 0 & 0 & 0 & 0 & 1 & 1 & 1 & 1 & 1 & 1 & 1 & 1 & 1 & 1 & 1 \\ 0 & 1 & 0 & 0 & 0 & 0 & 0 & 0 & 0 & 0 & 0 & 0 & 1 & 1 & 1 & 0 & 1 & 1 & 1 & 0 & 0 & 0 & 1 & 0 \\ 0 & 0 & 1 & 0 & 0 & 0 & 0 & 0 & 0 & 0 & 0 & 0 & 1 & 1 & 0 & 1 & 1 & 1 & 0 & 0 & 0 & 1 & 0 & 1 \\ 0 & 0 & 0 & 1 & 0 & 0 & 0 & 0 & 0 & 0 & 0 & 0 & 1 & 0 & 1 & 1 & 1 & 0 & 0 & 0 & 1 & 0 & 1 & 1 \\ 0 & 0 & 0 & 0 & 1 & 0 & 0 & 0 & 0 & 0 & 0 & 0 & 1 & 1 & 1 & 1 & 0 & 0 & 0 & 1 & 0 & 1 & 1 & 0 \\ 0 & 0 & 0 & 0 & 0 & 1 & 0 & 0 & 0 & 0 & 0 & 0 & 1 & 1 & 1 & 0 & 0 & 0 & 1 & 0 & 1 & 1 & 0 & 1 \\ 0 & 0 & 0 & 0 & 0 & 0 & 1 & 0 & 0 & 0 & 0 & 0 & 1 & 1 & 0 & 0 & 0 & 1 & 0 & 1 & 1 & 0 & 1 & 1 \\ 0 & 0 & 0 & 0 & 0 & 0 & 0 & 1 & 0 & 0 & 0 & 0 & 1 & 0 & 0 & 0 & 1 & 0 & 1 & 1 & 0 & 1 & 1 & 1 \\ 0 & 0 & 0 & 0 & 0 & 0 & 0 & 0 & 1 & 0 & 0 & 0 & 1 & 0 & 0 & 1 & 0 & 1 & 1 & 0 & 1 & 1 & 1 & 0 \\ 0 & 0 & 0 & 0 & 0 & 0 & 0 & 0 & 0 & 1 & 0 & 0 & 1 & 0 & 1 & 0 & 1 & 1 & 0 & 1 & 1 & 1 & 0 & 0 \\ 0 & 0 & 0 & 0 & 0 & 0 & 0 & 0 & 0 & 0 & 1 & 0 & 1 & 1 & 0 & 1 & 1 & 0 & 1 & 1 & 1 & 0 & 0 & 0 \\ 0 & 0 & 0 & 0 & 0 & 0 & 0 & 0 & 0 & 0 & 0 & 1 & 1 & 0 & 1 & 1 & 0 & 1 & 1 & 1 & 0 & 0 & 0 & 1 \end{bmatrix}.$$

Just due to this last feature, $\mathbf{A}^T = \mathbf{A}$ and a control matrix is

$$H = [\mathbf{A}^T, \mathbf{I}_{12}] = [\mathbf{A}, \mathbf{I}_{12}],$$

and the Golay code (24, 12) is self-dual (dual to itself).

Problem 6.6 The generator polynomial of cyclic code is

$$g(x) = 1 + x^2 + x^8.$$

(a) Explain the procedure for forming a sequence at the CRC encoder output if the information sequence i = (01100111) was emitted.
(b) Verify the remainder in decoder after division if there were no errors during the transmission.
(c) If during the transmission the errors at the second, at the last but one and at the last position occurred find the result in CRC decoder.
(d) Draw a code weight spectrum and comment its capability to detect the errors.

Solution

(a) A polynomial corresponding to the information sequence is $i(x) = x^7 + x^6 + x^5 + x^2 + x$, and the code sequence is

$$c(x) = i(x)x^{n-k} + \text{rem}\left\{\frac{i(x) \cdot x^{n-k}}{g(x)}\right\} = i(x)x^8 + r(x).$$

A procedure for finding the remainder $r(x)$ will be split into two phases. Firstly, a polynomial to be divided will be found and a corresponding binary sequence (bits corresponding to the coefficients)

$$i^*(x) = i(x)x^8 = x^9 + x^{10} + x^{13} + x^{14} + x^{15} \to i^* = (0000000001100111),$$

and the remainder after division of the found polynomial by the generator polynomial is

$$r(x) = \text{rem}\left\{\frac{i^*(x)}{g(x)}\right\} = 1 + x^4 + x^5 + x^6 \to r = (1000011),$$

therefore, a code polynomial and a corresponding code word are

$$c(x) = i(x) \times x^8 + r(x) = 1 + x^4 + x^5 + x^6 + x^9 + x^{10} + x^{13} + x^{14} + x^{15} \to c$$
$$= (1000111001100111).$$

(b) If during the transmission there were no errors, the code word was not changed. The decoder is performing a **Cyclic Redundancy Check** (CRC) to find the remainder after division of the received polynomial by a generator polynomial [35]. A procedure of division $c(x)$ by $g(x)$ is as follows (it starts from the highest power coefficient!):

$$\begin{array}{l}
1110011001110001 : 100000101 = 11100101\\
\underline{100000101}\\
110010011110001\\
\underline{100000101}\\
10010110110001\\
\underline{100000101}\\
0010100010001\\
\underline{100000101}\\
0100000101\\
\underline{100000101}\\
00000000
\end{array}$$

the obtained result is a polynomial $q(x) = 1 + x^2 + x^5 + x^6 + x^7$ (not important here) and a remainder equals zero. It is just the indicator that there were no errors during transmission.

(c) In this case the received message is

$$c = (1000011001100111)$$
$$e = (0100000000000011)$$
$$c' = c + e = (1\underline{1}000110011001\underline{00})$$

and a received word polynomial is $r(x) = 1 + x + x^5 + x^6 + x^9 + x^{10} + x^{13}$, while the remainder after division by a generator polynomial is

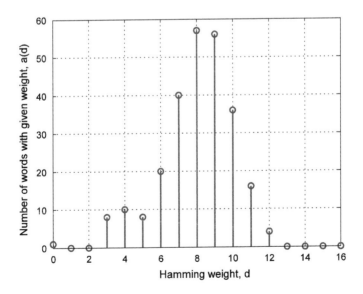

Fig. 6.6 CRC code (16, 8) weight spectrum

$$r(x) = \text{rem}\left\{\frac{c'(x)}{g(x)}\right\} = x^7 + x^6 + x^4 + x^3 + x^2 + 1$$

This remainder is different from zero, i.e. $r = (11011101) \neq (00000000)$, and the transmission error(s) is (are) detected, being the goal of CRC procedure. In this case the number of errors is not known, neither their positions, because this decoder is not designed to correct the errors.

A procedure for obtaining the code words, explained in the first part of this problem, can be repeated for any 8-bits code word. By a computer search all $2^8 = 256$ code words were generated and their Hamming weights were found, the result shown in Fig. 6.6.

As explained earlier, a linear block code cannot detect an error vector if it is the same as a code word (vector). The probability for this event can be found on the weight spectrum basis, and for $p \ll 1$ (e.g. $p < 10^{-2}$) the following approximation can be used

$$P_{e,d}(p) = \sum_{d=d_{\min}}^{n} a(d)p^d(1-p)^{n-d} \approx \sum_{d=d_{\min}}^{n} a(d)p^d \approx a(d_{\min})p^{d_{\min}}.$$

It is obvious that this probability would be greater if $d_{\min}$ is greater and $a(d_{\min})$ smaller. In this case $d_{\min} = 3$, and a code can detect all single and double errors, while more errors can be detected, if they do not correspond to the code word structures.

Problem 6.7 Binary message $i = (01100111)$ is transmitted through the system using CRC which has the generator polynomial $g(x) = x^4 + x^3 + 1$.

(a) Find a corresponding code word. Whether the same code word could be obtained using a nonsystematic procedure for some other information word? If yes, find it!
(b) Draw block schemes of encoder and decoder for error detection and explain how they work.
(c) Find a weight spectrum of a code and calculate the probability of the undetected error if the channel can be modeled as BSC with crossover probability $p = 10^{-3}$.
(d) Whether by using the same generator polynomial is possible to form linear block codes having parameters (7, 3) and (11, 4)? Are these codes cyclic?

Solution

(a) As mentioned in the previous problems, forming of a cyclic code can be conceived as multiplying information polynomial by a generator polynomial $c_{nesist}(x) = i(x)g(x)$, but the obtained code word is not a systematic one.

Problems

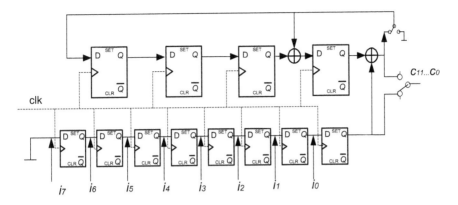

Fig. 6.7 Block scheme of CRC encoder for (12, 8) code

To form a systematic code, where a code word is composed by clearly separated information and parity-check bits, the word is formed on the basis of the following identity

$$c_{sist}(x) = i(x)x^{n-k} + \text{rem}\left(\frac{i(x)x^{n-k}}{g(x)}\right).$$

As shown in the previous problem, this word is divisible by $g(x)$ without remainder. It means that it can be obtained in the other way, multiplying a generator polynomial by some other information polynomial $i_2(x)$

$$c_{sist}(x) = i_2(x)g(x).$$

Combining two previous equations the following relation is obtained (addition and subtraction are equivalent)

$$i(x)x^{n-k} = i_2(x)g(x) + \text{rem}\left(\frac{i(x)x^{n-k}}{g(x)}\right),$$

from which it can be concluded that dividing a polynomial $i(x)x^{n-k}$ by $g(x)$ the result $i_2(x)$ is obtained, and the remainder is $r(x) = \text{rem}(i(x)x^{n-k}/g(x))$. In this example $i(x) = x + x^2 + x^5 + x^6 + x^7$, while a division by $g(x)$ is as follows

```
111001100000 :11001 =10110110
11001
 010111
 11001
  11100
  11001
   010100
   11001
    11010
    11001
     0011
```

It should be noted that a polynomial $i(x)x^{n-k}$ in a general case has n coefficients, while a generator polynomial is of order $n-k$, resulting in the following:

– a result obtained by division is a polynomial having k coefficients, and the information word looked for is

$$i_2(x) = x + x^2 + x^4 + x^5 + x^7 \Rightarrow i_2 = (0110111)$$

– a remainder after division in a general case is polynomial which has $n-k$ coefficients

$$r(x) = x + x^2.$$

Therefore, every code word obtained by a systematic procedure, can be obtained as well by starting from some other information word, using a nonsystematic procedure. It can be suitable for code spectrum calculating.

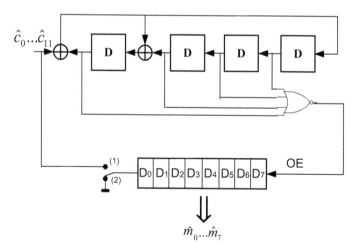

Fig. 6.8 Block scheme of CRC decoder for (12, 8) code

Table 6.8 Code words of a cyclic code (7, 3)

Information word	Code word
0 0 0	0 0 0 0 0 0 0
1 0 0	0 0 1 1 0 0 1
0 1 0	0 1 0 1 0 1 1
0 0 1	0 1 1 0 0 1 0
1 1 0	1 0 0 1 1 1 1
0 1 1	1 0 1 0 1 1 0
1 0 1	1 1 0 0 1 0 0
1 1 1	1 1 1 1 1 0 1

Instead of division a polynomial modulo polynomial, a polynomial multiplication modulo-2 can be used.

(b) Block scheme of a **CRC encoder** for (12, 8) code is shown in Fig. 6.7—during the first k shifts, the lower bits are red out, while at the upper part a division is performed. The encoder consists of two shift registers (usually realized using D-flip flops, as standard elements for delay). Generally, a lower register has k delay cells. The writing is parallel, but the reading (sending at the line) is serial, during k successive shifts. The upper register consists of $n - k$ delay cells and a feedback is defined by coefficients of generator polynomial. In this register, after k shifts, a remainder after division by a generator polynomial is obtained. It will be sent at the line after the information bits, because the switch will be at the upper position after the first k shifts.

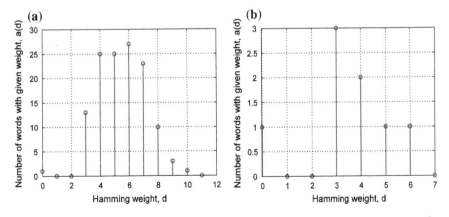

Fig. 6.9 Weight spectra of two CRC codes having the same generator polynomial $g(x) = x^4 + x^3 + 1$, but the different code words lengths—(11, 7) (**a**) and (7, 3) (**b**)

Table 6.9 Minimum code distances profile—optimum and for two considered codes

D_{min}	Optimum	CCITT, $n_c = 32767$	C_1, $n_c = 151$
17	17		
12	18		
10	19, ..., 21		17, ..., 20
9	22		
8	23, ..., 31		21, 22
7	32, ..., 35		
6	36, ..., 151		23, ..., 151
5	152, ..., 257		
4	258, ..., 32767	17, ..., 32767	
3	32768, ..., 65535		
2	≥ 65536	≥ 32768	≥ 152

Block scheme of **CRC decoder** is shown in Fig. 6.8, using simplified symbols for delay cells. When the switch is in position (1), the first k bits enter the division circuit and the shift register. Here the remainder after division by a generator polynomial, whose coefficients determine the feedback structure as well, is obtained. If the remainder differs from zero, the signal OE (*Output Enable*) is not active and the reading from the lower register is not permitted. If the remainder equals zero indicating that there were no errors during transmission, signal OE is active and the decoded information can be delivered to a user.

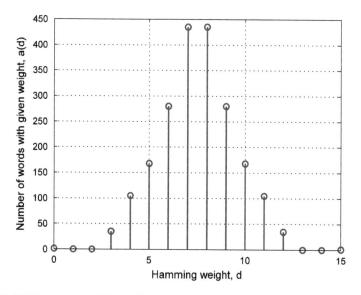

Fig. 6.10 Weight spectrum of the cyclic code (15, 11)

(c) CRC procedure as a result gives the code words of a code obtained by the *shortening a binary cyclic code*. The polynomial $g(x) = x^4 + x^3 + 1$ is an irreducible factor of a polynomial $x^{15} + 1$, CRC procedure based on this polynomial can generate all codes having the parameters $n \leq 15$ and $n - k = 4$. Codes having the parameters (11, 7) and (7, 3) fulfill these conditions and they can be generated using this polynomial. By a computer search all code words of these codes were found and their weight spectra are shown in Figs. 6.9a, b. In Table 6.8 all code words of a code (7, 3) are listed. It can be easily seen that this code is not cyclic, nor equivalent to the code from Problem 6.3.

It is clear that every obtained code, whose spectra were found, has a minimum Hamming distance $d_{min} = 3$, and could correct one error in a code word. However, here the CRC codes are considered where the aim is not the error correction. If $e_c = 0$, they can detect at most $e_d = 2$ errors in the code word. It should be noticed that even and a spectrum shape indicates that a code is cyclic. If a spectrum is symmetric (as for the case of non-shortened code (15, 11), spectrum shown in Fig. 6.10), a code is cyclic. From Fig. 6.9 it is clear that a code shortening usually disturbs this feature.

Problem 6.8 During CRC-8 procedure for error detecting, after the information word, CRC extension is included, where its forming is based on the generator polynomial $g(x) = x^8 + x^7 + x^6 + x^4 + x^2 + 1$.

(a) Find the code word corresponding to information word $i = (1000000)$ and find the code distances spectrum.
(b) Find the code word corresponding to information word $i = (10000000)$ and find the code distances spectrum.

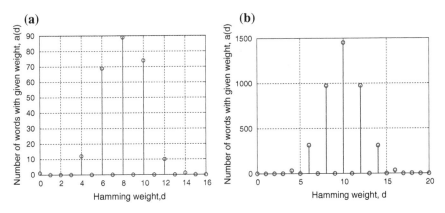

Fig. 6.11 Weight spectra for two CRC codes which have the same generator polynomial $g(x) = x^8 + x^7 + x^6 + x^4 + x^2 + 1$, but the different code word lengths—(16, 8) (**a**) and (20, 12) (**b**)

Solution

CRC code word is formed using the identity

$$c_{CRC}(x) = i(x)x^{n-k} + rem\left(\frac{i(x)x^{n-k}}{g(x)}\right),$$

and always consists of clearly separated information and parity-check bits. Where CRC-8 is applied, the remainder has always eight bits, while a length of the information part depends on the message to be transmitted.

(a) If the information sequence length is $k = 8$, a code word has length $n = 16$ and for the considered example $i(x)$ $i(x) = 1$ yielding

$$c^{(I)}(x) = x^8 + rem\left(\frac{x^8}{g(x)}\right) = 1 + x^2 + x^4 + x^6 + x^7 + x^8.$$

Here generator polynomial corresponds to the code word, and the code word is obtained by reading coefficients starting from x^0 to x^{15}

$$c^{(n=8)} = (1010101110000000).$$

A code spectrum found by a computer search is shown in Fig. 6.11a. In this case $d_{min} = 4$, the code detects all three-bit error combinations in a 16-bits block. It is obvious that there are no code words which have odd weights and the code will detect all sequences having an odd number of errors.

(b) Although the information sequence length here is $k = 10$, a procedure for obtaining a code word is practically unchanged, because here the following is satisfied as well

$$c^{(II)}(x) = g(x).$$

However, a code word in this case has the length $n = 18$ being obtained by reading coefficients starting from x^0 to x^{17}

$$c^{(n=10)} = (101010111000000000).$$

If the same information polynomial is used, but the code words have the various lengths, it is obvious that the resulting code polynomials will be unchanged, and under the same conditions, a longer code word relating to a shorter one has only zeros at the right side. The code spectrum is shown in Fig. 6.11b. In this case $d_{min} = 4$ as well, and a code can also detect all three-bit error combinations, but this time in the twenty-bit block. The code will detect all sequences with an odd number of errors as well.

Problems 277

Some general rules for the choice and analysis of CRC code performances are:

(1) By shortening a cyclic code, the code is obtained having minimum Hamming distance at least as the original code. In fact, by shortening a cyclic code, a parameter d_{min} cannot be decreased! Taking into account that code corrects at least the same number of errors as before a shortening, but has shorter code words, it is obvious that the capability for error detection is increased by code shortening. Of course, it is paid by decrease of code rate, because in shorter code word the number of parity-checks remains the same.

(2) A main advantage of code shortening is a possibility to adjust a code word length to a length of the message entering the encoder. In such way, the same generator polynomial can be used for encoding of information words of various lengths. However, for every generator polynomial there is an upper bound for a code word length, after which code performances become worse abruptly. It will be considered in the next problem.

(3) It is desirable for a code to have the capability to detect an odd number of errors. It can be achieved by a generator polynomial choice, if it has a factor $(x + 1)$. It is interesting that in this case a code can correct an odd number of errors although $x + 1$ is not a factor of generator polynomial!

(4) A spectrum of a code (20, 12) shown in Fig. 6.11b, was obtained by inspection of Hamming weights of all 4096 code words. In communication systems the information bits are grouped into still longer blocks, and to find a linear block code spectrum a long time is needed. In practice, the codes are usually used which have a maximum generator polynomial power 16 or 24 (wireless systems) up to $n - k = 32$ (e.g. *Ethernet*), while the information block lengths are a few hundreds and

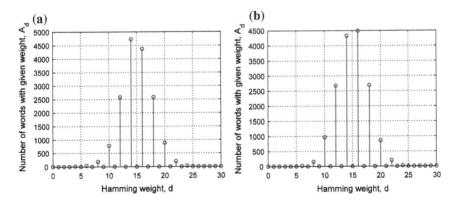

Fig. 6.12 Weight spectra for a two CRC codes, for the same code word length ($n = 30$), but with different generator polynomials $g_1(x) = x^{16} + x^{12} + x^5 + 1$ (**a**) and $g_2(x) = x^{16} + x^{13} + x^{12} + x^{11} + x^{10} + x^8 + x^6 + x^5 + x^2 + 1$ (**b**)

even thousands of bits. In this case a procedure to find a code spectrum used in this problem is useless (about 2^{1000} code words should be generated and analyzed!), but in this case a procedure of performance analysis can be done by observing a dual code spectrum.

Problem 6.9 A communication system uses CRC, where generator polynomials could be $g_1(x) = x^{16} + x^{12} + x^5 + 1$ or $g_2(x) = x^{16} + x^{13} + x^{12} + x^{11} + x^{10} + x^8 + x^6 + x^5 + x^2 + 1$.

(a) If a code word length is $n = 30$, draw spectra of both codes. How the spectrum shape changes and what are their possibilities to detect errors if the code word length is changed?
(b) If for the channel, a BSC model can be used (crossover probability $p = 10^{-3}$), draw a dependence of a probability that the errors are not detected versus a code word length (for $n \leq 300$). Find ranges of information word (message) length where it is possible to apply CRC procedure in both cases.
(c) If CRC procedure is applied for a word length $n = 120$, find the dependence not to detect the errors vs. crossover probability. The same procedure repeat for code word length $n = 200$. What are the advantages and the drawbacks of both codes?

Solution

(a) A code word length is relatively short and a direct method was used to find a spectrum by a computer search. Here $n = 30$ and $n - k = 16$, in both cases

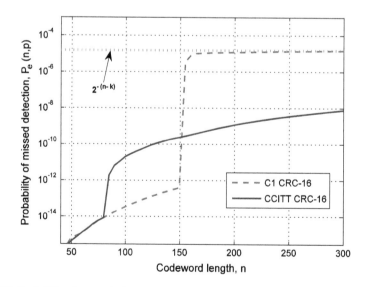

Fig. 6.13 Probability that a code does not detect error vs. the code word length for two analyzed codes

$2^{14} = 16384$ code words were generated, their Hamming weights and their spectra were found. For both polynomials they are shown in Figs. 6.12a, b. A code with generator polynomial $g_1(x) = x^{16} + x^{12} + x^5 + 1$ is a standard CCITT CRC-16 code having $d_{min} = 4$, and in a code there are $A(d_{min}) = 14$ words having four ones. A CRC-16 code using polynomial $g_2(x) = x^{16} + x^{13} + x^{12} + x^{11} + x^{10} + x^8 + x^6 + x^5 + x^2 + 1$ was proposed [36] as C1 code which has $d_{min} = 6$ and in a code there are $A(d_{min}) = 18$ words having six ones. Both codes can correct any odd number of errors. The reader should verify the divisibility of generator polynomials by factor $x + 1$.

As mentioned earlier, a shortening of a cyclic code cannot additionally decrease the minimum Hamming weight. If CCITT CTC-16 code is shortened to the code word length $n = 25$, $d_{min} = 4$ (the same), in the code there are less words with this weight, because $A(4) = 9$. Similarly, for a code whose generator polynomial is $g_2(x) = x^{16} + x^{13} + x^{12} + x^{11} + x^{10} + x^8 + x^6 + x^5 + x^2 + 1$ for $n = 25$, $d_{min} = 4$, but now $A(6) = 3$.

On the basis of the analysis from the previous chapter, the probability that the error was not detected is

$$P_{ed}(p,n) = \sum_{d=d_{min}(n)}^{n} A_d(n) p^d (1-p)^{n-d},$$

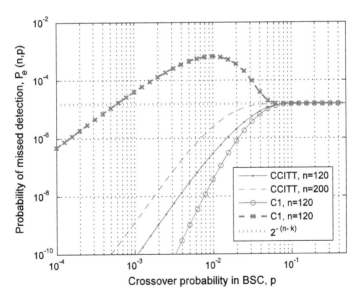

Fig. 6.14 Probability for a code not to detect error vs. crossover probability for two considered CRC codes and two code word lengths

where is was stressed that the minimum Hamming distance, as well as the weight spectrum coefficients, depend on the code word length (it is important how much a CRC code was shortened). CRC codes are generally used for the channels which have a low noise power which, except for rare packet errors, that can be modeled as a BSC where the crossover probability is sufficiently small ($p \ll 10^{-2}$), and the codes are optimized for such a channel type. Then, the first element of a series is dominant, and the following approximation can be used

$$P_{ed}(p,n) \approx A_{d_{\min}(n)}(n) p^{d_{\min}(n)}.$$

(b) The above used procedure is repeated for a code words where $n \leq 300$. Of course, here $n - k = 16$ and it is obvious that the code word length cannot be shorter than $n = 17$ and a typical example of such a code is the simple parity-check. For $n > 35$ the performances are found using a dual code spectrum (as described in Problem 5.9) and the results are shown in Fig. 6.13. It is obvious that a code with generator polynomial $g_2(x) = x^{16} + x^{13} + x^{12} + x^{11} + x^{10} + x^8 + x^6 + x^5 + x^2 + 1$ (here denoted as a C1 code), provides greater reliability for code word lengths $n \leq 151$, while CCITT code has much better performances for a longer code word lengths.

The above exposed results can be understand easier if one notice that a generator polynomial CCITT CRC-16 code $g_1(x) = x^{16} + x^{12} + x^5 + 1$ is a factor of polynomial $x^{32767} + 1$, while a generator polynomial of C1 CRC-16 code $g_2(x) = x^{16} + x^{13} + x^{12} + x^{11} + x^{10} + x^8 + x^6 + x^5 + x^2 + 1$ is a factor of polynomial $x^{151} + 1$. The code length of a non-shortened CCITT code is $n_c = 32767$ and a polynomial $g_1(x)$ can generate only codes being a shortened versions of a cyclic code (32767, 32751). Similarly, a polynomial $g_2(x)$ can generate shortened versions of a cyclic code (151, 135) because here $n_c = 151$. By using the polynomial $g_2(x)$ it is not possible to obtain a code for effective error detection for code word length $n > 151$. It was shown that a code

Table 6.10 Galois field GF(2^3)

Exponential equivalent	Polynomial equivalent			Binary equivalent	Minimal polynomial
α^0	1			1 0 0	$m_3(x) = x + 1$
α^1		α		0 1 0	$m_1(x) = x^3 + x^2 + 1$
α^2			α^2	0 0 1	$m_1(x) = x^3 + x^2 + 1$
α^3	1+		α^2	1 0 1	$m_2(x) = x^3 + x + 1$
α^4	1+	α+	α^2	1 1 1	$m_1(x) = x^3 + x^2 + 1$
α^5	1+	α		1 1 0	$m_2(x) = x^3 + x + 1$
α^6		α+	α^2	0 1 1	$m_2(x) = x^3 + x + 1$

becomes more effective as it is more shortened because of two effects, the minimum Hamming distance cannot be further shortened and $d_{min}(n) \geq d_{min}(n_c)$, the number of detectable errors is the same either it increases. Also, the code word length is shorter and the number of positions where the errors can occur is smaller.

For a CCITT CRC-16 code for any code word length from a range $17 \leq n \leq 32767$, $d_{min} = 4$, while A_{dmin} decreases with a code shortening. On the other hand, minimum Hamming distance for the code with generator polynomial $g_2(x)$ increases with the code shortening and in a range $17 \leq n \leq 20$ it is $d_{min} = 10$, in a range $21 \leq n \leq 22$ it is $d_{min} = 8$, while for $23 \leq n$ 151 it is $d_{min} = 6$. Because in this third group is the greatest number of code word lengths, it is said that this code is optimized for $d_{min} = 6$. Both codes have always a possibility to detect an odd number of errors, and it is obvious that a minimum Hamming distance is $d_{min} = 2$.

In [36] was analyzed in details a possibility to find ***minimum distance profile*** of CRC-16 codes. The analysis was based on an exhausting search and it was shown that a code which has maximum d_{min} for all code word lengths does not exists. The first two columns of Table 6.9 give the maximum possible minimum Hamming distance for some ranges of code words lengths. These distances are not achieved by any CRC-16 code, but to various code word lengths correspond the different optimal generator polynomials. In the third column the profile for a code CCITT CRC-16 is written and it is obvious that it is mainly optimum for a greater code word lengths ($n \geq 258$), while a code C1 CRC-16 defined by a polynomial $g_2(x)$ is approximate optimum for a range $36 \leq n \leq 151$.

It is interesting to notice that the probability not to detect an error for CRC code approaches to the bound $2^{-(n-k)}$ if a code word length approaches to infinity, i.e.

$$\lim_{n \to \infty} P_{ed}(p,n) = 2^{-(n-k)}.$$

Codes fulfilling as well for all values of n

$$P_{ed}(p,n) \leq 2^{-(n-k)}$$

and if their probability not to detect error monotonically decreases are called ***proper*** codes. It was shown that code C1 is proper for all code word lengths $n \leq n_c$, while CRC-CCITT code is not proper for all values $n \leq n_c$.

In Fig. 6.14 the probability for a code not to detect error versus BSC crossover probability is shown. It is obvious that CCITT CRC-16 code has such probability higher than C1 CRC-16 code, but this probability does increase significantly with the code word length increasing. On the other hand, a code defined by generator polynomial $g_2(x)$ provides better performances for a short code words, but when $n > 151$, performances are drastically worse because it is not more a code obtained by a cyclic code shortening. A basic advantage of CCITT code is, therefore, more flexibility concerning the code word length.

Problem 6.10 Field $GF(2^3)$ is generated by a primitive polynomial $p(x) = x^3 + x^2 + 1$. Find a BCH code constructed over this field and its distance spectrum, if one root of a generator polynomial is $\alpha_1 = \alpha$ under condition

(a) the code should correct one error in the code word,
(b) the code should correct two errors in the code word,
(c) the code should correct three errors in the code word,

Solution
Firstly, the root of *primitive polynomial* generating a field should be found. The polynomial $p(x)$ coefficients take the values from a binary field (addition is modulo-2), and it is obvious that binary zero and binary one are not polynomial roots because $p(0) = p(1) = 1$. An element α is introduced, for which

$$p(\alpha) = \alpha^3 + \alpha^2 + 1 = 0$$

It is obvious that this element is not from a binary field, but a relation $\alpha^3 = \alpha^2 + 1$ must formally be satisfied. It provides the expression of every power of α (the element of binary field extension) as a polynomial $b_0\alpha^0 + b_1\alpha^1 + b_2\alpha^2$, where the ordered coefficients (b_0 b_1 b_2) are a binary equivalent of a corresponding element from field extension $GF(2^3)$. E.g. it can be written $\alpha^4 = (\alpha^2 + 1)\alpha = \alpha^3 + \alpha = 1 + \alpha^2$ and a binary equivalent is (1,1,1). A complete Galois field $GF(2^3)$ is given in Table 6.10.

In this table the identity element for addition (combination "all zeros") is not included, it exists as the eighth field element, but it is not interesting for BCH codes

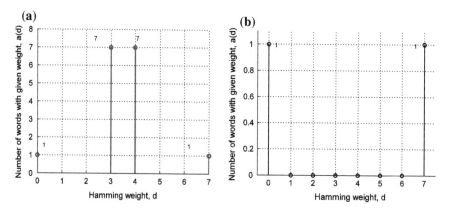

Fig. 6.15 Code distances spectrum of BCH(7, 4) (**a**) and BCH(7, 1) (**b**) codes

construction [37, 38]. It is obvious that $\alpha^7 = \alpha^0 = 1$ as well as $\alpha^l = \alpha^{l-7}$ for any natural number l.

For any arbitrary element β from GF(2^m), an **order of element** r can be defined. It is a minimum natural number equal to the element power yielding the identity element for multiplication, i.e. $\beta^r = \alpha^0 = 1$. An element is primitive if his order is maximum i/e/ being $r_{max} = 2^m - 1$. In other words, an element is primitive if by its successive powers all field elements can be obtained. From the extended field construction view, it is obvious that element α must always be a primitive one.

In this example, element α^5 is primitive, because by its successive powers yield the elements $(\alpha^5)^1 = \alpha^5$, $(\alpha^5)^2 = \alpha^3$, $(\alpha^5)^3 = \alpha$, $(\alpha^5)^4 = \alpha^5$, $(\alpha^5)^5 = \alpha^4$, $(\alpha^5)^6 = \alpha^2$, $(\alpha^5)^7 = \alpha^0 = 1$ and its order is seven. On the other hand, it is obvious that α^0 is not primitive because its order equals one. **Conjugate elements** of element β are found as all different powers β^2, β^4, β^8, β^{16}, ... existing in a considered field extension (power is always of the form 2^i, i being a natural number). In considered field GF(8) element α has conjugate elements α^2 and α^4, because the other powers are repeating $\alpha^8 = \alpha^{8-7} = \alpha$, $\alpha^{16} = \alpha^{16-2\times7} = \alpha^2$... Further, α^3 has the conjugate elements $(\alpha^3)^2 = \alpha^6$ and $(\alpha^3)^4 = \alpha^{12} = \alpha^5$, because $(\alpha^3)^8 = \alpha^{24} = \alpha^3$. The last field element is the identity element for multiplication, it has not conjugate elements, because its value does not change by raising to any power.

To every group of conjugate elements a **minimal polynomial** can be joined. To the groups of conjugated roots $(\alpha, \alpha^2 \text{ i } \alpha^4)$, $(\alpha^3, \alpha^5 \text{ and } \alpha^6)$ and α^0 correspond the following minimal polynomials, respectively

$$m_1(x) = (x+\alpha)(x+\alpha^2)(x+\alpha^4) = x^3 + (\alpha^4 + \alpha^2 + \alpha)x^2 + (\alpha^6 + \alpha^5 + \alpha^3)x + \alpha^7$$
$$= x^3 + x^2 + 1$$
$$m_2(x) = (x+\alpha^3)(x+\alpha^5)(x+\alpha^6) = x^3 + (\alpha^6 + \alpha^5 + \alpha^3)x^2 + (\alpha^8 + \alpha^9 + \alpha^{11})x + \alpha^{14}$$
$$= x^3 + x + 1$$
$$m_3(x) = (x+\alpha^0) = x+1$$

where for calculations, the relation $\alpha^l = \alpha^{l-7}$ was used as well as a corresponding equivalence between exponential and polynomial equivalents, given in Table 6.10.

Coefficients of minimal polynomials are by rule from a binary field. Further, in a general case, minimal polynomials are all irreducible factors of polynomial $x^n + 1$ where $n = 2^m - 1$, 2^m is the number of elements in extension field GF(2^m)

$$x^7 + 1 = (x^3 + x^2 + 1)(x^3 + x + 1)(x + 1).$$

This relation, previously written in Problem 6.1, can now be explained more precisely. To obtain a binary cyclic code it suffices that its generator polynomial is a product of minimal polynomials, and a code word polynomial is easily obtained multiplying information polynomial by code generator polynomial. Of course, all combinations of minimal polynomials will not result in equally effective error control codes.

On the other hand, **BCH code** construction guarantees to obtain a code where a capability to correct errors is given in advance. For every natural number $m \geq 3$ it is possible to construct BCH code where the code word length is $n = 2^m - 1$, correcting e_c errors, where $n - k \leq m e_c$. To provide these features, it is necessary and sufficient that a code generator polynomial has the following (successive) roots from a field extension

$$\alpha^{m_0}, \alpha^{m_0+1}, \alpha^{m_0+2}, \ldots, \alpha^{m_0+2e_c-1}.$$

If a starting element α^{m_0} is a primitive one, all its conjugate elements are primitive as well, and the corresponding minimal polynomial is primitive. BCH code, where a starting element is primitive, it is called a **primitive** code. As the element α is primitive by rule, usually as a generator polynomial the polynomial of a minimal power is chosen having $2e_c$ successive powers of α

$$\alpha, \alpha^2, \alpha^3, \ldots, \alpha^{2e_c}.$$

(a) To obtain a code which has code word length $n = 7$, capable to correct one error, a generator polynomial should have roots α i α^2. As both elements are roots of minimal polynomial $m_1(x)$, it is obvious

$$g_1(x) = m_1(x) = x^3 + x^2 + 1.$$

The generator polynomial highest power corresponds to the number of parity-check bits $n - k = 3$, and a code (7, 4) is obtained. A code spectrum can be easily found by noticing the following:

- to information polynomial $i_1(x) = 1$ corresponds code polynomial $c_1(x) = i_1(x)g_1(x) = 1 + x^2 + x^3$, i.e. the word (1 0 1 1 0 0 0). Because the code is cyclic, the six cyclic shifts of this code word yield six corresponding code words with Hamming weight $d = 3$.
- to information polynomial $i_2(x) = 1 + x$ corresponds code polynomial $c_2(x) = i_2(x)g_1(x) = 1 + x + x^2 + x^4$, i.e. a code word (1 1 1 0 1 0 0). This code word and its cyclic shifts (seven in total) has weight $d = 4$.

By noticing as well that in a linear block code there is always an "all zeros" code word, it is obvious that a code distances spectrum has a form shown in Fig. 6.15a. This code is equivalent to Hamming code, because their spectra are identical. Because a minimum Hamming distance is $d_{min} = 3$, the code can correct one error, as required.

(b) If the goal is to construct a code, having code word length $n = 7$ and correcting two errors, the generator polynomial should have roots α, α^2, α^3, α^4.

Elements α, α^2 and α^4 are roots of a minimal polynomial $m_1(x)$, while element α^3 is element of a minimal polynomial $m_2(x)$, therefore

$$g_2(x) = m_1(x)m_2(x) = (x^3 + x^2 + 1)(x^3 + x^2 + 1)$$
$$= 1 + x + x^2 + x^3 + x^4 + x^5 + x^6.$$

In this case $n - k = 6$, and a code (7, 1) is obtained. It is obviously sevenfold repetition code where to information polynomial $i_1(x) = 0$ corresponds code word "all zeros", while to information polynomial $i_2(x) = 1$ corresponds the word "all ones". These are the only two existing code words, the code distance spectrum can be easily found, as shown in Fig. 6.15b. It can be easily noticed that a minimum Hamming distance is $d_{min} = 7$, and a code can correct up to three errors, one more than required. A detailed explanation of this effect will be given in the next part of solution.

(c) To construct a code correcting three errors and having word code length $n = 7$, a generator polynomial should have the roots α, α^2, α^3, α^4, α^5, α^6. Elements α, α^2 i α^4 are roots of minimal polynomial $m_1(x)$, while the elements α^3, α^5 i α^6 are roots of minimal polynomial $m_2(x)$, and a generator polynomial is the same as in a previous case

$$g_3(x) = m_1(x)m_2(x) = 1 + x + x^2 + x^3 + x^4 + x^5 + x^6.$$

A generator polynomial completely determines a cyclic code, and BCH code correcting three errors is the same as in a previous case (sevenfold repetition) and its code distances spectrum is the same. As explained previously, the correcting of three errors is satisfied because the relation $d_{min} = 2e_c + 1 = 7$ holds.

It is known the BCH codes satisfy the relation $n - k \leq me_c$. In this case $m = 3$ and $n = 7$, and if a code corrects one error the information word length must be $k \geq 4$, while, when two errors are corrected a relation $k \geq 1$ holds, being satisfied with equality in both cases. It is interesting that for a three errors correction this condition is $k \geq -2$, is satisfied as well. This feature of BCH code gives a lower bound for an information word length, while a Singleton bound $n - k \geq 2e_c$ gives a fundamental upper bound for all block codes. In this case a Singleton bound is reduced to the condition $k \leq 1$, being satisfied with equality for a MDS code. Therefore, although code (7, 1) has a low code rate, it has a maximum possible value of code rate under these conditions.

From these examples it is obvious that the constructed BCH code corrects at least a number of errors it was projected for, but it could be possible that it corrects a larger number of errors. This effect appears, because a generator polynomial is formed as a product of the corresponding minimal polynomials,

Table 6.11 Galois field GF(2^4)

Exponential equivalent	Polynomial equivalent	Binary equivalent	Minimal polynomial
α^0	1	1 0 0 0	$m_5(x)$
α^1	α	0 1 0 0	$m_1(x)$
α^2	α^2	0 0 1 0	$m_1(x)$
α^3	α^3	0 0 0 1	$m_2(x)$
α^4	$1 + \alpha$	1 1 0 0	$m_1(x)$
α^5	$\alpha + \alpha^2$	0 1 1 0	$m_4(x)$
α^6	$\alpha^2 + \alpha^3$	0 0 1 1	$m_2(x)$
α^7	$1 + \alpha + \alpha^3$	1 1 0 1	$m_3(x)$
α^8	$1 + \alpha^2$	1 0 1 0	$m_1(x)$
α^9	$\alpha + \alpha^3$	0 1 0 1	$m_2(x)$
α^{10}	$1 + \alpha + \alpha^2$	1 1 1 0	$m_4(x)$
α^{11}	$\alpha + \alpha^2 + \alpha^3$	0 1 1 1	$m_3(x)$
α^{12}	$1 + \alpha + \alpha^2 + \alpha^3$	1 1 1 1	$m_2(x)$
α^{13}	$1 + \alpha^2 + \alpha^3$	1 0 1 1	$m_3(x)$
α^{14}	$1 + \alpha^3$	1 0 0 1	$m_3(x)$

and a whole polynomial enters into the product, regardless if it is necessary that only one or all its roots is a generator polynomial root. It seems that a feature of a code to correct more errors than required is suitable. However, in some cases, it is desirable that a number of errors is just as it was projected, if it can provide a greater code rate. E.g. it would be very useful if a code exists correcting just $e_c = 2$ errors for a code which has the code word length $n = 7$, being in the same time $k > 1$. However, it was found that such binary code cannot be constructed. But, using nonbinary codes it is possible to construct codes correcting just a required number of errors which have at the same time a maximum possible code rate. These are Reed-Solomon codes, which are in details described in one of the next problems.

Problem 6.11 Field GF(2^4) is generated by the primitive polynomial $p(x) = x^4 + x + 1$. Find BCH code constructed over this field under condition

(a) one root of generator polynomial is $\alpha_1 = \alpha$, and one error should be corrected, and draw a code spectrum distance;
(b) one root of generator polynomial is $\alpha_1 = \alpha$, and two errors should be corrected, and draw a code spectrum distance;
(c) one root of generator polynomial is $\alpha_1 = \alpha^4$, and two errors should be corrected, compare its code rate to the code rate of a code b);
(d) one root of generator polynomial is $\alpha_1 = \alpha^5$, and four errors should be corrected.

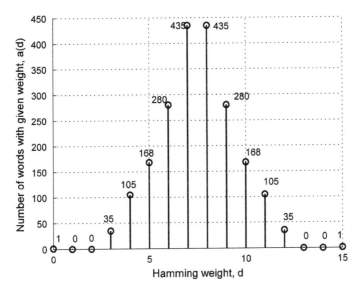

Fig. 6.16 BCH(15, 11) code distances spectrum

Solution

In this case a root of primitive polynomial is found from

$$p(\alpha) = \alpha^4 + \alpha + 1 = 0$$

and the relation $\alpha^4 = \alpha+1$ must hold, enabling to form the corresponding extended field GF(2^4) given in Table 6.11.

The minimal polynomials in GF(2^4) are obtained

$$m_1(x) = (x+\alpha)(x+\alpha^2)(x+\alpha^4)(x+\alpha^8) = x^4 + x + 1$$
$$m_2(x) = (x+\alpha^3)(x+\alpha^6)(x+\alpha^{12})(x+\alpha^9) = x^4 + x^3 + x^2 + x + 1$$
$$m_3(x) = (x+\alpha^7)(x+\alpha^{14})(x+\alpha^{13})(x+\alpha^{11}) = x^4 + x^3 + 1$$
$$m_4(x) = (x+\alpha^5)(x+\alpha^{10}) = x^2 + x + 1$$
$$m_5(x) = (x+\alpha^0) = x + 1$$

Although all polynomials are irreducible and the following is valid $x^{15} + 1 = m_1(x)m_2(x)m_3(x)m_4(x)m_5(x)$, it could be noticed that only these polynomials having primitive elements as a roots are primitive.

It is obvious that element α is primitive, as well as its conjugate elements and a polynomial $m_1(x)$ is primitive, the same being true and for a polynomial $m_3(x)$. However, it can be easily shown that the element α^3 has the order five, and the element α^5 has the order three. The corresponding minimal polynomials are not primitive [as well as polynomial $m_5(x)$]. Therefore, field GF(2^4) can be constructed only with polynomials $m_1(x)$ and $m_3(x)$, but any element can be used to find a BCH code polynomial. However, whether the obtained BCH code is primitive depends on a nature of generator polynomials roots.

(a) To construct a code correcting one error in the code word length $n = 15$, the generator polynomial should have the roots α and α^2, being here

$$g_1(x) = m_1(x) = x^4 + x + 1$$

and the obtained code is (15, 11). A code spectrum, shown in Fig. 6.16, was obtained by a computer search generating all $2^{11} = 2048$ information words, obtaining the corresponding code words and then finding their weights.
It is obvious that the code minimum Hamming distance is $d_{min} = 3$ providing the correction of all single errors, as required. This code is equivalent to Hamming (15, 11) code.

(b) To construct a code correcting two errors in the code word length $n = 15$, a generator polynomial should have the roots α, α^2, α^3, α^4, where α, α^2 and α^4 are roots of minimal polynomial $m_1(x)$, while α^3 is a root of $m_2(x)$, and generator polynomial is

$$\begin{aligned}g_2(x) &= m_1(x)m_2(x) = (x^4 + x + 1)(x^4 + x^3 + x^2 + x + 1) \\ &= x^8 + x^7 + x^6 + x^4 + 1\end{aligned}$$

A code (15, 7) spectrum, shown in Fig. 6.17, was obtained by a computer search generating all $2^7 = 128$ information words. In this case the minimum Hamming distance is $d_{min} = 5$ and a code can correct all double errors.

(c) In this case the goal is to construct a code correcting two errors in the code word length $n = 15$, but the roots are taken starting from the element α^4. Therefore, generator polynomial should have roots α^4, α^5, α^6, α^7, where α^4 is a root of minimal polynomial $m_1(x)$, α^5 is a root of minimal polynomial $m_4(x)$, α^6 is a root of minimal polynomial $m_2(x)$ and α^3 is a root of minimal polynomial $m_3(x)$ yielding

$$\begin{aligned}g_2(x) &= m_1(x)m_2(x)m_3(x)m_4(x) \\ &= (x^4 + x + 1)(x^4 + x^3 + x^2 + x + 1)(x^4 + x^3 + 1)(x^2 + x + 1)\end{aligned}$$

In this case $n - k = 14$, it is code (15,1), in fact, a 15 times repetition. Minimum Hamming distance is $d_{min} = 15$ and the code corrects up to seven errors (it is easy to prove that an n-times repetition code can correct up to $(n - 1)/2$ errors, and it is still MDS code). Therefore, a careless choice of the starting element for a series of generator polynomial roots can result in a code correcting substantially more errors than required, but it is paid by a great code rate reduction compared to the optimal choice. This effect can appear even and when the starting element is primitive, and can be avoided if a greater number

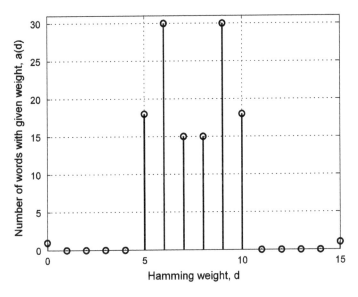

Fig. 6.17 BCH(15, 7) code distances spectrum

of generator polynomial elements are the roots of the same minimal polynomial.

(d) In this case, a goal is to construct a code correcting four errors in the code word length $n = 15$ bits, but the roots are taken starting from the element α^5. In this case a generator polynomial should have roots α^5, α^6, α^7, α^8, α^9, α^{10}, α^{11}, α^{12} and it is easy to conclude that a generator polynomial is

$$g_2(x) = m_1(x)m_2(x)m_3(x)m_4(x)$$
$$= (x^4 + x + 1)(x^4 + x^3 + x^2 + x + 1)(x^4 + x^3 + 1)(x^2 + x + 1)$$

Of course, it is the same code as in the above case, i.e. 15-fold repetition code correcting up to seven errors at a code word length. It is easy to conclude that this effect cannot be avoided by a suitable choice of the starting element for roots, and in this case, the requirement to construct a code correcting $e_c \geq 4$ errors will always result in a code (15, 1), correcting seven errors.

Problem 6.12 Field GF(2^4) is generated by the primitive polynomial $p(x) = x^4 + x + 1$. Find a BCH code constructed over this field, if one root of generator polynomial is $\alpha_1 = \alpha$ and a code should correct two errors in a code word.

(a) Find the code distances spectrum on the basis of spectrum of a dual code by using McWilliams identities.
(b) Find a code word corresponding to information polynomial $i(x) = 1$ and find a received word if the errors during transmission occurred at the second and the eighth position.

(c) Calculate syndrome values at the receiver end and verify whether the relations connecting syndromes and error locators in the previous example are fulfilled.
(d) Explain in short a procedure for finding locator error polynomial on the basis of syndrome and verify the procedure carried out in the previous example.
(e) Explain in short Peterson procedure for BCH codes decoding and verify its efficiency in a previous example.

Solution

Spectra of dual codes for BCH codes correcting double errors in the code word length $n = 2^m - 1$ (m even), can be found using identities given in Table 6.12 [25].

When $m = 4$, BCH code correcting double error in the code word length $n = 15$ bits has the parameters (15, 7). Dual code (15, 8) spectrum in this case has six nonzero components, and it is easy to calculate that in the dual code there are 1 word with weight 0, 15 words with weight 4, 100 words with weight 6, 75 words with weight 8, 60 words with weight 10 and 5 words with weight 12, what can be described by a relation

$$B(x) = 1 + 15x^4 + 100x^6 + 75x^8 + 60x^{10} + 5x^{12}.$$

BCH code (15, 7) spectrum can be now found by using the first McWilliams identity

$$A(x) = 2^{-8}(1+x)^{15} B\left(\frac{1-x}{1+x}\right)$$

$$= 2^{-8}(1+x)^{15}\left[1 + 15\left(\frac{1-x}{1+x}\right)^4 + 100\left(\frac{1-x}{1+x}\right)^6 + 75\left(\frac{1-x}{1+x}\right)^8\right.$$

$$\left. + 60\left(\frac{1-x}{1+x}\right)^{10} + 5\left(\frac{1-x}{1+x}\right)^{12}\right]$$

$$= 2^{-8}\left[(1+x)^{15} + 15(1-x)^4(1+x)^{11} + 100(1-x)^6(1+x)^9\right.$$

$$\left. + 75(1-x)^8(1+x)^7 + 60(1-x)^{10}(1+x)^5 + 5(1-x)^{12}(1+x)^3\right]$$

$$= 1 + 18x^5 + 30x^6 + 15x^7 + 15x^8 + 30x^9 + 18x^{10} + x^{15}$$

and the same result was obtained as in the previous problem, where a computer search was used (Fig. 6.17),

(b) In the previous problem it was shown that a generator polynomial for BCH (15, 7) code is

$$g_2(x) = m_1(x)m_2(x) = x^8 + x^7 + x^6 + x^4 + 1$$

In this case information polynomial is extremely simple (it is 1), a code polynomial is the same as a generator polynomial, and a code word is obtained

Problems

Table 6.12 Spectra of codes dual to BCH codes with $e_c = 2$, m even

Weight d	Number of code words having weight d, b_d
0	1
$2^{m-1} - 2^{(m+2)/2-1}$	$2^{(m-2)/2-1}(2^{(m-2)/2} + 1)(2^m - 1)/3$
$2^{m-1} - 2^{m/2-1}$	$2^{(m+2)/2-1}(2^{m/2} + 1)(2^m - 1)/3$
2^{m-1}	$(2^{m-2} + 1)(2^m - 1)$
$2^{m-1} + 2^{m/2-1}$	$2^{(m+2)/2-1}(2^{m/2} - 1)(2^m - 1)/3$
$2^{m-1} + 2^{(m+2)/2-1}$	$2^{(m-2)/2-1}(2^{(m-2)/2} - 1)(2^m - 1)/3$

by reading out its coefficients from the lowest to the highest power. Therefore, the code word, error vector and the received word are, respectively

$$c = (1000101110000000)$$
$$e = (0\underline{1}00000\underline{1}00000000)$$
$$r = (1\underline{1}00101\underline{0}10000000)$$

and the polynomial corresponding to the received word is

$$r(x) = 1 + x + x^4 + x^6 + x^8.$$

Syndrome values at the receiver end are calculated as polynomial values for generator polynomial roots which should exist for a case $e_c = 2$ being α, α^2, α^3, α^4. For the above polynomial, syndrome components are

$$S_1 = r(\alpha) = 1 + \alpha + \alpha^4 + \alpha^6 + \alpha^8 = 1 + \alpha + (1+\alpha) + (\alpha^2 + \alpha^3) + (1+\alpha^2)$$
$$= 1 + \alpha^3 = \alpha^{14},$$
$$S_2 = r(\alpha^2) = 1 + (\alpha^2) + (\alpha^2)^4 + (\alpha^2)^6 + (\alpha^2)^8$$
$$= 1 + \alpha^2 + (1+\alpha^2) + (1+\alpha+\alpha^2+\alpha^3) + \alpha = 1 + \alpha^2 + \alpha^3 = \alpha^{13},$$
$$S_3 = r(\alpha^3) = 1 + (\alpha^3) + (\alpha^3)^4 + (\alpha^3)^6 + (\alpha^3)^8 = 1 + \alpha^3 + (1+\alpha+\alpha^2+\alpha^3) + \alpha^3$$
$$+ (\alpha + \alpha^3) = \alpha^2,$$
$$S_4 = r(\alpha^4) = 1 + (\alpha^4) + (\alpha^4)^4 + (\alpha^4)^6 + (\alpha^4)^8 = 1 + (1+\alpha) + \alpha + (\alpha+\alpha^3) + \alpha^2$$
$$= \alpha + \alpha^2 + \alpha^3 = \alpha^{11}.$$

If there were no errors during the transmission, roots α, α^2, α^3, α^4 would have been zeros of the received word polynomial, because it is code word polynomial, divisible by a generator polynomial.

Syndromes can be obtained as well from error locators (locator $X_i = \alpha^k$ corresponds to ith error which occurred in the kth position in a code word) by following operations

$$S_1 = X_1 + X_2 = \alpha^1 + \alpha^7 = \alpha + (1 + \alpha + \alpha^3) = 1 + \alpha^3 = \alpha^{14},$$
$$S_2 = X_1^2 + X_2^2 = (\alpha^1)^2 + (\alpha^7)^2 = \alpha^2 + (1 + \alpha^3) = \alpha^{13},$$
$$S_3 = X_1^3 + X_2^3 = (\alpha^1)^3 + (\alpha^7)^3 = \alpha^3 + (\alpha^2 + \alpha^3) = \alpha^2,$$
$$S_4 = X_1^4 + X_2^4 = (\alpha^1)^4 + (\alpha^7)^4 = (1 + \alpha) + (1 + \alpha^2 + \alpha^3) = \alpha^{11}.$$

It is important to note that the syndrome values in two previous sets are obtained in different ways. In the first case, syndrome values are determined by a received word structure, while in the second, syndrome values are determined by error positions. Of course, the receiver does not know the error positions (it has to find them!) but it just gives a possibility for a decoder construction.

(d) Consider firstly a general case, i.e. BCH code for correcting e_c errors in the code word. Let during a transmission at v bits errors occurred. Syndrome values can be found as polynomial values for roots $\alpha, \alpha^2, \ldots, \alpha^{2e_c}$ and an equations system can be formed to find unknown error locators $X_1, X_2, \ldots, X_v$

$$\begin{aligned} S_1 &= X_1 + X_2 + \cdots + X_v, \\ S_2 &= X_1^2 + X_2^2 + \cdots + X_v^2, \\ S_3 &= X_1^3 + X_2^3 + \cdots + X_v^3, \\ &\vdots \\ S_{2e_c} &= X_1^{2e_c} + X_2^{2e_c} + \cdots + X_v^{2e_c}. \end{aligned}$$

These equations are called **power-sum symmetric functions**, they form a system of nonlinear algebraic equations of more variables and they are not suitable for solution by a direct procedure [39].

Peterson showed that, by introducing a notion of error locator polynomial, these equations can be transformed into a series of linear equations. **Error locator polynomial** is a polynomial having as the roots the error locators, i.e. [40].

$$\Lambda(x) = \prod_{i=1}^{v} (1 + X_i x) = \Lambda_0 + \Lambda_1 x + \Lambda_2 x^2 + \ldots + \Lambda_v x^v$$

It is obvious that error locations can be easily found if the coefficients of error locator polynomial are known. On the other hand, it was shown that syndrome components and coefficients of error locator polynomial are connected by a system of linear equations

$$S_1 + \Lambda_1 = 0 \qquad\qquad S_v + \Lambda_1 S_{v-1} + \Lambda_2 S_{v-2} + \cdots + \Lambda_{v-1} S_1 + v\Lambda_v = 0$$
$$S_2 + \Lambda_1 S_1 + 2\Lambda_2 = 0 \qquad\qquad S_{v-1} + \Lambda_1 S_v + \Lambda_2 S_{v-1} + \cdots + \Lambda_v S_1 = 0$$
$$S_3 + \Lambda_1 S_2 + \Lambda_2 S_1 + 3\Lambda_3 = 0 \quad \vdots$$
$$\vdots \qquad\qquad S_{2e_c} + \Lambda_1 S_{2e_c-1} + \Lambda_2 S_{2e_c-2} + \cdots + \Lambda_v S_{2e_c-v} = 0$$

In the special case, when all operations are in a binary field and when there are $v = e_c$ errors, the system is additionally simplified and can be written as

$$A\Lambda = \begin{bmatrix} 1 & 0 & 0 & 0 & \cdots & 0 & 0 \\ S_2 & S_1 & 1 & 0 & \cdots & 0 & 0 \\ S_4 & S_3 & S_2 & S_1 & \cdots & 0 & 0 \\ S_6 & S_5 & S_4 & S_3 & \cdots & 0 & 0 \\ \vdots & \vdots & \vdots & \vdots & \ddots & \vdots & \vdots \\ S_{2e_c-4} & S_{2e_c-5} & S_{2e_c-6} & S_{2e_c-7} & \cdots & S_{e_c-2} & S_{e_c-3} \\ S_{2e_c-2} & S_{2e_c-3} & S_{2e_c-4} & S_{2e_c-5} & \cdots & S_{e_c} & S_{e_c-1} \end{bmatrix} \begin{bmatrix} \Lambda_1 \\ \Lambda_2 \\ \Lambda_3 \\ \Lambda_4 \\ \vdots \\ \Lambda_{e_c-1} \\ \Lambda_{e_c} \end{bmatrix} = \begin{bmatrix} -S_1 \\ -S_3 \\ -S_5 \\ -S_7 \\ \vdots \\ -S_{2e_c-3} \\ -S_{2e_c-1} \end{bmatrix}.$$

This system has a unique solution if and only if syndrome matrix A is not singular. Peterson showed that, due to the limitations for syndromes, when a code is binary one, matrix A has a determinant different from zero if there are e_c or $e_c - 1$ errors in a received word. Solutions of a system of equations for $e_c = 1$, $e_c = 2$, $e_c = 3$ and $e_c = 4$ are given in Table 6.13.

In the previous example the case of a code correcting $e_c = 2$ errors was considered and on the basis of the received word the syndrome values were calculated $S_1 = \alpha^{14}$, $S_2 = \alpha^{13}$, $S_3 = \alpha^2$, $S_4 = \alpha^{11}$. System of equations connecting syndrome and coefficients of error locator polynomial is for this case

$$A\Lambda = \begin{bmatrix} 1 & 0 \\ S_2 & S_1 \end{bmatrix} \begin{bmatrix} \Lambda_1 \\ \Lambda_2 \end{bmatrix} = \begin{bmatrix} S_1 \\ S_3 \end{bmatrix},$$

and it is easily found $\Lambda_1 = S_1 = \alpha^{14}$ as well as $S_2\Lambda_1 + S_1\Lambda_2 = S_3$, from which follows

$$\Lambda_2 = (S_3 - S_2 S_1)/S_1 = (\alpha^2 - \alpha^{13}\alpha^{14})/\alpha^{14} = -(1+\alpha+\alpha^3)/\alpha^{14} = -\alpha^7/\alpha^{14}$$
$$= \alpha^{-7+15} = \alpha^8,$$

and a same result could be obtained using relation $\Lambda_2 = (S_3 + S_1^3)/S_1$ and error locator polynomial is

Table 6.13 Solutions of a system of equations for $e_c = 1$, $e_c = 2$, $e_c = 3$ and $e_c = 4$

$e_c = 1$:	$e_c = 2$:
$\Lambda_1 = S_1$	$\Lambda_1 = S_1$, $\Lambda_2 = (S_3 + S_1^3)/S_1$
$e_c = 3$:	$e_c = 4$:
$\Lambda_1 = S_1$	$\Lambda_1 = S_1$
$\Lambda_2 = \dfrac{S_1^2 S_3 + S_5}{S_1^3 + S_3}$	$\Lambda_2 = \dfrac{S_1(S_7 + S_1^7) + S_3(S_1^5 + S_5)}{S_3(S_1^3 + S_3) + S_1(S_1^5 + S_5)}$
$\Lambda_3 = (S_1^3 + S_3) + S_1\Lambda_2$	$\Lambda_3 = (S_1^3 + S_3) + S_1\Lambda_2$
	$\Lambda_4 = ((S_5 + S_1^2 S_3) + (S_1^3 + S_3)\Lambda_2)/S_1$

$$\Lambda(x) = 1 + \alpha^{14}x + \alpha^8 x^2.$$

The same result can be obtained from error locators

$$\Lambda(x) = \prod_{i=1}^{2}(1 + X_i x) = (1 + \alpha x)(1 + \alpha^7 x) = 1 + (\alpha + (1 + \alpha + \alpha^3))x + \alpha^8 x^2$$
$$= 1 + \alpha^{14}x + \alpha^8 x^2,$$

confirming the validity of the exposed procedure.

(e) **Peterson algorithm** for decoding of a binary BCH code correcting e_c errors consists of a few phases as follows [40]:

1. Find syndromes S_j as values of the received word polynomial $r(\alpha^j)$, $j = 1, 2, \ldots, 2e_c$.
2. Form syndrome matrix A and calculate its determinant. If determinant differs from zero, go immediately to the step 4.
3. Form a new syndrome matrix by omitting last two columns and the lower two rows from the old syndrome matrix. Go back to the step 2.
4. Solve a linear equations system, on the basis of known syndrome values find coefficients of error locator polynomial and form $\Lambda(x)$.
5. Find roots of $\Lambda(x)$. If the roots are not obvious or $\Lambda(x)$ has not roots in a given field, go to the step 8.
6. Invert bits pointed out by polynomial $\Lambda(x)$ roots. If a number of corrected errors is smaller than e_c, verify whether the obtained word satisfy all $2e_c$ syndrome equations. If not, go to step 8.
7. Send at the output a correct word and stop algorithm.
8. Signal a decoding error and stop algorithm.

Procedure defined by steps 1–3 is explained in details in the previous text. However, the finding of error locator polynomial roots is not a trivial task. To find them a **Chien search** can be used, being a systematic procedure to find the error locator polynomial in a field $GF(2^m)$ [39].

In the previous example it was found that the error locator polynomial is $\Lambda(x) = 1 - \alpha^{14}x + \alpha^8 x^2$. Its factorization yields

$$\Lambda(x) = (1 + X_1 x)(1 + X_2 x) = 1 + (X_1 + X_2)x + X_1 X_2 x^2,$$

resulting in the system of equations for solving

$$X_1 + X_2 = \alpha^{14}, \quad X_1 X_2 = \alpha^8.$$

In this case even without a Chien search, the values of error locator can be easily found $X_1 = \alpha^1$, $X_2 = \alpha^7$, and a correction is carried out by inversion of the first and the eighth bit.

Polynomial corresponding to the corrected received word is now

$$\hat{r}(x) = 1 + x^4 + x^6 + x^7 + x^8,$$

and syndrome components for the corrected polynomial have the values

$$\hat{S}_1 = \hat{r}(\alpha) = 1 + \alpha^4 + \alpha^6 + \alpha^7 + \alpha^8 = 1 + (1+\alpha) + (\alpha^2 + \alpha^3) + (1+\alpha+\alpha^3)$$
$$+ (1+\alpha^2) = 0,$$
$$\hat{S}_2 = \hat{r}(\alpha^2) = 1 + (\alpha^2)^4 + (\alpha^2)^6 + (\alpha^2)^7 + (\alpha^2)^8 = 1 + (1+\alpha^2)$$
$$+ (1+\alpha+\alpha^2+\alpha^3) + (1+\alpha^3) + \alpha = 0,$$
$$\hat{S}_3 = \hat{r}(\alpha^3) = 1 + (\alpha^3)^4 + (\alpha^3)^6 + (\alpha^3)^7 + (\alpha^3)^8 = 1 + (1+\alpha+\alpha^2+\alpha^3)$$
$$+ \alpha^3 + (\alpha^2+\alpha^3) + (\alpha+\alpha^3) = 0,$$
$$\hat{S}_4 = \hat{r}(\alpha^4) = 1 + (\alpha^4)^4 + (\alpha^4)^6 + (\alpha^4)^7 + (\alpha^4)^8 = 1 + \alpha + (\alpha+\alpha^3)$$
$$+ (1+\alpha^2+\alpha^3) + \alpha^2 = 0.$$

All syndrome components are equal to zero and it can be concluded that roots α, α^2, α^3, α^4 are polynomial $\hat{r}(x)$ zeros. Otherwise speaking, this polynomial is divisible by a generator polynomial and its coefficients define a valid code word. In this case it can be considered that the decoding was successful, because decoder corrected both errors in a code word.

Problem 6.13 Field GF(2^5) is generated by the primitive polynomial $p(x) = x^5 + x^2 + 1$. Find a BCH code constructed over this field, if one generator polynomial root is $\alpha_1 = \alpha$ and a code has to correct up to three errors in a code word.

(a) If the polynomial corresponding to received word is $r(x) = x^{10}$, explain a decoding procedure by using Peterson algorithm.
(b) Explain Berlekamp algorithm for BCH codes decoding and apply it to the previous example.
(c) If a polynomial corresponding to the received word is $r(x) = 1 + x^9 + x^{11} + x^{14}$, explain decoding by applying Berlekamp algorithm.

Solution
In this case a primitive polynomial root is found on the basis

$$p(\alpha) = \alpha^5 + \alpha^2 + 1 = 0,$$

and the relation $\alpha^5 = \alpha^2 + 1$ is satisfied, allowing to form a corresponding extended field GF(32).

To construct a code correcting three errors at a code word length $n = 31$ bit, a generator polynomial should have roots α, α^2, α^3, α^4, α^5 i α^6. As given in Table 6.14, in the field there are seven conjugated roots groups, where to the generator polynomial roots correspond the groups $(\alpha, \alpha^2, \alpha^4, \alpha^8, \alpha^{16})$, $(\alpha^3, \alpha^6, \alpha^{12}, \alpha^{17}, \alpha^{24})$, $(\alpha^5, \alpha^9, \alpha^{10}, \alpha^{18}, \alpha^{20})$. Differently to the previous cases, identity element

Table 6.14 Galois field GF(2^5)

Exponential equivalent	Binary equivalent	Group	Exponential equivalent	Binary equivalent	Group
0	0 0 0 0 0	–	α^{15}	1 1 1 1 1	G
α^0	1 0 0 0 0	A	α^{16}	1 1 0 1 1	B
α^1	0 1 0 0 0	B	α^{17}	1 1 0 0 1	C
α^2	0 0 1 0 0	B	α^{18}	1 1 0 0 0	D
α^3	0 0 0 1 0	C	α^{19}	0 1 1 0 0	E
α^4	0 0 0 0 1	B	α^{20}	0 0 1 1 0	D
α^5	1 0 1 0 0	D	α^{21}	0 0 0 1 1	F
α^6	0 1 0 1 0	C	α^{22}	1 0 1 0 1	F
α^7	0 0 1 0 1	E	α^{23}	1 1 1 1 0	G
α^8	1 0 1 1 0	B	α^{24}	0 1 1 1 1	C
α^9	0 1 0 1 1	D	α^{25}	1 0 0 1 1	E
α^{10}	1 0 0 0 1	D	α^{26}	1 1 1 0 1	F
α^{11}	1 1 1 0 0	F	α^{27}	1 1 0 1 0	G
α^{12}	0 1 1 1 0	C	α^{28}	0 1 1 0 1	E
α^{13}	0 0 1 1 1	F	α^{29}	1 0 0 1 0	G
α^{14}	1 0 1 1 1	E	α^{30}	0 1 0 0 1	G

for addition is included, always existing in a field, but not important for a BCH code construction.

The code (31, 16) generator polynomial is obtained as a product of three minimal polynomials

$$g_1(x) = x^{15} + x^{11} + x^{10} + x^9 + x^8 + x^7 + x^5 + x^3 + x^2 + x + 1.$$

(a) Polynomial corresponding to the received word $r(x) = x^{10}$, in the case when the received word consists of all zeros, except at the 11th position, where is binary one. The code corrects at least three errors, the guaranteed minimum Hamming distance is $d = 2 \times 3 + 1 = 7$ and in a code there is not a code word having less than seven ones (except "all zeros" word). It is obvious that the most probable is that the received word originates just from this word consisting of all zeros, i.e. it is the most probable that the error occurred at the eleventh position. It will be now verified by the Peterson algorithm [40].
The first step is the syndrome values determining

$$S_1 = r(\alpha) = \alpha^{10}, \ S_2 = r(\alpha^2) = (\alpha^{10})^2 = \alpha^{20}, \ S_3 = r(\alpha^3) = (\alpha^{10})^3 = \alpha^{30},$$
$$S_4 = r(\alpha^4) = (\alpha^{10})^4 = \alpha^9, \ S_5 = r(\alpha^5) = (\alpha^{10})^5 = \alpha^{19}, \ S_6 = r(\alpha^6) = (\alpha^{10})^6 = \alpha^{29}.$$

Further, the syndrome matrix is constructed

$$A = \begin{bmatrix} 1 & 0 & 0 \\ \alpha^{20} & \alpha^{10} & 1 \\ \alpha^{9} & \alpha^{30} & \alpha^{20} \end{bmatrix}$$

and it is verified whether it is non-singular. The third row of matrix **A** equals to the second row raised at the power α^{29}, therefore the second and the third column should be omitted as well as two lower rows. A unity matrix (dimensions 1×1) is obtained and a single error locator polynomial coefficient is $\Lambda_1 = \alpha^{10}$, the error locator being $X_1 = \alpha^{10}$. Therefore, the conclusion is drawn that the eleventh bit was inverted and the word which has all zeros is decoded.

(b) To apply Berlekamp algorithm, firstly the syndrome polynomial should be defined

$$S(x) = S_1 x + S_2 x^2 + \ldots + S_{2e_c} x^{2e_c},$$

where, as earlier, $S_k = r(\alpha^k)$, $k = 1, 2, \ldots, 2e_c$.
Further, the **error magnitude polynomial** is formed

$$\begin{aligned}\Omega(x) &= [1 + S(x)]\Lambda(x) \\ &= (1 + S_1 x + S_2 x^2 + \ldots + S_{2e_c} x^{2e_c})(1 + \Lambda_1 x + \Lambda_2 x^2 + \ldots + \Lambda_\nu x^\nu) \\ &= 1 + \Omega_1 x + \Omega_2 x^2 + \ldots\end{aligned}$$

In this (binary) case coefficients of $\Omega(x)$ having an odd index must be zero, and a decoding reduces to finding the polynomial $\Lambda(x)$ having a power smaller or equal to e_c satisfying

$$[1 + S(x)]\Lambda(x) \equiv (1 + \Omega_2 x^2 + \Omega_4 x^4 + \ldots + \Omega_{2e_c} x^{2e_c}) x^{2e_c + 1}$$

Berlekamp algorithm can now be defined as a procedure for finding coefficients of polynomial $\Lambda(x)$, by solving a series of smaller problems which have the form

$$[1 + S(x)]\Lambda^{(2k)}(x) \equiv (1 + \Omega_2 x^2 + \Omega_4 x^4 + \ldots + \Omega_{2k} x^{2k}) x^{2k + 1}, \ k = 1, 2, \ldots, e_c$$

Firstly, it is verified whether $\Lambda^{(0)}(x) = 1$ is a solution for a case $k = 1$. If it is true, a counter k is incremented, but if it is not satisfied, a correction factor is calculated and added to $\Lambda^{(0)}$, yielding a new solution $\Lambda^{(2)}(x)$. The algorithm ingenuity is in the correction factor calculating, which is chosen so as that a new solution is not valid only for a current case, but for all previous k values. **Berlekamp algorithm** for binary BCH codes decoding can now be formulated with a few important steps [41].

1. Ser the initial conditions: $\Lambda^{(-1)}(x) = \Lambda^{(0)}(x) = 1$, $d_{-1} = 1$, $d_0 = S_1$, $l_{-1} = l_0 = 1$.
2. An improved error locator polynomial is calculated

    a. if $d_k = 0$, then $\Lambda^{(k+1)}(x) = \Lambda^{(k)}(x)$
    b. if $d_k \neq 0$, then $\Lambda^{(k+1)}(x) = \Lambda^{(k)}(x) + d_k\left[d_\rho^{-1}x^{k-\rho}\cdot\Lambda^{(\rho)}(x)\right]$

3. Find d_k as coefficient of x^{k+1} in the product $\Lambda^{(k)}(x)[1 + S(x)]$, i.e.

$$d_k = S_{k+1} + \Lambda_1^{(k)}S_k + \ldots + \Lambda_1^{(k)}S_{k+1-l_k}$$

4. Find $l_{k+1} = \max(l_k + k - \rho)$
5. Set $k = k + 1$. If $k < 2e_c$, go to step 2.
6. Find roots of $\Lambda(x) = \Lambda^{(2t)}(x)$, correct the corresponding locations in the received word and stop algorithm.
7. If the previous step cannot be carried out, signal a decoding error.

an efficient implementation. A complexity of Peterson technique increases as the square of the number of corrected errors, yielding an efficient application for binary BCH decoders correcting a small number of errors. On the other hand, the Berlekamp algorithm complexity increases linearly, allowing forming the efficient decoders correcting more tens of errors.

Table 6.15 Berlekamp algorithm for a Problem 6.13b

K	$\Lambda^{(k)}(x)$	d_k	l_k
−1	1	1	0
0	1	α^{10}	0
1	$1 + \alpha^{10}x$	0	1
2	$1 + \alpha^{10}x$	0	1
3	$1 + \alpha^{10}x$	0	1
4	$1 + \alpha^{10}x$	0	1
5	$1 + \alpha^{10}x$	0	1
6	$1 + \alpha^{10}x$		

Table 6.16 Berlekamp algorithm for a Problem 6.13c

K	$\Lambda^{(k)}(x)$	d_k	l_k
−1	1	1	0
0	1	1	0
1	$1 + x$	0	1
2	$1 + x$	α^3	1
3	$1 + x + \alpha^3x^2$	0	2
4	$1 + x + \alpha^3x^2$	α^{20}	2
5	$1 + x + \alpha^{16}x^2 + \alpha^{17}x^3$	0	3
6	$1 + x + \alpha^{16}x^2 + \alpha^{17}x^3$		

For the previously considered example, polynomial corresponding to the received word is $r(x) = x^{10}$ and syndrome components are $S_1 = \alpha^{10}$, $S_2 = \alpha^{20}$, $S_3 = \alpha^{30}$, $S_4 = \alpha^9$, $S_5 = \alpha^{19}$, $S_6 = \alpha^{29}$. Syndrome polynomial is $S(x) = \alpha^{10}x + \alpha^{20}x^2 + \alpha^{30}x^3 + \alpha^9 x^4 + \alpha^{19}x^5 + \alpha^{29}x^6$, and algorithm continues as shown in Table 6.15.

Error locator polynomial is $\Lambda(x) = 1 + \alpha^{10}x$ pointing to error at the 11th position, and the word "all zeros" is decoded. As a difference to Peterson algorithm, $v < e_c$ errors are successfully corrected without the additional procedure decoding modifications.

(c) For a polynomial corresponding to the received word $r(x) = 1 + x^9 + x^{11} + x^{14}$, syndrome polynomial is $S(x) = x + x^2 + \alpha^{29}x^3 + x^4 + \alpha^{23}x^5 + \alpha^{27}x^6$, and a decoding procedure is shown in details in Table 6.16.

Error locator polynomial now can be factored

$$\Lambda(x) = 1 + x + \alpha^{16}x^2 + \alpha^{17}x^3 = (1 + \alpha^{13}x)(1 + \alpha^{16}x)(1 + \alpha^{19}x),$$

pointing to errors at the positions corresponding to α^{13}, α^{16} and α^{19} and the correct received word is

$$r(x) = 1 + x^9 + x^{11} + x^{13} + x^{14} + x^{16} + x^{19} = (x^4 + x + 1)g(x).$$

Problem 6.14 GF(2^3) is generated by a primitive polynomial $p(x) = x^3 + x^2 + 1$. Find Reed-Solomon code constructed over this field, write one code word, find the code rate, minimum Hamming distance and draw code distance spectrum, if one generator polynomial root is $\alpha_1 = \alpha$ and if

(a) the code can correct one error in the code word,
(b) the code can correct two errors in the code word,
(c) the code can correct three errors in the code word.

Solution Galois field GF(2^3) construction was explained in one of the previous problems and field structure is given in Table 6.10. As previously explained, a procedure for BCH construction guarantees the code correcting at least the number of errors as required. However, it might happen that a code in some cases corrects more errors than defined in advance, as illustrated in Problems 6.1 and 6.2. This BCH feature, sometimes undesirable, is a consequence of the fact that generator polynomial has a greater number of successive roots than directly required by code construction procedure (conjugate roots are included). In this problem it will be illustrated how this drawback can be avoided, if the requirement for a binary code is given up.

Reed-Solomon codes are a special class of nonbinary BCH codes. For a difference from binary BCH codes, here the code symbols and generator polynomial roots as well are from the same field. Because of that, generator polynomial of Reed-Solomon code for correcting e_c errors is not a minimal polynomials product, but a direct product of factors $\alpha^{m_0}, \alpha^{m_0+1}, \alpha^{m_0+2}, \ldots, \alpha^{m_0+2e_c-1}$ where usually $m_0 = 1$ is chosen, because the element α is primitive by rule [42]. In such a way the code having word length $n = 2^m - 1$ symbols is obtained where the number of redundant symbols is always $n - k = 2e_c$.

(a) To construct a code to correct one symbol at the code word having length $n = 7$ nonbinary symbols, a generator polynomial should have roots α and α^2. Number of redundant symbols is $n - k = 2$, and the obtained code is (7, 5). A generator polynomial is obtained directly

$$g(x) = (x+\alpha)(x+\alpha^2) = \alpha^3 + x(\alpha + \alpha^2) + x^2 = \alpha^3 + \alpha^6 x + x^2,$$

One possible information word is 10001 and information polynomial is $i(x) = 1 + x^4$. Corresponding code polynomial is $c(x) = i(x)g(x) = \alpha^3 + \alpha^6 x + x^2 + \alpha^3 x^4 + \alpha^6 x^5 + x^6$ and the resulting code word is $\alpha^3 \alpha^6 1 0 \alpha^3 \alpha^6 1$. Every polynomial coefficient in GF(8) can be represented by one three-bit and it can be said that the considered Reed-Solomon encoder transformed information sequence (100,000,000,000,100) into the code sequence (101,011,100,000,101,011,100).

Encoder transforms 5 information bits (represented by 15 bits) into 7 code symbols (represented by 21 bit) and it is obvious that code rate is $R = 5/7$. It is interesting to note that an increasing of the numbers of bits in the code word (relating to BCH) is additionally justified by the fact that this code can correct as well double and triple errors occurred in one nonbinary symbol.

Every Reed-Solomon code is MDS code, meaning that a Singleton bound is satisfied with equality. In this case $d_{\min} = n - k + 1 = 3$, and a code can correct one error. It is known that for any MDS code defined over a field GF (q) a code distances spectrum can be found by using the relation [27]

$$A_i = \binom{n}{i}(q-1) \sum_{j=0}^{i-d_{\min}} (-1)^j \binom{i-1}{j} q^{i-d_{\min}-j}, \quad i=d_{\min}, d_{\min}+1, \ldots, n$$

where n is a code word length, $d_{\min}$ is minimum Hamming length while A_i denotes a number of code words having i nonzero symbols. In this case the relation becomes

$$A_i = 7\binom{7}{i} \sum_{j=0}^{i-3} (-1)^j \binom{i-1}{j} 8^{i-3-j}, \quad i=3, 4, \ldots, 7.$$

It is interesting to note that at any word position can be one from $q = 8$ symbols and the code words number is $q^k = 8^5 = 32768$. More, it should be also noticed that there is always one code word comprising binary zeros only

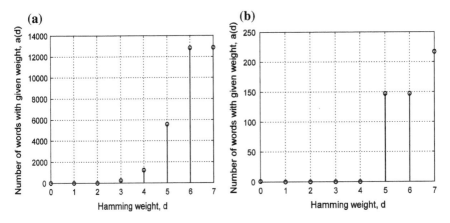

Fig. 6.18 Code distances spectrum of Reed-Solomon codes RS(7, 5) (a) and RS(7, 3) (b)

and it is obvious that code distances spectrum has a shape as shown in Fig. 6.18a. Although the code has the same Hamming distance as BCH code constructed for $e_c = 1$ in Problem 6.1, its code rate is increased for some amount, what is in any case desirable.

(b) To construct a RS code correcting two symbol errors in the code word length $n = 7$, roots of generator polynomial should be α, α^2, α^3, α^4, and for the code (7, 3) it is obtained

$$g(x) = (x+\alpha)(x+\alpha^2)(x+\alpha^3)(x+\alpha^4) = \alpha^3 + x + \alpha^3 x^2 + \alpha^2 x^3 + x^4.$$

Code polynomial corresponding to information word 100 is $c(x) = g(x) = \alpha^3 + x + \alpha^3 x^2 + \alpha^2 x^3 + x^4$ and the obtained code word is $\alpha^3 1 \alpha^3 \alpha^2 100$. Every polynomial coefficient in GF(8) can be represented by one three-bit word and a corresponding code sequence is (101,100,101,001,100,000,000). At the encoder output $8^3 = 512$ various code words can appear, code rate is $R = 3/7$, minimum Hamming distance is $d_{min} = n - k + 1 = 5$, the code spectrum is shown in Fig. 6.18b.

(c) To construct a code correcting three errors in the code word length $n = 7$, roots of generator polynomial should be α, α^2, α^3, α^4, α^5, α^6, and it is obtained

$$g(x) = (x+\alpha)(x+\alpha^2)(x+\alpha^3)(x+\alpha^4)(x+\alpha^5)(x+\alpha^6)$$
$$= 1 + x + x^2 + x^3 + x^4 + x^5 + x^6.$$

The coefficients of generator polynomial are binary and a code is a binary one. It is obvious that a code RS(7, 1) reduces to BCH code (7, 1) equivalent to sevenfold repetition code, where $d_{min} = 7$, the corresponding spectrum shown in Fig. 6.15b.

Problem 6.15 Field GF(2^4) is generated by the primitive polynomial $p(x) = x^4 + x + 1$. Construct RS code over this field and find code distances spectrum under conditions that a code corrects three errors as well as that a code word length is a maximum possible. Analyze code properties after shortening where two most significant information symbols are omitted.

Solution

Galois field GF(2^4) construction by using polynomial $p(x) = x^4 + x + 1$ is explained in Problem 6.2. To construct a code correcting three errors in code word length $n = 15$, roots of generator polynomial should be $\alpha, \alpha^2, \alpha^3, \alpha^4, \alpha^5, \alpha^6$, and it is obtained

$$g(x) = (x+\alpha)(x+\alpha^2)(x+\alpha^3)(x+\alpha^4)(x+\alpha^5)(x+\alpha^6)$$
$$= \alpha^6 + \alpha^9 x + \alpha^6 x^2 + \alpha^4 x^3 + \alpha^{14} x^4 + \alpha^{10} x^5 + x^6,$$

and the generating matrix of the corresponding code (15, 9) is

$$G = \begin{bmatrix} \alpha^6 & \alpha^9 & \alpha^6 & \alpha^4 & \alpha^{14} & \alpha^{10} & 1 & 0 & 0 & 0 & 0 & 0 & 0 & 0 & 0 \\ 0 & \alpha^6 & \alpha^9 & \alpha^6 & \alpha^4 & \alpha^{14} & \alpha^{10} & 1 & 0 & 0 & 0 & 0 & 0 & 0 & 0 \\ 0 & 0 & \alpha^6 & \alpha^9 & \alpha^6 & \alpha^4 & \alpha^{14} & \alpha^{10} & 1 & 0 & 0 & 0 & 0 & 0 & 0 \\ 0 & 0 & 0 & \alpha^6 & \alpha^9 & \alpha^6 & \alpha^4 & \alpha^{14} & \alpha^{10} & 1 & 0 & 0 & 0 & 0 & 0 \\ 0 & 0 & 0 & 0 & \alpha^6 & \alpha^9 & \alpha^6 & \alpha^4 & \alpha^{14} & \alpha^{10} & 1 & 0 & 0 & 0 & 0 \\ 0 & 0 & 0 & 0 & 0 & \alpha^6 & \alpha^9 & \alpha^6 & \alpha^4 & \alpha^{14} & \alpha^{10} & 1 & 0 & 0 & 0 \\ 0 & 0 & 0 & 0 & 0 & 0 & \alpha^6 & \alpha^9 & \alpha^6 & \alpha^4 & \alpha^{14} & \alpha^{10} & 1 & 0 & 0 \\ 0 & 0 & 0 & 0 & 0 & 0 & 0 & \alpha^6 & \alpha^9 & \alpha^6 & \alpha^4 & \alpha^{14} & \alpha^{10} & 1 & 0 \\ 0 & 0 & 0 & 0 & 0 & 0 & 0 & 0 & \alpha^6 & \alpha^9 & \alpha^6 & \alpha^4 & \alpha^{14} & \alpha^{10} & 1 \end{bmatrix}.$$

In general case a Reed-Solomon code word is found as

$$c(x) = i(x)g(x) = (i_0 + i_1 x + i_2 x^2 + \ldots + i_k x^k)g(x) = c_0 + c_1 x + c_2 x^2 + \ldots + c_n x^n,$$

and a code shortening is carried out by omitting some number (in general case—s) information bits (symbols) usually at the most significant positions.

$$c'(x) = (i_0 + i_1 x + i_2 x^2 + \ldots + i_{k-s} x^{k-s})g(x) = c_0 + c_1 x + c_2 x^2 + \ldots + c_{n-s} x^{n-s}.$$

As ith row of generator matrix corresponds to polynomial $x^i g(x)$, it is easy to show that a shortening operation corresponds to omitting last s rows and columns of

Table 6.17 Galois field GF(2^3)

Exponential equivalent	Polynomial equivalent			Binary equivalent
α^0	1			1 0 0
α^1		α		0 1 0
α^2			α^2	0 0 1
α^3	1+	α		1 1 0
α^4		$\alpha+$	α^2	0 1 1
α^5	1+	$\alpha+$	α^2	1 1 1
α^6	1+		α^2	1 0 1

a matrix **G**. The generator polynomial is not changed and a minimum Hamming distance (code capability for error correction) is not changed by shortening as well. By omitting two last information symbols, code word length becomes $n' = 13$, generator matrix dimensions become 13 × 7, code rate is $R' = 7/13$ (<$R = 9/15$) but $d_{min} = 7$ does not change, and $e_c = 3$.

Problem 6.16 Field GF(2^3) is generated by the primitive polynomial $p(x) = x^3 + x + 1$. Find a Reed-Solomon code constructed over this field, if one generator polynomial root is $\alpha_1 = \alpha$ and a code has to correct two errors in the code word.

(a) Explain a procedure for obtaining systematic Reed-Solomon code and draw the block scheme of a corresponding encoder.
(b) Starting from a nonsystematic code version form a shortened RS(5, 2) code and draw its spectrum.

Solution
Galois field GF(8) construction by using polynomial $p(x) = x^3 + x + 1$ is given in Table 6.17 (not the same as in Problem 6.1!). Minimal polynomials corresponding to conjugate roots are not given, because they are not interesting for RS code construction.

(a) Reed-Solomon code correcting two errors in the code word length $n = 7$ has the generator polynomial

$$g(x) = (x+\alpha)(x+\alpha^2)(x+\alpha^3)(x+\alpha^4) = \alpha^6 + \alpha^5 x + \alpha^5 x^2 + \alpha^2 x^3 + x^4.$$

Let the input sequence is (110010101). From Table 6.17 it can be concluded that the corresponding symbols are $\alpha^3 \, \alpha^1 \, \alpha^6$.

Procedure to obtain the remainder after division by a generator polynomial can be symbolically written as follows:

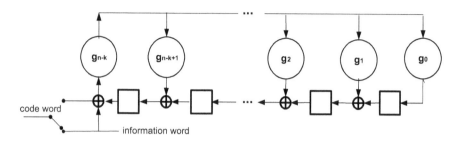

Fig. 6.19 Reed-Solomon systematic encoder

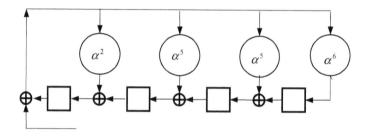

Fig. 6.20 Division circuits as a main element of RS(7, 3) systematic encoder

$$\alpha^5\alpha^0\alpha^5\ 0\ 0\ 0\ 0\ :\ \alpha^0\alpha^2\alpha^5\alpha^5\alpha^6 = \alpha^5\alpha^2$$
$$\underline{\alpha^5\alpha^0\alpha^3\alpha^3\alpha^4}$$
$$\alpha^2\alpha^3\alpha^4\ 0\ 0$$
$$\underline{\alpha^2\alpha^4\alpha^0\alpha^0\alpha^1}$$
$$\alpha^6\alpha^5\alpha^0\alpha^1$$

and the corresponding cyclic code word is ($\alpha^5\ \alpha^0\ \alpha^5\ \alpha^6\ \alpha^5\ \alpha^0\ \alpha^1$), the binary equivalent is (111001111101111001010). Block scheme of **systematic Reed-Solomon encoder** for any generator polynomial is given in Fig. 6.19. The functioning is the same as for binary codes—at a link the information bits are send firstly and after them the remainder polynomial coefficients found in delay cells of the circuit for division by polynomial $g(x)$. Circuit for division, as a main encoder element, is shown in details in Fig. 6.20 for a considered case.

(b) Code polynomial of the nonsystematic (7, 3) code is

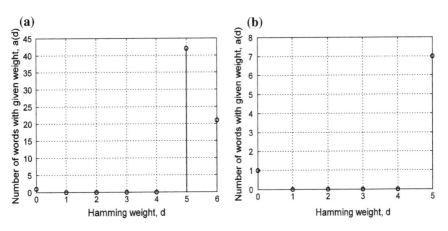

Fig. 6.21 Code distance spectrum of the shortened codes RS(6, 1) (a) and RS(5, 1) (b)

$$c(x) = i(x)g(x) = (i_0 + i_1 x + i_2 x^2)(\alpha^6 + \alpha^5 x + \alpha^5 x^2 + \alpha^2 x^3 + x^4),$$

as every code word symbol can take one from eight field values, total number of code words is $8^3 = 512$.

The omitting of the last information symbol yields the code (6, 2) having $8^2 = 64$ code words, the word length $n = 6$, with the corresponding polynomial

$$c'(x) = (i_0 + i_1 x)(\alpha^6 + \alpha^5 x + \alpha^5 x^2 + \alpha^2 x^3 + x^4),$$

while the omitting the last two information symbols yields the code (5, 1), with the corresponding polynomial

$$c''(x) = i_0(\alpha^6 + \alpha^5 x + \alpha^5 x^2 + \alpha^2 x^3 + x^4).$$

Code distance spectra of shortened codes RS(6, 2) and RS(5, 1) are shown in Figs. 6.21a, b. It is obvious that a minimum Hamming distance for both codes is $d_{min} = 5$, and they can, as the code RS(7, 3), by which shortening they were obtained, correct two errors in the code word length.

Problem 6.17 Consider a primitive RS code correcting double errors, constructed over the field GF(8), generated by the primitive polynomial $p(x) = x^3 + x + 1$.

(a) Write the equations connecting syndrome components and error locator polynomial coefficients, as well as the equations connecting syndrome components, error positions and error magnitudes. Verify whether these relations are satisfied if the transmitted word "all zeros" was sent and the errors occurred at the first and the fifth code word symbols, having magnitudes α^4 i α^5, respectively.

(b) By using Peterson-Gorenstein-Zierler algorithm decode a received word whose polynomial is $r(x) = \alpha^5 x^2 + x^3 + \alpha^2 x^4 + \alpha^2 x^6$.
(c) By using Peterson-Gorenstein-Zierler algorithm decode a received word whose polynomial is $r(x) = \alpha^5 x^2 + \alpha^3 x^3 + \alpha x^4 + \alpha^5 x^5 + \alpha^2 x^6$.

Solution

(a) The relation connecting syndrome components and error locator polynomial coefficients, when during the transmission a number of errors occurred being just equal to the number of errors ($v = e_c$) correctable by Reed-Solomon code is [40]

$$\mathbf{A'\Lambda} = \begin{bmatrix} S_1 & S_2 & S_3 & S_4 & \cdots & S_{e_c-1} & S_{e_c} \\ S_2 & S_3 & S_4 & S_5 & \cdots & S_{e_c} & S_{e_c+1} \\ S_3 & S_4 & S_5 & S_6 & \cdots & S_{e_c+1} & S_{e_c+2} \\ S_4 & S_5 & S_6 & S_7 & \cdots & S_{e_c+2} & S_{e_c+3} \\ \vdots & \vdots & \vdots & \vdots & \ddots & \vdots & \vdots \\ S_{e_c-1} & S_{e_c} & S_{e_c+1} & S_{e_c+2} & \cdots & S_{2e_c-3} & S_{2e_c-2} \\ S_{e_c} & S_{e_c+1} & S_{e_c+2} & S_{e_c+3} & \cdots & S_{2e_c-2} & S_{2e_c-1} \end{bmatrix} \begin{bmatrix} \Lambda_{e_c} \\ \Lambda_{e_c-1} \\ \Lambda_{e_c-2} \\ \Lambda_{e_c-3} \\ \vdots \\ \Lambda_2 \\ \Lambda_1 \end{bmatrix} = \begin{bmatrix} S_{e_c+1} \\ S_{e_c+2} \\ S_{e_c+3} \\ S_{e_c+4} \\ \vdots \\ S_{2e_c-1} \\ S_{2e_c} \end{bmatrix}.$$

It can be shown that matrix $\mathbf{A'}$ is nonsingular if in a received word there are just e_c errors, while $\mathbf{A'}$ is singular if less than e_c errors occurred. If matrix $\mathbf{A'}$ is singular, then the last right column and the lower row are omitted and the determinant of new obtained matrix is calculated, the procedure repeating until the resulting matrix becomes nonsingular. Error locator polynomial coefficients are then found by using standard linear algebraic techniques and the corresponding polynomial roots define error positions X_i. Finally, error magnitudes e_{Xi} can be found from the matrix relation

$$\mathbf{Be} \begin{bmatrix} X_1 & X_2 & \cdots & X_v \\ X_1^2 & X_2^2 & \cdots & X_v^2 \\ \vdots & \vdots & \ddots & \vdots \\ X_1^v & X_2^v & \cdots & X_v^v \end{bmatrix} \begin{bmatrix} e_{\log_\alpha X_1} \\ e_{\log_\alpha X_2} \\ \vdots \\ e_{\log_\alpha X_v} \end{bmatrix} = \begin{bmatrix} S_1 \\ S_2 \\ \vdots \\ S_v \end{bmatrix}.$$

In the considered example $X_1 = \alpha^0$, $e_1 = \alpha^4$, $X_2 = \alpha^4$, $e_2 = \alpha^5$ and $\Lambda(x) = (1 + X_1 x)(1 + X_2 x) = \alpha^4 x^2 + \alpha^5 x + 1$ and $r(x) = \alpha^4 + \alpha^5 x^4$. Because of $e_c = 2$, syndrome components are $S_1 = r(\alpha) = \alpha^4 + \alpha^2 = \alpha$, $S_2 = r(\alpha^2) = \alpha^4 + \alpha^6 = \alpha^3$, $S_3 = r(\alpha^3) = \alpha^4 + \alpha^3 = \alpha^6$ and $S_4 = r(\alpha^4) = \alpha^4 + \alpha^0 = \alpha^5$, and it is easy to verify the following

$$\mathbf{A'\Lambda} = \begin{bmatrix} \alpha & \alpha^3 \\ \alpha^3 & \alpha^6 \end{bmatrix} \begin{bmatrix} \alpha^4 \\ \alpha^5 \end{bmatrix} = \begin{bmatrix} \alpha^6 \\ \alpha^5 \end{bmatrix}, \quad \mathbf{Be} = \begin{bmatrix} \alpha^4 & \alpha^5 \\ \alpha & \alpha^3 \end{bmatrix} \begin{bmatrix} \alpha^0 \\ \alpha^4 \end{bmatrix} = \begin{bmatrix} \alpha \\ \alpha^3 \end{bmatrix}.$$

(b) **Peterson-Gorenstein-Zierler algorithm** is based on a few steps [40, 43]:
  1. Find syndromes for a received word: $\{S_l\} = \{r(\alpha^l)\}$, $l = 1, 2, 3, \ldots, 2e_c$, then construct syndrome matrix $\mathbf{A'}$.
  2. Calculate syndrome matrix determinant. If it equals zero, construct a new matrix by omitting the last right and the lowest row from the previous syndrome matrix. Shorten Λ for one position by erasing the coordinate Λ_t. Repeat this step until the determinant becomes not equal to zero.
  3. Find polynomial $\Lambda(x)$ coefficients and its roots as well. If the roots are not obvious or $\Lambda(x)$ has not roots in the observed field go to step 7.
  4. Construct matrix $\mathbf{B}$ and solve it to find error magnitudes.
  5. Subtract error magnitudes from the values in the corresponding positions of the received word.
  6. Deliver the corrected word and stop algorithm.
  7. Signal decoding error and stop algorithm.

In this case the received polynomial $r(x) = \alpha^5 x^2 + x^3 + \alpha^2 x^4 + \alpha^2 x^6$ determines syndrome components $S_1 = \alpha^6$, $S_2 = \alpha^3$, $S_3 = \alpha^4$, $S_4 = \alpha^3$ and the following relation is valid

$$\mathbf{A'}\Lambda \begin{bmatrix} \alpha^6 & \alpha^3 \\ \alpha^3 & \alpha^4 \end{bmatrix} \begin{bmatrix} \Lambda_2 \\ \Lambda_1 \end{bmatrix} = \begin{bmatrix} \alpha^4 \\ \alpha^3 \end{bmatrix}.$$

Matrix $\mathbf{A'}$ is nonsingular, a method of simple substitution yields error locator polynomial coefficients $\Lambda_2 = \alpha$, $\Lambda_1 = \alpha^2$ and $\Lambda(x) = \alpha x^2 + \alpha^2 x + 1 = (1 + \alpha^3 x)(1 + \alpha^5 x)$, the errors occurred at the fourth and the sixth code word symbol. Error magnitudes are then found from the following system of equations

$$\mathbf{B}e = \begin{bmatrix} \alpha^3 & \alpha^5 \\ \alpha^6 & \alpha^3 \end{bmatrix} \begin{bmatrix} e_3 \\ e_5 \end{bmatrix} = \begin{bmatrix} \alpha^6 \\ \alpha^3 \end{bmatrix} \Rightarrow \begin{cases} \alpha^3 e_3 + \alpha^5 e_5 = \alpha^6 \\ \alpha^6 e_3 + \alpha^3 e_5 = \alpha^3 \end{cases}$$

from which by a simple substitution it is found that the error at the fourth position has the magnitude α, while the error at the sixth symbol has the magnitude α^5. Finally, error polynomial becomes

$$e(x) = e_{\log_\alpha X_1} x^{\log_\alpha X_1} + e_{\log_\alpha X_2} x^{\log_\alpha X_2} = \alpha x^3 + \alpha^5 x^5,$$

and a nearest code word is found by adding error polynomial and the received polynomial

$$c(x) = r(x) + e(x) = (\alpha^5 x^2 + x^3 + \alpha^2 x^4 + \alpha^2 x^6) + (\alpha x^3 + \alpha^5 x^5)$$
$$= \alpha^5 x^2 + \alpha^3 x^3 + \alpha^2 x^4 + \alpha^5 x^5 + \alpha^2 x^6.$$

Finally, it is suitable to verify do the obtained result is a code word. Generator RS polynomial correcting double errors at the code word length $n = 7$ is

$$g(x) = (x - \alpha)(x - \alpha^2)(x - \alpha^3)(x - \alpha^4) = x^4 + \alpha^3 x^3 + x^2 + \alpha x + \alpha^3,$$

and it is easy to verify that the code word can be written as $c(x) = \alpha^2 x^2 g(x)$.

(c) A Peterson-Gorenstein-Zierler algorithm is applied as well, but in this case the received polynomial is $r(x) = \alpha^2 x^6 + \alpha^5 x^5 + \alpha x^4 + \alpha^3 x^3 + \alpha^5 x^2$ and syndrome components are $S_1 = \alpha$, $S_2 = \alpha^5$, $S_3 = \alpha^2$, $S_4 = \alpha^4$. This time one starts from a matrix

$$A' = \begin{bmatrix} \alpha & \alpha^5 \\ \alpha^5 & \alpha^2 \end{bmatrix},$$

which is singular, because the second row equals to the first multiplied by α^4. A matrix equation reduces to scalar equality $\mathbf{A''\Lambda} = \alpha \Lambda_1 = \alpha^5$ from which easily follows $\Lambda_1 = \alpha^4$ and $\Lambda(x) = 1 + \alpha^4 x$. Therefore, an error occurred at the fifth code word symbol.

Error magnitude is found using

$$\mathbf{B}e = \alpha^4 e_4 = \alpha,$$

from which easily follows that the error at the fourth symbol has the magnitude α^4 and the error polynomial becomes

$$e(x) = e_{i1} x^{\log_\alpha X_1} = \alpha^4 x^4,$$

and a nearest code word is found by adding error polynomial and the received polynomial

$$c(x) = r(x) + e(x) = (\alpha^2 x^6 + \alpha^5 x^5 + \alpha x^4 + \alpha^3 x^3 + \alpha^5 x^2) + (\alpha^4 x^4)$$
$$= \alpha^2 x^6 + \alpha^5 x^5 + \alpha^2 x^4 + \alpha^3 x^3 + \alpha^5 x^2.$$

Although the fifth symbol of received word is α and the fifth code word symbol is α^2 it should be noted that the error magnitude is a very large being α^4. However, it is only a consequence of the way for defining addition in a field GF(8) shown in Table 6.15.

Problem 6.18 Consider a primitive RS code correcting double errors, constructed over the field GF(8), generated by the primitive polynomial $p(x) = x^3 + x + 1$.

(a) Explain in short Berlekamp-Massey algorithm for RS codes decoding.
(b) By applying Berlekamp-Massey algorithm decode the word $r(x) = \alpha^5 x^2 + \alpha^3 x^3 + \alpha x^4 + \alpha^5 x^5 + \alpha^2 x^6$.

Solution

(a) **Berlekamp-Massey algorithm** for RS codes decoding starts from the fact that the error position polynomial does not depend on error magnitudes in a code word. It makes possible to use Berlekamp algorithm to find the error position, and later on to determine error magnitudes and to do their correction. A complete algorithm is divided into a few steps as follows [41, 44]:

1. Find syndromes for a received word: $\{S_l\} = \{r(\alpha^l)\}$, $l = 1, 2, 3, \ldots, 2e_c$
2. Set initial conditions $\Lambda^{(-1)}(x) = \Lambda^{(0)}(x) = 1$, $d_{-1} = 1$, $d_0 = S_1$, $l_{-1} = l_0 = 1$.
3. Calculate improved error locator polynomial

   a. if $d_k = 0$ then $\Lambda^{(k+1)}(x) = \Lambda^{(k)}(x)$
   b. if $d_k \neq 0$ then $\Lambda^{(k+1)}(x) = \Lambda^{(k)}(x) + d_k \left[d_\rho^{-1} x^{k-\rho} \Lambda^{(\rho)}(x) \right]$

4. Calculate d_k as a coefficient of the power x^{k+1} in the product $\Lambda^{(k)}(x)[1 + S(x)]$, i.e.

$$d_k = S_{k+1} + \Lambda_1^{(k)} S_k + \ldots + \Lambda_{lk}^{(k)} S_{k+1-lk}$$

5. Calculate $l_{k+1} = \max(l_k, l_\rho + k - \rho)$
6. Set $k = k + 1$. If $k < 2e_c$, go to the step 3.
7. Find the roots of $\Lambda(x) = \Lambda^{(2t)}(x)$, correct corresponding locations in a received word and stop the algorithm.
8. If the previous step is not possible, signal decoding error.
9. Form error magnitude polynomial—a main equation for RS code decoding:

$$\begin{aligned}\Omega(x) &= [1 + S(x)]\Lambda(x) x^{2e_c+1} \\ &= 1 + (S_1 + \Lambda_1)x + x^2(S_2 + S_1\Lambda_1 + \Lambda_2) + \ldots \\ &\quad + x^{e_c}(S_{e_c} + S_{e_c-1}\Lambda_1 + S_{e_c-2}\Lambda_2 + \ldots + S_{e_c})\end{aligned}$$

10. If v is a number of occurred errors, error magnitude at the position $\log_\alpha X_i$ is

$$e_{\log_\alpha X_i} = z(X_i^{-1}) / \prod_{\substack{k=1 \\ k \neq i}}^{v} (1 + X_k X_i^{-1})$$

11. Form error polynomial $e(x) = e_{\log_\alpha X_1} x^{\log_\alpha X_1} + \ldots + e_{\log_\alpha X_2} x^{\log_\alpha X_2}$, add it to the received word polynomial to find the nearest code word $c'(x)$.

Table 6.18 Berlekamp-Massey algorithm for Problem 6.18c

k	$\Lambda^{(k)}(x)$	d_k	l_k
-1	1	1	0
0	1	α	0
1	$1 + \alpha x$	α^3	1
2	$1 + \alpha^4 x$	0	1
3	$1 + \alpha^4 x$	0	1
4	$1 + \alpha^4 x$	–	1

12. If $c'(x)$ is not divisible by $g(x)$ signal a decoding error.

It is obvious that the essential difference relating to Berlekamp algorithm for BCH codes decoding are the last four steps, allowing finding the error magnitude and its correction.

(b) In this case the received polynomial is $r(x) = \alpha^5 x^2 + \alpha^3 x^3 + \alpha x^4 + \alpha^5 x^5 + \alpha^2 x^6$, corresponding syndromes are $S_1 = \alpha$, $S_2 = \alpha^5$, $S_3 = \alpha^2$, $S_4 = \alpha^4$. The first eight algorithm steps are illustrated in Table 6.18.

Syndrome polynomial is $S(x) = \alpha x + \alpha^5 x^2 + \alpha^2 x^3 + \alpha^4 x^4$, error locator polynomial $\Lambda(x) = 1 + \alpha^4 x$ (to which corresponds error locator $X_1 = \alpha^4$) and the error magnitude polynomial can be formed

$$\Omega(x) = \Lambda(x)[1 + S(x)] x^5 = (1 + \alpha^4 x)(1 + \alpha x + \alpha^5 x^2 + \alpha^2 x^3 + \alpha^4 x^4) x^5$$

and a part of error magnitude polynomial up to a power $e_c = 2$ is $\Omega(x) = 1 + \alpha^2 x$.

Error magnitudes are

$$e_{\log_\alpha X_i} = \frac{\Omega(X_i^{-1})}{1 + X_k X_i^{-1}}, i = 1, 2; k \neq i.$$

Although a code was designed to correct $e_c = 2$ errors, in this case only one error occurred ($v = 1$) and its magnitude is

$$e_4 = \Omega(\alpha^{-4})/1 = 1 + \alpha^2(\alpha^3) = 1 + \alpha^5 = \alpha^4$$

and error polynomial is $e(x) = \alpha^4 x^4$, the nearest code word is found by adding error polynomial and received polynomial

$$c(x) = r(x) + e(x) = (\alpha^5 x^2 + \alpha^3 x^3 + \alpha x^4 + \alpha^5 x^5 + \alpha^2 x^6) + (\alpha^4 x^4)$$
$$= \alpha^5 x^2 + \alpha^3 x^3 + \alpha^2 x^4 + \alpha^5 x^5 + \alpha^2 x^6 = \alpha^2 x^2 g(x).$$

Problems

Problem 6.19 Consider a primitive RS code correcting double errors, constructed over a field GF(8), generated by the primitive polynomial $p(x) = x^3 + x + 1$.

(a) Explain extended Euclidean algorithm for finding the greatest common divisor of two integers.
(b) Explain shortly Euclidean algorithm for RS codes decoding.
(b) Decode the received word $r(x) = \alpha^5 x^2 + x^3 + \alpha^2 x^4 + \alpha^2 x^6$ using Euclidean algorithm.
(c) Decode the received word $r(x) = \alpha x^2 + \alpha^5 x^4$ using Euclidean algorithm.

Solution

(a) The Euclidean algorithm is an efficient method for finding the greatest common divisor (GCD) for a group of elements. It can be expressed as a linear combination of these elements. Extended form of this algorithm, besides the finding GCD, makes possible to find the corresponding coefficients.

Let a GCD for two elements (a,b) should be found. If the initial conditions $r_{-1} = a$, $r_0 = b$, $s_{-1} = 1$, $s_0 = 0$, $t_{-1} = 0$, $t_0 = 1$ are given, the algorithm continues according to the following recursive relations

$$r_i = r_{i-2} - q_i r_{i-1}$$
$$s_i = s_{i-2} - q_i s_{i-1}$$
$$t_i = t_{i-2} - q_i t_{i-1}$$

where in every step the parameter q_i is chosen providing that the relation $r_i < r_{i-1}$ is satisfied. Algorithm stops when the remainder $r_n = 0$. The remainder r_{n-1} is GCD(a,b). Recursive relations are chosen so as that in every (ith) iteration provide

$$s_i a + t_i b = r_i$$

Table 6.19 illustrates the procedure for finding GCD(322, 115) and coefficients in linear combination for every iteration. It is obvious that GCD for 322 and 115 is 23, and the in every iteration the above relation is satisfied.

Table 6.19 Euclidean extended algorithm for finding the greatest common divisor for two positive integers

i	q_i	r_i	s_i	t_i	$s_i a + t_i b$
−1	–	a = 322	1	0	–
0	–	b = 115	0	1	–
1	2	92	1	−2	1 × 322 − 2 × 115 = 92
2	1	23	−1	3	−1 × 322 + 3 × 115 = 23
3	4	0	5	−14	5 × 322 − 14 × 115 = 0

(b) ***Euclidean algorithm*** for RS codes decoding is defined by the following steps [39]

1. Find syndrome polynomial $S(x) = \sum_{j=1}^{n-k} s_j x^{j+1}$. If $S(x) = 0$, the received vector corresponds to a code word. If it is not true, go to the next step.
2. Set the following initial conditions: $\Omega_{-1}(x) = x^{2e_c+1}$, $\Omega_0(x) = 1 + S(x)$, $\Lambda_{-1}(x) = 0$, $\Lambda_0(x) = 1$, $i = 1$.
3. By applying the extended algorithm find successive remainders $r_i(x)$ and the corresponding polynomials $t_i(x)$

$$\Omega_i(x) = \Omega_{i-2}(x) - q(x)\Omega_{i-1}(x)$$
$$\Lambda_i(x) = \Lambda_{i-2}(x) - q(x)\Lambda_{i-1}(x)$$

where polynomial $q_i(x)$ is chosen so as that the power of remainder polynomial (denoted by $r_i(x)$) is minimum.
4. Repeat the previous step until $deg[\Omega_n(x)] \leq e_c$. Remainder $\Omega_n(x)$ is GCD for $\Omega_{-1}(x)$ and $\Omega_0(x)$.
5. Calculate polynomial $\Lambda_n(x)$ roots, defining the error positions.
6. Find error magnitudes.

It is obvious that this algorithm is a modification of Euclidean algorithm for polynomials (instead of integers). Its modification is possible because the main equation for BCH and Reed-Solomon codes decoding

$$[1 + S(x)]\Lambda(x) \bmod x^{2e_c+1} = \Omega(x),$$

can be written in a slightly different form

$$\Theta(x) x^{2e_c+1} + \Lambda(x)[1 + S(x)] = \Omega(x).$$

Now the extended Euclidean algorithm for finding GCD for x^{2t+1} and $[1 + S(x)]$, can be used to obtain a complete of solutions $(\Lambda^{(i)}(x), \Omega^{(i)}(x))$ enabling that in ith iteration the following is satisfied

$$\Theta^{(i)}(x) x^{2e_c+1} + \Lambda^{(i)}(x)[1 + S(x)] = \Omega^{(k)}(x).$$

During RS codes decoding, finding $\Theta(x)$ is not of interest, but of a great interest is a pair of solutions $(\Lambda^{(i)}(x), \Omega^{(i)}(x))$. Solution corresponding to error locator and magnitude polynomials is obtained when a power of $\Omega^{(i)}(x)$ is less or equal to the power of the polynomial $\Lambda^{(i)}(x)$.

It is interesting that in the literature can be found a variant where this problem is formulated in slightly different form (the result is the same)

Table 6.20 Euclidean algorithm for Problem 6.19c

k	$\Omega_k(x)$	$q_k(x)$	$\Lambda_k(x)$
−1	x^5	–	0
0	$1 + \alpha^6 x + \alpha^5 x^2 + \alpha x^3 + \alpha x^4$	–	1
1	$\alpha^6 + \alpha x + x^2 + \alpha^5 x^3$	$\alpha^6 + \alpha^6 x$	$\alpha^6 + \alpha^6 x$
2	$\alpha^3 + \alpha x + \alpha^6 x^2$	$\alpha^2 + \alpha^3 x$	$\alpha^3 + \alpha^4 x + \alpha^2 x^2$

$$\Theta^{(i)}(x) x^{2e_c} + \Lambda^{(i)}(x) S(x) = \Omega^{(k)}(x).$$

Berlekamp-Massey algorithm is more efficient than Euclidean algorithm, but the difference between these two is not as dramatic as the difference between Euclidean approach and a direct solving in Peterson-Gorenstein-Zierler algorithm. A basic advantage of Euclidean algorithm is that is easy to understand and to apply this algorithm.

(b) Here the code RS(7, 3) is considered, correcting double error and Euclidean algorithm is applied, but syndromes are calculated starting from the element α^0 i.e. from $S_i = r(\alpha^{i-1})$. Here $r(x) = \alpha x^2 + \alpha^5 x^4$ and syndrome components are $S_1 = \alpha^6$, $S_2 = \alpha^5$, $S_3 = \alpha$, $S_4 = \alpha$. Initial conditions are $\Omega_{-1}(x) = x^5$, $\Omega_0(x) = 1 + S(x) = 1 + \alpha^6 x + \alpha^5 x^2 + \alpha x^3 + \alpha x^4$, $\Lambda_{-1}(x) = 0$, $\Lambda_0(x) = 1$. The next steps of algorithm are given in Table 6.20.

The obtained results are polynomials

$$\Omega(x) = \alpha^3 + \alpha x + \alpha^6 x^2 = \alpha^3(1 + \alpha^5 x + \alpha^3 x^2) \text{ and } \Lambda(x) = \alpha^3 + \alpha^4 x + \alpha^2 x^2$$
$$= \alpha^3(1 + \alpha x + \alpha^6 x^2),$$

satisfying the equation

$$(1 + \alpha x + \alpha^6 x^2)(1 + \alpha^6 x + \alpha^5 x^2 + \alpha x^3 + \alpha x^4) = (1 + \alpha^5 x + \alpha^3 x^2) x^5.$$

Brief Introduction to Algebra II

The aim of this subsection is to facilitate the study of the cyclic codes. Practically all proofs are omitted. The corresponding mathematical rigor can be found in many excellent textbooks [25, 27, 34].

It is in fact, the continuation of the introduction to algebra from the previous chapter.

Finite (Galois) Field Arithmetic

Losely speaking, rings can be regarded as an initial step when introducing the fields. The ring is a commutative group under addition, while the multiplication is closed, associative and distributive over addition. The existence of an identity element for multiplication is not supposed (although, it may exist), Even, if the identity element exists (it must be unique) all elements of the ring need not have the inverses. If all ring elements (with the exception of identity element for addition—0) have inverses and the multiplication is commutative, the field is obtained. Here it will be entered more deeply to the theory of finite fields.

Ideals, Residue Classes, Residue Class Ring

The normal (invariant) subgroup is very important notion in the theory of groups. The corresponding notion in the ring theory is *ideal*.

Axioms for the ideal:

I.1. Ideal I is the subgroup of the ring's (R) additive group
I.2. $\forall i \in I, \forall r \in R : ir \wedge ri \in I$.

Example

1. In the ring of all integers, the set of multiples of any particular integer is an ideal.
2. In the ring of polynomials in one variable with integer coefficients, the set of polynomials which are multiples of any particular polynomial is an ideal.

Cosets can be formed, ideal being a subgroup. These cosets are here called *residue classes*. The elements of the ideal form the first row (starting from 0). The procedure is the same as being explained for the coset decomposition.

$i_1 = 0$	i_2	i_3	.	.	.
$r_1 = r_1 + i_1$	$r_1 + i_2$	$r_1 + i_3$	.	.	.
$r_2 = r_2 + i_1$	$r_2 + i_2$	$r_2 + i_3$	.	.	.
.	.	.	.	.	.
.	.	.	.	.	.
.	.	.	.	.	.

The ideal is a normal (invariant) subgroup and the cosets (residue classes) form a group (*factor group*) under addition defined as follows

$$\{r_i\} + \{r_j\} = \{r_i + r_j\},$$

where $\{r\}$ denotes the residue class containing r. If the multiplication is defined as follows

$$\{r_i\}\{r_j\} = \{r_i r_j\},$$

it can be shown that the residue classes of a ring with respect to an ideal form also a ring. This ring is called *residue class ring*.

Example In the ring of all integers, the integers which are multiples of 3 form an ideal. The residue classes are {0}, {1} and {2}. They form the residue class ring—modulo 3 addition and multiplication. In fact, they are elements of GF(3).

Ideals and Residue Class of Integers

As explained earlier, the set of all integers forms a ring under addition and multiplication. Let start with some definitions (only integers are considered):

1. If $rs = t$, it is said that t is *divisible* by r (and by s as well), or that r (s) *divides t*, or that r and s are *factors* of t.
2. A positive integer $p > 1$ divisible only by $\pm p$ or ± 1 is called a *prime* (integer).
3. The *greatest common divisor* of two integers, GCD(r, s), is the largest positive integer dividing both of them.
4. The *least common multiple* of two integers, LCM(r, s), is the smallest positive integer divisible by both of them.
5. Two integers are *relatively prime* if their greatest common divisor equals 1.

For every pair of integers t and d (non-zero integers) there is a unique pair of integers q (the *quotient*) and r (the *remainder*), such that

$$t = dq + r, \ 0 \leq r < |d|.$$

This is well known *Euclid division algorithm*. In the following the remainder will be written as

$$r = R_d[t]$$

or

$$r = t(\bmod d).$$

In fact, the last relation means that r and t have the same remainder modulo d (both can be greater than d)—it is called *congruence*.

Some interesting properties of the remainders are

$$R_d[t + id] = R_d[t], \text{ for any } i,$$
$$R_d[t] = R_{-d}[t], \text{ only quotient changes sign.}$$

For two distinct non-zero integers (r, s) their GCD can always be expressed in the form

$$\text{GCD}(r, s) = ar + bs,$$

and computed by an iterative application of the division algorithm.

Example Find GCD(105, 91)

$$105 = 1 \times 91 + 14$$
$$91 = 6 \times 14 + 7$$
$$7 = 1 \times 7 + 0$$

Therefore GCD(105, 91) = 7 and, starting at the bottom we obtain

$$7 = 91 - 6 \times 14$$
$$7 = 91 - 6 \times (105 - 91)$$
$$7 = 91 - 6 \times 105 + 6 \times 91$$
$$7 = -6 \times 105 + 7 \times 91$$

Note that the representation is not unique, also

$$7 = -19 \times 105 + 22 \times 91 = 7 \times 105 - 8 \times 91.$$

Two more interesting properties of remainders are:

$$R_d(n_1 + n_2) = R_d(n_1) + R_d(n_2),$$
$$R_d(n_1 n_2) = R_d[R_d(n_1) R_d(n_2)].$$

Therefore, if one is interested only in remainders, to avoid big integers, they can be replaced by their remainders at any point in computation.

It can be shown also that every positive integer can be written as a product of prime number powers.

Using previously explained procedure it can be shown that residue class ring modulo p is a field if (and only if) p is a prime number. Therefore, the arithmetic in GF(p) (p—prime number) can be described as addition and multiplication modulo p. These fields are called also *prime fields*.

In such a way, the arithmetic in prime fields is defined. To define the arithmetic in GF(p^n) the polynomial rings must be introduced.

Ideals and Residue Classes of Polynomials
Some definitions:

1. A *polynomial* with one variable x with the coefficients (f_i) from ("over") a field GF(q) is of the following form

$$f(x) = f_{n-1} x^{n-1} + \cdots + f_1 x + f_0$$

2. The polynomial is *monic* if $f_{n-1} = 1$.
3. The zero polynomial is $f(x) = 0$.

In the following exposition the monic polynomials would be supposed if not otherwise stated.

The polynomials over GF(q) form a commutative ring if the operations are defined as the "usual" addition and multiplication of polynomials. The obtained

Brief Introduction to Algebra II

results will also have the coefficients from the field because they are obtained by addition and multiplication of the field elements. The inverse element for addition is just the same polynomial having as the coefficients the corresponding inverse elements from the field. Generally, the inverse elements for multiplication do not exist, but the division is possible. To obtain the field special conditions should be imposed.

Some more definitions:

1. A polynomial $p(x)$ of degree n is *irreducible* if it is not divisible by any polynomial of degree less than n, but is greater than 0.
2. A monic irreducible polynomial of degree of at least 1 is called a *prime polynomial*.
3. The *greatest common divisor* of two polynomials is the monic polynomial of largest degree which divides both of them.
4. Two polynomials are *relatively prime* if their greatest common divisor is 1.
5. The *least common multiple* of two polynomials is the monic polynomial of smallest degree divisible by both of them.

Examples (for the irreducible polynomials):

1. $x^2 + x + 1$ in GF(2), but $x^2 + 1 = (x+1)^2$ is not irreducible.
2. $x^2 - 3$ in the field of rational numbers.
3. $x^2 + 1$ in the field of real numbers.

The *Euclidian division algorithm* for polynomials can be stated as follows. For every pair of polynomials $t(x)$ and $d(x)$ (non equal to zero) there is a unique pair of polynomials $q(x)$ (quotient polynomial) and $r(x)$ (remainder polynomial), such that

$$t(x) = d(x)q(x) + r(x),$$

where the degree of $r(x)$ is less than the degree of $d(x)$. The remainder polynomial can be written

$$r(x) = R_{d(x)}[t(x)].$$

It can also be called a *residue* of $t(x)$ when divided by $d(x)$. The corresponding *congruence* relation is

$$t(x) \equiv r(x)((d(x)),$$

where the degree of $r(x)$ can be greater than the degree of $d(x)$.

Two important properties of remainders are:

$$R_{d(x)}[a(x) + b(x)] = R_{d(x)}[a(x)] + R_{d(x)}[b(x)]$$
$$R_{d(x)}[a(x)b(x)] = R_{d(x)}\{R_{d(x)}[a(x)]R_{d(x)}[b(x)]\}.$$

Further, the greatest common divisor of two polynomials $r(x)$ and $s(x)$ can always be expressed in the form

$$\text{GCD}(r(x), s(x)) = a(x)r(x) + b(x)s(x),$$

where $a(x)$ and $b(x)$ are polynomials (they are not unique!). They can be obtained in an analogous way as for the corresponding integer relation.

Therefore, there is a parallel between the ring of integers and rings of polynomial over a field. The word "integer" can be changed to word "polynomial", "$a < b$" to "deg $a(x)$ < deg $b(x)$", "prime number" to "irreducible polynomial".

The nonzero polynomial over a field has a unique factorisation into a product of prime polynomials (like the factorisation of integers into a product of primes). The irreducible (prime) polynomial cannot be factored further, because it has no field elements as the zeros.

Having the preceding note in view, the next conclusion is evident:

The set of polynomials which are multiples of any particular polynomial $f(x)$ is an ideal. The residue class ring formed from this ideal is called *polynomial ring modulo $f(x)$*.

One more note. The number of residue classes in the integer ring modulo d is just d ($\{0\}, \{1\}, \ldots, \{d-1\}$). The number of residue classes in the polynomial ring modulo $d(x)$ of the degree n over the GF(q) equals the number of all possible remainders—all possible polynomials—with the degree less equal to $n-1$, i.e. there are q^n residue classes.

The Algebra of Polynomial Residue Classes

It can be easily proved that the residue classes of polynomials modulo polynomial $f(x)$ form a commutative linear algebra of a dimension n over the corresponding coefficients field GF(q).

Denoting the field element with a, the scalar multiplication is defined as

$$a\{r(x)\} = \{ar(x)\}.$$

The other axioms are also easily verified. Among q^n residue classes, only n are linearly independent spanning the vector space. They are, for example

$$\{1\}, \{x\}, \{x^2\}, \ldots, \{x^{n-1}\}.$$

Every residue class contains the polynomial of degree less n, therefore

$$\{a_0 + a_1 x + \ldots + a_{n-1} x^{n-1}\} = a_0\{1\} + a_1\{x\} + \ldots + a_{n-1}\{x^{n-1}\}.$$

Of course,

$$\{f(x)\} = \{0\} = 0$$
$$\{xf(x)\} = 0$$
$$\{x + f(x)\} = \{x\}.$$

Brief Introduction to Algebra II

In the algebra of polynomials modulo $f(x)$ there are ideals—the residue classes which are multiples of some monic polynomial (class) $g(x)$. But the polynomial $g(x)$ must divide $f(x)$. It is called *generator (generating) polynomial* of the ideal. In fact, every factor of $f(x)$ is a generator polynomial of an ideal. The factors can be irreducible polynomials or their products as well. There are no other ideals. The ideal is a subgroup of the additive group. It is the basis of the corresponding subspace. Let the degree of $f(x)$ is n and the degree of $g(x) - n - k$. Then, the dimension of the subspace is k and the linearly independent residue classes are

$$\{g(x)\}, \{xg(x)\}, \ldots, \{x^{n-k-1}g(x)\}.$$

Multiplying $g(x)$ by x^{n-k}, the polynomial of degree n is obtained, having to be reduced modulo $f(x)$.

Let

$$f(x) = g(x)h(x),$$

then the dimension of subspace corresponding to $h(x)$ is $n - k$ (the degree of $h(x)$ being k).

The next relation also holds

$$\{g(x)\}\{h(x)\} = \{g(x)h(x)\} = \{f(x)\} = 0.$$

It can be said also that the corresponding subspaces generated by $g(x)$ and $h(x)$ are *null-spaces* of each other.

In the following exposition, the parenthesis "{.}" will be usually omitted.

The "value" of the polynomial for any field element can be easily computed. Let be $p_1(x) = 1 + x + x^2$ and $p_2(x) = 1 + x$ over GF(2). then

$$p_1(0) = 1 + 0 + 0^2 = 1, p_1(1) = 1 + 1 + 1^2 = 1; p_2(0) = 1 + 0 = 1, p_2(1) = 1 + 1 = 0,$$

and "1" is the zero (root) of the polynomial $p_2(x)$.

Vectors and Polynomials

Consider the correspondence between the n-tuples (vectors) and polynomials modulo $f(x)$ of degree n. While the vector is an ordered sequence of the field elements $(a_0, a_1, \ldots, a_{n-1})$, using polynomial the field elements are ordered by the degrees of x, i.e. the corresponding polynomial is $a_0 + a_1 x + \ldots + a_{n-1}x^n$. The multiplication by scalar (the field element) gives in both cases the same corresponding result as well as the addition. In fact, an n-tuple and corresponding polynomial can be considered to be only a different ways of representing the same element of the algebra. Therefore, both names will be used accordingly.

There is one problem. The requirement for the orthogonality of vectors (inner product equals 0) differs from the requirement that the product of polynomials be 0.

However, a very suitable connection is obtained if $f(x) = x^n - 1$ (in GF(2) – $x^n + 1$). In this case

$$\{x^n - 1\} = 0 \Rightarrow \{x^n\} = \{1\},$$

or, by omitting the parenthesis $x^n = 1$.

Let

$$a(x) = a_0 + a_1 x + \ldots + a_{n-1} x^n$$
$$b(x) = b_0 + b_1 x + \ldots + b_{n-1} x^n,$$

then

$$c(x) = a(x)b(x) = c_0 + c_1 x + \ldots + c_{n-1} x^n,$$

because

$$x^{n+j} = x^j.$$

The corresponding coefficient is

$$c_j = (a_0 b_j + a_1 b_{j-1} + \ldots + a_j b_0) + (a_{j+1} b_{n-1} + a_{j+2} b_{n-2} + \ldots + a_{n-1} b_{j+1}).$$

The first part of the expression corresponds to the terms whose sum of the indexes is j and the second part to the terms whose sum is $n + j$. The coefficient c_j can be rewritten as an inner product

$$c_j = (a_0, a_1, \ldots, a_j, a_{j+41}, a_{j+2}, \ldots, a_{n-1})(b_j, b_{j-1}, \ldots, b_0, b_{n-1}, b_{n-2}, \ldots, b_{j+1}).$$

It should be noted that the second vector is obtained taking the coefficients of $b(x)$ in reverse order and shifted cyclically $j + 1$ positions to the right. Therefore, if $a(x)b(x) = 0$, then the vector corresponding to $a(x)$ is orthogonal to the vector corresponding to $b(x)$ with the order of its components reversed as well as to every cyclic shift of this vector. Of course, the multiplication is commutative, therefore, $a(x)b(x) = 0 \Rightarrow b(x)a(x) = 0$ and the same rule can be applied keeping components of $b(x)$ in normal order and reverting and shifting the components of $a(x)$.

Having the cyclic codes in view, the residue classes of polynomials modulo polynomial $x^n - 1$ will be almost exclusively considered in the following.

Example

Let $f(x) = 1 + x^3 = (1 + x)(1 + x + x^2)$ over GF(2). The residue classes are (in the first column the corresponding vectors are written):

(000)	$\{0\}$	$1 + x^3$	$x(1 + x^3) = x + x^4$	.	.	.
(100)	$\{1\}$	x^3	$1 + x + x^4$	.	.	.

(continued)

Brief Introduction to Algebra II

(continued)

(000)	{0}	$1 + x^3$	$x(1 + x^3) = x + x^4$	.	.	.
(010)	{x}	$1 + x + x^3$	x^4	.	.	.
(110)	{1 + x}	$x + x^3$	$1 + x^4$	.	.	.
(001)	{x^2}	$1 + x^2 + x^3$	$x + x^2 + x^4$	.	.	.
(101)	{1 + x^2}	$x^2 + x^3$	$1 + x + x^2 + x^4$	.	.	.
(011)	{$x + x^2$}	$1 + x + x^2 + x^3$	$x^2 + x^4$	.	.	.
(111)	{1 + $x + x^2$}	$x + x^2 + x^3$	$1 + x^2 + x^4$	.	.	.

The first row is the ideal, the other rows are the residue classes. The elements of any row have the same remainder after dividing by $f(x)$. The corresponding basis is $\{1\}, \{x\}$ and $\{x^2\}$. $f(x)$ is the product of two factors—$1 + x$ and $1 + x + x^2$. Let $g(x) = 1 + x$. The elements of the corresponding ideal are $\{0\}, \{1 + x\}, \{xg(x)\} = \{x + x^2\}$ as well as their sum $\{(1 + x)g(x)\} = \{1 + x\} + \{x + x^2\} = \{1 + x^2\}$. This ideal is the null-space for the ideal generated by $h(x) = 1 + x + x^2$ whose elements are $\{0\}$ and $\{1 + x + x^2\}$. The corresponding orthogonality can be easily verified.

Galois Fields

Let $p(x)$ is a polynomial with coefficients in a field F. The algebra of polynomials modulo $p(x)$ is a field if (and only if) $p(x)$ is irreducible in F. If $p(x)$ is of degree n, the obtained field is called the *extension field of degree n over F*, while the field F is called the *ground field*. The extension field contains as the residue classes all the elements of the ground field. It is said that extension field contains the ground field. (Note: some authors instead of irreducible polynomials consider only *monic* irreducible polynomials called a *prime polynomials*—in the case of GF(2) there is no any difference).

The residue classes modulo prime number p form GF(p) of p elements. The ring of polynomials over any finite field has at least one irreducible polynomial of every degree. Therefore, the ring of polynomials over GF(p) modulo an irreducible polynomial of degree n is the finite field of p^n elements—GF(p^n).

The existence of inverse elements in the field GF(p^n) is proved in an analogous way as in the field GF(q), using now the relation

$$1 = a(x)r(x) + b(x)s(x).$$

There are no other finite fields. Moreover, the finite fields with the same number of elements are *isomorphic*—they have the unique operation tables. The difference is only in the way of naming their elements.

The fields GF(p^n) are said to be fields of *characteristic p*. In the GF(p), $p = 0$ (mod p), therefore, in an extension field GF(p^n), $p = 0$ as well. Then

$$(a+b)^p = \sum_{i=0}^{p} \binom{p}{i} a^i b^{p-i} = a^p + b^p,$$

because all binomial coefficients except $\binom{p}{0}$ and $\binom{p}{p}$ have $p(=0)$ as a factor.

A few definitions and theorems follow (without proof!):

1. Let F be a field and α let be the element of an extension field of F. The irreducible polynomial $m(x)$ of the smallest degree over F with $m(\alpha) = 0$, is called the *minimal polynomial* of α over F.
2. The minimal polynomial always exists and it is unique.
3. Every element of the extension field has a minimal polynomial.
4. If $f(x)$ is the polynomial over F and if $f(\alpha) = 0$, then $f(x)$ is divisible by the corresponding minimal polynomial $m(x)$.

For example, the minimal polynomial for the "imaginary unit" j from the field of complex numbers (being extension of the field of real numbers) is $x^2 + 1$, whose coefficients are from the field of real numbers.

The Multiplicative Group of a Galois Field

Let G be any finite group. Consider the set of elements formed by any element (g) and its powers

$$g, gg = g^2, gg^2 = g^3, \ldots$$

There is a finite number of elements. Therefore, after some finite number of exponentiations there must be the repetition, i.e.

$$g^i = g^j \Rightarrow 1 = g^{j-i}.$$

The conclusion is that some power of g equals 1. Let e be the smallest such positive integer ($g^e = 1$). Then e is called the *order* of the element g. Obviously, the set of elements

$$1, g, g^2, \ldots, g^{e-1}$$

forms a subgroup. There are e elements in the subgroup—the order of any field element divides the order of the group (Lagrange's theorem). A group consisting of all the powers of one of its elements is called a *cyclic group*. Such an element is called a *primitive* element. There can be more distinct primitive elements in the group.

The multiplicative group in the Galois field is cyclic—all the elements of the field (except 0) are the powers of an element (primitive). Therefore, the Galois field has a primitive element. The order of the primitive element equals the number of elements of the multiplicative group—$p^n - 1$.

The following definition will be useful:

A *primitive polynomial* $p(x)$ (also called *primitive irreducible polynomial*) over Galois field is a prime polynomial over the same field with the property that in the extension field obtained modulo $p(x)$, the field element represented by x is primitive. The conclusion is that all the field elements can be obtained by the corresponding exponentiation. Because $\{p(x)\} = 0$, one can say that a primitive polynomial is a prime polynomial having a primitive element as a zero (root). Let α is a primitive element, then $p(\alpha) = 0$.

It can be shown that the primitive polynomials of every degree exist over every Galois field.

Example The GF(2^4) may be formed as a field of polynomials modulo *primitive polynomial* $x^4 + x + 1$ over GF(2) (as the prime numbers, the irreducible polynomials are found by search!). According to the previous definition of the primitive element ("the field element represented by x is primitive"), denoted by α, it is the root of the primitive polynomial ($\alpha^4 + \alpha + 1 = 0$). All 15 nonzero field elements are represented by the corresponding powers of α. The table is obtained dividing the corresponding power of x by the primitive polynomial and taking the remainder. For α^7, the procedure is the following

$$x^7 : (x^4+x+1) = x^3+1$$
$$\underline{x^7+x^4+x^3}$$
$$x^4+x^3$$
$$\underline{x^4+x+1}$$
$$x^3+x+1 \Rightarrow \alpha^7 = \alpha^3+\alpha+1$$

The corresponding vectors are also shown.

(1000)	$\alpha^0=$	1			(1010)	$\alpha^8=$	1	$+\alpha^2$
(0100)	$\alpha^1=$		α		(0101)	$\alpha^9=$	α	$+\alpha^3$
(0010)	$\alpha^2=$			α^2	(1110)	$\alpha^{10}=$	$1+\alpha+\alpha^2$	
(0001)	$\alpha^3=$				α^3	(0111)	$\alpha^{11}=$	$\alpha+\alpha^2+\alpha^3$
(1100)	$\alpha^4=$	$1+\alpha$			(1111)	$\alpha^{12}=$	$1+\alpha+\alpha^2+\alpha^3$	
(0110)	$\alpha^5=$		$\alpha+\alpha^2$		(1011)	$\alpha^{13}=$	1	$+\alpha^2+\alpha^3$
(0011)	$\alpha^6=$			$\alpha^2+\alpha^3$	(1001)	$\alpha^{14}=$	1	$+\alpha^3$
(1101)	$\alpha^7=$	$1+\alpha$		$+\alpha^3$	(1000)	$\alpha^{15}(\alpha^0)=1$		

It should be noted that $x^4 + x + 1$ is a factor of $x^{15} + 1$ as well as that α is the root of the both polynomials. Therefore,

$$\{x^{15}+1\} = 0 \Rightarrow \alpha^{15}+1 = 0 \Rightarrow \alpha^{15} = 1,$$

and the last row in the table is added to show the cyclic structure only.

The addition in the Galois field GF(p^n) can be easily carried out by adding the corresponding coefficients in GF(p). On the other hand, the multiplication is more complicated. After multiplying the polynomials (residue classes) the obtained result should be divided by the corresponding primitive polynomial and the remainder taken as a final result. The representation of the field elements by the powers of the primitive element will ease the task. For the preceding example, when multiplying $\{x^2 + x\} = \alpha^5$ by $\{x^3 + x^2 + x\} = \alpha^{11}$, the result is $\alpha^5 \alpha^{11} = \alpha^{16} = \alpha^{15} \alpha = 1\alpha = \alpha \ (=\{x\})$. The addition using primitive element is also possible by introducing Zech logarithms.

The next theorems are also important:

1. All the $p^n - 1$ nonzero elements of GF(p^n) are roots of the polynomial $x^{p^n-1} - 1$. The element 0 can be included by considering the polynomial $(x - 0)(x^{p^n-1} - 1) = x^{p^n} - x$.
2. If $m(x)$ is the minimal polynomial of degree r over GF(p) of an element $\beta \in$ GF(p^n), then $m(x)$ is also minimal polynomial of β^p.

In fact, m elements

$$\beta, \beta^p, \beta^{p^2}, \ldots, \beta^{p^{r-1}}$$

are all the roots of $m(x)$. These elements are, by analogy with the fields of real and complex numbers, called *conjugates*. All conjugates have the same order. Let e denotes the order of β, i.e. $\beta^e = 1$. Then

$$\left(\beta^{p^i}\right)^e = \beta^{ep^i} = (\beta^e)^{p^i} = 1^{p^i} = 1.$$

There are no the others roots of $m(x)$, because β is also the root of $x^{p^r-1} - 1$, i.e. $\beta^{p^r-1} - 1 = 0 \Rightarrow \beta^{p^r-1} = 1$. Therefore, $\left(\beta^{p^{r-1}}\right)^p = \beta^{p^r} = \beta \cdot \beta^{p^r-1} = \beta \cdot 1 = \beta$. Let remind that the coefficients of $m(x)$ are from the ground field GF(p).

Let α be a primitive element in a Galois field, then all the conjugates are also the primitive elements because they all have the same order.

Example Find the minimal polynomial for $\alpha \in$ GF(2^4) over GF(2). The conjugates are

$$\alpha, \alpha^2, \alpha^4, \alpha^8 (\alpha^{16} = \alpha^{15} \alpha = 1\alpha = \alpha),$$

therefore (in GF(2) $-\alpha = \alpha$)

$$m_1(x) = (x + \alpha)(x + \alpha^2)(x + \alpha^4)(x + \alpha^8) = \ldots = x^4 + x + 1,$$

where the corresponding table shown earlier is used.

The verification is as follows (still using the same table)

$$m_1(\alpha) = \alpha^4 + \alpha + 1 = 1 + \alpha + \alpha + 1 = 0$$
$$m_1(\alpha^2) = \alpha^8 + \alpha^2 + 1 = 1 + \alpha^2 + \alpha + 1 = 0$$
$$m_1(\alpha^4) = \alpha^{16} + \alpha^4 + 1 = \alpha + 1 + \alpha + 1 = 0 \ (\alpha^{15} = 1 \Rightarrow \alpha^{16} = \alpha)$$
$$m_1(\alpha^8) = \alpha^{32} + \alpha^8 + 1 = \alpha^2 + 1 + \alpha^2 + 1 = 0 \ (\alpha^{30} = 1 \Rightarrow \alpha^{32} = \alpha^2).$$

The minimal polynomial for α^3 is

$$m_2(x) = (x+\alpha^3)(x+\alpha^6)(x+\alpha^{12})(x+\alpha^{24}) = (x+\alpha^3)(x+\alpha^6)(x+\alpha^{12})(x+\alpha^9)$$
$$= x^4 + x^3 + x^2 + x + 1.$$

In a similar way the other minimal polynomials are obtained

$$m_3(x) = x^4 + x^3 + 1 \text{ (for } \alpha^7, \alpha^{14}, \alpha^{28} = \alpha^{13}, \alpha^{56} = \alpha^{11})$$
$$m_4(x) = x^2 + x + 1 \text{ (for } \alpha^5, \alpha^{10})$$
$$m_5(x) = x + 1 \text{ (for } \alpha^0 = 1).$$

According the above cited theorem

$$x^{15} + 1 = (x+1)(x^4+x+1)(x^4+x^3+x^2+x+1)(x^4+x^3+1)(x^2+x+1)$$

and all the nonzero elements GF(2^4) are the roots of $x^{15} + 1$ [$\alpha^0 = 1$ is the element both of GF(2) and GF(2^4)].

It should be noted that α^7 and its conjugates are also the primitive elements:

$$\alpha^7, (\alpha^7)^2 = \alpha^{14}, (\alpha^7)^3 = \alpha^{21} = \alpha^6, (\alpha^7)^4 = \alpha^{28} = \alpha^{13}, \alpha^5, \alpha^{12}, \ldots, (\alpha^7)^{14} = \alpha^{98}$$
$$= \alpha^8, (\alpha^7)^{15} = \alpha^{105} = \alpha^0 = 1.$$

Therefore, the GF(2^4) can be obtained also modulo $x^4 + x^3 + 1$ [starting form $\alpha^7 (= \beta)$]. The obtained table will appear different, but the results of addition and multiplication of field elements will be always the same. As stated earlier, all finite fields of the same order are isomorphic. When forming the new table only the field elements are denoted by different symbols.

It should be noted also that conjugate non-primitive elements form a corresponding subgroup. The multiplicative group in GF(2^4) has the following subgroups (the order of the subgroup is divisor of the order of the group—$15 = 1 \times 3 \times 5$):

$$\alpha^0 = 1$$
$$\alpha^3, \alpha^6, \alpha^9, \alpha^{12}, \alpha^{15} = \alpha^0 = 1$$
$$\alpha^5, \alpha^{10}, \alpha^{15} = \alpha^0 = 1.$$

Chapter 7
Convolutional Codes and Viterbi Algorithm

Brief Theoretical Overview

Parallelly to the block codes, the ***convolutional codes*** are the second important family of error correcting codes. They are the most important subset of so called ***tree codes***, where some additional limitations are imposed (finite memory order, time invariance, linearity). They were first introduced in the "systematic" form by Elias in 1955 [46] as an alternative to block codes.

During block encoding, a block of k information symbols is represented by a code word of length n. In the case of systematic code, information symbols are not changed and $n - k$ symbols are added. If information bits (symbols) are statistically independent the obtained code words will be statistically independent as well. From theory, it is known that good results can be obtained if the code words are relatively long. On the other hand, for a longer code word encoders, and especially decoders, are more and more complex. During convolutional encoding, to obtain n channel bits, parity checks added do not depend on the considered k information bits only, but as well on m previous k-tuples of information bits. Therefore, the statistical dependence is introduced not for n channel bits, but for $(m + 1)n$ channel bits. Otherwise speaking, if the channel bits are observed using "window" of length $(m + 1)n$, the statistical dependence will be found. In the next step, the window "slides" for n bits, and the observed bits are statistically dependent as well. The encoding procedure can be interpreted as follows. The input signal (bits) consisting of blocks of k bits enters into the linear system whose impulse response is $m + 1$ blocks "long". At its output, the blocks of n symbols are generated and the output signal is the convolution of the input signal and the system impulse response. The name convolutional code originates from this interpretation. In such way, and for smaller values of k and n, the longer "code words" can be obtained, i.e. one would expect that good performance will be obtained. Often, the codes with $k = 1$ are used, i.e. with ***code rate*** $R = 1/n$. The convolutional codes can be ***systematic*** as well as ***nonsystematic*** ones.

Generally, the convolutional encoding can be accomplished according to the block scheme shown in Fig. 7.1. Convolutional codes can be generated using more than two symbols, but in the following only the name "bit" will be used, although at every place the name "symbol" can be put. Information bits (k bits) from buffer enter parallelly into the memory keeping in the same time $m \cdot k$ previous information bits. The memory consists of k shift registers, of the length m. New k bits enter into the block *logic* as well. In this block, according the chosen procedure, n bits are formed, written parallelly into the output register. From register, they are sent serially into the channel. For a systematic encoding, k current information bits enter unchanged into output register. Convolutional codes are usually denoted with (n, k, m), or only the *code rate* is given ($R = k/n$). It should be mentioned that at the beginning of the encoding the zeros are in the memory and in the output register.

Sometimes, some number of "zeros" (or some other sequence) is added after the information bits to reset the encoder. In this case, an ***effective code rate*** can be defined as

$$R_{\text{eff}} = \frac{kL}{n(L+m-1)} < R,$$

where kL is a total number of encoded information bits.

Important parameter as well is the ***constraint length*** $v = (m+1)n$. There are more definitions [27]. For (n, k, m) convolutional code some authors define it as

$$v = n(m+1)$$

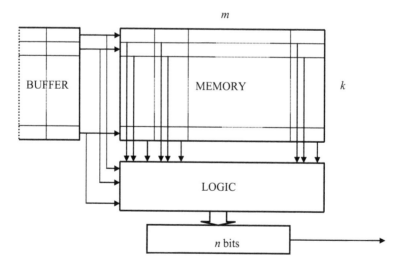

Fig. 7.1 General block scheme of a convolutional encoder

Brief Theoretical Overview

where *m* is length of the longest shift register in memory (there is no need for all shift registers to be the same length). It can be interpreted as the maximum number of encoder output bits that can be affected by a single information bit.

As said previously, if the incoming information bits are not changed during encoding, a **systematic** code is obtained. In Fig. 7.2 a **nonsystematic** convolutional encoder is shown. If there is not a feedback in the encoder, it is not **recursive** (as in Fig. 7.2) (Problems 7.3, 7.8). The parameters are $k = 1$, $n = 2$, $m = 2$, $R = 1/2$ and $v = 6$. The encoder input is denoted as x and outputs as y_1 and y_2.

The outputs are obtained by adding the bit which is at the encoder input and bits at the corresponding cells of shift register. They are denoted as s_1, s_2 being the components of **state vector**, i.e. the **encoder state** is $S = [s_1 s_2]$. This encoder has four states corresponding to dibits or to its decimal equivalents ($00 = 0$, $01 = 1$, $10 = 2$, $11 = 3$). Convolutional encoder can be considered as a finite state automate. Its functioning can be represented by table or by state diagram. If the state vector components in the next moment are denoted as s'_1 i s'_2, the corresponding equations are

$$s'_1 = x, \quad s'_2 = s_1, \quad y_1 = x \oplus s_1 \oplus s_2, \quad y_2 = x \oplus s_2.$$

on the basis of which Table 7.1 is obtained.

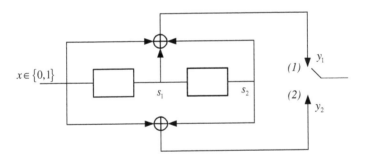

Fig. 7.2 Convolutional encoder for (2,1,2) nonsystematic code

Table 7.1 The functioning of encoder from Fig. 7.2

x	s_1	s_2	s'_1	s'_2	y_1	y_2
0	0	0	0	0	0	0
1	0	0	1	0	1	1
0	0	1	0	0	1	1
1	0	1	1	0	0	0
0	1	0	0	1	1	0
1	1	0	1	1	0	1
0	1	1	0	1	0	1
1	1	1	1	1	1	0

Fig. 7.3 State diagram for encoder from Fig. 7.2

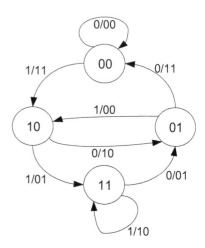

The corresponding *state diagram* (Problems 7.1–7.3) is shown in Fig. 7.3 (state diagram is explained in Chap. 2). The same encoder is used in Problem 7.3, but there Viterbi algorithm is illustrated.

Along the branches in state diagram, the bit entering the encoder is denoted as well as the bits at the encoder output. Historically, for the first description of convolutional encoder the *code tree* was used (code tree is explained in Chap. 2). In Fig. 7.4 a corresponding code tree for the considered example is shown, and the path (heavy lines) corresponding to sequence 1010 at the encoder input (when the encoder is reset). The corresponding states sequence is 00-10-01-10-01-..., while at the encoder output, the sequence is 11100010.... Therefore, to every input (and output) sequence corresponds a unique path in the state diagram.

The main drawback of the code tree is that the number of nodes grows exponentially. To follow the system "behavior" some kind of "dynamic" state diagram can be used—*trellis* (explained as well in Chap. 2). The corresponding trellis is shown in Fig. 7.5.

There exist the other useful descriptions of convolutional encoders. Consider firstly the *polynomial description* (Problem 7.1). Every memory cell is a part of delay circuit. By introducing delay operator D (*"delay"*) or z^{-1}, denoting one shift delay, the following can be written

$$s_1 = Dx, \; s_2 = Ds_1,$$
$$y_1 = x \oplus s_1 \oplus s_2 = x(1+D+D^2),$$
$$y_2 = x \oplus s_2 = x(1+D^2).$$

Therefore, the encoder can be completely described using two *generator polynomials* [47]

Brief Theoretical Overview

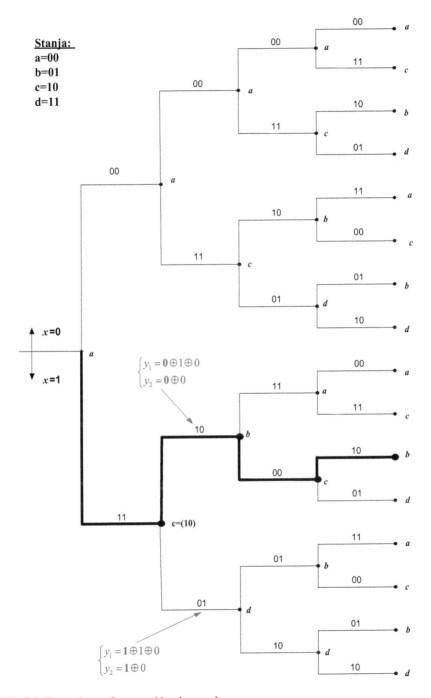

Fig. 7.4 The code tree for a considered example

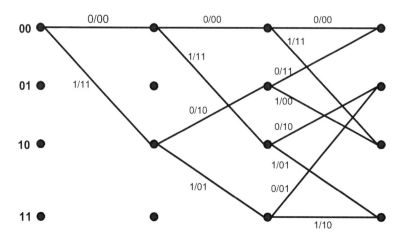

Fig. 7.5 Trellis for the considered example

$$g_1(D) = 1 + D + D^2, g_2(D) = 1 + D^2,$$

and the output sequence can be obtained by multiplying polynomial corresponding to the encoder input (***information polynomial***) and generator polynomials. These polynomials can be put together as a ***transfer function matrix*** (Problem 7.1)

$$G(D) = [g_1(D), g_2(D)] = [1 + D + D^2, 1 + D^2]$$

describing completely the convolutional encoder structure.

Consider the following example. Let the input of the encoder shown in Fig. 7.2 is bit sequence $x = 101$. The corresponding information polynomial (the most significant coefficient corresponds to the last bit) is

$$x(D) = 1 \cdot D^0 + 0 \cdot D^1 + 1 \cdot D^2 = 1 + D^2.$$

By multiplying the polynomials, the following is obtained

$$y_1(D) = x(D)g_1(D) = (1+D^2)(1+D+D^2) = 1 \cdot D^0 + 1 \cdot D^1 + 0 \cdot D^2 + 1 \cdot D^3 + 1 \cdot D^4$$
$$y_2(D) = x(D)g_2(D) = (1+D^2)(1+D^2) = 1 \cdot D^0 + 0 \cdot D^1 + 0 \cdot D^2 + 0 \cdot D^3 + 1 \cdot D^4,$$

and the corresponding output sequence, after multiplexing the obtained sequences is

$$y = 11\,10\,00\,10\,11.$$

In the previous example the information sequence is transformed into polynomial, further multiplied by generator polynomials and later again transformed back into the output sequence. This procedure can be simplified if instead of polynomials, the corresponding ***vectors*** are used, in this case

Brief Theoretical Overview

$$g_1 = (111), \quad g_2 = (101).$$

In the literature, sometimes **octal equivalent** is used. Sets of three-bit (from left to right) are written using octal symbols. If the number of bits is not divisible by three, at the sequence beginning the corresponding number of zeros is added (some authors put the zeros at the sequence end!). In the considered example, generator matrix is $G_{OKT} = [7, 5]$.

The "impulse response" can be used as well. For the considered example for a single 1 at the input, at the first output 111 will be obtained, and at the second one 101.

Generally, for the input sequence

$$i = (i_0, i_1, \ldots, i_l, \ldots),$$

the corresponding output sequences (encoder with two outputs) are the corresponding convolutions

$$v^{(1)} = \left(v_0^{(1)}, v_1^{(1)}, \ldots, v_l^{(1)}, \ldots\right), v^{(2)} = \left(v_0^{(2)}, v_1^{(2)}, \ldots, v_l^{(2)}, \ldots\right).$$

By introducing vectors according to encoder configuration, the output sequences are

$$v^{(1)} = i * g^{(1)}, v^{(2)} = i * g^{(2)}.$$

For the previous example (Fig. 7.2) for input sequence $x = 101$, the output sequence is $y = 1110001011$:

Input i	Output (in time)				
1	11	10	11		
0		00	00	00	
1			11	10	11
Sum:	11	10	00	10	11

It should be noted as well that shift registers are not uniformly drawn in literature. For example, two encoders shown in Fig. 7.6 are equivalent.

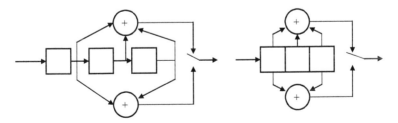

Fig. 7.6 Different drawings of the same encoder

Now, the decoding problem will be considered. The following procedures are of interest:

- majority logic decoding;
- sequential decoding;
- Viterbi algorithm.

It should be mentioned that majority logic decoding and Viterbi algorithm can be used for block codes decoding as well (Chaps. 8 and 9). Majority logic decoding and sequential decoding are specific for error control, while Viterbi decoding is universal. *Viterbi decoding* and *sequential decoding* use trellis.

Majority logic decoding is a special case of **threshold decoding** elaborated in details by Massey [48] having in mind convolutional encoding. However, this approach was used earlier, especially for Reed-Muller codes [49, 50] (Chap. 5). Hard decision and soft decision can be applied as well. Threshold decoding can be applied for some codes classes only. Further, majority logic decoding is possible only if hard decision is used.

Sequential decoding can be used for a very long constraint lengths. It was proposed by Wozencraft [51]. Fano [52] proposed a new algorithm now known as **Fano algorithm**. Later, Zigangirov [53] and Jelinek [54] (separately) proposed a new version of decoding procedure called now **stack algorithm**. The idea of sequential decoding is very simple. Decoder slides over only one path in trellis measuring distance (hard or soft) of the received sequence and this path. If it is "satisfying", the received sequence is decoded and the process starts anew. If the distance grows up over some threshold, decoder returns to some node and starts considering the other branch. Threshold should be adjusted during the decoding. This adjusting and mode of sliding form in fact the corresponding algorithm. Convolutional codes and sequential decoding were used in some space missions (Pioneer Program, Voyager Program, Mars Pathfinder).

However, the **Viterbi algorithm** is the most interesting and will be exclusively considered and explained in details in this chapter. Viterbi proposed his algorithm more as a pedagogical device, but it has found many applications. Here the important code characteristic is called *free distance* (d_{free}) (Problem 7.2). It is a minimum weight of encoded sequence whose first information block is nonzero. Viterbi algorithm is a maximum likelihood decoding. It moves through the trellis keeping only the best path (*survivor*) (Problem 7.3) to each state. It means that the corresponding metric is used—usually Hamming distance (Problems 7.3–7.6) or Euclidean (squared) distance (Problems 7.5–7.7). The signal used for metric calculation can be quantized. If there are only two levels it results in a Hamming distance (hard decision), otherwise the Euclidean metric is used (soft decision).

The state diagram can be modified by splitting the initial state into two states (Problem 7.2) to find a *modified transfer (generator) function*, from which the corresponding weight spectrum can be found. It can be used to find the bounds of residual error probability. Consider convolutional encoder shown in Fig. 7.2, which has *transfer function matrix* $G(D) = [g_1(D), \ g_2(D)] = [1 + D + D^2, \ 1 + D^2]$.

Brief Theoretical Overview

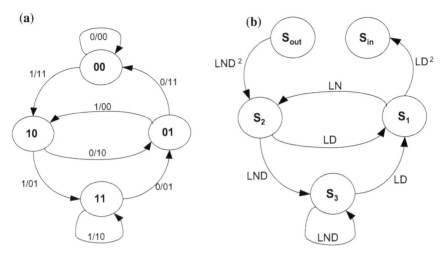

Fig. 7.7 Original (**a**) and modified (**b**) state diagram of the considered encoder

To find a *modified transfer (generator) function*, the corresponding state diagram is obtained by splitting the initial state into two (zero) states—from one it is possible only to start (S_{out}) and to other where it is possible only to enter (S_{in}). The system cannot stay in this state(s). In. Fig. 7.7 original (a) and modified (b) state diagram of the encoder are shown.

The marks on the branches of a modified state diagram are chosen as follows: every branch having n units at the input is marked with N^n, every branch having d units at the output is marked with D^d. One time interval corresponds to the transmission of one information bit, is symbolically denoted by L. These marks are multiplied yielding the corresponding system of equations for modified state diagram

$$S_2 = S_{out}LND^2 + S_1NL; \quad S_1 = S_2LD + S_3DL;$$
$$S_3 = S_2LND + S_3LND; \quad S_{in} = S_1LD^2,$$

By eliminating S_1, S_2 and S_3, the corresponding relation between two zero states is obtained

$$S_{in} = D^2L \frac{LD}{1-LND} \frac{LND^2(1-LND)}{1-NLD-NL^2D} S_{out},$$

yielding a *modified transfer (generator) function*

$$T(L,N,D) = \frac{S_{in}}{S_{out}} = \frac{L^3ND^5}{1-NLD(1+L)}.$$

By using identity

$$\frac{1}{1-x} = \sum_{i=0}^{\infty} x^i = 1 + x + x^2 + x^3 + x^4 + \cdots$$

modified generator function can be represented as an infinite series

$$\begin{aligned}T(L,N,D) &= L^3 ND^5 \left[1 + NLD(1+L) + N^2 L^2 D^2 (1+L)^2 + \ldots\right] \\ &= L^3 ND^5 + L^4 N^2 D^6 + L^5 N^2 D^6 + L^5 N^3 D^7 + 2L^6 N^3 D^7 + L^7 N^3 D^7 + \cdots\end{aligned}$$

where the element $m \, L^l N^n D^d$ shows that there are m weight paths, d formed for l entering bits, from which in this entering sequence there are n ones.

Denoting with $a_{d,n}$ the number of paths having weight d, generated by an input of the weight n, a modified generating function for $L = 1$ can be written in a form

$$T(N,D) = \sum_{d=d_{free}}^{\infty} \sum_{n=1}^{\infty} a_{d,n} N^n D^d,$$

where the total number of input ones on the paths having a weight d, denoted by c_d, can be found from

$$\frac{\partial T(D,N)}{\partial N}\bigg|_{N=1} = \sum_{d=d_{free}}^{\infty} \sum_{n=1}^{\infty} n a_{d,n} D^d = \sum_{d=d_{free}}^{\infty} c_d D^d \Rightarrow c_d = \sum_{n=1}^{\infty} n a_{d,n}.$$

Trellis corresponding to the considered encoder is shown in Fig. 7.8, while the weight spectrum is given in Table 7.2.

The first nonzero component is obtained for $d = 5$, therefore $d_{free} = 5$. It corresponds as well to the above obtained infinite series where the first nonzero term is

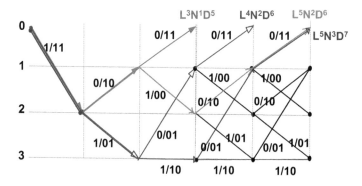

Fig. 7.8 Trellis corresponding to the considered encoder

Brief Theoretical Overview

Table 7.2 Weight spectrum of the considered encoder

d	1	2	3	4	5	6	7	8	9	10	11	12	13	14
a_d	0	0	0	0	1	2	4	8	16	32	64	128	256	512
c_d	0	0	0	0	1	4	12	32	80	192	448	1024	2304	5120

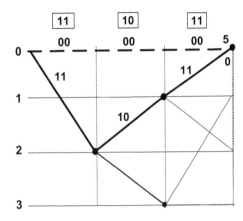

Fig. 7.9 Illustration of decision error

L^3ND^5. In this code $k = 1$, minimum distance after the first $(i + 1)$ information bits $(d_i - i = 0,1,2,...)$ is $d_0 = 0$, $d_1 = 0$, $d_2 = 5$, $d_3 = 5,\ldots$. Therefore $d_m = d_2 = 5$, and the code can correct up to $e_c = (d_{\text{free}}-1)/2 = 2$ errors on one constraint length (6 bits).

The next question is what will happen if the number of errors is greater than code correcting capability. If there are no errors at further received bits, the decoder will return to the correct path. The path part comprising errors is called "**error event**". However, some bits during the error event will be correctly decoded. The standard measure of error correcting capability is BER ("*Bit Error Rate*").

Performance of convolutional code can be found using the weight spectrum. For the above example suppose that the encoder emits all zeros sequence and that the error vector transforms this sequence into some possible sequence. Let this sequence is 111011. According to Fig. 7.9 this sequence is valid. Generally, the decision error will appear if three or more bits of the received sequence differ from the emitted sequence. It should be noted that the error at the fourth bit (0 for both paths) does not affect the decision. Because of that the summing is up to $d = 5$. For BSC with crossover probability p, the probability to choose the path with Hamming weight $d = 5$ (denoted as P_5)

$$P_5 = \sum_{e=3}^{5} \binom{5}{e} p^e (1-p)^{5-e}$$

For $d = 6$, 3 bit errors will yield the same probability for right and wrong decision, while 4 and more bit errors will always result in wrong decision yielding

$$P_6 = \frac{1}{2}\binom{6}{3}p^3(1-p)^3 + \sum_{e=4}^{6}\binom{6}{e}p^e(1-p)^{6-e}.$$

Generally, **path error event probability** for path having the weight d is

$$P_d = \begin{cases} \sum_{e=(d+1)/2}^{d}\binom{d}{e}p^e(1-p)^{d-e}, & d \text{ neparno} \\ \frac{1}{2}\binom{d}{d/2}p^{d/2}(1-p)^{d/2} + \sum_{e=d/2+1}^{d}\binom{d}{e}p^e(1-p)^{d-e}, & d \text{ parno} \end{cases}.$$

This probability can be limited from the upper side (for both even and odd d values) [55]

$$P_d = \sum_{e=(d+1)/2}^{d}\binom{d}{e}p^e(1-p)^{d-e} < \sum_{e=(d+1)/2}^{d}\binom{d}{e}p^{d/2}(1-p)^{d/2} = 2^d p^{d/2}(1-p)^{d/2}.$$

Taking into account the number of paths which have the weight $d(a_d)$, the **average probability for wrong path choice** does not depend on the node where the decision is made, being

$$P_e \leq \sum_{d=d_{\text{free}}}^{\infty} a_d P_d < \sum_{d=d_{\text{free}}}^{\infty} a_d \times [2\sqrt{(1-p)p}]^d = T(D)\Big|_{D=2\sqrt{(1-p)p}}.$$

Having in mind that p is usually a small number ($p \ll 0.5$), previous sum can be calculated on the basis of its first term as follows:

$$P_e < a_{d_{\text{free}}} 2^{d_{\text{free}}} [p(1-p)]^{d_{\text{free}}/2} \approx a_{d_{\text{free}}} 2^{d_{\text{free}}} p^{d_{\text{free}}/2}$$

This analysis can used for BER calculation. In fact, for every wrong decision, the number of bits in error equals the number of ones at the wrong part. Denoting this number with c_d (the total number of ones at all sequences with weight d) and denoting with k the number of information bits entering the encoder in one time unit, BER is upper-limited as follows

$$P_b < \frac{1}{k}\sum_{d=d_{\text{free}}}^{\infty} c_d P_d.$$

Brief Theoretical Overview

By using the above expression for P_d, this expression can be further simplified

$$P_b < \frac{1}{k}\sum_{d=d_{\text{free}}}^{\infty} c_d P_d = \frac{1}{k}\sum_{d=d_{\text{free}}}^{\infty} c_d \left[2\sqrt{p(1-p)}\right]^d = \frac{1}{k}\frac{\partial T(D,N)}{\partial D}\bigg|_{N=1, D=2\sqrt{p(1-p)}}$$

and if it is supposed that the first term in the sum is dominant, a simple expression for hard decision Viterbi algorithm is obtained [25] (Problem 7.5)

$$P_b < \frac{1}{k} c_{d_{\text{free}}} 2^{d_{\text{free}}} p^{d_{\text{free}}/2}.$$

For the above considered example $d_{\text{free}} = 5$, $a_{d_{\text{free}}} = 1$ and $c_{d_{\text{free}}} = 1$, and for $p = 10^{-2}$ and $k = 1$ it is obtained

$$P_e \approx P_b < 2^5 p^{5/2} = 3.2 \times 10^{-4}.$$

Consider the system using BPSK and coherent demodulation with binary quantization of demodulator output modeled as BSC. Denoting by E_b/N_0 the ratio of energy per information bit and average power density spectrum, error probability without encoding is

$$p = \frac{1}{2}\text{erfc}\left(\sqrt{\frac{E_b}{N_0}}\right) \approx \frac{1}{2} e^{-E_b/N_0},$$

for hard decision Viterbi decoding ($k = 1$) this probability is

$$P_b < \frac{1}{2} c_{d_{\text{free}}} 2^{d_{\text{free}}} e^{-(d_{\text{free}}/2)(E_b/N_0)}.$$

By comparing these two expressions the asymptotic coding gain is obtained (R is the code rate)

$$g_{\text{tvrdo}} = 10 \cdot \log_{10}(R d_{\text{free}}/2).$$

For soft decision Viterbi algorithm the following is obtained [25]

$$P_e \leq \frac{1}{2}\sum_{d=d_{\text{free}}}^{\infty} a_d \text{erfc}\left(\sqrt{R_c d \frac{E_b}{N_0}}\right) \approx T(D)\bigg|_{D=1/2 \cdot e^{-RE_b/N_0}},$$

and

$$P_b \leq \frac{1}{2}\sum_{d=d_{\text{free}}}^{\infty} c_d \text{erfc}\left(\sqrt{\frac{E_b}{N_0} R_c d}\right) \approx \frac{1}{k}\frac{\partial T(D,N)}{\partial D}\bigg|_{N=1, D=1/2 \cdot e^{-RE_b/N_0}}.$$

The following approximation can be used (Problem 7.5)

$$P_d \le \frac{a_d}{2} erfc(\sqrt{R_c d_{free} \frac{E_b}{N_0}}) \approx \frac{a_d}{2} e^{-R_c d_{free} \frac{E_b}{N_0}}$$

$$P_b \le \frac{c_d}{2} erfc(\sqrt{R_c d_{free} \frac{E_b}{N_0}}) \approx \frac{c_d}{2} e^{-R_c d_{free} \frac{E_b}{N_0}}.$$

Asymptotic coding gain in this case is

$$g_{meko} = 10 \cdot \log_{10}(R d_{free}).$$

If the initial encoder state is not known (Problem 7.6), the algorithm starts from all possible states. If the synchronization is lost, decoder and encoder can be resynchronized after some steps, if there were no channel errors. One of drawbacks of Viterbi algorithm is variable rate (Problem 7.4) of the decoded bits at the decoder output. It can be overcomed by so called *fixed delay* (Problem 7.6). The decoding is performed until some trellis depth (defined by fixed delay) and then the decisions are made. An interesting class are **punctured convolutional codes** (Problem 7.7), where some output bits of the originating code are omitted ("punctured"). It can result in the encoder having the same code rate and better error correcting capability.

There are two ways to approach the limiting performance of digital communication systems. In the case when the signal-to-noise ratio is limited (i.e. where the average signal power is limited), the performance can be improved by using binary error control coding including the adequate frequency band broadening. When the frequency band is limited, performance can be improved using multilevel signaling and the corresponding more powerful transmitter. For the last case an interesting procedure is TCM (**Trellis Coded Modulation**) (Problem 7.8) being a combination of convolutional encoder and modulator whose parameters are simultaneously optimized (e.g. the mapping information bit combinations into a modulated signal levels). For decoding the Euclidean metric is used. The first mention of this possibility was in 1974 by Massey [56], while Ungerboeck later, starting from 1976, elaborated the idea in many articles [57–60]. Coding and modulation are jointed here.

The next step is the substitution of code symbols by the channel signals. In Fig. 7.10 some one-dimensional and two-dimensional signal constellations are shown. One dimension corresponds to the baseband transmission, but two dimensions (complex numbers) corresponds to modulation (multilevel PSK, QAM). To compare the efficiency of coding schemes the notion of **asymptotic coding gain** is often used. For soft Viterbi algorithm decoding it is (Problem 7.8)

$$G_a = 20 \cdot \log\left(\frac{d_{free}}{d_{ref}}\right) \text{ [dB]},$$

where d_{free} is free Euclidean distance and d_{ref} is minimal Euclidean distance without the error control coding (for the same signal-to-noise ratio).

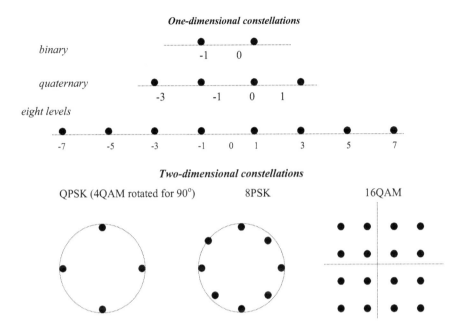

Fig. 7.10 Some one-dimensional and two-dimensional signal constellations

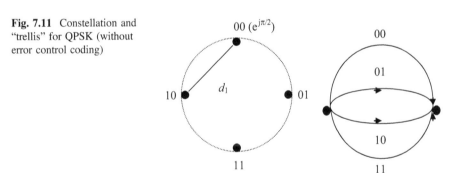

Fig. 7.11 Constellation and "trellis" for QPSK (without error control coding)

Consider now digital communication system using QPSK where every level (carrier phase) corresponds to two information bits. Let the error probability is higher than allowed. What to do? One way is to use the higher transmission power. However, there are also some other ways. One is to use convolutional code $R = 2/3$, with the same power. Error probability will be smaller, but the information rate will be smaller as well. The other way is to use 8PSK and the same convolutional code without changing the power nor bandwidth. In Fig. 7.11 the constellation for QPSK signal (without error coding) is shown and the corresponding "degenerated trellis". Gray code is used. There is no any memory in the signal and all four branches enter into the same node. Error probability is determined by referent (minimal) distance $d_1 = \sqrt{2}$.

Let augment the number of levels (phases) to eight. Every phase corresponds to three bits. In Fig. 7.12 the corresponding constellation is shown with the corresponding distances denoted.

Consider now 8PSK and convolutional coding $R = 2/3$. Two equivalent encoder configurations are shown in Fig. 7.13 and the part of the corresponding trellis. The encoder is systematic (easily seen from the second variant). From every node four branches (paths) are going out (corresponding to different information bits combinations—last two bits in tribit) and four paths enter into every node. It should be noticed that every nonsystematic convolutional encoder [if it is not catastrophic (Problem 7.4)] can be made equivalent to a systematic encoder, if feedback is allowed.

The corresponding TCM modulator is shown in Fig. 7.14.

By comparing the paths going from node (0) to node (0) one can find that the distance of other paths from the all zeros path, over the node (1) is $d_{(1)} = \sqrt{d_0^2 + d_1^2}$, over the node (2) $d_{(2)} = \sqrt{d_2^2 + d_3^2}$ and over the node (3) $d_{(3)} = \sqrt{d_3^2 + d_0^2}$. Therefore, the free distance is $d_{\text{free}} = d_{(1)} = \sqrt{d_0^2 + d_1^2}$, and the code gain is

$$G_a = 10 \cdot \log \frac{d_0^2 + d_1^2}{d_1^2} = 10 \cdot \log \frac{2 - \sqrt{2} + 2}{2} = 1{,}1 \text{ dB}.$$

This gain is modest, but it can be increased by optimal constellation mapping and by choice of optimal convolutional encoder for a given constellation.

If during Viterbi decoding some errors occur, they can propagate. The *catastrophic error propagation* can occur as well, but only if the corresponding

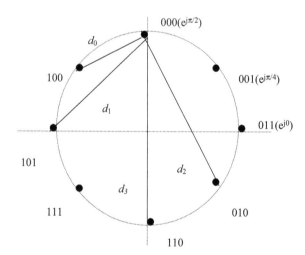

Fig. 7.12 8PSK signal constellation, bit combinations and Euclidean distances. ($d_0 = \sqrt{2 - \sqrt{2}}, d_1 = \sqrt{2}, d_2 = \sqrt{2 + \sqrt{2}}, d_3 = 2$)

Brief Theoretical Overview

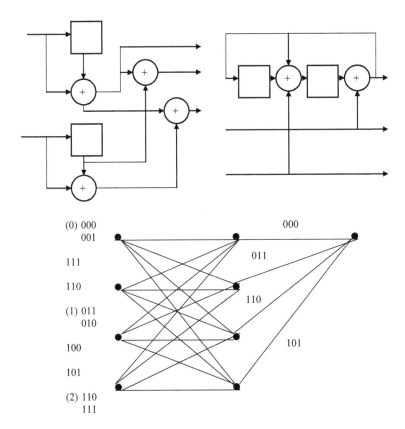

Fig. 7.13 Two equivalent scheme of convolutional encoder and part of trellis

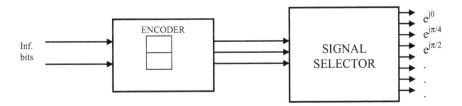

Fig. 7.14 TCM modulator

generator polynomials are not suitably chosen, i.e. if they have the common factors. At the end of convolutional decoding a few bits non carrying information (*tail bits*) (Problem 7.8) are added to provide the trellis termination in the initial state. However, this number is very small comparing to the number of bits usually transferred in one file. Convolutional codes are practically used for error correction only. They can be used as well to correct bursts of errors. Block codes decoding by using trellis will be considered in the next chapter.

Problems

Problem 7.1 The convolutional encoder is shown in Fig. 7.15.

Write polynomial and matrix description of the encoder, find the outputs and the next states corresponding to all possible initial states and to all inputs.

Solution

Polynomial form describes how the inputs (and their delayed values as well) yield the encoder outputs

$$y_1 = x_1 \oplus s_1 \oplus s_2 = (1+D+D^2)x_1,$$
$$y_2 = x_1 \oplus s_3 = x_1 + Dx_2,$$
$$y_3 = x_2 = x_2,$$

also described in a matrix form (**transfer function matrix**), having as the elements **polynomials, binary or octal numbers** [61]

$$G_{POL}(D) = \begin{bmatrix} 1+D+D^2 & 1 & 0 \\ 0 & D & 1 \end{bmatrix} \Rightarrow (y_1 \, y_2 \, y_3) = (x_1 \, x_2) \times G_{POL},$$

$$G_{BIN} = \begin{bmatrix} 111 & 1 & 0 \\ 0 & 10 & 1 \end{bmatrix} = \begin{bmatrix} 111 & 001 & 000 \\ 000 & 010 & 001 \end{bmatrix} \Rightarrow G_{OCT} = \begin{bmatrix} 7 & 1 & 0 \\ 0 & 2 & 1 \end{bmatrix},$$

where, in binary form a corresponding number of zeros should be added from the left (or from right, it is a matter of convention which is not unique in the literature) and after that every three-bit is transformed into a corresponding octal number.

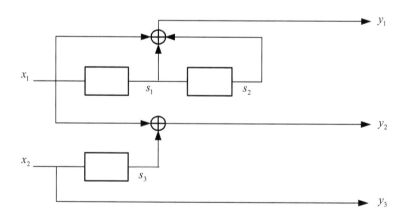

Fig. 7.15 The convolutional encoder structure

Corresponding trellis parameters are determined directly from the encoder structure:

- the encoder has in total $m = 3$ delay cells, the number of states is $2^m = 8$;
- the encoder has in total $k = 2$ inputs (number of rows in G_{OCT}), number of branches leaving every node is $2^k = 4$;
- the encoder has in total three outputs (number of columns in G_{OCT}), number of outgoing bits for every branch is $n = 3$.

Transitions into the next states are given by $s'_1 = x_1$, $s'_2 = s_1$, $s'_3 = x_2$. The encoder is described in details by Table 7.3, while the corresponding *state diagram* is shown in Fig. 7.16. For a better insight the corresponding combinations of input and output bits are omitted.

It can be shown that with the code memory increasing, its correcting capability increases as well, the code rate remaining constant (the same number of inputs and outputs), but with the number of delay cells the number of states grows exponentially, resulting in more complicate decoding procedure.

On the other hand, all convolutional codes which have the same code rate and the memory order have not the same error correction capability. A code quality depends on the encoder structure and is primarily determined by a corresponding free distance, as shown in the next problem.

Problem 7.2

(a) Find the modified generating function and the weight spectrum of the convolutional code defined by matrix $G(D) = [1 + D, D]$;
(b) Draw a few first steps of a trellis diagram needed to find the weight spectrum of the convolutional code defined by matrix $G(D) = [1 + D+D^2, 1]$.

Solution

(a) The encoder block scheme is shown in Fig. 7.17, and the corresponding state diagram is shown in Fig. 7.18a. Modified state diagram is shown in Fig. 7.18b.

Table 7.3 Description of encoder functioning

s_1	s_2	s_3	$x_1 = 0, x_2 = 0$				$x_1 = 0, x_2 = 1$				$x_1 = 1, x_2 = 0$				$x_1 = 1, x_2 = 1$			
			s'_1	s'_2	s'_3	y_1y_2	s'_1	s'_2	s'_3	y_1y_2	s'_1	s'_2	s'_3	y_1y_2	s'_1	s'_2	s'_3	y_1y_2
0	0	0	0	0	0	000	0	0	1	001	1	0	0	110	1	0	1	111
0	0	1	0	0	0	010	0	0	1	011	1	0	0	100	1	0	1	101
0	1	0	0	0	0	100	0	0	1	101	1	0	0	010	1	0	1	011
0	1	1	0	0	0	110	0	0	1	111	1	0	0	000	1	0	1	001
1	0	0	0	1	0	100	0	1	1	101	1	1	0	010	1	1	1	011
1	0	1	0	1	0	110	0	1	1	111	1	1	0	000	1	1	1	001
1	1	0	0	1	0	000	0	1	1	001	1	1	0	110	1	1	1	111
1	1	1	0	1	0	010	0	1	1	011	1	1	0	100	1	1	1	101

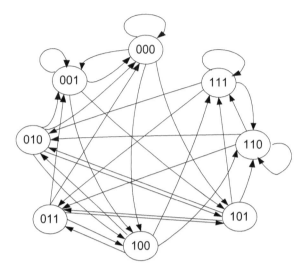

Fig. 7.16 State diagram

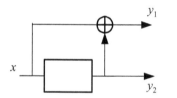

Fig. 7.17 Convolutional encoder structure

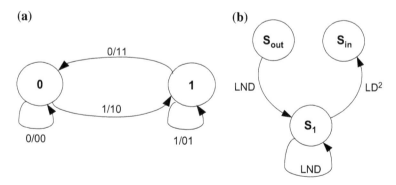

Fig. 7.18 Original (**a**) and modified (**b**) state diagram of the encoder shown in Fig. 7.17

The marks on the branches of a modified state diagram are standard, every branch that has n units at the input is marked with N^n, every branch that has d units at the output is marked with D^d. One time interval corresponds to the transmission of one information bit, it is symbolically denoted with L. The corresponding system of equations is

$$S_1 = S_{out}LND + S_1LND; \quad S_{in} = S_1LD^2,$$

and it is easy to find a relation connecting the initial and the final zero state, yielding a *modified generation function* of a code [47]

$$S_{in} = LD^2 \frac{LND}{1-LND} S_{out} \Rightarrow T(L,N,D) = \frac{S_{in}}{S_{out}} = \frac{L^2ND^3}{1-LND}.$$

As explained in the introductory part of this chapter, by using

$$\frac{1}{1-x} = \sum_{i=0}^{\infty} x^i = 1 + x + x^2 + x^3 + x^4 + \cdots$$

the modified generating function can be represented as an infinite series

$$T(L,N,D) = L^2ND^3\left[1 + LND + L^2N^2D^2 + L^3N^3D^3\ldots\right]$$
$$= L^2ND^3 + L^3N^2D^4 + L^4N^3D^5 + L^2N^4D^6 + \cdots$$

where the element $m\,L^lN^nD^d$ shows that there are m weight paths, d formed for l entering bits, from which in this entering sequence there are n ones. Denoting with $a_{d,n}$ a number of paths having weight d, generated by an input of the weight n, a modified generating function for $L = 1$ can be written in a form [47]

$$T(N,D) = \sum_{d=d_{free}}^{\infty} \sum_{n=1}^{\infty} a_{d,n} N^n D^d,$$

where the total number of input ones on the paths having a weight d, denoted by c_d, can be found from

$$\frac{\partial T(D,N)}{\partial N}\Big|_{N=1} = \sum_{d=d_{free}}^{\infty} \sum_{n=1}^{\infty} n a_{d,n} D^d = \sum_{d=d_{free}}^{\infty} c_d D^d \Rightarrow c_d = \sum_{n=1}^{\infty} n a_{d,n},$$

while, a simplified generating code function $T(D)$ is

$$T(D) = T(L, N, D)|_{L=N=1} = \sum_{d=d_{free}}^{\infty} a_d D^d = D^3 + D^4 + D^5 + \ldots$$

Coefficients $a_{d,n}$ and c_d determine weight spectrum components of the analyzed code. For a considered case these coefficients are given in Table 7.4, and they can be determined as well from a ***modified trellis*** shown in Fig. 7.19. For this encoder, there is one path for every weight of sequence at the encoder output, where the number of bits in the input sequence is always for two smaller than its length, therefore, $a_d = 1$ and $c_d = d-2$ for $d \geq 3$. A free code distance is here $d_{free} = 3$, it is relatively small, as well as a code capability to correct the errors.

(b) Block scheme of the encoder is given in Fig. 7.20 and the corresponding state diagram in Fig. 7.21a. Modified state diagram is shown in Fig. 7.21b.
From the trellis shown in Fig. 7.22 it can be noticed that there are two paths with weight $d = 4$, four paths with weight $d = 6$ etc. Code free distance is $d_{free} = 4$, and this code is better than the code analyzed in (a).

Table 7.4 Spectrum of a code which state diagram is shown in Fig. 7.19

d	1	2	3	4	5	6	7	8	...
a_d	0	0	1	1	1	1	1	1	...
c_d	0	0	1	2	3	4	5	6	...

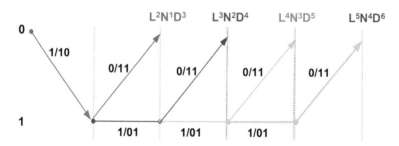

Fig. 7.19 Modified trellis for finding weight spectrum for a code from (a)

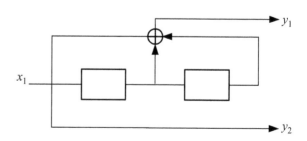

Fig. 7.20 Convolutional encoder structure

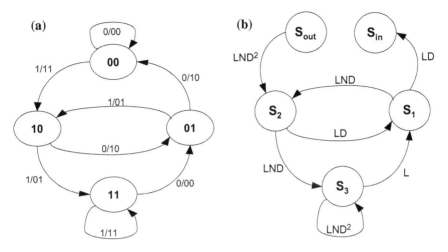

Fig. 7.21 Original (a) and modified (b) state diagram of the encoder shown in Fig. 7.20

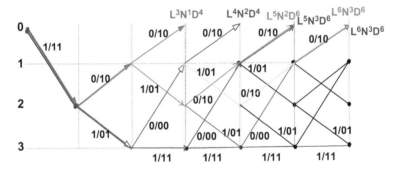

Fig. 7.22 Trellis paths determining a code spectrum

Problem 7.3 One digital communication system for error control coding use the convolutional encoder and Viterbi decoding. A convolutional encoder has one input, two outputs, defined by matrix

$$\mathbf{G}_{POL}(D) = \begin{bmatrix} 1+D+D^2, & 1+D^2 \end{bmatrix}$$

The initial convolutional encoder state is 00.

(a) Draw the convolutional encoder block scheme, find the table describing the encoder functioning and draw a state diagram.
(b) If the decoder input sequence is

$$00 \quad 11 \quad 10 \quad 11 \quad 11 \quad 0$$

decode the sequence using Viterbi algorithm with Hamming metric.
(c) What result would be obtained if decoding starts from the second received bit?

Solution

Using matrix $G_{POL}(D)$ a convolutional encoder structure is firstly obtained, shown in Fig. 7.23, where D denotes a delay corresponding to the duration of one information symbol. This symbol is used because D flop-flops are usually used as delay cells. There are $m = 2$ delay cells, a total number of states is $2^m = 4$ being determined by all combinations of cell delay outputs, denoted by s_1 and s_2. Number m is usually called **memory order**, and $v = (m + 1)n$ is a **constraint length**, showing a number of output bits depending on one input bit. There is only one input and from every state $2^k = 2$ paths start. The number of encoder outputs is $n = 2$, its **code rate** is $R = k/n = 1/2$, meaning that a binary rate at the output is two times greater than information rate at its input,

(a) If the series of encoder input bits is denoted by x_n, states are determined by previously emitted bits in this sequence, i.e.

$$x = x_n, \quad s_1 = x_{n-1}, \quad s_2 = x_{n-2}$$

where s_1, s_2 define a current encoder state $S = (s_1 s_2)$, they are usually denoted by the corresponding decimal numbers (00 = 0, 01 = 1, 10 = 2, 11 = 3). On the other hand, the next state $S' = (s'_1 s'_2)$ and the output are determined by encoder structure and by the previous state:

$$s'_1 = x, \quad s'_2 = s_1,$$
$$y_1 = x \oplus s_1 \oplus s_2, \quad y_2 = x \oplus s_2.$$

For all possible combinations of x, s_1 and s_2 values, using the above relations, Table 7.5 is formed. Corresponding state diagram and **trellis** are shown in Fig. 7.24a, b. The transitions corresponding to '0' information bits are drawn by dashed lines and those corresponding to '1' information bits are drawn by full lines. It can be noticed that from two paths going out of a considered state, the branch going "up" (relatively to the other one) corresponds to '0' information bit and vice versa. This rule is always valid when there is not a

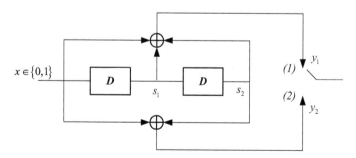

Fig. 7.23 Convolutional encoder block scheme

Problems

Table 7.5 The encoder functioning

x	s_1	s_2	s_1'	s_2'	$y_1 y_2$
0	0	0	0	0	00
0	0	1	0	0	11
0	1	0	0	1	10
0	1	1	0	1	01
1	0	0	1	0	11
1	0	1	1	0	00
1	1	0	1	1	01
1	1	1	1	1	10

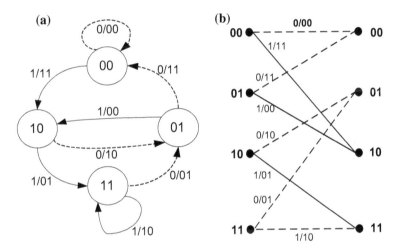

Fig. 7.24 State diagram (a) and trellis (b) of convolutional encoder from Fig. 7.23

feedback in the encoder (i.e. when the encoder is not *recursive*). In the following it will be shown that this feature makes the decoding process easier.

(b) On the basis of input information bits the encoder emits one of the possible sequences, denoted by *i*. The input sequence uniquely determines the output sequence *c* transmitted through the channel. If the sequence at the channel output, i.e. at the decoder input (denoted by *r*), is known, the input encoder sequence can be uniquely reconstructed. Therefore, at the receiving end, one should estimate which of possible sequences was emitted resulting directly in a decoding sequence.

The encoder functioning can be described by a semi-infinite tree starting from one node and branching infinitely, and the decoding is equivalent to finding the corresponding tree branch. To every output sequence corresponds a unique path through a state diagram—trellis diagram. An example for a trellis corresponding to the considered convolutional encoder, if the initial state was 00

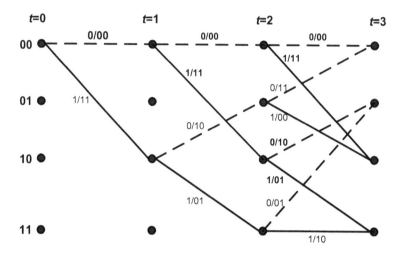

Fig. 7.25 Trellis corresponding to a sequence for the first three information bits at the encoder input

(encoder was reset at the beginning) is shown in Fig. 7.25. The transient regime is finished when in every of states the same number of branches enters as started from the initial state at the depth $t = 0$. In the case of encoder having one input, the transient regime duration equals the number of delay cells.

In the case of ***Viterbi decoder with Hamming metric*** the rules are as follows [62]:

1. To every branch a ***branch metric*** is adjoined being equal to the number of different outgoing bits from the branch relating to the received bits in this step.
2. To the nodes a metric is adjoined equal to the sum of the ***node metric*** of the node from which the incoming branch starts and a branch metric of this branch. In this problem in every node two branches are coming, resulting in two different metrics. From these two the branch is chosen which has a smaller metric, the corresponding path is ***survived***. The branch with a larger Hamming distance is ***eliminated*** *(overwritten) and **erased from a trellis*** (the corresponding data are erased from the memory). In such way the path with smaller Hamming distance is always chosen.

   If at some trellis depth both branches outgoing from some state (a trellis node) are eliminated, then a branch (i.e. all the branches) entering into this state are eliminated.
3. A unique path from depth $t = 0$ till depth $t = N$, determines N decoded bits, having the same values as the first N bits at the convolutional encoder input, to which corresponds this path.
4. If the Hamming distance for two paths is the same, any of them can be chosen. However, to accelerate the decoding, usually one of the following criteria is applied:

- a "more suitable" path is chosen, so as to eliminate a branch making "more trouble", i.e. the branch allowing the elimination of more branches at the trellis.
- always the upper branch is chosen—although suboptimal, this approach yields a good results if successively repeated many times.

For a considered case the decoder input sequence is:

$$00 \quad 11 \quad 10 \quad 11 \quad 11 \quad 01 \quad 01 \quad 11 \quad 00 \quad 11 \quad 01 \quad 10 \quad 1.$$

At the beginning, the encoder was reset, and decoding starts from a state 00 as well. For every next step (for every new information bit entering the encoder) a new state diagram is drawn, but only the states are connected where the transitions are possible in this step.

1. Step—transient regime

As shown in Fig. 7.26 and in Table 7.6, in this case the transition regime ends at the depth $t = 3$, because in this step two branches are entering into the every state. The possible sequences at the convolutional encoder output are compared to the received sequence and for every pair of paths entering the same node that one is chosen differing in smaller number of bits, i.e. having smaller Hamming distance [for a depth t, it is denoted by $d(t)$]. Survived path goes from state 00 into the state 00. This change of states in convolutional encoder is caused by the input bit 0, easily seen from the state diagram (encoder output is 00). In the case of correct transmission encoder and decoder change the states simultaneously, and the decoder performs an inverse operation, i.e. to the encoder input corresponds the decoder output. Therefore, the decoder output is $i_1 = 0$.

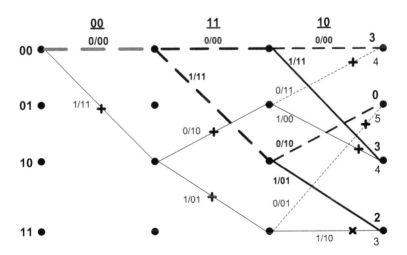

Fig. 7.26 Decoding after the transient regime

354 7 Convolutional Codes and Viterbi Algorithm

Table 7.6 Detailed data concerning decoding during the transient regime

	Corresponding path		Decoder input/ encoder output	$d(3)$
Received sequence			00 11 10	
Possible paths, corresponding emitted bits at the convolutional encoder output and Hamming distance to the received sequence	$0 \to 0$	0-0-0-0	00 00 00	3
		0-2-1-0	11 10 11	4
	$0 \to 1$	**0-0-2-1**	**00 11 10**	**0**
		0-2-3-1	11 01 01	5
	$0 \to 2$	0-0-0-2	00 00 11	3
		0-2-1-2	11 10 00	4
	$0 \to 3$	**0-0-2-3**	**00 11 01**	**2**
		0-2-3-3	11 01 10	3

2. Step—stationary regime

All overwritten paths are eliminated and only the survived paths are withheld, i.e. the paths having a smaller Hamming distance. The Hamming distance of a survived path becomes the Hamming distance of a node where the path terminates. At the depth $t = 4$ in state 00 enter two paths—one with $d(4) = 5$ and the other with $d(4) = 0$. Therefore, the probability that a code sequence 00 00 00 00 was emitted is very small, because the sequence 00 11 10 11 in this case could be received only if there were five bit errors (bits 3, 4, 5, 7 and 8). The decision made is more reliable if the metric difference between two competing paths is larger, and the decisions for states 01, 10 and 11 are less reliable than that one for a state 00.

For further calculation it is taken into account the previous Hamming distance of every node (from which the branches start) and a trellis structure is the same between trellis depths $t = 2$ and $t = 3$. It is clear that after this step the path to trellis depth $t = 2$ survived, and the decoded bits are $i_1 = 0$ and $i_2 = 1$ (see the Fig. 7.27

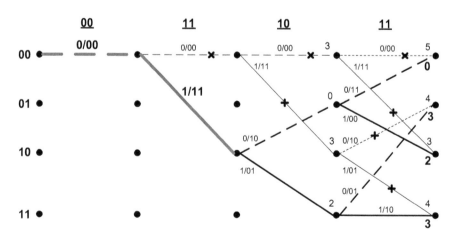

Fig. 7.27 Decoding after eight received bits

Problems

Table 7.7 Detailed data concerning decoding after eight received bits

	Corresponding path $t = 3 \to t = 4$		$d(3)$	Decoder input/ encoder output	$d = d(3) + d$ $(3 - 4)$
Received sequence				11	
Possible paths, corresponding emitted bits at the convolutional encoder output and Hamming distance to the received sequence	$0 \to 0$	0-0	3	00	$3 + 2 = 5$
		1-0	0	11	$0 + 0 = 0$
	$0 \to 1$	2-1	3	10	$3 + 1 = 4$
		3-1	2	01	$2 + 1 = 3$
	$0 \to 2$	0-2	3	11	$3 + 0 = 3$
		1-2	0	00	$0 + 2 = 2$
	$0 \to 3$	2-3	3	01	$3 + 1 = 4$
		3-3	2	10	$2 + 1 = 3$

and Table 7.7). In this step the second information bit was decoded because i_1 was decoded in a previous step.

3. Step—stationary regime continuation

The principle is the same as in the previous case, but the node metrics have the other values. All overwritten paths are omitted and only survived paths are withheld, i.e. the paths which have a smaller Hamming distance. Hamming distance of survived paths becomes the node metric where the path ends.

From the trellis shown in Fig. 7.28 and from Table 7.8, it is clear that in this step the third information bit is decoded and decoded bits are $i_1 = 0$, $i_2 = 1$ and $i_3 = 0$. It should be noticed that the accumulated Hamming distance of survived path till the depth $t = 5$ is $d(5) = 0$, meaning that during the transmission of the first ten bits over the channel there were no errors. Therefore, with a great probability it can be supposed that in the future the survived path will follow the path where node

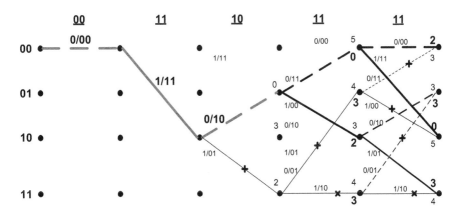

Fig. 7.28 Decoding after ten received bits

Table 7.8 Detailed data concerning decoding after ten received bits

	Corresponding path $t = 4 \rightarrow t = 5$		$d(4)$	Decoder input/ encoder output	$d = d(4) + d$ $(4 - 5)$
Received sequence				11	
Possible paths, corresponding emitted bits at the convolutional encoder output and Hamming distance to the received sequence	$0 \rightarrow 0$	0-0	0	00	$0 + 2 = 2$
		1-0	3	11	$3 + 0 = 3$
	$0 \rightarrow 1$	2-1	2	10	$2 + 1 = 3$
		3-1	3	01	$3 + 1 = 4$
	$0 \rightarrow 2$	0-2	0	11	$0 + 0 = 0$
		1-2	3	00	$3 + 2 = 5$
	$0 \rightarrow 3$	2-3	2	01	$2 + 1 = 3$
		3-3	3	10	$3 + 1 = 4$

metrics are equal to zero, i.e. that the next decoded bits are $i_4 = 0$, $i_5 = 1$. However, these two bits are not still decoded!

Because the transmission is without errors, bit sequence at the encoder output is the same as the sequence at the decoder input (until this moment—ten transmitted bits). Therefore, the decoder output sequence must be identical to the convolutional encoder input sequence. However, for a five bits at the encoder input, at the decoder output appeared only three. Reader should think about this phenomenon.

(c) If the decoding starts from the second bit of the received sequence, it can be considered that the decoder input is

$$01 \quad 11 \quad 01 \quad 11 \quad 10$$

The transient regime for this case is shown in Fig. 7.29, and it is clear that at the depth $t = 3$ no one bit was decoded. The same situation is repeated and after two additional received bits (Fig. 7.30)—no one information bit was decoded and the path with Hamming distance equals zero does not exists.

After two more received bits the situation is unchanged, only the node metrics have other values (shown in Fig. 7.31). Also, at the trellis depth $t = 5$, the paths entering into two nodes have the same Hamming distance. In spite of here applied approach where a more suitable path was chosen, i.e. that one was eliminated yielding a successive elimination of a more paths in previous steps, the noticeable branch shortening was not achieved, and after three steps no one bit was detected (decoded)!

There can be two possible reasons:

1. Large number of errors during the transmission
2. Bad synchronization.

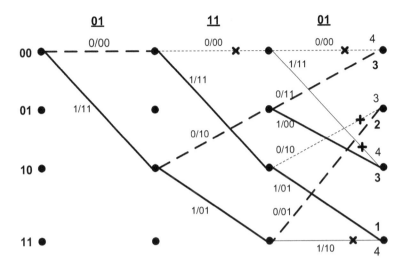

Fig. 7.29 Decoding after the transient regime

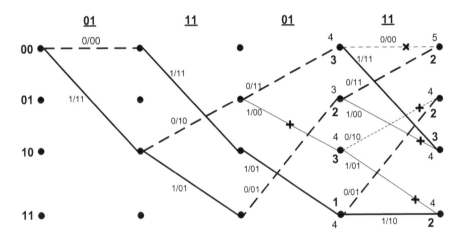

Fig. 7.30 Decoding after eight received bits

Because all paths during the first bit decoding are shortened, it is obviously the bad synchronization. It can be concluded as well and during the further decoding, because a minimum Hamming distance for any state at some large trellis depth will attain the large value, and a distance of two paths entering the same node will differ for a small values.

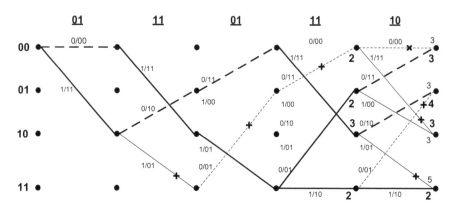

Fig. 7.31 Decoding after ten received bits

Problem 7.4 One digital communication system for error control coding use the convolutional encoder and Viterbi decoding. Convolutional encoder has one input and three outputs defined by

$$y_{1n} = x_n, \quad y_{2n} = x_n \oplus x_{n-2}, \quad y_{3n} = x_n \oplus x_{n-1} \oplus x_{n-2}.$$

(a) Draw the block scheme of the convolutional encoder and find its generator matrix.
(b) Find the code rate and draw the state diagram. Is the encoder a systematic one?
(c) If it is known that the decoding starts from the first received bit and that the initial state of convolutional encoder is 01, decode the first two information bits if at the input of Viterbi decoder is the sequence

$$100 \quad 110 \quad 010 \quad 100 \quad 000 \quad 000 \quad 111$$

Whether in this moment some further information bit was decoded? If yes, how many?
(d) Comment the code capability to correct the errors.

Solution

(a) Encoder block scheme is shown in Fig. 7.32, and the code generator is

$$G(D) = \begin{bmatrix} 1 + D^2, 1 + D, D^2 \end{bmatrix}.$$

(b) Code rate equals to the ratio of inputs and outputs of encoder—$R = k/n = 1/3$. It is a systematic encoder because the first input is directly sent to the third output. State diagram is shown in Fig. 7.33. Because there is no a loop containing all zeros at the output for a nonzero input sequence, the encoder is not a catastrophic one.

Fig. 7.32 Encoder block scheme

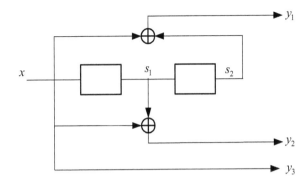

Fig. 7.33 State diagram of the encoder

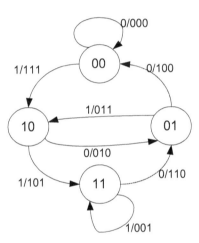

(c) According to problem text, the encoder *initial state* is 01. The decoding starts from this state. At the decoder input at every moment enter $n = 3$ bits, the trellis has $2^m = 4$ states. As shown in Fig. 7.34, after the transient regime, the path survived at the trellis depth $t = 1$, and one information bit is decoded— $i_1 = 0$. After the second step (see Fig. 7.35) there are no new decoded bits. The decoding should be continued, because two information bits are not yet decoded.

In the next step, shown in Fig. 7.36, two additional bits are decoded, $i_2 = 1$ and $i_3 = 0$. At this moment the condition that at least two information bits are decoded is fulfilled, the decoding procedure should be stopped, and there is no need to use other bits at the decoder input.

The Viterbi algorithm has just a feature that the rate at the decoder output is not constant. It can be explained by the fact that this algorithm does not make the decision instantly, but in some way it "ponders" until the moment when the

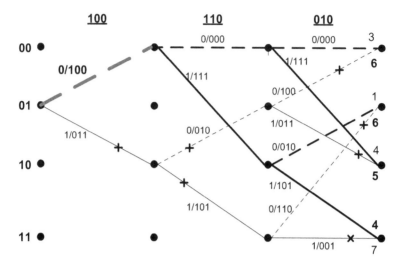

Fig. 7.34 Viterbi algorithm result after the transient regime

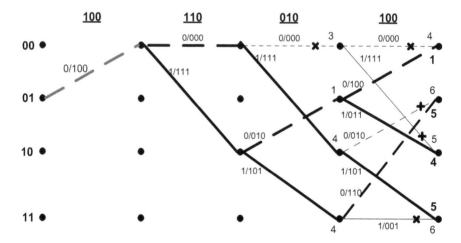

Fig. 7.35 Viterbi algorithm result after twelve received bits

similarity of the received sequence is "sufficiently large" to correspond to one of emitted sequences at the convolutional encoder output, i.e. to one of the trellis paths. The delay during decoding is of variable duration by itself, and at the moment when the decision is finally made, a greater number of information bits can be decoded. Because the delay can be very large, decoded signal rate variations can be lessen if at the decoder output a buffer is inserted having relatively large capacity.

Problems

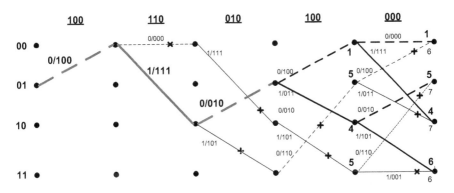

Fig. 7.36 Viterbi algorithm result after fifteen received bits

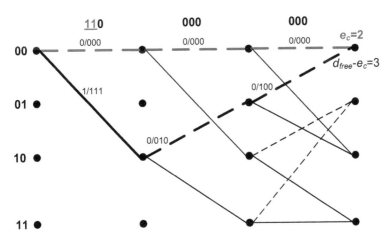

Fig. 7.37 The influence of parameter d_{free} on the decoding capability

Suppose at a moment that the encoder at the beginning was reset and that at his output the sequence of all zeros is emitted, and that the channel introduces the errors at the first $e_c = 2$ bits. On the basis of Fig. 7.37 it is clear that the decoder will nevertheless made a good decision, i.e. it will chose the path corresponding to the emitted sequence. If $e_c = 3$ the decision will be wrong as this decoder can correct at most $e_c = 2$ errors if they are sufficiently separated. It can be shown that by using Viterbi algorithm, regardless the initial state and the sequence at the encoder input, one can correct up to $(d_{free} - 1)/2$ bits (errors) on the constraint length $v = (m + 1)n$ bits.

Problem 7.5 One digital communication system for error control coding uses the convolutional encoder and Viterbi decoding. The encoder is shown in Fig. 7.38.

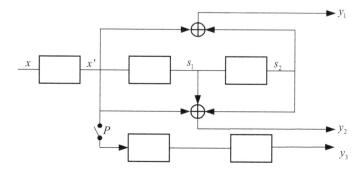

Fig. 7.38 Convolutional encoder block scheme

(a) Do the encoder structure can be presented in a form more suitable for decoding?
(b) Find the bit sequence at the encoder output if the switch is closed, and at his input is the binary sequence $i = (0110\ldots)$.
(c) When the samples of channel noise are $n = (0.2, -0.1, -0.5, -0.15, -0.6, 0.6, -0.1, 0.15, 0.4, 0.1, 0.3, 0.1)$, find the decoder output sequence if Hamming metric is used, after that repeat the same procedure if the decoder uses Euclidean metric.
(d) If the switch is open decode a sequence by decoder using Euclidean metric.
(e) If the switch is open calculate the approximate values of error probability for both ways of decoding (hard and soft) if $E_b/N_0 = 5$ [dB].

Solution

(a) It is clear that the delay cell at the encoder input will not substantially influence the decoding result—sequence x is obtained at the cell output becoming x' after only one step. Therefore, during the decoding it is sufficient to reconstruct sequence x', and this cell can be omitted. On the other hand, output y_3 corresponds to the input sequence delayed for three bits, denoted by s_2 (corresponding to the second state component), and the output is active only when the switch is closed. Now, it is obvious that this encoder can be completely described by a **simplified block scheme** shown in Fig, 7.39, i.e. by the state diagram having four states.

Fig. 7.39 Simplified convolutional encoder structure (switch closed)

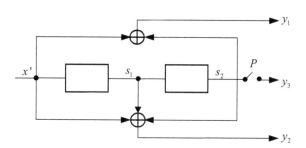

(b) When the switch is closed, the corresponding code rate is $R = 1/3$, and the code is systematic because the input is practically sent to the third output (only the delay of two bit intervals is introduced). One can suppose that the encoder was reset at the beginning, because nothing contrary was said in the problem text. Corresponding trellis and state diagram are similar to these ones in Problem 7.3, but the first and the second output bit change the places, while the third output bit is determined by the first bit of the state from which the corresponding branch at the trellis starts. For a sequence $i = (0110)$ at the encoder input, the following transitions in the Fig. 7.40 can be observed:

- the first input bit 0 causes the transition from the state 00 into the state 00, and 000 is emitted at the output,
- the second input bit 1 causes the transition from the state 00 into the state 10, and 110 is emitted at the output,
- the third input bit 1 causes the transition from the state 10 into the state 11, and 100 is emitted at the output,
- the fourth input bit 0 causes the transition from the state 11 into the state 01, and 101 is emitted at the output,

yielding the channel input sequence

$$c = (000110100101).$$

(c) If for transmission the unipolar pulses are used, the corresponding voltage levels being 1 and 0, then $x = c$, and for a known input sequence and noise samples the following is obtained

$$y = x + n = (0.2, -0.1, -0.5, 0.85, 0.4, 0.6, 0.9, 0.15, 0.4, 1.1, 0.3, 1.1).$$

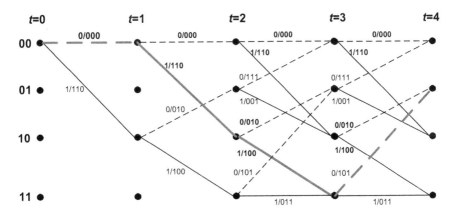

Fig. 7.40 Trellis for the simplified convolutional encoder structure (switch closed)

364 7 Convolutional Codes and Viterbi Algorithm

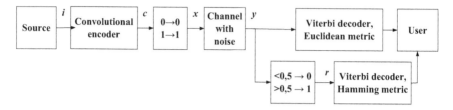

Fig. 7.41 Block scheme of a complete system, both ways of decoding are illustrated

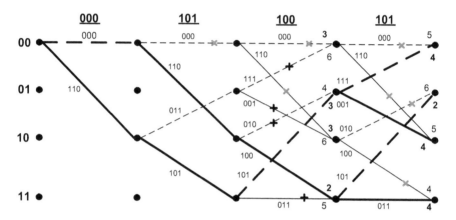

Fig. 7.42 Viterbi decoding using Hamming metric, switch *closed*

The further procedure is illustrated in Fig. 7.41. When the hard decoding is used, between the channel output and the decoder input, a decision block should be inserted (a comparison to a threshold), while for soft decoding there is no need for it. If the decision is carried out on one sample basis, it is understood that the symbols are equally probable, and a threshold is at 0.5.
At the trellis corresponding to the hard decoding, shown in Fig. 7.42, the input bits corresponding to branches are not denoted (from two branches leaving one node, the "upper" branch always corresponds to input bit 0 and it is denoted by dashed line). The branches eliminated after the transient regime are denoted with '+' and those eliminated in the next step with '×'. After the first two steps of decoding no one bit was decoded, but it is clear that if the first decoded bit is $i_1 = 0$ the second must be $i_2 = 1$, while, the alternative combination is $i_1 = 1$, $i_2 = 1$. Most likely to be detected is a sequence $i = (0110)$, because for a corresponding path $d = 2$, but it is not still the algorithm result. In Fig. 7.43 the procedure of soft decoding is shown, when the metrics are calculated as a square of ***Euclidean distance*** between the received sequence y and a supposed code word x, i.e. [63]

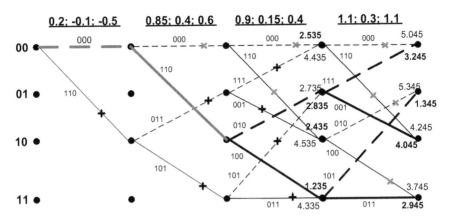

Fig. 7.43 Viterbi decoding using Euclidean metric, the switch is closed

$$d_E^2(\mathbf{x},\mathbf{y}) = \sum_{i=1}^{n}(x_i - y_i)^2.$$

Here two bits were successfully decoded—$i_1 = 0$ and $i_2 = 1$. It is rather certain that the next two decoded bits will be $i_3 = 1$ and $i_4 = 0$ (look at the metrics relation at depth $t = 4!$). If the numerical metrics values are compared, the Hamming metrics can be considered as a rounded off values of Euclidean metrics, where the error of rounding off propagates resulting in non-optimum decision.

(d) When the switch is open, the encoder is no more a systematic one, and its code rate is increased to $R = 1/2$. This encoder is not catastrophic (as well as in the case of closed switch) because the polynomial corresponding to the second output is not divisible with the polynomial corresponding to the third output [D^2 is not a factor of polynomial $1 + D + D^2$ over the field GF(2)].

In this case the trellis structure is changed in such a way that the third encoder output is not taken into account. Code word is now $\mathbf{c} = (0,0, 1,1, 1,0, 1,0)$, and at the decoder input is the sequence obtained by superimposition of the first eight noise samples on the code word, i.e. $\mathbf{y} = (0.2; -0.1; 0.5; 0.85; 0.4; 0.6; 0.9; 0.15)$. Decoding procedure is illustrated in Fig. 7.44, and one bit is correctly decoded ($i_1 = 0$). Although the transmitted sequence $\mathbf{i} = (0110)$ is most likely to be reconstructed, at this moment even its second bit is not yet decoded, because the noise influence is stronger than in previous case (e.g. it is not the same case if noise sample -0.5 is superimposed on 0 and 1!), and because of the increased code rate, the correcting capability is decreased.

(e) Error probability registered by a user when hard decoding is used, for implemented encoder having one input ($k = 1$) is [25, 26]

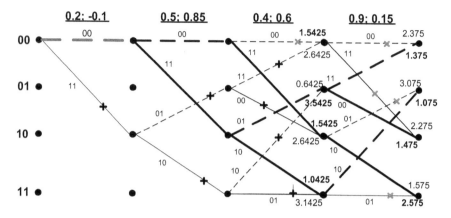

Fig. 7.44 Viterbi decoding using Euclidean metric, switch open

$$P_{b,HM} < \sum_{d=d_{free}}^{\infty} c_d \left[2\sqrt{p(1-p)}\right]^d \approx c_{d_{free}} 2^{d_{free}} p^{d_{free}/2}$$

$$= c_{d_{free}} 2^{d_{free}} \left(\frac{1}{2}\text{erfc}(\sqrt{\frac{E_b}{N_0}})\right)^{d_{free}/2},$$

while for a Viterbi algorithm using soft decision is

$$P_{b,EM} \leq \frac{1}{2} \sum_{d=d_{free}}^{\infty} c_d \text{erfc}(\sqrt{\frac{E_b}{N_0}} Rd) \approx \frac{c_{d_{free}}}{2} \text{erfc}(\sqrt{Rd_{free} \frac{E_b}{N_0}}).$$

where c_d denotes the total number of input binary ones at all sequences having the weight d, while p denotes the probability of error at the binary symmetric channel (crossover probability) when error control coding is not applied. For the first case $R = 1/3$, $d_{free} = 7$, $c_7 = 1$, and in the case when the switch is open $R = 1/2$, $d_{free} = 5$, $c_5 = 1$. If it is known that before the error control application the signal-to-noise ratio was $E_b/N_0 = 10^{5/10}$ (it is a value per information bit, while after the encoding the ratio of energy per code word bit and noise spectral power density it decreases to $E_c/N_0 = E_b/N_0 \times R$). The reader should find the numerical values for every considered case (switch closed or open, Hamming or Euclidean metric).

Problem 7.6

(a) If at the input of the convolutional encoder defined by the generator $G(D) = [1 + D+D^2, 1 + D^2]$ is the sequence $i = (010010)$ find the corresponding code word and explain the effect of quantization on the system performances. The system uses BPSK modulation, Viterbi decoding and

Euclidean metric. Find a sequence for decoding if two and four quantization regions are used and if the real noise samples are:

$$n = (-0.51, 0.1, -0.95, -1.1, -0.2, -0.9, 2.34, 0.68, -0.42, 0.56, 0.08, -0.2).$$

(b) Using the same example explain the decoding with a fixed delay where the decision about transmitted bit is made after three steps. What are the advantages and the drawbacks of this decoding type? Decode the first four information bits if at the receiver input is the same sequence as above, but if eight-level quantization is used.

Solution

(a) Soft decision in practice can never be applied exactly because the Viterbi algorithm is usually implemented by software or hardware, and the samples of received signal are represented by a limited number of bits. Therefore, the received signal should be *quantized with the finite number of levels*, and it is suitable that this number is a power of two.

Let BPSK signal has a unity power (baseband equivalent is +1/−1) while the received signal is uniformly quantized. The positive and negative values of additive noise are equally probable (easily concluded from a constellation diagram of BPSK signal and noise), it is logical that the thresholds are symmetric with respect to the origin.

- Quantization into only $Q = 2$ regions is trivial reducing to the hard decision, where only one threshold is used for the comparison. Centers of quantization regions can be denoted using one bit (1-*bit quantization*), and samples at the decoder input are denoted by −1 and +1, as shown in Fig. 7.45a.
- In the case of quantization using $Q = 4$ decision regions, it is logical that regions in the negative part are symmetric in respect to the points corresponding to transmitted signals, in the same time not disturbing the symmetry regarding to the origin. In this case *two* bits are used for quantization, regions borders have values −1, 0, 1 and quantization regions centers are points −1.5, −0.5, 0.5 and 1.5, as shown in Fig. 7.45b.

To the encoder input sequence $i = (010010)$ corresponds the code word $c = (001110111110)$, and the baseband equivalent of BPSK signal becomes

$$x = (-1, -1, 1, 1, 1, -1, 1, 1, 1, 1, 1, -1).$$

Taking into account the noise samples, signal at the receiver input is

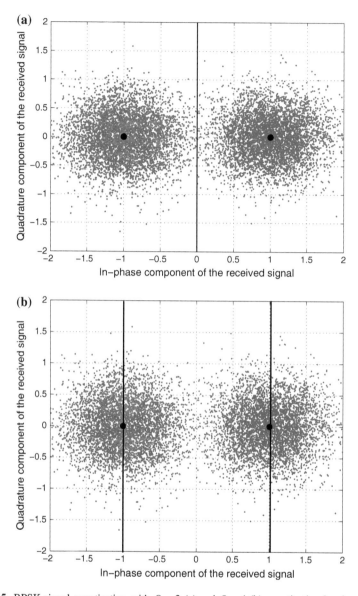

Fig. 7.45 BPSK signal quantization with $Q = 2$ (a) and $Q = 4$ (b) quantization levels

$$y = x + n = (-1.51,\ -0.90,\ 0.05,\ -0.10,\ 0.80,\ -1.90,\ 3.34,\ 1.68,\ 0.58,\ 1.56,\ 1.08,\ -1.20),$$

while the signal values at the quantizer output for $Q = 2$, $Q = 4$ are $Q = 8$ are as follows:

$y_{q2} = (-1.00, -1.00, 1.00. -1.00, 1.00, -1.00, 1.00, 1.00, 1.00, 1.00, 1.00, -1.00)$,
$y_{q4} = (-1.5, -0.5, 0.5, -0.5, 0.5, -1.5, 1.5, 1.5, 0.5, 1.5, 1.5, -1.5)$,
$y_{q8} = (-1.75, -0.75, 0.25, -0.25, 0.75, -1.75, 1.75, 1.75, 0.75, 1.75, 1.25, -1.25)$.

It is proposed to reader to carry out a complete decoding procedure for non-quantized and quantized signal and to draw the corresponding conclusions. In the second part of solution the quantized signal decoding procedure will be described in details when the initial decoder state is not known, and during decoding a fixed delay is used.

Of course, the number of regions can be further increased, but it is shown that the optimum relation between performances and memory capacity is achieved already for $Q = 8$ quantization levels (3-*bit quantization*). In Fig. 7.46 the results giving residual error probability per bit for $Q = 2, 4, 8, 16$ quantization levels are shown, as well as for the case without quantization. The results are obtained by Monte Carlo simulation for a sample size of $N = 100/P_e$ bits, where P_e denotes the estimated error probability. As mentioned earlier, the case $Q = 2$ corresponds to decision using Hamming metric, and a case without quantization corresponds to decision using Euclidean metric, both being considered in a previous problem.

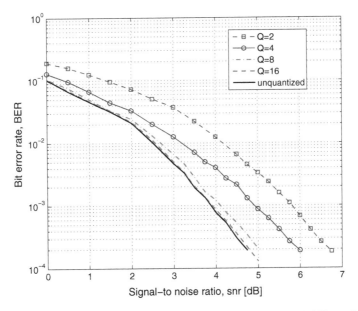

Fig. 7.46 The influence of the number of quantization levels on the error probability registered by user, encoder with $G = [7,5]$, BPSK modulation, coherent detection

(c) When the decoding is carried out using the *fixed delay*, the rules are as follows [25]:

1. For the first step a classic Viterbi algorithm is applied, using a chosen metric. After the transition regime, in every state a path having smaller metric is chosen, i.e. path having larger metric is eliminated.
2. If **the initial state is not known**, the decoding should start from all states, considering that the accumulated distance in these states are equal to zero. Duration of transient regime in this case is only one step and the decision can be done already at a trellis depth $t = 1$.
3. The procedure from the previous steps is repeated until the trellis depth determined by a fixed delay duration (registers containing the metrics have a limited length determining this numerical value!). When this depth is achieved, the survived paths are found and the state is identified at the largest depth to whom corresponds the smaller metric. For the considered example the decoding procedure until this step is shown in Fig. 7.47, fixed delay duration is three, the survived paths are denoted by heavy lines and the minimal metric value at the depth $t = 3$ is 3.375 corresponding to the state 01.
4. At the trellis the path is found connecting the state having a minimal metric (at the depth corresponding to a fixed delay) with some of candidates for the initial state. If a fixed delay is sufficiently large and if during the decoding a sufficient number of branches was eliminated, this path is unique. In Fig. 7.48 it is denoted by the strongest lines, with arrows pointing to the left. In such a way, the initial state is found (in this case 00).
5. Now, it can be considered that the first bit corresponding to this path is decoded. Therefore, although for the depth $t = 1$ all paths are not

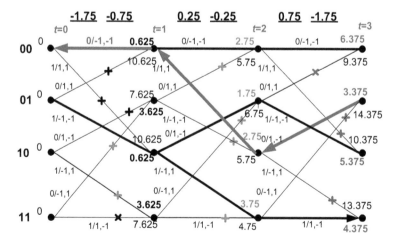

Fig. 7.47 Decoding using a fixed delay, the first step

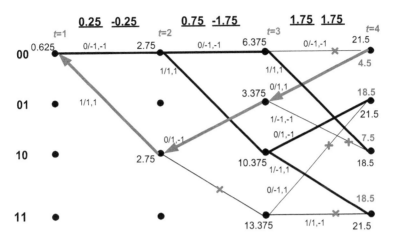

Fig. 7.48 Decoding using a fixed delay, the second step

eliminated except one, it will be considered that the decoded bit is $i_1 = 0$, because to the denoted path at this depth corresponds the transition from state 00 into the state 00.

6. After the first bit decoding, the trellis part till the depth $t = 1$ is erased (from memory) because there is no need to use it further and the trellis part from depth $t = 3$ to depth $t = 4$ is drawn. It is considered that at the depth $t = 1$, one must start from the state 00 and that all paths starting at this depth from other states are erased, the corresponding trellis shown in Fig. 7.48.

One more fact should be noted. Because the decision was made that at the depth $t = 1$ one should start from state 00, some branches being eliminated in the previous step now should be taken into account again. For example, it was previously decided that at depth $t = 3$ in the state 11 is optimum to enter from state 11 (distance 4.375), and not from the state 10 (distance 13.375). However, now state 11 can be entered from 11 only if at the depth $t = 1$ one starts from the state 10, now not being a valid option. Therefore, if from the depth $t = 1$ one starts from the state 00, in the state 11 at depth $t = 3$ can be entered only from the state 10 at depth $t = 2$, what previously seemed suboptimal.

Also, the following should be noted:

- The branch elimination in steps 1-3 had a goal only to determine the initial state and to decode the first information bit. After that, these decisions can be changed
- By the decision change, the metric difference between the path corresponding to the encoder output ("code word") and other paths increases, and the decisions become more reliable.

- The previous claim is valid only in a case if the decision about the first information bit was made correctly. Then the uncertainty about the result of decoding really becomes smaller, because this bit value influences to a series of next bits emitted by encoder. However, if the decision is incorrect, the consequences for a further decoding procedure are catastrophic.
- Of course, the decision about the first information bit is more reliable if a fixed delay has a larger value.

7. Now, the trellis is formed from depth $t = 1$ to depth $t = 4$, starting from a state 00, where some branches are back, and metrics are corrected as shown in Fig. 7.48 The procedure from 1-6 is now repeated:

    - metrics are calculated, some branches are eliminated and survived paths are denoted;
    - at the depth $t = 4$ the state having the smallest metric is identified (it is 10, the corresponding distance 4.5);
    - the path is found using the survived branches from this state to the initial state (being now known);
    - the second information bit is decoded, being here $i_2 = 1$;
    - part of a trellis till the depth $t = 2$ is erased, and a new part from the depth $t = 4$ to the depth $t = 5$ is formed, while a new initial state is 10;
    - the metrics from the previous step are taken, with a correction described in 6. (being here applied to the states 01, 10 and 11).

8. The previous procedure is repeated until the end of decoding.

The next step is shown in Fig. 7.49, where the decoded bit is $i_3 = 0$. It should be noted that not here neither in the previous steps, there would not be a correct

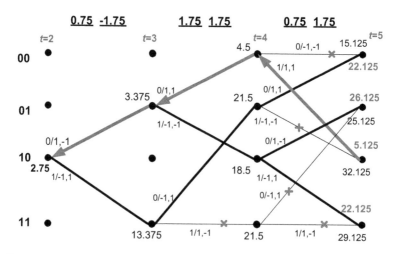

Fig. 7.49 Decoding using fixed delay, the third step

decoding if the decision was not made "by force", because the branch elimination was not resulted in a unique path till the depth $t = 3$. Because of that, after this step, a trellis part till the depth $t = 3$ is erased, new initial state is 01, and (insignificant) metrics correction is carried out only for a state 01.

The last considered step is shown in Fig. 7.50. For a difference to the previous cases, here already a standard procedure of branches elimination, to which correspond larger distances, yields the decoding of the fourth information bit ($i_4 = 1$). After this step one should erase only the trellis part till the depth $t = 4$, new initial state is now 00, the trellis part till the depth $t = 7$ is drawn and a decoding procedure is continued.

It can be noted that in this example in every step only one information bit was decoded. However, it is a consequence of the fixed delay small value—it is here minimum and corresponds to the encoder transient state duration, to simplify the procedure illustration. Usually, a fixed delay is 15–20 bits and it is clear that, in such case, in one window more bits can be decoded, and that a "truncation", i.e. a decision "by force" is applied only when in a standard decoding procedure there is some delay.,

It was shown earlier that in this example the decoded information bits are $i_1 = 0$, $i_2 = 1$, $i_3 = 0$, $i_4 = 0$, corresponding completely to the sequence at the encoder input (see the first part of the solution). One can verify that by using a classic decoding procedure at least after the first three steps no one bit would be detected (it is recommended to a reader to verify it!). If the decoding was continued, if there were no more transmission errors, all bits would be decoded. However, the decoder would not emit anything for some longer time, but after he would emit a large number of bits.

In addition to lessening the need for a registers length where the metrics are stored, the decoding with a fixed delay is used just to lessen the variable rate at the

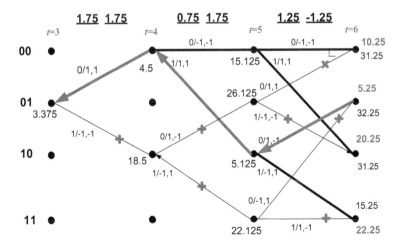

Fig. 7.50 Decoding using fixed delay, the fourth step

decoder output as well. However, one should keep in the mind that the decoding will be correct only if a delay is sufficiently long to "clear up" the path, i.e. that it has no branches for a number of steps equal to the delay.

Problem 7.7

(a) Draw a block scheme of a convolutional encoder defined by the generator $G(D) = [1 + D, D, 0; 0, D, 1 + D]$ and a corresponding trellis with denoted transitions $(x_1x_2/y_1y_2y_3)$. Find the code rate and the free distance.
(b) Design the encoder with one input which combined with a suitable puncturing provides the same d_{free} and the same code rate as the encoder above.
(c) Decode using Viterbi algorithm with Euclidean metric if at the input of encoder from (b) the sequence

$$i = (1101),$$

is entering, represented by unit power pulses, the channel noise samples are

$$n = (0.3, -0.7, 0.4, -0.3, -0.8, +1.1).$$

Solution

(a) The encoder block scheme is shown in Fig. 7.51, and a corresponding trellis for a transient regime is shown in Fig. 7.52. Code rate is $R = 2/3$. Free distance is $d_{free} = 3$.
(b) According to the problem condition, a ***punctured convolutional code*** should be obtained which has the code rate $R = 2/3$ and free distance at least $d_{free} = 3$, where for starting convolutional encoder is chosen one having one input. The simplest way to realize it is to start from the best possible encoder having two inputs and two delay cells ($R_m = 1/2$). It is well known that the corresponding generator is $G(D) = [1 + D+D^2, 1 + D^2]$ (or $G(D) = [1 + D^2, 1 + D+D^2]$), which has $d_{free} = 5$, which after the puncturing will be additionally decreased.

Fig. 7.51 Encoder block scheme

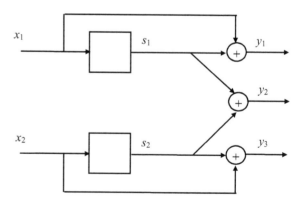

Fig. 7.52 Trellis corresponding to a transient regime, paths to find d_{free} are heavier

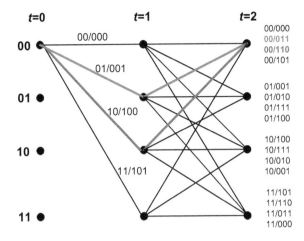

Later, it will be verified do by using this starting encoder the wished performances can be obtained.

It is known that the code rate after the puncturing equals the ratio of puncturing period (denoted by P) and a number of ones in the puncturing matrix (denoted by a) [1]

$$R_P = R_m \times (nP/a) = P/a,$$

and that the simplest *puncturing matrices* providing code rate $R = 2/3$ are as follows

$$\Pi_1 = \begin{bmatrix} 0 & 1 \\ 1 & 1 \end{bmatrix}, \Pi_2 = \begin{bmatrix} 1 & 1 \\ 0 & 1 \end{bmatrix}, \Pi_3 = \begin{bmatrix} 1 & 0 \\ 1 & 1 \end{bmatrix}, \Pi_4 = \begin{bmatrix} 1 & 1 \\ 1 & 0 \end{bmatrix}.$$

From Fig. 7.53 the weights of paths returning in the state 00 after three and four steps can be found, the results are shown in Table 7.9 (it can be considered that there are no longer paths having smaller weights). All matrices satisfy the problem conditions, but the encoder will have better correction capabilities if matrices Π_3 are Π_4 are used yielding $d_{\text{free}} = 4$. Therefore, by puncturing the originating code having one input, to which corresponds a simpler encoder block scheme and a lower complexity trellis, it is possible to construct an encoder which has the same code rate as well as a higher capability to correct errors [64].

(c) In what follows the puncturing matrix Π_3 will be used, the complete block scheme of corresponding encoder is shown in Fig. 7.54. To information sequence $i = (1101)$ corresponds the code word $c = (11010100)$. However, at the transmission line a polar equivalent of this sequence is emitted where (because of puncturing) third and seventh bit are omitted, and the following sequences are at the channel input i.e. output

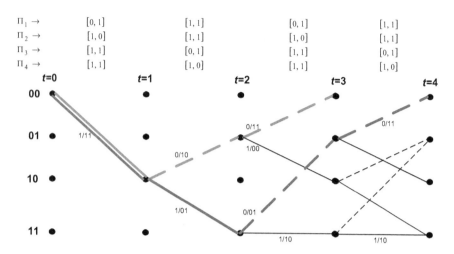

Fig. 7.53 Trellis corresponding to punctured convolutional code which has the code rate $R = 2/3$ and the free distance $d_{\text{free}} = 3$

Table 7.9 Weights of two observed paths for puncturing matrices

Path 00 → 00	Π_1	Π_2	Π_3	Π_4
Path terminates in $t = 3$	$d = 3$	$d = 3$	$d = 4$	$d = 5$
Path terminates in $t = 4$	$d = 5$	$d = 4$	$d = 5$	$d = 4$

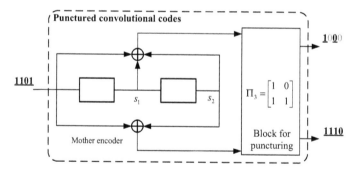

Fig. 7.54 Convolutional encoder structure corresponding to a chosen punctured code

$$x = (+1, +1, +1, -1, +1, -1),$$
$$y = x + n = (1.3, 0.3, 1.4, -1.3, 0.2, 0.1).$$

Decoding procedure using Euclidean metric (without quantization) is shown in Fig. 7.55. The difference in respect to previously considered cases is that instead of symbols, being erased, neutral values are inserted (symbol 0 has the same distance from −1 and +1). Although the channel noise is quite high, in

Problems

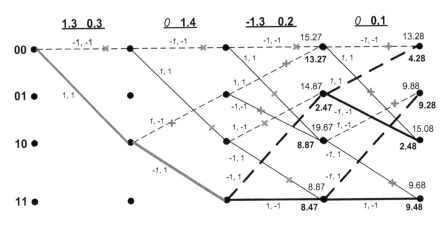

Fig. 7.55 Decoding procedure illustration

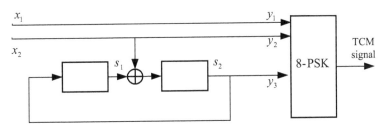

Fig. 7.56 TCM modulator structure

every of two decoding steps, one bit of information sequence was successfully reconstructed ($i_1 = 1$, $i_2 = 1$), and there is a high probability that the next two decoded bits will be $i_3 = 0$ and $i_4 = 1$, because the corresponding path Euclidean distance is substantionaly smaller with respect to the other paths.

Problem 7.8 The systematic recursive convolutional encoder in Fig. 7.56 is shown. It is used for TCM procedure combined with 8-PSK modulation.

(a) Draw state diagram and trellis corresponding to convolutional encoder and find its free distance.
(b) Find which mapping (binary or Gray) at 8-PSK constellation provides a greater value of minimum Euclidean distance.
(c) Find the TCM modulator output for the input sequence $x = (01, 00, 10)$. If at the obtained signal in the channel a narrowband Gaussian noise is superimposed which has complex samples (corresponding to a baseband)

$$n = (0.1 - 0.2j, 0, 0.3 + 0.1j, 0, 0.2j)$$

describe in details a procedure for information bits reconstruction in TCM demodulator.

Solution

(a) In TCM (*Trellis Coded Modulation*) transmitting end consists from encoder and modulator, their parameters are jointly optimized [65]. State diagram of *recursive systematic convolutional encoder* is shown in Fig. 7.57, where dashed lines correspond to zero value of the second information bit. The first information bit does not influence the path shape, but only the first output value (it is the reason for parallel paths).

In Fig. 7.58 complete trellis at the depth $t = 1$ and a trellis part to the depth $t = 3$ are shown. It is clear that the free code distance is $d_{\text{free}} = 1$. Parallel to all

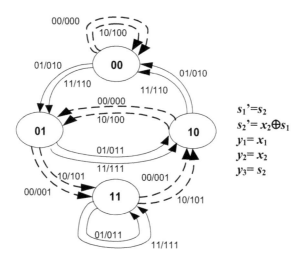

Fig. 7.57 State diagram and encoder (Fig. 7.58) working rule

$s_1' = s_2$
$s_2' = x_2 \oplus s_1$
$y_1 = x_1$
$y_2 = x_2$
$y_3 = s_2$

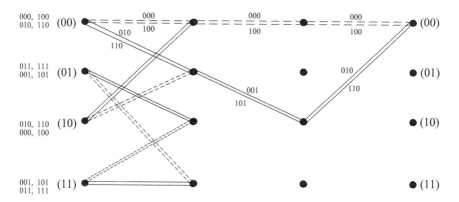

Fig. 7.58 Trellis corresponding to the encoder shown in Fig. 7.48

zeros path is a path which in $t = 0$ leaves state 00, and in $t = 1$ returns into the same state containing only one output bit different from one.

(b) Furthermore, two ways of constellation diagram mapping (binary and Gray) will be considered, for a complex form of modulated signal which has a normalized amplitude, for every three bit combination at the modulator input.

1. **Binary mapping**:

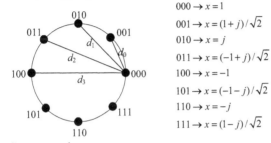

$000 \to x = 1$
$001 \to x = (1+j)/\sqrt{2}$
$010 \to x = j$
$011 \to x = (-1+j)/\sqrt{2}$
$100 \to x = -1$
$101 \to x = (-1-j)/\sqrt{2}$
$110 \to x = -j$
$111 \to x = (1-j)/\sqrt{2}$

2. **Gray mapping**:

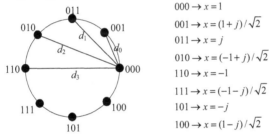

$000 \to x = 1$
$001 \to x = (1+j)/\sqrt{2}$
$011 \to x = j$
$010 \to x = (-1+j)/\sqrt{2}$
$110 \to x = -1$
$111 \to x = (-1-j)/\sqrt{2}$
$101 \to x = -j$
$100 \to x = (1-j)/\sqrt{2}$

From the trellis in Fig. 7.59 the paths can be noticed which at depth $t = 0$ leave the state 00 and in some next moment return to this state, i.e. they are competing for a path having a minimum distance. Euclidean distance is calculated step by step, as a distance of a point at constellation diagram to the

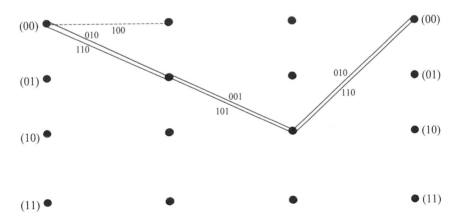

Fig. 7.59 Trellis part for finding minimum Euclidean distance for TCM modulator

point corresponding to all zeros combination, and a squares of distances of single steps are added to obtain a total distance.

It can be noticed that there are nine competing paths where one stops at the depth $t = 1$, and the others at $t = 3$ (it is supposed that the longer paths are not of interest) [65]:

1. $t = 1$:

$$c = (110) \rightarrow d^2_{(I),bin} = d^2_3, d^2_{(I),Grej} = d^2_0$$

2. $t = 3$:

$$c = (010, 001, 010) \rightarrow d^2_{(II),bin} = d^2_0 + 2d^2_1, d^2_{(II),Grej} = 2d^2_2 + d^2_0$$
$$c = (010, 101, 110) \rightarrow d^2_{(III),bin} = d^2_0 + 2d^2_1, d^2_{(III),Grej} = d^2_2 + d^2_0 + d^2_3$$
$$c = (010, 001, 010) \rightarrow d^2_{(IV),bin} = d^2_2 + 2d^2_1, d^2_{(IV),Grej} = 2d^2_2 + d^2_1$$
$$c = (010, 101, 110) \rightarrow d^2_{(V),bin} = d^2_2 + 2d^2_1, d^2_{(V),Grej} = d^2_2 + d^2_1 + d^2_3$$
$$c = (110, 001, 010) \rightarrow d^2_{(VI),bin} = d^2_0 + 2d^2_1, d^2_{(VI),Grej} = d^2_3 + d^2_0 + d^2_2$$
$$c = (110, 001, 110) \rightarrow d^2_{(VII),bin} = d^2_0 + 2d^2_1, d^2_{(VII),Grej} = 2d^2_3 + d^2_0$$
$$c = (110, 101, 010) \rightarrow d^2_{(VIII),bin} = d^2_2 + 2d^2_1, d^2_{(VIII),Grej} = d^2_3 + d^2_1 + d^2_2$$
$$c = (110, 101, 110) \rightarrow d^2_{(IX),bin} = d^2_2 + 2d^2_1, d^2_{(IX),Grej} = 2d^2_3 + d^2_1$$

where $d_0 = \sqrt{2 - \sqrt{2}}, d_1 = \sqrt{2}, d_2 = \sqrt{2 + \sqrt{2}}, d_3 = 2$. It is easy to verify that in both cases the path terminating at the depth $t = 1$ is the more critical, because is this case the Euclidean distance is the smallest. Corresponding **code gains** are

$$G_{bin} = 10 \log \frac{d^2_3}{d^2_1} = 10 \log \frac{4}{2} = 3.01 \text{ dB},$$

$$G_{Grej} = 10 \log \frac{d^2_0}{d^2_1} = 10 \log \frac{2 - \sqrt{2}}{2} = -5.33 \text{ dB},$$

and it is obvious that for this encoder binary mapping is optimal.

(c) From the trellis shown in Fig. 7.59 it is obvious that to the information sequence at the encoder input $x = (01,00,10)$ corresponds the code sequence $y = (010,001,100)$ at its output. For the applied binary mapping, this sequence of bits in 8-PSK modulator is represented by modulated symbols

$$u = (j, (1+j)/\sqrt{2}, -1),$$

and after noise superimposing at the decoder input the following symbols enter the decoder

$$v = u + n = (0.1 + 0.8j, (1+j)/\sqrt{2}, -0.7 + 0.1j).$$

TCM demodulator functioning is shown in details in Fig. 7.60. It is obvious that in this case demodulation and decoding are performed simultaneously—instead that a demodulation is performed by observing the octant where the received signal "enter", and thus obtained decisions send to a Viterbi decoder, the decoding is performed using Euclidean metrics where in every step, the metrics are calculated in a two-dimensional space (dimensions correspond to so called I and Q received signal components) and from all paths entering in one node, this one is selected which has a minimum Euclidean distance in respect to a received sequence. For a better insight in this problem only, the eliminated paths are shown, denoted by dashed lines.

In this example, after the transient regime ending, it was found that the encoder terminated in the state 00, corresponding to the code sequence 010, i.e. to symbol j at the modulator output (denoting a signal whose real part equals zero and imaginary part equals one, i.e. the phase shift is $\pi/2$). The encoder is systematic and the first two bits in the code sequence 010 correspond to information bits at the decoder input, therefore, at this moment decoded bits are $x_1 = 0$ and $x_2 = 1$.

The bits most likely to be decoded further are $x_3 = 0$, $x_4 = 0$, $x_5 = 1$ and $x_6 = 0$, but the final decision is not yet made because even in an optimum regime the last mk bits are not decoded (m is the encoder memory order and k is a number of input bits). However, it can be overcomed if during the encoding after the information bit sequence mk "tail bits" are added, do not bearing any information, but providing the

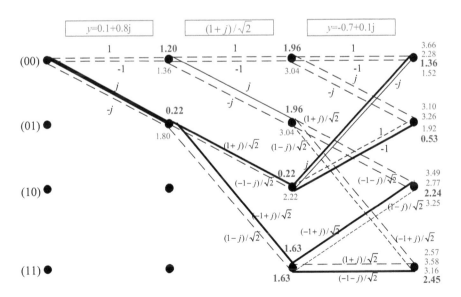

Fig. 7.60 Procedure of demodulation/decoding of TCM signal, transient regime

trellis termination in state 00. In considered case trellis should return from the state 01 (where the encoder was after emitting a code sequence c) into the state 00, which is provided if in the fourth step $x_2 = 1$ and in the fifth $x_2 = 1$. The corresponding values of x_1 are not interesting, it will be supposed that they equal zero.

A complete sequence at the encoder input is now $x = (01,00,10,01,01)$, the corresponding code sequence is $y = (010,001,100,011,010)$, while the modulated symbols are

$$u = (j, (1+j)/\sqrt{2}, -1, (-1+j)/\sqrt{2}, j),$$

and after the superimposition of noise at the decoder input is the following sequence

$$v = u + n = (0.1 + 0.8j, (1+j)/\sqrt{2}, -0.7 + 0.1j, (-1+j)/\sqrt{2}, 1.2j).$$

Trellis after two more decoding steps is shown in Fig. 7.61. In this case, the decoding of all emitted information bits is terminated, i.e. the sequence $x = (01,00,10)$ was reconstructed correctly. Trellis has still more branches, and the fourth and fifth bit from the encoder input are not decoded. But, these bits do not carry any information and their reconstruction is not important. Although the metric corresponding to the state 00 at the depth $t = 5$ is the smallest, the corresponding numerical values at the last two depth of a trellis are not written, because they are not interesting, the decoding being finished (the reader should find them, repeating the procedure from the previous steps).

In this case for transmitting six information bits even four tail bits are used, and it may seem that a trellis termination procedure have extremely low efficacy. Of course, it is not a case, because the information bits block to be transmitted can be extremely long—typical files lengths downloaded from Internet are greater than 1 [MB], and after their transmission typically 4–16 tail bits are added. For a turbo

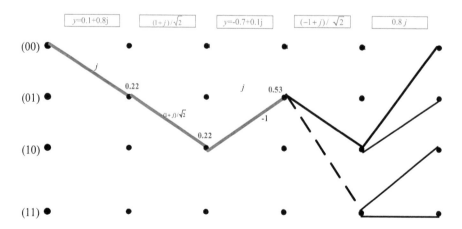

Fig. 7.61 Demodulation/decoding of TCM signal, trellis termination

codes, a block of a few thousands of information bits is entered into parallel recursive convolutional encoders and with trellis termination it is provided that every block is considered separately, i.e. by using the convolutional codes a linear block code is obtained, having the possibility to be decoded by using a trellis as well. These procedure will be analyzed in details in the Chap. 2

Chapter 8
Trellis Decoding of Linear Block Codes, Turbo Codes

Brief Theoretical Overview

During the "long" history of error control coding, especially taking into account that the proof of the Second Shannon theorem is not a "constructive" one, the search for constructive codes to approach the "promised" limits have taken a long period of time. Algebraic coding theory flourished. However, even the best cyclic codes (RS codes) had not good performance. Namely, they reduce the error probability for relatively high signal-to-noise ratios and have high code rate, but they are far from the Shannon limit. Furthermore, the practical coding schemes were developed allowing the better reliability for lower signal-to-noise ratios. These were convolutional codes having a moderate constraint length decoded using Viterbi algorithm, and convolutional codes having a long constraint length decoded using sequential decoding. Next, cascade schemes were developed where block and convolutional codes were combined to maximize the coding gain. However, such codes were suitable only for systems having limited power and unlimited bandwidth. For systems having limited bandwidth, the spectral efficiency was augmented by combining multilevel signal constellations and punctured convolutional codes, where the modulation (constellation) and coding (code rate) were adaptive. Later it was conceived that the better performance can be achieved by joined coding and modulation (trellis coded modulation). The next step was iterative decoding allowing at last coming into "the land of promise". In fact, there is no need to make the difference between block codes and convolutional codes, because the last ones can be terminated (by adding non information bits) and considered as block codes.

In the previous chapter, the Viterbi algorithm was introduced having in view mainly convolutional codes. Even if the information bits are statistically independent, the dependence is introduced into a linear block codes codewords as well, by adding parity-check bits. Therefore, they can be regarded as generated by some finite automat. The corresponding trellis can be constructed which has clearly defined the starting and the finishing state.

A method to find the trellis for linear block codes was proposed in [66] and later Wolf [67] introduced identical trellis and applied Viterbi algorithm for ML block codes decoding. More details can be found in [68, 69]. The procedure will be firstly explained for binary code (Problems 8.1, 8.2 and 8.4), but it can be similarly applied for linear block codes over $GF(q)$ (Problem 8.3).

Consider a linear block code (n, k) over $GF(2)$. Its control matrix has the dimensions $(n - k) \times n$ and can be written in the form

$$H = [h_1 h_2 \ldots h_j \ldots h_n],$$

where h_j are vector columns which have dimensions $(n - k) \times 1$. Vector (row) v with dimensions $n \times 1$ is a code vector if the following is true

$$vH^T = 0.$$

The trellis will start at the depth 0 and will terminate at the depth n. Suppose the systematic code where the first k bits are information bits, followed by $n - k$ parity checks. State at the depth j will be denoted as S_j and numerically expressed using its ordinal number. It can be denoted as the second index, i.e. as $S_{j,\text{ord.numb}}$. Maximal value of these ordinal numbers is $2^{n-k} - 1$. There are 2^{n-k} possible values (states). They can be represented by vectors having length $n - k$. If the information vector is denoted as $i(i_1, i_2, \ldots, i_k)$ a new state for the depth j can be found (second index denoting the ordinal number is omitted) as

$$S_j = S_{j-1} + x_j h_j,$$

where $x_j = i_j$ ($j \leq k$) and for $j = k + 1, \ldots, n x_j$ is the corresponding parity check bit obtained on the basis of the previous information bits using generator matrix G. In such way the terminating state S_n will be equal to the starting one—S_0. In this case all paths in trellis will correspond to the possible code words.

Consider (7, 4) code where the corresponding generator and control matrices are

$$G = \begin{bmatrix} 1 & 0 & 0 & 0 & 1 & 1 & 1 \\ 0 & 1 & 0 & 0 & 1 & 1 & 0 \\ 0 & 0 & 1 & 0 & 1 & 0 & 1 \\ 0 & 0 & 0 & 1 & 0 & 1 & 1 \end{bmatrix} = [I_4 P],$$

$$H = [PI_3] = \begin{bmatrix} 1 & 1 & 1 & 0 & 1 & 0 & 0 \\ 1 & 1 & 0 & 1 & 0 & 1 & 0 \\ 1 & 0 & 1 & 1 & 0 & 0 & 1 \end{bmatrix}.$$

The corresponding trellis has $2^{n-k} = 8$ states (i.e. three bit vectors). It starts from depth 0 going to the depth $n = 7$. Total number of paths equals the number of code words $2^k = 2^4 = 16$. If index of S denotes only the trellis depth and state index is written in the parenthesis, the calculations are as follows (for $S_0 = (000)$):

Brief Theoretical Overview

$j = 1$:
$$\mathbf{S}_1 = \mathbf{S}_0 \oplus x_k \mathbf{h}_1 = (0\ 0\ 0) + x_1(1\ 1\ 1)$$
$$= \begin{cases} (0\ 0\ 0), & x_1 = 0 \\ (1\ 1\ 1), & x_1 = 1 \end{cases}$$

$j = 2$:
$$\mathbf{S}_2 = \mathbf{S}_1 \oplus x_2 \mathbf{h}_2 = \mathbf{S}_1 + x_2(1\ 1\ 0)$$
$$\mathbf{S}_2 = (0\ 0\ 0) + x_2(1\ 0\ 1) = \begin{cases} (0\ 0\ 0), & x_2 = 0 \\ (1\ 0\ 1), & x_2 = 1 \end{cases}$$
$$\mathbf{S}_2 = (1\ 1\ 1) + x_2(1\ 1\ 0) = \begin{cases} (1\ 1\ 1), & x_2 = 0 \\ (0\ 0\ 1), & x_2 = 1 \end{cases}$$

$j = 3$:
$$\mathbf{S}_3 = \mathbf{S}_2 \oplus x_3 \mathbf{h}_3 = \mathbf{S}_2 + x_3(1\ 0\ 1)$$
$$\mathbf{S}_3 = (0\ 0\ 0) + x_3(1\ 0\ 1) = \begin{cases} (0\ 0\ 0), & x_3 = 0 \\ (1\ 0\ 1), & x_3 = 1 \end{cases}$$
$$\mathbf{S}_3 = (0\ 0\ 1) + x_3(1\ 0\ 1) = \begin{cases} (0\ 0\ 1), & x_3 = 0 \\ (1\ 0\ 0), & x_3 = 1 \end{cases}$$
$$\mathbf{S}_3 = (1\ 1\ 0) + x_3(1\ 0\ 1) = \begin{cases} (1\ 1\ 0), & x_3 = 0 \\ (0\ 1\ 1), & x_3 = 1 \end{cases}$$
$$\mathbf{S}_3 = (1\ 1\ 1) + x_3(1\ 0\ 1) = \begin{cases} (1\ 1\ 1), & x_3 = 0 \\ (0\ 1\ 0), & x_3 = 1 \end{cases}$$

$j = 4$:
$$\mathbf{S}_4 = \mathbf{S}_3 \oplus x_4 \mathbf{h}_4 = \mathbf{S}_3 + x_4(101)$$
$$\mathbf{S}_4 = (0\ 0\ 0) + x_4(0\ 1\ 1) = \begin{cases} (0\ 0\ 0), & x_3 = 0 \\ (0\ 1\ 1), & x_3 = 1 \end{cases} \quad \mathbf{S}_4 = (1\ 0\ 0) + x_4(0\ 1\ 1) = \begin{cases} (1\ 0\ 0), & x_4 = 0 \\ (1\ 1\ 1), & x_4 = 1 \end{cases}$$
$$\mathbf{S}_4 = (0\ 0\ 1) + x_4(0\ 1\ 1) = \begin{cases} (0\ 0\ 1), & x_3 = 0 \\ (0\ 1\ 0), & x_3 = 1 \end{cases} \quad \mathbf{S}_4 = (1\ 0\ 1) + x_4(0\ 1\ 1) = \begin{cases} (1\ 0\ 1), & x_4 = 0 \\ (1\ 1\ 0), & x_4 = 1 \end{cases}$$
$$\mathbf{S}_4 = (0\ 1\ 0) + x_4(0\ 1\ 1) = \begin{cases} (0\ 1\ 0), & x_3 = 0 \\ (0\ 0\ 1), & x_3 = 1 \end{cases} \quad \mathbf{S}_4 = (1\ 1\ 0) + x_4(0\ 1\ 1) = \begin{cases} (1\ 1\ 0), & x_4 = 0 \\ (1\ 0\ 1), & x_4 = 1 \end{cases}$$
$$\mathbf{S}_4 = (0\ 1\ 1) + x_4(0\ 1\ 1) = \begin{cases} (0\ 1\ 1), & x_3 = 0 \\ (0\ 0\ 0), & x_3 = 1 \end{cases} \quad \mathbf{S}_4 = (1\ 1\ 1) + x_4(0\ 1\ 1) = \begin{cases} (1\ 1\ 1), & x_4 = 0 \\ (1\ 0\ 0), & x_4 = 1 \end{cases}$$

Now, the trellis is completely developed and there are 16 paths going to 8 nodes (two paths into every node). These paths correspond to 16 possible information bit sequences (of the length 4). It is supposed that the information bits are independent and that all these combinations are equally probable. But, now the parity check bits come. They depend on previous path bits, i.e. bits x_5, x_6 and x_7 depend on the state (node). The state now determines completely the next parity check bits, as shown in Fig. 8.1. E.g. in a considered case:

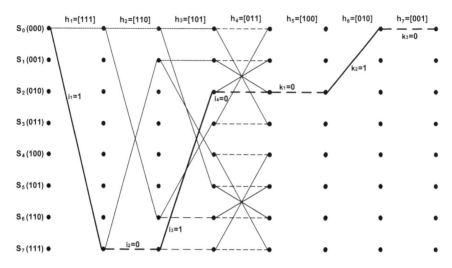

Fig. 8.1 Trellis after the first four steps

$$S_4 = S_3 + x_4 h_4 = (S_2 + x_3 h_3) + x_4 h_4 = ((S_1 + x_2 h_2) + x_3 h_3) + x_4 h_4$$
$$= (((S_0 + x_1 h_1) + x_2 h_2) + x_3 h_3) + x_4 h_4 = S_0 + (i_1 h_1 + i_2 h_2 + i_3 h_3 + i_4 h_4)$$
$$= S_0 + (i_1 H_{11} + i_2 H_{21} + i_3 H_{31} + i_4 H_{41} \, i_1 H_{12} + i_2 H_{22} + i_3 H_{32} + i_4 H_{42} \, i_1 H_{13}$$
$$+ i_2 H_{23} + i_3 H_{33} + i_4 H_{43})$$
$$= S_0 + (i_1 P_{11} + i_2 P_{12} + i_3 P_{13} + i_4 P_{14} \, i_1 P_{21} + i_2 P_{22} + i_3 P_{23} + i_4 P_{24} \, i_1 P_{31}$$
$$+ i_2 P_{32} + i_3 P_{33} + i_4 P_{34})$$
$$= (000) + (k_1 k_2 k_3) = (k_1 k_2 k_3).$$

From Fig. 8.1 it is easy to see that to the branch (path) which corresponds to the input $i_1 = 1$, $i_2 = 0$, $i_3 = 1$ and $i_4 = 0$, correspond as well the same output bits (systematic encoder). This path after the fourth step is in state $S_2 = (010)$, and the next three control bits correspond to the bits in parenthesis, i.e. $k_1 = 0$, $k_2 = 1$ and $k_3 = 0$. Final state for the considered path is

$$S_7 = S_6 + x_7 h_7 = (S_5 + x_6 h_6) + x_7 h_7 = ((S_4 + x_5 h_5) + x_6 h_6) + x_7 h_7$$
$$= [k_1 k_2 k_3] + k_1 [100] + k_2 [010] + k_3 [001] = [000].$$

The complete trellis is shown in Fig. 8.2. Parts of the path corresponding to symbol $x_j = 0$ are dashed, and parts corresponding to symbol $x_j = 1$ are denoted by full lines.

Consider now linear block code (5, 3) which has the generator matrix

Brief Theoretical Overview

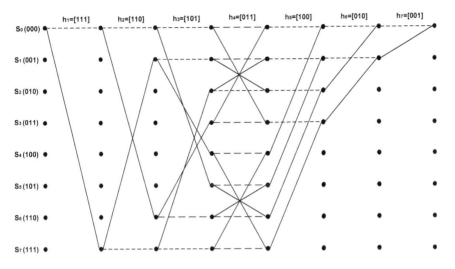

Fig. 8.2 Trellis structure for the considered example

$$G = \begin{bmatrix} 1 & 0 & 1 & 0 & 0 \\ 0 & 1 & 0 & 1 & 0 \\ 0 & 0 & 1 & 1 & 1 \end{bmatrix}$$

To obtain the trellis, the control matrix should be found. Code is not systematic and firstly the generator matrix of the equivalent systematic code has to be found. It can be achieved by column permutation (here 1-2-3-4-5 → 3-4-1-2-5) or by addition of corresponding rows (here the third row of systematic code matrix G' can be obtained by adding all three rows of matrix G). Consider submatrix of this matrix corresponding to parity checks P (here first two columns of G). Using the standard procedure the control matrix of systematic code is obtained:

$$G' = \begin{bmatrix} 1 & 0 & 1 & 0 & 0 \\ 0 & 1 & 0 & 1 & 0 \\ 1 & 1 & 0 & 0 & 1 \end{bmatrix} = [\mathbf{P}, \mathbf{I}_3], \quad H' = [\mathbf{I}_2, \mathbf{P}^T] = \begin{bmatrix} 1 & 0 & 1 & 0 & 1 \\ 0 & 1 & 0 & 1 & 1 \end{bmatrix}.$$

Of course, the relation $G'(H')^T = 0$ holds but as well as the relation $(GH')^T = 0$.

Corresponding trellis is shown in Fig. 8.3. It should be noticed that in this example the used generator matrix G generates the code words where the control bits are at the beginning, followed by information bits. Corresponding code words are $c_1 = (00000)$, $c_2 = (11001)$, $c_3 = (01010)$, $c_4 = (10011)$, $c_5 = (10100)$, $c_6 = (01101)$, $c_7 = (11110)$, $c_8 = (00111)$. Minimal Hamming weight determines the minimal Hamming distance is here $d_{min} = 2$.

Therefore, to construct trellis, the corresponding parity-check matrix should be found providing the trellis construction. When parity-check matrix of linear block code is not known, a trellis structure can be determined from the generator matrix, if

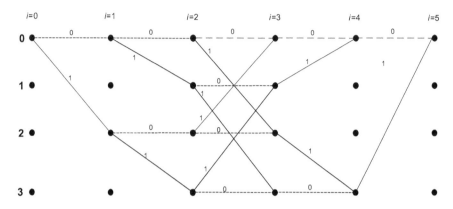

Fig. 8.3 Trellis structure for the considered example (code (5, 3))

it is written in a ***Trellis Oriented Form*** (TOF). For a generator matrix G it is said that is TOF if the next two conditions are satisfied [70, 71]: the leading one in every matrix G row is before the leading one in the every next row, i.e. the row below it and no two rows have the last ones in the same column. In fact, ***trellis oriented generator matrix*** G_{TOGM} is obtained by generator matrix transformation as done in the previous examples. The generator matrix of cyclic codes has always trellis oriented form.

The trellis provides for application of hard or soft Viterbi algorithm. Path metrics are calculated as Hamming or (squared) Euclidean distances of the received sequence u from the sequences v that could have been transmitted, taking into account the trellis structure. Therefore, the decoding can be hard [by using Hamming metric (Problem 8.5)] or soft [by using Euclidean metric with or without the quantization (Problems 8.5 and 8.6)]. Accumulated squared metric of every state is calculated as follows

$$d_E(u, v) = \sum_{i=1}^{n} (u_i - v_i)^2.$$

If ML rule is applied by using the Viterbi algorithm, a ***generalized Viterbi algorithm*** is obtained (Problem 8.6).

Consider now the first example in this section. Let the samples at the decoder input are

$$0.4, 0.2, 0.5, 0.1, 0.2, 0.1 \text{ and } 0.2,$$

forming one code word.

Decoding procedure is shown in Table 8.1. The procedure starts at the moment when the paths joint for the first time (here for depth 4). It is supposed that bit 0 corresponds to the sample value 0.0, and bit 1 to the sample value 1.0. In the first

Brief Theoretical Overview

Table 8.1 Viterbi decoding for the trellis shown in Fig. 8.2

	First step	Second step	Third step	Fourth step
Samples→	0.4 0.2 0.1 0.5	0.2	0.1	0.2
0–0:	0000: **0.46** 0111: 1.86	0: $0.46 + 0.2^2 =$ **0.5** 1: $0.66 + 0.8^2 = 1.3$	0: $0.5 + 0.1^2 =$ **0.51** 1: $1.3 + 0.9^2 = 2.31$	0: $0.51 + 0.2^2 =$ **0.55** 1: $1.31 + 0.8^2 = 1.95$
0–1:	1100: **1.26** 1011: 1.46	0: $1.26 + 0.2^2 =$ **1.3** 1: $1.06 + 0.8^2 = 1.7$	0: $1.3 + 0.1^2 =$ **1.31** 1: $0.5 + 0.9^2 = 1.31$	
0–2:	1101: **1.26** 1010: 1.46	0: $1.26 + 0.2^2 =$ **1.3** 1: $1.06 + 0.8^2 = 1.7$		
0–3:	0001: **0.46** 0110: 1.86	0: $0.46 + 0.2^2 =$ **0.5** 1: $0.36 + 0.8^2 = 1.3$		
0–4:	1110: 2.06 1001: **0.66**			
0–5:	0010: 1.26 0101: **1.06**			
0–6:	0011: 1.26 0100: **1.06**			
0–7:	1111: 2.06 1000: **0.66**			

step between nodes 0 and 0 (the first row in table) corresponding paths are 0000 and 0111.

For path S_0-S_0-S_0-S_0-S_0 the corresponding metric is

$$d = (0.4 - 0)^2 + (0.2 - 0)^2 + (0.5 - 0)^2 + (0.1 - 0)^2 = 0.46,$$

and for path S_0-S_0-S_6-S_3-S_0

$$d = (0.4 - 0)^2 + (0.2 - 1)^2 + (0.5 - 1)^2 + (0.1 - 1)^2 = 1.86.$$

The survived path has the distance 0.46 (fat numbers). The procedure continues for the next step where into state S_0 enter paths from S_0 and S_4 and a path is chosen which has smaller accumulated distance taking as well into account metrics of the nodes from which the part of path starts. In the third step there is a case where both distances are equal (1.31) and arbitrarily, one path is chosen. However, and even if the other path was chosen, it would not influence the last step. Therefore, all zeros path (0000000) has survived and the information bits are (0000).

Soft decision is optimum if at the channel output the samples are from continuous set. In real systems it is not the case because samples are quantized. Therefore, it may be conceived that at the channel output there are specific (finite) number of discrete values. For binary transmission, the channel can be usually represented as discrete channel with two inputs and more outputs. The case with two outputs is a binary channel and corresponds to hard decision. For a higher number of outputs the limiting case is soft decision. ML rule can be applied for such channels and if

Viterbi algorithm is used, this procedure is usually named *generalized Viterbi algorithm* (Problem 8.6) [72]. The procedure is as follows:

1. *Path metrics* are defined by trellis structure and by transient discrete channel probabilities

$$\gamma_i(\mathbf{S}^{(i-1)}, \mathbf{S}^{(i)}) = P(\mathbf{S}^{(i)}|\mathbf{S}^{(i-1)})P(Y_i|X_i = x).$$

where $P(X_i = x|\mathbf{S}^{(i-1)}\mathbf{S}^{(i)})$ denotes the probability that during transition from state $\mathbf{S}^{(i-1)}$ in state $\mathbf{S}^{(i)}$ at the encoder output, x is emitted ith symbol. It can be equal 1 (if that transition exists at the trellis or 0, if it is not allowed). Probability that at the channel output at ith step the symbol Y_i appears if at the channel input is symbol x is denoted with $P(Y_i|X_i = x)$. The number of symbols is finite and these probabilities can be calculated and given by the table.
2. *Node metrics* are obtained in the following way—at depth $i = 0$ only to one node (usually corresponding to state $\mathbf{S}^{(0)} = 0$) the unity metric is joined ($\alpha_0(\mathbf{S}^{(0)}) = 1$), while for other nodes the metric equals zero. At ith trellis depth node metrics are calculated as follows:

$$\alpha_i(\mathbf{S}^{(i)}) = \alpha_{i-1}(\mathbf{S}^{(i-1)})\gamma_i(\mathbf{S}^{(i-1)}, \mathbf{S}^{(i)}) = \prod_{j=1}^{i} \gamma_j(\mathbf{S}^{(j-1)}, \mathbf{S}^{(j)}), \quad i = 1, 2, \ldots, n$$

There are more paths to enter into a specific node and the corresponding products are calculated for all possible paths. For node metric the highest one is chosen.

Consider the encoder defined in the second example in this section. Let the probability of binary symbols is 0.5. The encoded sequence is transmitted over discrete channel without memory. Let channel output is quantized yielding symbols 0, 0^+, 1^- i 1. Two output symbols are more reliable (0 i 1) and two are less reliable (0^+, 1^-). Channel graph is shown in Fig. 8.4.

Let message $\mathbf{m} = (000)$ is transmitted, the corresponding code word is $c_1 = (00000)$. If at the channel output $\mathbf{y} = (0^+, 0, 1^-, 0, 0)$ is obtained, apply generalized Viterbi algorithm. Transition probabilities are given in Table 8.2.

Using trellis shown in Fig. 8.3 the corresponding path metrics are found and for the transient regime shown in Fig. 8.5. The metrics can be easily found. E.g.

$$\alpha_3(2)' = \gamma_1(0,0)\gamma_2(0,0)\gamma_3(0,2) = P(0^+|0)P(0|0)P(1^-|1)$$
$$= 0.3 \times 0.5 \times 0.3 = 0.045^*,$$
$$\alpha_3(2)'' = \gamma_1(0,2)\gamma_2(2,2)\gamma_3(2,2) = P(0^+|1)P(0|0)P(1^-|0)$$
$$= 0.15 \times 0.5 \times 0.15 = 0.01125.$$

The path which has a larger metric is chosen. These pats are denoted with heavy lines as well as their values.

Brief Theoretical Overview 393

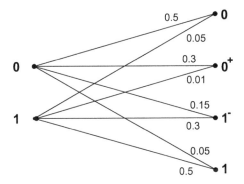

Fig. 8.4 Graph of discrete channel (two inputs, four outputs)

Table 8.2 Transition probabilities for the received message

i	1	2	3	4	5
Y_i	0^+	0	1^-	0	0
$(P(Y_i/0); P(Y_i/1))$	(0.3, 0.15)	(0.5, 0.05)	(0.15, 0.3)	(0.5, 0.05)	(0.5, 0.05)

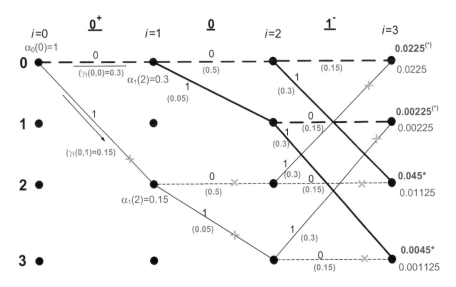

Fig. 8.5 Trellis part corresponding to the transient regime

It can be noticed that after the third step, the choice was between two paths entering state 0 (both metrics are 0.0225) and as well between two pats entering state 1 (both 0.00225). In Fig. 8.5 the "upper" paths were chosen in both cases. The complete corresponding trellis is shown in Fig. 8.6 where heavy dashed line is decoded, i.e. the word (00000). If the "lower" paths were chosen for both cases, the corresponding complete trellis is shown in Fig. 8.7, i.e. the word (10100) resulting in error event. It should be stressed both results are equally reliable.

As pointed above, generalized Viterbi algorithm is based on *Maximum Likelihood* (ML) rule. This rule does not take into account a priori probabilities of symbols at the channel input. Similarly, Viterbi algorithm is not an optimal procedure if the symbol probabilities at the encoder input are unequal. To achieve the optimal performance, the decisions should be based on *Maximum A Posteriori Probability* (MAP). Such an algorithm was proposed by Bahl, Cocke, Jelinek and Raviv [73] known today as **BCJR algorithm**. It takes into account the symbol probabilities at the channel input. Further, this algorithm guarantees an optimal decision for every system for which trellis can be constructed.

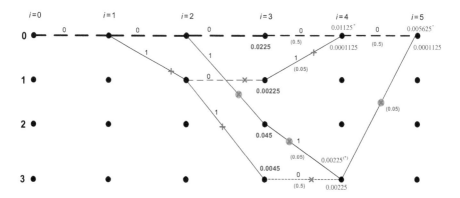

Fig. 8.6 Complete trellis, the first variant

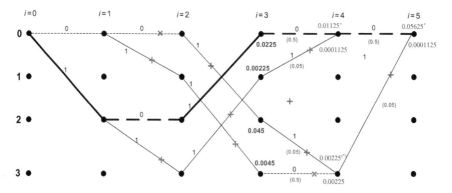

Fig. 8.7 Complete trellis, the second variant

Here *path (branch) metric* is modified as follows:

$$\gamma_i(S^{(i-1)}, S^{(i)}) = \sum_{x \in A_x} P(S^{(i)}|S^{(i-1)})P(X_i = x|S^{(i-1)}S^{(i)})P(Y_i|X_i = x).$$

In this expression, the new parameter is $P(S^{(i)}|S^{(i-1)})$, the conditional probability that system in ith step from the state $S^{(i-1)}$ goes into the state $S^{(i)}$. For binary codes, this probability depends on the probabilities of 0 and 1 at the encoder input (not taken into account in Viterbi algorithm). For the equiprobable input symbols for every state $P(S^{(i)}|S^{(i-1)}) = 0,5$. However, sometimes the path from state $S^{(i-1)}$ must go into only one state (then $P(S^{(i)}|S^{(i-1)}) = 1, 0$. The second important difference relating to the Viterbi algorithm is that node metrics are calculated in two ways—"forward" (with increasing trellis depth) and "backward" (from the end of trellis).

1. *Forward path metrics* are calculated as follows:
   - at depth $i = 0$, only to one node (usually corresponding to state $S^{(0)} = 0$) unit metric is adjoined, while for other nodes the metric equals zero.
   - at ith trellis depth node metrics are calculated as follows

$$\alpha_i(S^{(i)}) = \sum_{S^{(i-1)} \in S_x} \alpha_{i-1}(S^{(i-1)}) \gamma_i(S^{(i-1)}, S^{(i)}), \quad i = 1, 2, \ldots, n$$

where S_x is a set of all possible states. After finding all possible transitions from the previous state depth to the considered state the metric of the starting states is multiplied by the metric of the corresponding branch and all obtained results are summed to obtain the node metric of the specific node.

2. *Backward path metrics* are calculated as follows:
   - at depth $i = n$, only to one node (usually corresponding to state $S^{(n)} = 0$) unit metric is adjoined, while for other nodes the metric equals zero.
   - at the depth $i - 1$ node metrics are calculated as follows:

$$\beta_{i-1}(S^{(i-1)}) = \sum_{S^{(i)} \in S_x} \beta_i(S^{(i)}) \gamma_i(S^{(i-1)}, S^{(i)}), \quad i = n, n-1, \ldots, 1$$

where S_x is a set of all possible states. Here the metric of node where the branch enters is multiplied by the metric of corresponding branch and all obtained results are summed to obtain the node metric of the specific node.

Consider the previous example where $y = (0^+, 0, 1^-, 0, 0)$. Corresponding trellis is shown in Fig. 8.3. E.g. for node at the depth $i = 1$, the following is obtained:

$$\gamma_1(0,0) = \sum_{x \in \{0,1\}} P(S^{(1)} = 0|S^{(0)} = 0)P(X_i = x|S^{(0)} = 0, S^{(1)} = 0)P(Y_1|X_1 = x)$$
$$= 0.5 \times 1 \times P(0|0) + 0.5 \times 0 \times P(0|1) = 0.15$$
$$\gamma_1(0,2) = \sum_{x \in \{0,1\}} P(S^{(1)} = 2|S^{(0)} = 0)P(X_i = x|S^{(0)} = 0, S^{(1)} = 2)P(Y_1|X_1 = x)$$
$$= 0.5 \times 0 \times P(0|0) + 0.5 \times 1 \times P(0|1) = 0.075,$$

while $\gamma_1(0,1) = 0$ and $\gamma_1(0,3) = 0$, because the transitions from state 0 into the states 1 and 3 do not exist.

Node metrics (presented in Table 8.3) are calculated according to the following example:

$$\alpha_1(2) = \alpha_0(0)\gamma_1(0,2) = 1 \times 0.075 = 0.075,$$
$$\beta_4(3) = \beta_5(0)\gamma_5(3,0) = 1 \times 0.05 = 0.05$$

shown in Fig. 8.8. In the case when more branches enter into one node, the metrics are added only. They are not compared neither only one branch is chosen, as in Viterbi algorithm case.

Now, the further calculation is needed:

For every branch entering alone in the specific node, as well for the node, parameter $\lambda_i(S^{(i)})$ is calculated as

$$\lambda_i(S^{(i)}) = \alpha_i(S^{(i)})\beta_i(S^{(i)}).$$

From the trellis one finds to which symbol of encoded sequence (at that depth) corresponds this transition (i.e. node) and the probability is calculated that in this step at the encoder output the corresponding symbol has been emitted. If at the considered depth into every node only one branch enters, the probability that at this (*i*th) depth binary zero is decoded is

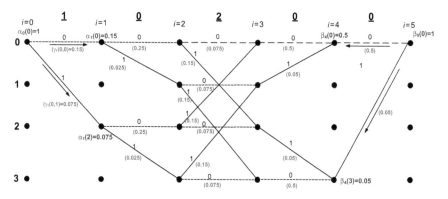

Fig. 8.8 Node metrics calculation "forward" and "backward"

Brief Theoretical Overview

Table 8.3 List of metrics for all nodes "forward" and "backward"

		$t = 0$	$t = 1$	$t = 2$	$t = 3$	$t = 4$	$t = 5$
Forward	$\alpha_t(0)$	1	0.15	0.0375	0.005625	0.002841	0.001455
	$\alpha_t(1)$	0	0	0.00375	0.0005625	0	0
	$\alpha_t(2)$	0	0.075	0.01875	0.007031	0	0
	$\alpha_t(3)$	0	0	0.001875	0.0007031	0.0007031	0
Backward	$\beta_t(0)$	0.001455	0.004922	0.01912	0.25	0.5	1
	$\beta_t(1)$	0	0	0.05625	0.025	0	0
	$\beta_t(2)$	0	0.009562	0.03769	0.0025	0	0
	$\beta_t(3)$	0	0	0.05625	0.025	0.05	0

$$P_i(0) = \frac{\sum_{A_{x0}} \lambda_j^{(x=0)}(\mathbf{S}^{(i)})}{\sum_{A_x} \lambda_j^{(x=0)}(\mathbf{S}^{(i)})},$$

where the summation in numerator is over all node metrics to which in this step corresponds binary zero ($x_i = 0$), and summation in denominator is over metric of all branches entering into any of states.

If two branches enter the node, the following metric is calculated

$$\sigma_i(\mathbf{S}^{(i-1)}, \mathbf{S}^{(i)}) = \alpha_{i-1}(\mathbf{S}^{(i-1)})\gamma_i(\mathbf{S}^{(i-1)}, \mathbf{S}^{(i)})\beta_i(\mathbf{S}^{(i)}).$$

From trellis (at that depth) one finds to which symbol of encoded sequence correspond the transitions and the probability is calculated that in this step at the encoder output the corresponding symbol has been emitted. The probability that at this (*i*th) depth binary zero is decoded is

$$P_i(0) = \frac{\sum_{A_{x0}} \sigma_i(\mathbf{S}^{(i-1)}, \mathbf{S}^{(i)})}{\sum_{A_x} \sigma_i(\mathbf{S}^{(i-1)}, \mathbf{S}^{(i)})},$$

where the summation in numerator is over metrics of all branches to which in this step corresponds binary zero ($x_i = 0$), and summation in denominator is over metrics of all branches entering into any of states.

For the considered example, at depths 1 and 2 these probabilities are

$$P_1(0) = \frac{\lambda_1(0)}{\lambda_1(0) + \lambda_1(2)} = 0.5072,$$

$$P_2(0) = \frac{\lambda_2(0) + \lambda_2(2)}{\lambda_2(0) + \lambda_2(2) + \lambda_2(1) + \lambda_2(3)} = 0.9783$$

meaning that at the first step probability of zero is 0.5072, and at the second—0.9783.

Corresponding probabilities at depths three and four are

$$P_3(0) = \frac{\sigma_3(0,0) + \sigma_3(1,1) + \sigma_3(2,2) + \sigma_3(3,3)}{\sigma_3(0,0) + \sigma_3(1,1) + \sigma_3(2,2) + \sigma_3(3,3) + \sigma_3(0,2) + \sigma_3(1,3) + \sigma_3(2,0) + \sigma_3(3,1)}$$
$$= 0.4927,$$

$$P_4(0) = \frac{\sigma_4(0,0) + \sigma_4(3,3)}{\sigma_4(0,0) + \sigma_4(3,3) + \sigma_4(1,0) + \sigma_4(2,3)} = 0.9783.$$

Values $(S^{(i)})$ and $\sigma_i(S^{(i-1)}, S^{(i)})$ are shown in Fig. 8.9 where in the parentheses are given the probabilities that zero is emitted in every step. On the basis of these values BCJR decoder could decide that sequence (0, 0, 1, 0, 0) was emitted (wrongly!). However, the trellis branches corresponding to the above sequence (fat lines in figure) do not form the path in trellis.

BCJR algorithm is completed, but still the decision was not made. Now, the valid code sequence should be found which is the nearest to the above estimation (00100). One way to do it is the comparison of this estimation to all eight possible code words and to choose one having the minimum Hamming distance from it. In this case there are two such words:

- code word (00000) (Hamming distance $d = 1$) and according to BCJR its estimation of the third bit being 1 is not so reliable (only 0.5072)
- code word (10100) (Hamming distance $d = 1$) and according to BCJR its estimation of the first bit being 1 is not so reliable (only 0.5072).

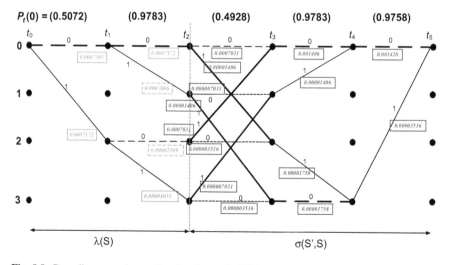

Fig. 8.9 Decoding procedure and estimation probabilities

Both estimations are equally unreliable and the final decision about the emitted sequence has the same probability.

The obtained result should be expected because the Viterbi algorithm is optimum when the symbols at the discrete channel input are equally probable, what is here the case. Because of that, there is no sense to use BCJR here. However, BCJR has many advantages:

- when the input symbols are not equally probable, Viterbi algorithm (ML algorithm) is not optimum. But, BCJR algorithm uses MAP, is an optimum in this case
- BCJR algorithm is suitable for iterative decoding, where on the basis of additional information, the estimation of symbols can be changed, until they become sufficiently large to make a reliable estimation.

Decoding using MAP algorithm needs large memory and a substantial number of operations including multiplication and exponentiation. As in the case of Viterbi algorithm, here instead of metrics, their logarithms can be used yielding summation instead of multiplication. Now, **Log-MAP algorithm** [74] is obtained (Problem 8.9), which has the same performance as BCJR. However, in this case the expressions of the type $\ln(e^{L(d_1)} + e^{L(d_2)})$ appear. Here, the following exact expression can be used

$$\ln(e^{\delta_1} + e^{\delta_2}) = \max(\delta_1, \delta_2) + \ln(1 + e^{-|\delta_1 - \delta_2|}),$$

where the second term should be successively calculated. If the second term is neglected, the suboptimum procedure is obtained—**Max-Log-MAP algorithm** [75] (Problem 8.9), having smaller complexity. There exist also further modifications. One is **Constant-Log-MAP** [76], where the approximation

$$\ln(e^{\delta_1} + e^{\delta_2}) \approx \max(\delta_1, \delta_2) + \begin{cases} 0, & |\delta_1 - \delta_2| > T \\ C, & |\delta_1 - \delta_2| < T \end{cases},$$

is used and where the value of threshold T is found on the basis of a specific criteria. The other is **Linear-Log-MAP**, where the following approximation is used

$$\ln(e^{\delta_1} + e^{\delta_2}) \approx \max(\delta_1, \delta_2) + \begin{cases} 0, & |\delta_1 - \delta_2| > T \\ a|\delta_1 - \delta_2| + b, & |\delta_1 - \delta_2| < T \end{cases},$$

and where constants **a** and **b** are determined on the basis of a specific criteria.

The application of MAP (and Log-MAP) implies the knowledge of signal-to-noise ratio at the receiver input. It is needed for metric calculation and for the corresponding scaling of information during the decoders communication in the case of iterative decoding.

The Viterbi algorithm output is always "hard". Only the SOVA (*Soft Output Viterbi Algorithm*) [77, 78] (Problem 8.7) algorithm makes possible at the decoder output, parallelly with a series of decoded bits, to find an estimation of the reliability of the corresponding decisions. A variant that use Euclidean metric is used as well as a variant that use correlation metric. SOVA overcomes the drawback of classical Viterbi algorithm where the estimations of reconstruction reliability of code sequence were not available. This procedure uses a good feature of Viterbi algorithm that a decoded word must correspond to a complete trellis path, i.e. it must be a codeword (BCJR has not this feature). It is ML procedure, and algorithm is still suboptimum with respect to BCJR. SOVA algorithm can be implemented and for unquantized signal values and can be used for channels which are not discrete.

Similarly to BCJR, SOVA is Viterbi algorithm modification where the complete received sequence is considered and decisions are based on forward and backward recursions. The algorithm output are not only the hard decisions, but as well the estimations of a posteriori probabilities of encoded symbols x_i or their logarithmic ratio—LLR(*log likelihood ratio*) [67]

$$\Lambda(x_i) = \log \frac{\Pr\{x_i = 1|r\}}{\Pr\{x_i = 0|r\}} = \begin{cases} \mu_{i,c} - \mu_{min}, & x_i = 1 \text{ in ML path} \\ \mu_{min} - \mu_{i,c}, & x_i = 0 \text{ in ML path} \end{cases}$$

where μ_{min} is the metric (squared Euclidean distance) of complete survived path and $\mu_{t,c}$ is the metric of path of the strongest concurrent (*competitor*) in step i.

Consider systematic linear block code (4, 3) yielding simple parity checks. Generator and control matrices are

$$G = \begin{bmatrix} 1 & 0 & 0 & 1 \\ 0 & 1 & 0 & 1 \\ 0 & 0 & 1 & 1 \end{bmatrix}, \quad H = [1 \ 1 \ 1 \ 1].$$

Corresponding trellis is shown in Fig. 8.10. In the same figure the decoding of sequence (0.8, 0.7, 0.1, 0.2) is shown using soft Viterbi algorithm, as well as branch metrics and the survived path. The decoded sequence is $(i_1, i_2, i_3) = (1, 1, 0)$.

SOVA algorithm is identical to the classic Viterbi algorithm until the moment when the path is found which has the minimum Euclidean distance. To find LLR, the metric of every branch that separated from this path using expression

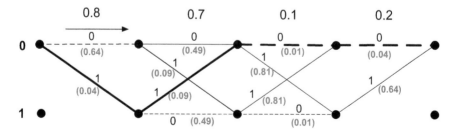

Fig. 8.10 "Classic" Viterbi algorithm *zapete*

$$\mu_{i,c} = \min_{l,l'}\left\{\mu_{i-1}^{f}(l') + v_i(l',l) + \mu_i^{b}(l)\right\}$$

is found, where $\mu_{i-1}^{f}(l')$ denotes the metric "forward" of the state from which the concurrent branch started, $v_i^{i-1}(l',l)$ branch metric at the depth i and $\mu_i^{b}(l)$ metric "backward" of state where enters the concurrent branch. Minimization is performed over all concurrent branches. At the first step, the first term does not exist, and at the last one, the third term does not exist. Therefore, taking into account that the decoded sequence is 1100:

$i = 1$: $\mu_{1,c} = 0.64 + 0.54 = 1.18$ and $\Lambda(x_1) = \mu_{1,c} - \mu_{\min} = 1.18 - 0.14 = 1.04$,
$i = 2$: $\mu_{2,c} = 0.04 + 0.49 + 0.65 = 1.18$ and $\Lambda(x_2) = \mu_{2,c} - \mu_{\min} = 1.18 - 0.14 = 1.04$,
$i = 3$: $\mu_{3,c} = 0.13 + 0.81 + 0.64 = 1.58$ and $\Lambda(x_3) = \mu_{3,c} - \mu_{\min} = 0.14 - 1.58 = -1.44$,
$i = 4$: $\mu_{4,c} = 0.94 + 0.64 + 0 = 1.58$ and $\Lambda(x_4) = \mu_{4,c} - \mu_{\min} = 0.14 - 1.58 = -1.44$.

Metrics according SOVA algorithm are shown in Fig. 8.11.

It is considered that SOVA algorithm is about 1.5 times more complex than classical Viterbi algorithm. For very long code words, the decision based on the whole received sequence needs too much decoder memory. However, it can be substantially reduced by using "*sliding window*" [68].

As mentioned earlier to obtain the results promised by the Second Shannon theorem, the codes having long codewords should be used. However, the decoding procedure for such codes can be very complex. In 1966, Forni proposed **concatenated codes** where two encoders in series are used. Decoding procedure can be performed *iteratively*, where the decoders interchange information. Further, in 1993 **turbo codes** [79] (Problem 8.10) were proposed where the feedback between the decoders is used to obtain a better estimation of every bit a posteriori probability during the iterations. Here, the feedback between decoders is used [80]. This principle was previously applied in turbo motors, and the name *turbo codes* was proposed. In the following only the basics of turbo coding will be explained, while the numerical example is given in Problem 8.10.

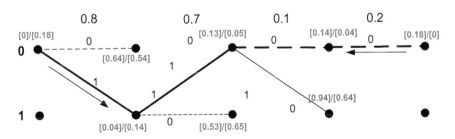

Fig. 8.11 SOVA algorithm—metrics forward and backward

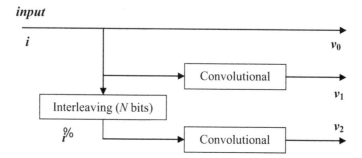

Fig. 8.12 Turbo encoding principle

One possible way to implement turbo encoder is shown in Fig. 8.12. Recursive systematic convolutional codes are supposed. Generally, different codes can be used. The same information sequence is led into both encoders, but the interleaving is applied in front of one. Interleaving is introduced to avoid the encoding of the same sequence by both encoders. In such way the packets of errors will be deinterleaved. After the information bit sequence some "non information" bits are used to reset encoders usually into all zeros state ("*tail biting*"). The corresponding trellises are terminated, and the decoding of the next information bit sequence is independent of the previous sequence. Therefore, turbo encoder is linear block coder generating long "code words", being one if the goals in construction of error control codes.

Let both encoders have code rate 1/2. Information bits are directly transmitted as well as parity checks of the first encoder. From the second encoder only the parity checks are send, but for interleaved sequence. In such way, an equivalent code is obtained which has code rate 1/3. It should be noted that the second encoder takes into account only the information bits and not control bits of the first encoder. It is the difference in regard to serial *concatenated* encoding, where the outer encoder "works" considering the complete output of the inner encoder.

In this example (code rate $R = 1/3$) for a sequence of N information bits, the encoded sequence has $3N$ bits. From all possible 2^{3N} binary sequences, only 2^N are valid code words. Therefore, it is linear $(3N, N)$ block code (N is the interleaver length).

Consider turbo encoder shown in Fig. 8.13. The interleaver work is described by ($N = 16$ bits)

$$I\{.\} = \begin{pmatrix} 1 & 2 & 3 & 4 & 5 & 6 & 7 & 8 & 9 & 10 & 11 & 12 & 13 & 14 & 15 & 16 \\ 1 & 5 & 9 & 13 & 2 & 6 & 10 & 14 & 3 & 7 & 11 & 15 & 4 & 8 & 12 & 16 \end{pmatrix}$$

where the second row denotes bit positions at the interleaver output and the first—bit positions in the original sequence. The puncturing (explained in the previous chapter) is supposed as well and let Π is the puncturing matrix.

Because of puncturing the code rate is $R = 1/2$ and the corresponding code has parameters (32, 16) as illustrated in Fig. 8.14.

Brief Theoretical Overview

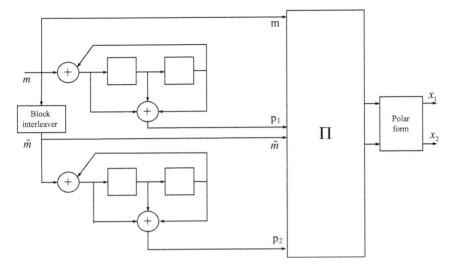

Fig. 8.13 Complete turbo encoder structure

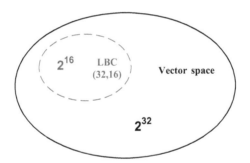

Fig. 8.14 Turbo code as a linear block code

Table 8.4 Sequences at some points of encoder from Fig. 8.13

i	1	2	3	4	5	6	7	8	9	10	11	12	13	14	15	16
m	1	1	0	1	1	0	1	1	0	1	1	1	0	1	0	0
m'	1	1	0	0	1	0	1	1	0	1	1	0	1	1	1	0
p_1	1	0	1	0	1	0	1	0	1	0	1	1	1	1	0	0
p_2	1	0	1	1	0	0	0	0	0	0	0	0	0	0	1	0
x_1	1	−1	1	−1	1	−1	1	−1	1	−1	1	1	1	1	−1	−1
x_2	1	−1	1	1	−1	−1	−1	−1	−1	−1	−1	−1	−1	−1	1	−1

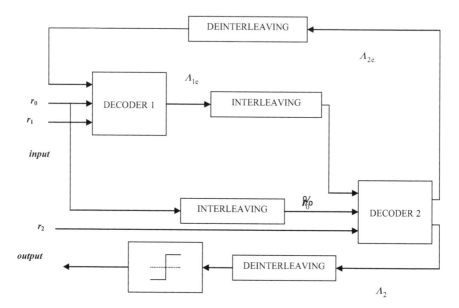

Fig. 8.15 Turbo decoding principle

For information sequence of length $m = 16$, control bits and output bits (in polar format) are given in Table 8.4. It is obvious that by varying interleaver length different codes can be obtained. Code rate does not change, but longer code words will provide better error correction capabilities.

Turbo decoder (Fig. 8.15) has to make best use of redundancy introduced by encoder. The used decoders work on "their own" information bit sequences and the corresponding deinterleaving and interleaving are used. The first decoder on the base of its own input and control bits estimate the information bits (e.g. MAP) and sends estimations to the second decoder (here interleaving is needed). The second decoder, on the base of these estimations and its own control bits sends to the first decoder its own estimations (here the deinterleaving is needed). First decoder now makes new estimations taking into account these estimations as well control bits and sends them to the second decoder. Therefore, the iterative procedure starts. In such way a try is made to extract all possible information introduced by the encoders. After a few iterations, the second decoder makes hard decisions and sends to output (after deinterleaving) decoded information bits. Number of iterations and interleaving procedure are the subjects of investigation.

In Problem 8.10 for above considered turbo encoder (and decoder) numerical example is given for interleaver length $N = 5$.

Turbo codes are an interesting example of the specific area in information theory. They were firstly introduced in praxis, and a theoretical explanation (and even understanding) came later [81]. Generally, it can be said [82] that during turbo decoding, the a posteriori probability for every bit of information sequence is calculated in iterations, to give the final decision at the end.

Problems

Problem 8.1 Hamming code generator matrix is

$$G = \begin{bmatrix} 1 & 1 & 1 & 0 & 0 & 0 \\ 1 & 0 & 0 & 1 & 1 & 0 \\ 0 & 1 & 0 & 1 & 0 & 1 \end{bmatrix},$$

(a) Is it possible to write a generator matrix in a systematic form, so as that the code is selfdual?
(b) Describe in details a procedure to form trellis diagram corresponding to a systematic code.

Solution
One of methods to construct a *linear block code trellis* based on its parity-check matrix structure is described in introductory part of this chapter. This method is applied to construct a trellis for the code whose matrix is given above. Only the explanations necessary for the understanding of the applied procedure are given here.

(a) The code has parameters (6, 3), and it can be noticed that it is a code whose construction is described in Problem 5.8, obtained by Hamming code (7, 4) shortening. Generator matrix of systematic version can be easily found by reordering column positions, but the solution is not a unique one. The reader should think over how many possible matrices exist for this code. Some of possible solutions are

$$G_{s1} = \begin{bmatrix} 1 & 0 & 0 & 0 & 0 & 0 \\ 0 & 1 & 0 & 1 & 0 & 1 \\ 0 & 0 & 1 & 0 & 1 & 1 \end{bmatrix}, G_{s2} = \begin{bmatrix} 1 & 0 & 0 & 0 & 1 & 1 \\ 0 & 1 & 0 & 1 & 1 & 0 \\ 0 & 0 & 1 & 1 & 0 & 1 \end{bmatrix},$$

$$G_{s3} = \begin{bmatrix} 1 & 0 & 0 & 1 & 0 & 1 \\ 0 & 1 & 0 & 1 & 1 & 0 \\ 0 & 0 & 1 & 0 & 1 & 1 \end{bmatrix}, \ldots$$

By a suitable matrix choice, the relation $P^T = P$ (matrix P is symmetric) can be achieved, the code described by a generator matrix G_s is *self-dual*, and it is easy to find the parity-check matrix

$$G_s = \begin{bmatrix} 1 & 0 & 0 & 0 & 1 & 1 \\ 0 & 1 & 0 & 1 & 0 & 1 \\ 0 & 0 & 1 & 1 & 1 & 0 \end{bmatrix} = [I_3 P] \Rightarrow H_s = [P, I_3] = \begin{bmatrix} 0 & 1 & 1 & 1 & 0 & 0 \\ 1 & 0 & 1 & 0 & 1 & 0 \\ 1 & 1 & 0 & 0 & 0 & 1 \end{bmatrix}.$$

General rule to form a trellis by using a parity-check matrix is given by

$$S_t = S_{t-1} \oplus x_t h_t,$$

where h_t denotes t the tth parity-check matrix column transposed and x_t can take values 0 or 1. Trellis part to the depth $t = 3$ is formed by using equations given below.

$t = 1$:
$$S_1 = S_0 \oplus x_1 h_1 = (0\,0\,0) + x_1(0\,1\,1)$$
$$= \begin{cases} (0\,0\,0), & x_1 = 0 \\ (0\,1\,1), & x_1 = 1 \end{cases}$$

$t = 2$:
$$S_2 = S_1 \oplus x_2 h_2 = S_1 + x_2(1\,0\,1)$$
$$(0\,0\,0) + x_2(1\,0\,1) = \begin{cases} (0\,0\,0), & x_2 = 0 \\ (1\,0\,1), & x_2 = 1 \end{cases}$$
$$(0\,1\,1) + x_2(1\,0\,1) = \begin{cases} (0\,1\,1), & x_2 = 0 \\ (1\,1\,0), & x_2 = 1 \end{cases}$$

$t = 3$:
$$S_3 = S_2 \oplus x_3 h_3 = S_2 + x_3(110)$$
$$(0\,0\,0) + x_3(110) = \begin{cases} (0\,0\,0), & x_3 = 0 \\ (1\,1\,0), & x_3 = 1 \end{cases}$$
$$(1\,0\,1) + x_3(110) = \begin{cases} (1\,0\,1), & x_3 = 0 \\ (0\,1\,1), & x_3 = 1 \end{cases}$$
$$(0\,1\,1) + x_3(110) = \begin{cases} (0\,1\,1), & x_3 = 0 \\ (1\,0\,1), & x_3 = 1 \end{cases}$$
$$(1\,1\,0) + x_3(110) = \begin{cases} (1\,1\,0), & x_3 = 0 \\ (0\,0\,0), & x_3 = 1 \end{cases}$$

In the introductory part of this chapter it is shown that for linear block codes, information bits at the encoder input determines uniquely the structure of the rest of trellis. If one starts from state $S_0 = (000)$, the following should be satisfied

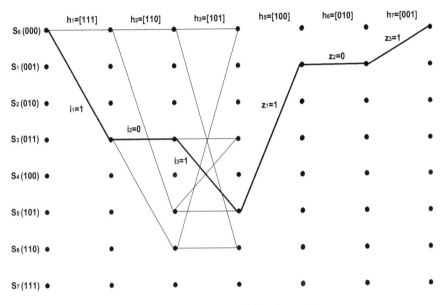

Fig. 8.16 Trellis structure and process of parity-check bits determining

$$S_3 = S_2 + x_3h_3 = ((S_1 + x_2h_2) + x_3h_3 = S_0 + (i_1h_1 + i_2h_2 + i_3h_3) = (z_1z_2z_3).$$

In Fig. 8.16 trellis for the information bits $i_1 = 1, i_2 = 0, i_3 = 1$ is shown.
To information bits $i_1 = 1, i_2 = 0, i_3 = 1$ the following trellis state is obtained

$$S_3 = (000) + 1 \times (011) + 0 \times (101) + 1 \times (110) = (101),$$

but to information bits $i_1 = 0, i_2 = 1, i_3 = 0$ the same state corresponds as well

$$S_3 = (000) + 0 \times (011) + 1 \times (101) + 0 \times (110) = (101).$$

As the states are the same after three steps for both sequences, it should be noticed as well that the parity-check bits are the same for these information sequences

$$c_1 = (101) \otimes G = (101101), \quad c_2 = (010) \otimes G = (010101).$$

The codeword has a structure $c = (i_1\ i_2\ i_3\ z_1\ z_2\ z_3)$, therefore, the rest of path is uniquely determined

$$S_4 = S_3 \oplus z_1h_4, \quad S_5 = S_4 \oplus z_2h_5$$
$$S_6 = S_3 \oplus z_1h_5 \oplus z_2h \oplus z_3h_7 = (z_1z_2z_3) \oplus z_1(100) \oplus z_2(010) \oplus z_3(001) = (000)$$

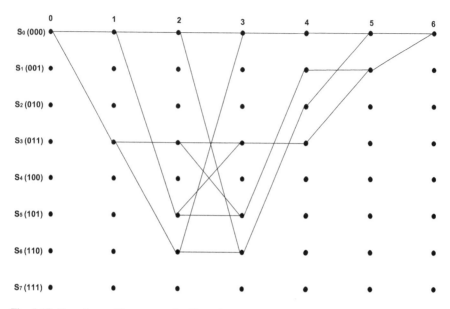

Fig. 8.17 Complete trellis structure for Hamming code (6, 3)

and trellis must terminate in all zeros state.

A complete Hamming code (6, 3) trellis structure is shown in Fig. 8.17. It should be noticed that a number of trellis states is determined by a number of parity-check (redundant) bits in a codeword, being in general case 2^{n-k}.

Problem 8.2 Determine the trellis structure for the Hamming code (8, 4).

Solution

Generator and parity-check matrix of systematic Hamming code (8, 4) have (one possible) form

$$G = \begin{bmatrix} 1 & 0 & 0 & 0 & 0 & 1 & 1 & 1 \\ 0 & 1 & 0 & 0 & 1 & 0 & 1 & 1 \\ 0 & 0 & 1 & 0 & 1 & 1 & 0 & 1 \\ 0 & 0 & 0 & 1 & 1 & 1 & 1 & 0 \end{bmatrix}, \quad H = \begin{bmatrix} 0 & 1 & 1 & 1 & 1 & 0 & 0 & 0 \\ 1 & 0 & 1 & 1 & 0 & 1 & 0 & 0 \\ 1 & 1 & 0 & 1 & 0 & 0 & 1 & 0 \\ 1 & 1 & 1 & 0 & 0 & 0 & 0 & 1 \end{bmatrix}$$

By applying procedure described in the previous problem, the trellis structure to the depth $t = 4$ is obtained and shown in Fig. 8.18, corresponding to codeword information bits. Trellis state where the path is at this depth determines uniquely parity-check bits and the rest of every single path at the trellis.

Problem 8.3 Form a trellis corresponding to Reed-Muller first order code, if the codeword length is eight bits.

(a) Starting from parity-check matrix,
(b) Starting from generator matrix given in trellis oriented form.

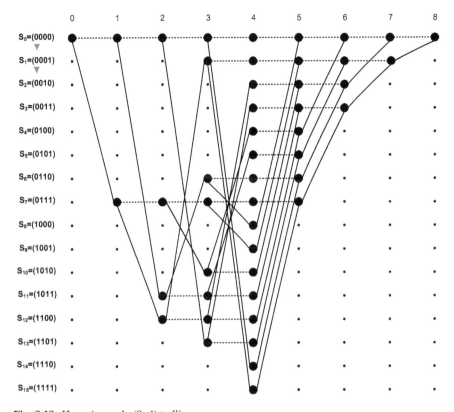

Fig. 8.18 Hamming code (8, 4) trellis

Solution

On the basis of the theory exposed in Chap. 5 it is clear that it is Reed-Muller code RM(8, 4), which can be described by a generator matrix [29, 30]

$$G = \begin{bmatrix} 1 & 1 & 1 & 1 & 1 & 1 & 1 & 1 \\ 0 & 0 & 0 & 0 & 1 & 1 & 1 & 1 \\ 0 & 0 & 1 & 1 & 0 & 0 & 1 & 1 \\ 0 & 1 & 0 & 1 & 0 & 1 & 0 & 1 \end{bmatrix}.$$

The easiest way to form a trellis is to start from a fact that RM(8, 4) code is self-dual, i.e. its generator matrix is its parity-check matrix as well

$$H = \begin{bmatrix} 1 & 1 & 1 & 1 & 1 & 1 & 1 & 1 \\ 0 & 0 & 0 & 0 & 1 & 1 & 1 & 1 \\ 0 & 0 & 1 & 1 & 0 & 0 & 1 & 1 \\ 0 & 1 & 0 & 1 & 0 & 1 & 0 & 1 \end{bmatrix}.$$

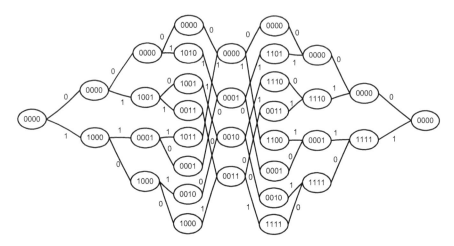

Fig. 8.19 Trellis diagram with the depth 8 for RM code (8, 4), the states are denoted by using parity-check matrix

Following the procedure described in previous problems the trellis is formed on the generator matrix basis, shown in Fig. 8.19. Trellis diagram of the depth eight was not drawn in a rectangular form, but a notation similar to a state diagram was used. However, the horizontal axis here corresponds to successive time intervals as well.

(b) When parity-check matrix of linear block code is not known, trellis can be formed on the generator matrix basis, if it is written in a **Trellis Oriented Form** (TOF). For a generator matrix G it is said that is TOF if the following conditions are satisfied [70, 71]:

1. The leading one in every matrix G row is before the leading one in the every next row, i.e. the row below it.
2. No two rows have the last ones in the same column.

The first element in non zero binary n-tuple $v = (v_0, v_1, \ldots, v_{n-1})$, different from zero, is a leading one, and the last element different from zero is the last one.

Matrix G is not TOF. Changing the second and the fourth row, matrix G' is obtained

$$G' = \begin{bmatrix} 1 & 1 & 1 & 1 & 1 & 1 & 1 & 1 \\ 0 & 1 & 0 & 1 & 0 & 1 & 0 & 1 \\ 0 & 0 & 1 & 1 & 0 & 0 & 1 & 1 \\ 0 & 0 & 0 & 0 & 1 & 1 & 1 & 1 \end{bmatrix}.$$

If the fourth row of matrix G' is added to first, second and the third row, the TOF matrix is obtained, called **trellis oriented generator matrix** G_{TOGM}

Problems

$$G_{\text{TOGM}} = \begin{bmatrix} g_0 \\ g_1 \\ g_2 \\ g_3 \end{bmatrix} = \begin{bmatrix} 1 & 1 & 1 & 1 & 0 & 0 & 0 & 0 \\ 0 & 1 & 0 & 1 & 1 & 0 & 1 & 0 \\ 0 & 0 & 1 & 1 & 1 & 1 & 0 & 0 \\ 0 & 0 & 0 & 0 & 1 & 1 & 1 & 1 \end{bmatrix}.$$

bit position $\longrightarrow$ 0 1 2 3 4 5 6 7

Now for every row the intervals from the first one to the last one can be found, measured with respect to the bit position in a row (starting from a zeroth position). If the first one is in the ith position, and the last in the jth position, then the active time intervals are $[i+1, j]$ if $j > i$ or ϕ for $j = i$, and the following is obtained

$$\tau_a(g_0) = [1, 3], \tau_a(g_1) = [2, 6], \tau_a(g_2) = [3, 5], \tau_a(g_3) = [5, 7].$$

At the moment $0 \leq i \leq n$ a matrix G_i^s can be defined consisting of these G_{TOGM} rows whose active time interval contains the moment i (in a previous problem the moments when the states change are denoted by dashed lines). This matrix is a basis for trellis forming because:

- ordinal number of a row appearing in matrix G_i^s at the moment $i+1$ (and was not at the i-moment) defines the ordinal number of current input information bit, denoted by a^*,
- ordinal number of a row existing at the ith moment in matrix G_i^s but being omitted at the next moment defines the ordinal number of the oldest information bit stored in the encoder memory at the moment, denoted by a^*.
- subset of information bits corresponding to the matrix G_i^s rows, denoted by A_i^s, where information bits from A_i^s define the encoder state at the moment i.

In Table 8.5 all previously defined notions from RM code (8, 4) are given, for the every moment when the state changes. Corresponding trellis is shown in Fig. 8.20, where the states are denoted according to the last column of this table,

Table 8.5 The sets defining states and marks for trellis which has the depth eight for RM code (8, 4)

Time I	G_i^s	a^*	a^0	A_i^s	State mark
0	ϕ	a_0	–	ϕ	(0000)
1	$\{g_0\}$	a_1	–	$\{a_0\}$	($a_0$000)
2	$\{g_0, g_1\}$	a_2	–	$\{a_0, a_1\}$	($a_0\,a_1$00)
3	$\{g_0, g_1, g_2\}$	–	a_0	$\{a_0, a_1, a_2\}$	($a_0\,a_1\,a_2$0)
4	$\{g_1, g_2\}$	a_3	–	$\{a_1, a_2\}$	(0 $a_1\,a_2$0)
5	$\{g_1, g_2, g_3\}$	–	a_2	$\{a_1, a_2, a_3\}$	(0 $a_1\,a_2\,a_3$)
6	$\{g_1, g_3\}$	–	a_1	$\{a_1, a_3\}$	(0 $a_1$0 a_3)
7	$\{g_3\}$	–	a_3	$\{a_3\}$	(000 a_3)
8	ϕ	–	–	ϕ	(0000)

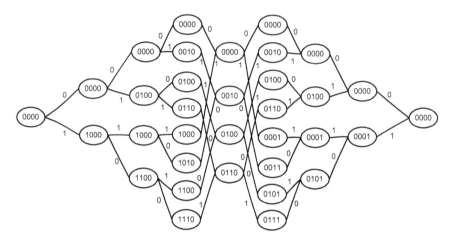

Fig. 8.20 Trellis diagram for RM code (8, 4), the states denoted by information bits set

Table 8.6 Data defining states and their marks for BCH code (7, 4)

i	G_i^s	a^*	a^0	A_i^s	State
0	ϕ	a_0	–	ϕ	(000)
1	$\{g_0\}$	a_1	–	$\{a_0\}$	$(a_0 00)$
2	$\{g_0, g_1\}$	a_2	–	$\{a_0, a_1\}$	$(a_0\ a_1 0)$
3	$\{g_0, g_1, g_2\}$	a_3	a_0	$\{a_0, a_1, a_2\}$	$(a_0\ a_1\ a_2)$
4	$\{g_1, g_2, g_3\}$	–	a_1	$\{a_1, a_2, a_3\}$	$(a_1\ a_2\ a_3)$
5	$\{g_2, g_3\}$	–	a_2	$\{a_1, a_3\}$	$(a_2\ a_3\ 0)$
6	$\{g_3\}$	–	a_3	$\{a_3\}$	$(a_3\ 00)$
7	ϕ	–	–	ϕ	(0000)

and the branches between two states contain the information about the input information bit by which the state mark is supplemented or about the oldest bit being omitted from the state mark. Although the marks are different from those in the Fig. 8.19 it is clear that both trellises have the same shape, i.e. they are equivalent.

Problem 8.4 Form the trellis for BCH code which corrects one error, the codeword length seven bits.

Solution

It is BCH code (7, 4) which has the generator polynomial $g(x) = x^3 + x + 1$, analyzed previously in Problem 6.2. It should be noticed that a cyclic codes generator matrix has always trellis oriented form

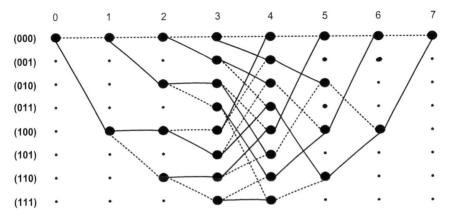

Fig. 8.21 Trellis diagram for BCH code (7, 4)

$$G = \begin{bmatrix} g_0 \\ g_1 \\ g_2 \\ g_3 \end{bmatrix} = \begin{bmatrix} g(x) \\ xg(x) \\ x^2 g(x) \\ x^3 g(x) \end{bmatrix} = \begin{bmatrix} 1 & 1 & 0 & 1 & 0 & 0 & 0 \\ 0 & 1 & 1 & 0 & 1 & 0 & 0 \\ 0 & 0 & 1 & 1 & 0 & 1 & 0 \\ 0 & 0 & 0 & 1 & 1 & 0 & 1 \\ 0 & 1 & 2 & 3 & 4 & 5 & 6 \end{bmatrix}$$

Furthermore, the procedure for determining the active time rows intervals is used, which for cyclic codes always differ for one bit $\tau_a(g_0) = [1, 3]$, $\tau_a(g_1) = [2, 4]$, $\tau_a(g_2) = [3, 5]$, $\tau_a(g_3) = [4, 6]$. Based on the data given in Table 8.6, trellis diagram for BCH code (7, 4) is shown in Fig. 8.21.

Problem 8.5 At the output of systematic Hamming code (6, 3) encoder is the codeword $c = (011011)$. This sequence is transmitted as a polar signal, while in channel the additive Gaussian noise is superimposed yielding at receiving end the signal $y = (-1.1; -0.2; -0.1; -0.8; +0.9; +1.2)$ which is decoded and delivered to the user.

(a) Draw the block scheme of the transmission system.
(b) Decode the received codeword applying Viterbi algorithm by using Hamming metric.
(c) Decode the received codeword applying Viterbi algorithm by using Euclidean metric, supposing that a quantization error can be neglected.
(d) Decode the received codeword applying Viterbi algorithm by using Euclidean metric, if the uniform quantizer with $q = 4$ levels is used $(-1.5; -0.5; 0.5; 1.5)$.

Solution

(a) The binary sequence generated by a source is firstly encoded by Hamming code with parameters (6, 3), for which the trellis is formed in Problem 8.1.

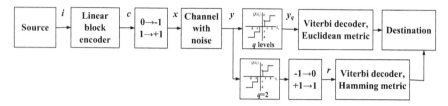

Fig. 8.22 Block diagram of Viterbi decoding for linear block codes

Encoding is performed multiplying all possible three-bit combinations by code generator matrix, given as well in the same problem.

Encoded sequence is converted into a polar signal adjoining the pulse with amplitude +1 V to binary one, and with amplitude −1 V to binary zero. At the transmitted signal x a white Gaussian noise is superimposed having zero average value. To the received signal corresponds the real numbers sequence

$$y = (-1.1, -0.2, -0.1, -0.8, +0.9, +1.2).$$

In the case of soft decoding, the sequence is first quantized (q levels) and then the Viterbi algorithm is applied by using Euclidean metric, and a decoded sequence is delivered to user. When the quantizer levels are (−1.5, −0.5, 0.5, 1.5) the boundaries of quantizing intervals are (−∞, −1, 0, 1, ∞), yielding

$$y_q = (-1.5, -0.5, -0.5, -0.5, 0.5, 1.5).$$

In the case of hard decoding, two level quantization is performed (threshold is at zero). Quantized signal is then converted into a binary sequence, being in this case

$$r = (0, 0, 0, 0, 1, 1),$$

and after that sent into the Viterbi decoder using Hamming metric. Complete system block scheme is shown in Fig. 8.22.

(b) In the case of **Viterbi algorithm** decoding by **using Hamming metric**, it is understood that at the receiver input there is decision block which has one threshold (corresponding to two-level quantizer). Therefore, at this block output only two voltage levels can appear, corresponding to binary one and to binary zero.

The decoding is then performed at a bit level, where a notion of Hamming distance is used, defined as the number of positions where two binary sequences differ. Therefore, the Hamming distance between the received sequence r and emitted codeword c is

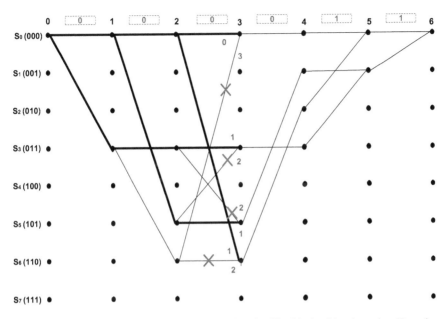

Fig. 8.23 The first step during code (6, 3) decoding by Viterbi algorithm by using Hamming metric

$$d(\mathbf{c},\mathbf{r}) = \sum_{t=1}^{n} (c_t \oplus r_t),$$

where $\oplus$ is a sign for modulo-2 addition, and summation over t is carried out in decimal system.

Trellis makes possible to find the Hamming distances between the received sequence and all possible codewords, with a minimal complexity by using a maximum likelihood criterion. In the first step, the part of trellis is observed to the moment when the transient regime is finished, in this example at the trellis depth $t = 3$. Part of a received sequence until this moment is compared to the corresponding paths, and from two paths entering the same state, that one is chosen having a smaller Hamming distance. The same procedure is continued until all information bits are received, as shown in Fig. 8.23. Obviously, these steps fully correspond to convolutional codes Viterbi decoding by using hard decision [67].

In the next steps, shown in Fig. 8.24, the previously overwritten branches are eliminated and a procedure continues in the trellis part where there is the rest of codeword bits determined by the state in which is the path after the receiving of information bits. Because of that, the number of states is reduced and as a result the path is obtained starting from all zeros state and terminating in it. In this case the sequence is decoded corresponding to states sequence $S_0 \to S_3 \to S_3 \to S_3 \to S_3 \to S_1 \to S_0$, and reconstructed codeword and decoded information word are

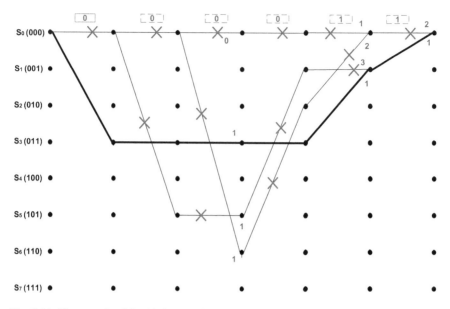

Fig. 8.24 The second and the third step during code (6, 3) decoding by Viterbi algorithm by using Hamming metric

$$\hat{c} = (100011) \Rightarrow \hat{i} = (100).$$

(c) The same example is considered, but decoding is performed by Viterbi **algorithm using Euclidean metric**. Euclidean distance is [63]

$$d_E(y, x) = \sqrt{\sum_{t=1}^{n} (y_t - x_t)^2},$$

but the square root operation is not often performed and the square of Euclidean distance is considered (it does not influence to the relation of the two distances). Because the quantization error is negligible, it will be supposed that the quantization was not performed. The procedure of decoding is shown in Fig. 8.25, where the branches overwritten in the first step are denoted by using the sign '×', while those overwritten in the second and the third step are denoted by '+' and 'o', respectively.

The procedure of calculating squared Euclidean distance is given in Table 8.7 where in bold font the metrics corresponding to survived paths are shown. As for the algorithm by using Hamming metric, the first decision is at the trellis depth $t = 3$ (here the transient regime is finished) and the second decision is at the depth

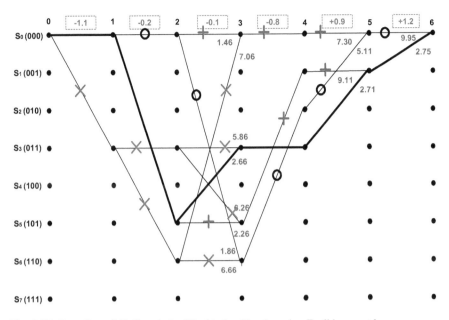

Fig. 8.25 Decoding of (6, 3) code by Viterbi algorithm by using Euclidean metric

Table 8.7 The squared Euclidean distances calculation, no quantization

		$t = 3$	$t = 5$	$t = 6$
		−1.1 −0.2 −0.1	−0.8 0.9	1.2
0–0:	S_0-S_0-S_0-S_0	−1 − 1 −1 **1.46**	−1 − 1: 1.46 + 3.65 = 7.30	−1: 5.11 + 4.84 = 9.95
	S_0-S_3-S_6-S_0	+1 +1 +1 7.06	+1 +1: 1.86 + 3.25 = **5.11**	+1: 2.71 + 0.04 = **2.75**
0–3:	S_0-S_3-S_3-S_3	+1 − 1 −1 5.86	+1 − 1: 2.26 + 6.85 = 9.11	
	S_0-S_0-S_5-S_3	−1 +1 +1 **2.66**	−1 +1: 2.66 + 0.05 = **2.71**	
0–5:	S_0-S_3-S_3-S_5	+1 − 1 +1 6.26		
	S_0-S_0-S_5-S_5	−1 +1−1 **2.26**		
0–6:	S_0-S_0-S_0-S_6	−1 − 1 +1 **1.86**		
	S_0-S_3-S_6-S_6	+1 +1−1 6.66		

$t = 5$. For a difference from hard decoding, by soft decoding a correct decision is made about the emitted sequence, as

$$\hat{c} = (011011) \quad \Rightarrow \quad \hat{i} = (011).$$

(d) If before the decoding by applying the Viterbi algorithm the **quantization** was performed, Euclidean distance is defined as

Table 8.8 Euclidean metrics calculation, $q = 4$ level calculation

		$t = 3$	$t = 5$	$t = 6$
		−1.5 −0.5 −0.5	−0.5 1.0	1.5
0–0:	S_0-S_0-S_0-S_0	−1 − 1 −1 **0.75**	−1 − 1: 0.75 + 2.50 = **3.25**	−1: 3.25 + 6.25 = 9.50
	S_0-S_3-S_6-S_0	+1 +1 +1 10.75	+1 +1: 2.75 + 2.50 = 5.25	+1: 5.25 + 2.75 = **8.00**
0–3:	S_0-S_3-S_3-S_3	+1 − 1 −1 6.75	+1 − 1: 2.75 + 4.50 = 7.25	
	S_0-S_0-S_5-S_3	−1 +1 +1 **4.75**	−1 +1: 4.75 + 0.50 = **5.25**	
0–5:	S_0-S_3-S_3-S_5	+1 − 1 +1 8.75		
	S_0-S_0-S_5-S_5	−1 +1−1 **2.75**		
0–6:	S_0-S_0-S_0-S_6	−1 − 1 +1 **2.75**		
	S_0-S_3-S_6-S_6	+1 +1−1 8.75		

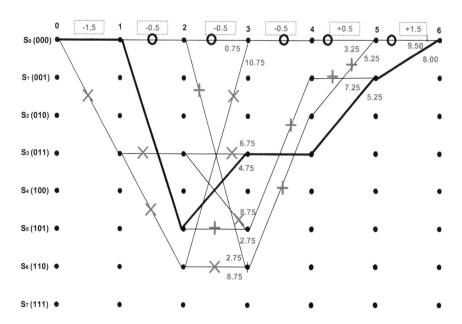

Fig. 8.26 Decoding by Viterbi algorithm by using Euclidean metric, $q = 4$ level quantization

$$d_E^2(y_q, x) = \sum_{t=1}^{n} (y_{qt} - x_t)^2,$$

the procedure in this case is given in Table 8.8, and the trellis and the order of paths elimination is shown in Fig. 8.26. It is obvious that the decoding is performed successfully although at the depth $t = 5$ one error was made during the decision, but it was corrected in the next step. Because the metric difference is smaller than in a previous case, it is clear that the decision made was less reliable.

Problems

Problem 8.6 Binary memoryless source emitting binary zero with the probability $P_b(0) = 0.8$ sends bits to Hamming encoder defined by the generator matrix

$$G = \begin{bmatrix} 1 & 0 & 0 & 0 & 1 & 1 & 0 \\ 0 & 1 & 0 & 0 & 1 & 0 & 1 \\ 0 & 0 & 1 & 0 & 0 & 1 & 1 \\ 0 & 0 & 0 & 1 & 1 & 1 & 1 \end{bmatrix}.$$

The source emits a bit sequence (0010) and a corresponding output encoder sequence enters a line encoder which generates a polar signal having amplitudes ± 1 V. In the channel at the transmitted signal a noise is superimposed. At the receiver input, before the decoder, there is a four level quantizer $(-1.5; -0.5; 0.5; 1.5$ V). The channel is memoryless and the transition probabilities of input symbols to the quantizer output levels are given in Fig. 8.27.

At the receiver input (before the quantizer) the following signal is detected $y = (1.1, -1.2, 2.1, -1.8, 0.9, 1.9, 1.2)$.

(a) Reconstruct the bit sequence emitted by a source, if the decoding is performed by using a generalized Viterbi algorithm.
(b) Repeat the previous procedure if the decoding is performed by BCJR algorithm.
(c) Decode the received sequence by BCJR algorithm if the probabilities of source symbols are not known.

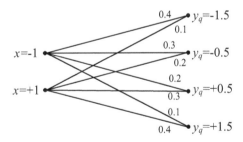

Fig. 8.27 Transition probabilities for channel and quantizer

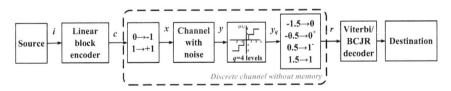

Fig. 8.28 Block diagram for linear block codes decoding by using trellis

Table 8.9 Transition probabilities for discrete memoryless channel

P(y/x)				
x/y	0	0⁺	1⁻	1
0	0.4	0.3	0.2	0.1
1	0.1	0.2	0.3	0.4

Solution

Similarly to a previous problem, it is suitable to draw system block scheme (shown in Fig. 8.28). The Hamming code (7, 4) is used and a generator matrix is known, therefore to the information sequence $i = (0010)$ corresponds the codeword $c = (0010011)$.

At the encoder output two binary symbols can appear (0 and 1) while at the input of Viterbi/BJCR decoder one from four levels obtained by quantization can appear. If to the amplitude levels the logic symbols 0, 0^+, 1^- and 1 are joined, line encoder, channel and quantizer can be considered as an equivalent discrete memoryless channel. At this channel output two symbols more reliable (0 and 1) and two symbols less reliable (0^+, 1^-) can appear, the corresponding transition probabilities are given in Table 8.9. From a known channel output sequence one obtains

$$y = (1.1, -1.2, 2.1, -1.8, 0.9, 1.9, 1.2) \rightarrow y_q = (1.5, -1.5, 1.5, -1.5, 0.5, 1.5, 1.5)$$

and the sequence at the decoder input is $r = (1, 0, 1, 0, 1^-, 1, 1)$.

Because the generator matrix is given in a systematic form, the parity-check matrix is easily obtained

$$H = \begin{bmatrix} 1 & 1 & 0 & 1 & 1 & 0 & 0 \\ 1 & 0 & 1 & 1 & 0 & 1 & 0 \\ 0 & 1 & 1 & 1 & 0 & 0 & 1 \end{bmatrix},$$

the trellis shown in Fig. 8.29 is easily obtained as well. It should be noticed that it is slightly different from that one shown in Fig. 8.2.

(a) When a **generalized Viterbi algorithm** is used, the distances accumulated in the nodes are calculated on the basis of relation [72]

$$\alpha_t(S^{(t)}) = \prod_{j=1}^{t} \gamma_j(S^{(j-1)}, S^{(j)}),$$

where $S^{(j-1)}$ denotes a starting state (trellis node at the depth $j - 1$) and $S^{(j)}$ denotes the state where the branch terminates (trellis node at the depth j). Branch metrics $\gamma_j(S^{(j-1)}, S^{(j)})$ are the same as the transition probabilities describing the channel and for a decoder input sequence are given in Table 8.10 (the value $P(y_j|0)$ corresponds to the horizontal segments, denoted by a dashed line, while the probability $P(y_j|1)$ corresponds to the segments

Problems

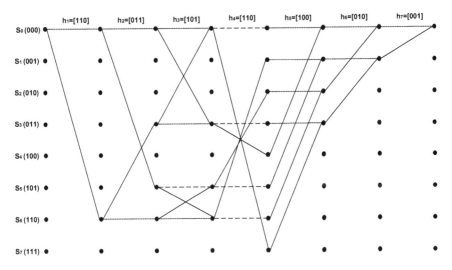

Fig. 8.29 Trellis of a systematic Hamming code (7, 4)

Table 8.10 Transition probabilities for a discrete memoryless channel

t	1	2	3	4	5	6	7
y_t	1	0	1	0	1⁻	1	1
$((P(y_t\|0); P(y_t\|1))$	(0.1, 0.4)	(0.4, 0.1)	(0.1, 0.4)	(0.4, 0.1)	(0.2, 0.3)	(0.1, 0.4)	(0.1, 0.4)

denoted by a full line). From the two paths entering into one node, this one is chosen which has the greater metric. Procedure of decoding by using generalized Viterbi algorithm is shown in Fig. 8.30. The survived path, i.e. the procedure result, corresponds to a codeword $c' = (1010101)$ and to information word $i' = (1010)$. It is obvious that the used algorithm did not result in a correct decoding of the source emitted sequence. In the continuation of the solution, it will be shown that, by applying BCJR algorithm, the successful decoding of is still possible.

(b) **BCRJ algorithm** procedure is based on the following branch calculation [73]

$$\gamma_t(S^{(t-1)}, S^{(t)}) = \sum_{x \in A_x} P(S^{(t)}|S^{(t-1)})P(X_t = x|S^{(t-1)}S^{(t)})P(Y_t|X_t = x)$$

depending besides on the channel transition probabilities and the trellis structure (described by the probabilities $P(S^{(t)}|S^{(t-1)})$) as well as on the probabilities of the encoder input symbols. It should be noticed that in this problem at the depths $t = 4$, $t = 6$ and $t = 7$ is $P(S^{(t)}|S^{(t-1)}) = 1, 0$. At the other trellis depths this probability is

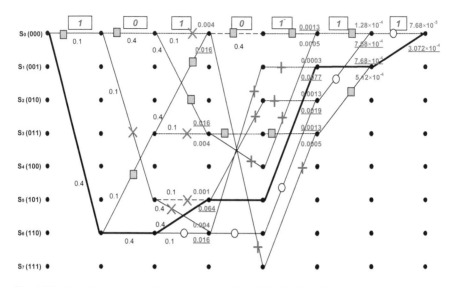

Fig. 8.30 Decoding procedure by using a generalized Viterbi algorithm

$$P(S^{(t)}|S^{(t-1)}) = \begin{cases} 0.8 & \text{if } S^{(t)} = S^{(t-1)}, \\ 0.2 & \text{if } S^{(t)} \neq S^{(t-1)}. \end{cases}$$

The metrics are calculated going forward and backward from

$$\alpha_t(S^{(t)}) = \sum_{S^{(t-1)} \in S_x} \alpha_{t-1}(S^{(t-1)}) \gamma_t(S^{(t-1)}, S^{(t)}), \quad t = 1, 2, \ldots, 7$$

$$\beta_{t-1}(S^{(t-1)}) = \sum_{S^{(t)} \in S_x} \beta_t(S^{(t)}) \gamma_t(S^{(t-1)}, S^{(t)}), \quad t = 7, 6, \ldots, 1$$

and the corresponding numerical values are given in Table 8.11. For the trellis depths $t = 1$, $t = 2$ and $t = 4$, the following is calculated

$$\lambda_t(S^{(t)}) = \alpha_t(S^{(t)}) \beta_t(S^{(t)}),$$

from which the estimations are obtained

$$P_1(0) = \frac{\lambda_1(0)}{\lambda_1(0) + \lambda_1(6)}, \quad P_2(0) = \frac{\lambda_2(0) + \lambda_2(6)}{\lambda_2(0) + \lambda_2(6) + \lambda_2(3) + \lambda_2(5)},$$

$$P_4(0) = \frac{\lambda_4(0) + \lambda_4(3) + \lambda_4(5) + \lambda_4(6)}{\lambda_4(0) + \lambda_4(3) + \lambda_4(5) + \lambda_4(6) + \lambda_4(1) + \lambda_4(2) + \lambda_4(4) + \lambda_4(7)}.$$

According to the procedure described in the introductory part of this, the following values are calculated

Table 8.11 Node metrics forward and backward for various trellis depths, BCJR algorithm

		$t=0$	$t=1$	$t=2$	$t=3$	$t=4$	$t=5$	$t=6$	$t=7$
Forward	$\alpha_t(0)$	1	0.08	0.0256	0.00217	6.96×10^{-4}	1.523×10^{-4}	1.023×10^{-4}	4.33×10^{-5}
	$\alpha_t(1)$	0	0	0	0	4.35×10^{-5}	2.176×10^{-4}	8.269×10^{-5}	0
	$\alpha_t(2)$	0	0	0	0	4.35×10^{-5}	2.176×10^{-4}	0	0
	$\alpha_t(3)$	0	0	0.0016	0.00217	6.96×10^{-4}	1.523×10^{-4}	0	0
	$\alpha_t(4)$	0	0	0	0	4.35×10^{-5}	0	0	0
	$\alpha_t(5)$	0	0	0.0016	0.00217	6.96×10^{-4}	0	0	0
	$\alpha_t(6)$	0	0.08	0.0256	0.00217	6.96×10^{-4}	0	0	0
	$\alpha_t(7)$	0	0	0	0	4.35×10^{-5}	0	0	0
Backward	$\beta_t(0)$	4.33×10^{-5}	3.17×10^{-4}	9.52×10^{-4}	0.0016	0.002	0.01	0.1	1
	$\beta_t(1)$	0	0	0	0	0.008	0.04	0.4	0
	$\beta_t(2)$	0	0	0	0	0.008	0.04	0	0
	$\beta_t(3)$	0	0	9.52×10^{-4}	0.0103	0.032	0.16	0	0
	$\beta_t(4)$	0	0	0	0	0.003	0	0	0
	$\beta_t(5)$	0	0	6.4×10^{-4}	0.004	0.012	0	0	0
	$\beta_t(6)$	0	2.24×10^{-4}	6.4×10^{-4}	0.004	0.012	0	0	0
	$\beta_t(7)$	0	0	0	0	0.048	0	0	0

$$\sigma_t(\mathbf{S}^{(t-1)}, \mathbf{S}^{(t)}) = \alpha_{t-1}(\mathbf{S}^{(t-1)})\gamma_t(\mathbf{S}^{(t-1)}, \mathbf{S}^{(t)})\beta_t(\mathbf{S}^{(t)}),$$

and on this basis the values of other bits are estimated

$$P_3(0) = \frac{\sigma_3(0,0) + \sigma_3(3,3) + \sigma_3(5,5) + \sigma_3(6,6)}{\sigma_3(0,0) + \sigma_3(3,3) + \sigma_3(5,5) + \sigma_3(6,6) + \sigma_3(0,3) + \sigma_3(3,0) + \sigma_3(5,6) + \sigma_3(6,5)},$$

$$P_5(0) = \frac{\sigma_5(0,0) + \sigma_5(1,1) + \sigma_5(2,2) + \sigma_5(3,3)}{\sigma_5(0,0) + \sigma_5(1,1) + \sigma_5(2,2) + \sigma_5(3,3) + \sigma_5(4,0) + \sigma_5(5,1) + \sigma_5(6,2) + \sigma_5(7,3)},$$

$$P_6(0) = \frac{\sigma_6(0,0) + \sigma_6(1,1)}{\sigma_6(0,0) + \sigma_6(1,1) + \sigma_6(2,0) + \sigma_6(3,1)}.$$

Numerical values of probabilities that the corresponding bit of the estimated encoded sequence equals to zero are

$$P_1(0) = 0.5865, P_2(0) = 0.9412, P_3(0) = 0.3071, P_4(0) = 0.9327,$$
$$P_5(0) = 0.5628, P_6(0) = 0.2362, P_7(0) = 0.2362,$$

from which the sequence emitted by the encoder is estimated $c' = (0010011)$. It is obvious that the second and the fourth bit are successfully estimated, while the estimations of the first and of the fifth bit have smaller reliability. In Fig. 8.31 it is shown that a path corresponding to the estimated sequence exists at the trellis and the decoder takes out the first four bits considering that a sequence $i' = (0010)$ is valid. In this case, the decoding was successful.

(c) When the decoder has not information about probabilities of symbols emitted by a source, it is assumed that these symbols are equiprobable. The decoding procedure was repeated and the numerical metrics values "forward" and "backward" are given in Table 8.12. The probability that to the ith trellis depth corresponds binary zero at the encoder input, denoted by $P_i(0)$, has a numerical value

$$P_1(0) = 0.3059, P_2(0) = 0.8, P_3(0) = 0.2, P_4(0) = 0.8,$$
$$P_5(0) = 0.4, P_6(0) = 0.4118, P_7(0) = 0.2,$$

and the estimated encoded sequence is $c' = (\underline{1010}111)$. It is easy to verify that such sequence does not exist at the trellis (trellis part shown by heavy line with the arrows shown in Fig. 8.31). It is obvious that at the depth $t = 6$ one cannot find a segment corresponding to binary one at the input, because the parity-check bits are fully defined by the information bits. Therefore, the decoded word is finally

Table 8.12 Node metrics forward and backward for BCJR algorithm

		$t=0$	$t=1$	$t=2$	$t=3$	$t=4$	$t=5$	$t=6$	$t=7$
Forward	$\alpha_t(0)$	1	0.05	0.01	0.0025	5×10^{-4}	1.375×10^{-4}	1.0625×10^{-4}	5.3125×10^{-5}
	$\alpha_t(1)$	0	0	0	0	1.25×10^{-4}	5.125×10^{-4}	1.0625×10^{-4}	0
	$\alpha_t(2)$	0	0	0	0	4.06×10^{-4}	2.3125×10^{-4}	0	0
	$\alpha_t(3)$	0	0	0.01	0.0025	5×10^{-4}	1.375×10^{-4}	0	0
	$\alpha_t(4)$	0	0	0	0	1.25×10^{-4}	0	0	0
	$\alpha_t(5)$	0	0	0.0025	0.0081	1.625×10^{-3}	0	0	0
	$\alpha_t(6)$	0	0.2	0.04	0.0025	5×10^{-4}	0	0	0
	$\alpha_t(7)$	0	0	0	0	1.25×10^{-4}	0	0	0
Backward	$\beta_t(0)$	5.31×10^{-5}	3.25×10^{-4}	1.45×10^{-4}	0.0028	0.002	0.01	0.1	1
	$\beta_t(1)$	0	0	0	0	0.008	0.04	0.4	0
	$\beta_t(2)$	0	0	0	0	0.008	0.04	0	0
	$\beta_t(3)$	0	0	8.875×10^{-4}	0.00655	0.032	0.16	0	0
	$\beta_t(4)$	0	0	0	0	0.003	0	0	0
	$\beta_t(5)$	0	0	7×10^{-4}	0.0028	0.012	0	0	0
	$\beta_t(6)$	0	1.84×10^{-4}	7×10^{-4}	0.0028	0.012	0	0	0
	$\beta_t(7)$	0	0	0	0	0.048	0	0	0

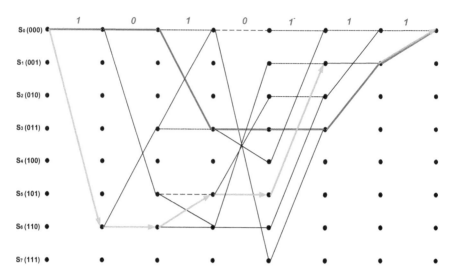

Fig. 8.31 Procedure for decoding by using BCJR algorithm for the known and the unknown a priori probabilities

$$c'' = (1010101),$$

identical to that one obtained by applying a generalized Viterbi algorithm.

Therefore, the following is shown:

(1) Viterbi algorithm is an optimal procedure for the case when the symbols at the encoder input are equiprobable (ML solution is equivalent to MAP solution).
(2) Viterbi algorithm does not yield the estimation of reliability for the decoded bits, for a difference from BCJR algorithm. This estimation (for every single bit) usually is described as *log likelihood ratio* (LLR)

$$\Lambda(x_t) = \ln \frac{\Pr\{i_t = 1 | r\}}{\Pr\{i_t = 0 | r\}},$$

and the estimation is more reliable as its absolute value is greater (negative value indicates that the zero was sent, and positive value indicates that one was sent).

(3) BCJR algorithm does not yield a correct solution if a sufficiently good estimation of a priori probabilities is not accessible. The estimations obtained as a result of BCJR algorithm are not optimal in this case.

Problem 8.7 The same transmission system from the previous problem is considered, but the quantizer output sequence enters the decoder whose procedure is based at SOVA algorithm with the correlation metric. Explain the decoding

Problems

procedure, reconstruct the transmitted codeword and find the corresponding estimations of reliability.

Solution

Block scheme is practically the same as shown in Fig. 8.28, with a difference that instead of BJCR decoder *Soft Output Viterbi Algorithm* (SOVA) is used [77]. When a classic Viterbi algorithm is applied, after decision which branches have been survived, a series of decisions about "better" paths results in the decoded bit sequence. Not taking into account whether the metrics are calculated by using Hamming or Euclidean distance (or the channel transition probabilities, as for a generalized algorithm version), i.e. not taking into account whether a hard or soft decoding is used, the Viterbi algorithm output is always "hard". Only SOVA algorithm makes possible at the decoder output, in parallel with a series of decoded bits, to find as well an estimation of the reliability of corresponding decisions.

The first step in SOVA algorithm is the same as in any Viterbi algorithm variant by using soft decision—on the basis of the branch metrics the survived branches are found. A variant that uses Euclidean metric is used usually or a variant that uses correlation metric calculated as follows [26]

$$d_K(c, y_q) = \sum_{t=1}^{n} (-1)^{\tilde{c}_t} y_{qt},$$

where $\tilde{c}_t$ represents the tth bit of the estimated codeword, and y_{qt} denotes the tth value of quantized received signal. The procedure, for this case, is shown in the Fig. 8.32.

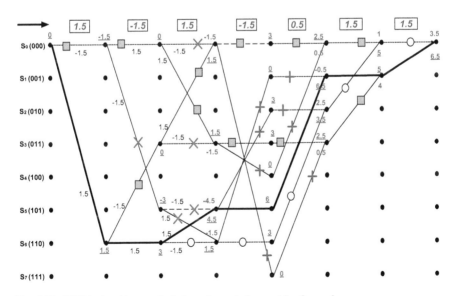

Fig. 8.32 SOVA algorithm—calculating of correlation metrics forward

By comparing Fig. 8.32 to the Fig. 8.30, it can be firstly noticed that the branches are overwritten following the same order and the decoded codeword in both cases is the same

$$c' = (1010101).$$

Although the metrics are calculated by using different formulas, it can be noticed that for a fixed trellis depth to the mutual same numerical metric values from Fig. 8.30, correspond the same mutual metric values in the Fig. 8.32. For example, at the depth $t = 3$ to the states (000), (011) and (110) in Fig. 8.30 corresponds the survived branch metric 0.016, while, in the Fig. 8.32 to the same states corresponds the metric 1.5. Similarly, at the depth $t = 5$ to the metric 0.013 for generalized Viterbi algorithm corresponds the correlation metric 2.5, and at the depth $t = 6$ to the metric 7.68×10^{-4} corresponds metric 5.

In the book [26] it was shown that the Euclidean metric and the correlation metric are equivalent concerning the making decisions when the channel noise is Gaussian, but the correlation metrics are more easy to calculate and usually have a smaller values. The advantage is obvious as well with the respect to calculate metrics when a generalized Viterbi algorithm is applied—in Fig. 8.30 the metrics had very small values already for a few trellis depths, and this approach for a codewords having long length yields the metric values being difficult to express with a sufficient exactness.

The second step in SOVA algorithm is the backward metric calculation, similarly as for BCJR procedure. As it is shown in Fig. 8.33, the node metrics in this case are accumulated from right to the left. In this example to every horizontal

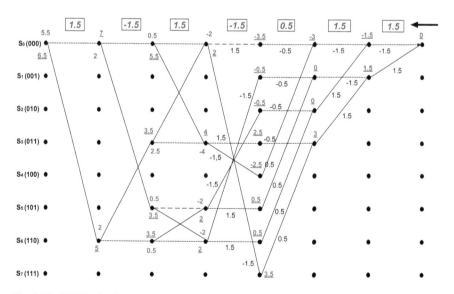

Fig. 8.33 SOVA algorithm, backward correlation metrics calculation

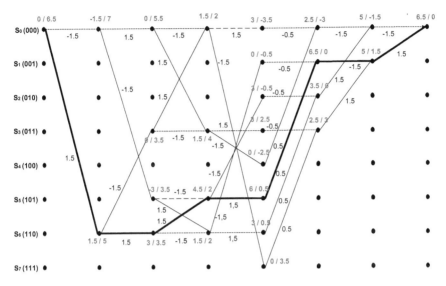

Fig. 8.34 SOVA algorithm, calculating the reliability estimation

transition corresponds the branch metric $-y_{qt}$, while to the other transitions correspond y_{qt} (it was valid and for the metrics "forward", shown in Fig. 8.32, as well). When calculating the metrics "backward" there is no need to make comparisons, decisions nor paths overwriting. The reader is advised to verify which path would survive in this case!

After the metrics calculation for both ways, it is suitable to write them at one trellis diagram, together with the branch metrics. For the considered example all elements needed for the next algorithm steps are shown in the Fig. 8.34. On the basis of this figure it is easy to find the reliability of the bit decoded at the trellis depth t by using the following procedure.

1. For every branch starting from the depth $t-1$ the starting node is noticed and its adjoined metric forward, denoted by d_{t-1}^{FW}, and the incoming node at the depth t and its adjoined metric backward, denoted by d_t^{BW}. If this branch metric is denoted by $d_{t-1,t}^{BM}$, a corresponding numerical value for this branch can be calculated

$$M_t = d_{t-1}^{FW} + d_{t-1,t}^{BM} + d_t^{BW},$$

it will be called the metric of a considered path.

2. All the branches at the trellis part from depth $t-1$ to the depth t corresponding to bits '0' (here these are horizontal branches) at the encoder input are considered, and that one is found which has the minimal value for a path metric, being further denoted as $M_t(0)$.

3. All the branches at the trellis part from depth $t-1$ to the depth t corresponding to bits '1' (the branches that are not horizontal) at the encoder input are considered, and that one is found which has the minimal value for a path metric, will be further denoted as $M_t(1)$.
4. If a bit '1' is decoded the reliability of estimation is as greater as $\Lambda_t = M_t(1) - M_t(0)$ is greater. If a bit '0' is decoded the reliability of estimation is as greater as $\Lambda_t = M_t(1) - M_t(0)$ is smaller.

In this case for a depth $t = 1$, the following is obtained

$$M_1(0) = 0 - 1.5 + 7 = 5.5; M_1(1) = 0 + 1.5 + 5 = 6.5 \Rightarrow \Lambda_1 = M_1(1) - M_1(0) = 1 > 0$$

and because Λ is positive, it is clear that the probability that the codeword first bit is binary one is greater.
At the depth $t = 2$

$$M_2(0) = \min\{-1.5 + 1.5 + 5.5; 1.5 + 1.5 + 3.5\} = 5.5$$
$$M_2(1) = \min\{-1.5 - 1.5 + 3.5; 1.5 - 1.5 + 3.5\} = 0.5$$
$$\Rightarrow \Lambda_2 = M_2(1) - M_2(0) = -5 < 0$$

and it is obvious that the second decoded bit is binary zero, and because the absolute value of parameter Λ_2 is substantially greater than for a first bit, the estimation of the second bit value is more reliable. It is in concordance with the results of the third part of a previous problem.
At the depth $t = 3$

$$M_3(0) = \min\{0 - 1.5 + 2; 0 - 1.5 + 4; -3 - 1.5 + 2; 3 - 1.5 + 2\} = -2.5$$
$$M_3(1) = \min\{0 + 1.5 + 4; 0 + 1.5 + 2; -3 + 1.5 + 2; 3 + 1.5 + 2\} = 0.5$$
$$\Rightarrow \Lambda_3 = M_3(1) - M_3(0) = 3 > 0$$

the third decoded bit is binary one, and this decision was made being more reliable than for the first codeword bit and less reliable than for the second codeword bit.
At the depth $t = 4$

$$M_4(0) = \min\{1.5 + 1.5 - 3.5; 1.5 + 1.5 + 2.5; 4.5 + 1.5 + 0.5; 1.5 + 1.5 + 0.5\} = -0.5$$
$$M_4(1) = \min\{1.5 - 1.5 + 3.5; 1.5 - 1.5 - 2.5; 4.5 - 1.5 - 0.5; 1.5 - 1.5 - 0.5\} = -2.5$$
$$\Rightarrow \Lambda_4 = M_4(1) - M_4(0) = -2 < 0$$

and according to this criterion, the decoded bit is binary zero, but the decision is relatively unreliable.

At the depth $t = 5$

$$M_5(0) = \min\{3 - 0.5 - 3; 0 - 0.5 + 0; 3 - 0.5 + 0; 3 - 0.5 + 3\} = -0.5$$
$$M_5(1) = \min\{0 + 0.5 - 3; 6 + 0.5 + 0; 3 + 0.5 + 0; 0 + 0.5 + 3\} = -2.5$$
$$\Rightarrow \Lambda_5 = M_5(1) - M_5(0) = -2 < 0$$

It is interesting that at this depth binary one is decoded, but the parameter Λ_5 value (negative one!) shows that this decision is very unreliable, and that it would be more logical that in this case binary zero is decoded, but that for some reason is not suitable (because of trellis structure).

At the depth $t = 6$

$$M_6(0) = \min\{2.5 - 1.5 - 1.5; 6.5 - 1.5 + 1.5\} = -0.5$$
$$M_6(1) = \min\{3.5 + 1.5 - 1.5; 2.5 + 1.5 + 1.5\} = 3.5$$
$$\Rightarrow \Lambda_6 = M_6(1) - M_6(0) = 4 > 0$$

and at this depth binary zero is decoded, but the great parameter Λ_6 value shows that this decision is extremely unreliable and almost surely wrong! Therefore, the possibility of this bit inversion should be considered.

Finally, at the depth $t = 7$

$$M_7(0) = 5 - 1.5 + 0 = 3.5; M_7(1) = 5 + 1.5 + 0 = 6.5$$
$$\Rightarrow \Lambda_7 = M_7(1) - M_7(0) = 3 > 0$$

and one can consider that in this case it is relatively reliable that binary one was sent.

It is obvious that the fifth and the sixth bit are decoded with the smallest reliability, but with only their inversion the obtained bit combination is not a codeword. By the first bit inversion (next according to the unreliability level) the codeword would be obtained that was really sent, i.e. $c = (0010011)$. Although the procedure for obtaining this word is simple, still it should be noticed that to the sent codeword corresponds a higher correlation level with the received word, than to this reconstructed one, as

$$d_K(c', y_q) = y_{q1} - y_{q2} + y_{q3} - y_{q4} + y_{q5} - y_{q6} + y_{q7} = 6.5,$$
$$d_K(c, y_q) = -y_{q1} - y_{q2} + y_{q3} - y_{q4} - y_{q5} + y_{q6} + y_{q7} = 5.5$$

i.e., the sent word is the next one, after above reconstructed, according the reliability level.

Now, some important features of SOVA algorithm can be noticed:

1. SOVA overcomes the drawback of classic Viterbi algorithm where the estimations of reconstruction reliability of code sequence were not available.

432   8 Trellis Decoding of Linear Block Codes, Turbo Codes

Therefore, this procedure is an advanced version of that one being explained in the first part of the proceeding problem.
2. This procedure uses a good feature of Viterbi algorithm that a decoded word must correspond to a complete trellis path, i.e. it must be a codeword (BCJR has not this feature). Just because of that a slightly better solution is obtained than in the third part of a previous problem.
3. Neither here, the a priori probabilities at the encoder input are not taken into account and it is ML procedure, and algorithm is still suboptimal in respect to BCJR (the obtained solution is worse than that one in the second part of the previous problem).
4. SOVA algorithm can be implemented and for unquantized signal values and can be used for channels which are not discrete. Later, it will be shown that the BCJR algorithm can be modified to be applied at such channels as well.

Problem 8.8 The system is considered whose transmitting side consists of the source emitting a series of equiprobable zeros and ones 010011..., a convolutional encoder whose structure is defined by generator $G(D) = [1 + D + D^2, 1 + D^2]$ and BPSK modulator. Signal power at the receiver input is $P_s = 1$ [μW] and signaling rate is $V_b = 1$ [Mb/s]. In the channel the white Gaussian noise with the average power density spectrum $N_0 = 10^{-12}$ [W/Hz] is added. The signal samples at the channel output are

$$y = (-0.95, -1.2, -0.1, -0.05, -1.2, 1.01, 0.3, 1.13, -0.4, -16),$$

while in the receiver a coherent BPSK demodulation is used as well as a soft decision decoding. Decode a transmitted information sequence if one block consists of three information bits for the following cases:

(a) BCJR algorithm is used for a previously given parameters.
(b) BCJR algorithm is used and it is supposed that in the channel the parameter N_0 changes as $N_0 = 10^{-13}$ [W/Hz], $N_0 = 10^{-11}$ [W/Hz] and $N_0 = 10^{-10}$ [W/Hz], for the same input sequence.
(c) BCRJ algorithm is used for $N_0 = 10^{-11}$ [W/Hz], but the probabilities of symbols emitted by a source are related as $P(0) = 9P(1)$ or $P(1) = 9P(0)$.

Solution

(a) The code rate is $R = 1/2$ and a decoding is performed by blocks consisting of three information bits, and the input decoder sequence is separated into the groups consisting of ten samples. In this case the signal entering the decoder is not quantized and a channel has not a finite number of output symbols (it is not discrete). Therefore, the transition probabilities cannot be found and the branch metrics are calculated according to the formula [73]

Problems

$$\gamma_t(S^{(t-1)}, S^{(t)}) = C_t \exp\left(\frac{u_t}{2} L(u_t)\right) \exp\left(\frac{L_c}{2} \sum_{l=1}^{n} x_{tl} y_{tl}\right)$$

where C_t does not affect the final result because it is canceled during the *Log-Likelihood Ratio* (LLR) coefficients calculating. $L(u_t)$ is a priori LLR for the *k*th bit at the encoder input ($u_t = 1$ for $i_t = 1$ and $u_t = -1$ for $i_t = 0$) and for equiprobable symbols it is obtained

$$L(u_t) = ld\frac{P(u_t = +1)}{P(u_t = -1)} = 0.$$

L_c is defined as

$$L_c = 4a\frac{E_c}{N_0} = 4aR\frac{E_b}{N_0},$$

where $E_c = RE_b$ denotes energy needed to transmit one bit of the codeword and E_b denotes energy needed to transmit one bit of the information word. Let the instantaneous value of amplification (corresponding to the channels with fading) is denoted by *a*. This value can change from one codeword to another (even from one to the other bit inside the codeword, if the fading is fast).

In this problem, the fading is not supposed and the value $a = 1$ is taken, and because C_k does not influence to the result, $C_t = 1$ will be used. For such numerical values it is obtained

$$\frac{E_c}{N_0} = \frac{P_s}{N_0 V_b} = 1 \Rightarrow L_c = 4.$$

The branch metrics are now considerably simplified becoming

$$\gamma_t(S^{(t-1)}, S^{(t)}) = \exp\left(2\sum_{l=1}^{n} x_{tl} y_{tl}\right)$$

where *n* denotes a number of convolutional encoder outputs (here $n = 2$), defining a number of samples at the decoder input in one step. The corresponding trellis is shown in Fig. 8.35, the corresponding metrics are

1. In the first step the following is received $y_{11} = -0.95$ and $y_{12} = -1.2$, yielding:

$$\gamma_1(0,0) = \exp(2[(-1) \times (-0.95) + (-1) \times (-1.2)]) = 73.6998,$$
$$\gamma_1(0,2) = \exp(2[(+1) \times (-0.95) + (+1) \times (-1.2)]) = 0.0136.$$

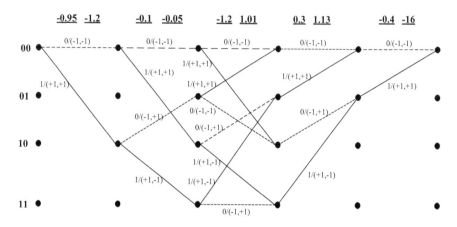

Fig. 8.35 Trellis corresponding to recursive convolutional encoder for a given input sequence

2. In the second step the following is received $y_{21} = -0.1$ and $y_{22} = -0.05$, yielding

$$\gamma_2(0,0) = \exp(2[(-1) \times (-0.1) + (-1) \times (-0.05)]) = 1.3499,$$
$$\gamma_2(0,2) = \exp(2[(+1) \times (-0.1) + (+1) \times (-0.05)]) = 0.7408,$$
$$\gamma_2(2,1) = \exp(2[(-1) \times (-0.1) + (+1) \times (-0.05)]) = 1.1052,$$
$$\gamma_2(2,3) = \exp(2[(+1) \times (-0.1) + (-1) \times (-0.05)]) = 0.9048.$$

3. In the third step the following is received $y_{31} = -1.2$ and $y_{32} = 1.01$, yielding:

$$\gamma_3(0,0) = 1.4623, \gamma_3(0,2) = 0.6839, \gamma_3(1,0) = 0.6839, \gamma_3(1,2) = 1.4623,$$
$$\gamma_3(2,1) = 83.0963, \gamma_3(2,3) = 0.012, \gamma_3(3,1) = 0.012, \gamma_3(3,3) = 83.0963.$$

4. In the fourth step the following is received $y_{41} = 0.3$ and $y_{42} = 1.13$, yielding:

$$\gamma_4(0,0) = 0.0573, \gamma_3(1,0) = 17.4615, \gamma_4(2,1) = 5.2593, \gamma_4(3,1) = 0.1901.$$

5. In the last, fifth, step the following is received $y_{51} = -0.4$ and $y_{52} = -16$, yielding:

$$\gamma_5(0,0) = 1.75 \times 10^{14}, \quad \gamma_5(1,0) = 5.69 \times 10^{-15}.$$

Forward metrics are calculated by using the same formula as in Problem 8.6, and from $\alpha_0(0) = 1$, it is easy to calculate

Problems

$$\alpha'_1(0) = \alpha_0(0)\gamma_1(0,0) = 73.6998, \quad \alpha'_1(2) = \alpha_0(0)\gamma_1(0,2) = 0.0136.$$

To reduce the overflow/underflow probability, the metrics forward are scaled so as that their sum equals one

$$\alpha_1(0) = \frac{\alpha'_1(0)}{\alpha'_1(0) + \alpha'_1(2)} = 0.9998, \quad \alpha_1(2) = \frac{\alpha'_1(2)}{\alpha'_1(0) + \alpha'_1(2)} = 0.0002.$$

The metrics corresponding to the depth $t = 2$ are

$$\alpha'_2(0) = \alpha_1(0)\gamma_2(0,0) = 1.3496, \quad \alpha'_2(1) = \alpha_1(2)\gamma_2(2,1) = 2.21 \times 10^{-4},$$
$$\alpha'_2(2) = \alpha_1(0)\gamma_2(0,2) = 0.7404, \quad \alpha'_2(3) = \alpha_1(2)\gamma_2(2,3) = 1.81 \times 10^{-4},$$

and the corresponding normalized forward metric values at the depth $t = 2$ are obtained dividing the previously obtained results by $\alpha'_2(0) + \alpha'_2(1) + \alpha'_2(2) + \alpha'_2(3) = 2.0904$, yielding

$$\alpha_2(0) = 0.6455, \alpha_2(1) = 1.06 \times 10^{-4}, \quad \alpha_2(2) = 0.3543; \alpha_2(3) = 0.87 \times 10^{-4}.$$

In a similar way, the backward metrics are calculated and $\beta_5(0) = 1$, for the depth $t = 4$ yielding

$$\beta'_4(0) = \gamma_5(0,0)\beta_5(0) = 1.75 \times 10^{14}, \quad \beta'_4(1) = \gamma_5(1,0)\beta_5(0) = 5.69 \times 10^{-15},$$

and the numerical values after normalization are (the values are exact to twenty ninth decimal!).

$$\beta_4(0) = \frac{\beta'_1(0)}{\beta'_4(0) + \beta'_4(1)} \approx 1, \quad \beta_4(1) = \frac{\beta'_4(1)}{\beta'_4(0) + \beta'_4(1)} \approx 0.$$

Backward metrics corresponding to depth $t = 3$ are

$$\beta'_3(0) = \gamma_1(0,0)\beta_4(0) = 0.0573, \quad \beta'_3(1) = \gamma_1(1,0)\beta_4(0) = 17.4615,$$
$$\beta'_3(2) = \gamma_1(2,1)\beta_4(1) = 0, \quad \beta'_3(3) = \gamma_1(3,1)\beta_4(1) = 0,$$

and the corresponding normalized backward metric values are

$$\beta_3(0) = 0.0033, \beta_3(1) = 0.9967, \beta_3(2) = 0, \beta_3(3) = 0,$$

the numerical values (normalized!) all forward and backward metrics are given in Table 8.13.

Table 8.13 Node forward and backward metrics for BCJR algorithm

		$t = 0$	$t = 1$	$t = 2$	$t = 3$	$t = 4$	$t = 5$
Forward	$\alpha_t(0)$	1	0.9998	0.6455	0.0306	0.9955	1
	$\alpha_t(1)$	0	0	0.0001	0.9547	0.0045	0
	$\alpha_t(2)$	0	0.0002	0.3543	0.0143	0	0
	$\alpha_t(3)$	0	0	0.0001	0.0004	0	0
Backward	$\beta_t(0)$	1	0.9998	0.0001	0.0033	1	1
	$\beta_t(1)$	0	0	0	0.9967	0	0
	$\beta_t(2)$	0	0.0002	0.9998	0	0	0
	$\beta_t(3)$	0	0	0.0001	0	0	0

It is interesting to notice the paths of maximal forward and backward metrics:

- Starting from the left to the right, the optimal states sequence would be $0 \to 0 \to 0 \to 1 \to 0 \to 0$. Because the transition $0 \to 1$ at the trellis at the third step is not possible, the nearest paths are $0 \to 0 \to 2 \to 1 \to 0 \to 0$ (corresponding to information sequence 010) and $0 \to 0 \to 0 \to 0 \to 0 \to 0$ (corresponding to information sequence 000). The metrics are in average larger for the first variant.
- When starting from the trellis end to the left, the optimal states sequence would be $0 \to 2 \to 2 \to 1 \to 0 \to 0$. However, the transition $2 \to 2$ at the trellis at the second step is not possible, the nearest paths are $0 \to 0 \to 2 \to 1 \to 0 \to 0$ (corresponding to information sequence 010) and the path $0 \to 2 \to 1 \to 0 \to 0 \to 0$ (corresponding to information sequence 100). The metrics are larger for the first variant.

However, the final decision is not made from forward metrics nor backward metrics, but on the basis of equality defined earlier in Problem 8.6

$$\sigma_t(\mathbf{S}^{(t-1)}, \mathbf{S}^{(t)}) = \alpha_{t-1}(\mathbf{S}^{(t-1)})\gamma_t(\mathbf{S}^{(t-1)}, \mathbf{S}^{(t)})\beta_t(\mathbf{S}^{(t)}),$$

where this value should be normalized so as that a sum over all transitions at a given depth is equal to one, to represent the transition probability from the state $\mathbf{S}^{(t-1)}$ into the state $\mathbf{S}^{(t)}$ for a known received sequence $\mathbf{y}$. The corresponding LLR for the tth information bit after the received sequence $\mathbf{y}$ can be found from the relation (non normalized values can be used)

$$L(u_t|\mathbf{y}) = \ln \frac{\sum_{R_1} \sigma_t(\mathbf{S}^{(t-1)}, \mathbf{S}^{(t)})}{\sum_{R_0} \sigma_t(\mathbf{S}^{(t-1)}, \mathbf{S}^{(t)})},$$

where R_1 denotes the transitions in the trellis corresponding to information bit '1' ($u_t = +1$), while R_0 denotes the transitions corresponding to information bit '0' ($u_t = -1$). Because the general trellis structure is the same in every step, it can be written

$$L(u_t|\mathbf{y}) = \ln\frac{\sigma_t(0,2)+\sigma_t(1,0)+\sigma_t(2,3)+\sigma_t(3,1)}{\sigma_t(0,0)+\sigma_t(1,2)+\sigma_t(2,1)+\sigma_t(3,3)},$$

and if in some steps (for some trellis depths) some transitions do not exist, the corresponding probabilities are equal to zero and a previous equality is further simplified.

Finally, it is obtained

$$L(u_1|\mathbf{y}) = \ln\frac{\sigma_1(0,2,\mathbf{y})}{\sigma_1(0,0,\mathbf{y})} = \ln\frac{\alpha_0(0)\gamma_1(0,2)\beta_1(2)}{\alpha_0(0)\gamma_1(0,0)\beta_1(0)} = -17.035,$$

$$L(u_2|\mathbf{y}) = \ln\frac{\sigma_2(0,2)+\sigma_2(2,3)}{\sigma_2(0,0)+\sigma_2(2,1)} = 9.16,$$

$$L(u_3|\mathbf{y}) = \ln\frac{\sigma_3(0,2)+\sigma_3(1,0)+\sigma_3(2,3)+\sigma_3(3,1)}{\sigma_3(0,0)+\sigma_3(1,2)+\sigma_3(2,1)+\sigma_3(3,3)} = -17.035,$$

while for a terminating part of trellis the expressions are slightly different, but as a rule simplified (the corresponding encoder inputs are uniquely determined by information sequence, but are not necessarily equal to all zeros sequence) yielding here

$$L(u_4|\mathbf{y}) = \ln\frac{\sigma_4(1,0)+\sigma_4(4,2)}{\sigma_4(0,0)+\sigma_4(3,2)} = 9.16, \quad L(u_5|\mathbf{y}) = \ln\frac{\sigma_5(1,0,\mathbf{y})}{\sigma_5(0,0,\mathbf{y})} = -\infty,$$

LLR values and the branch metrics with a denoted path corresponding to a codeword are shown in Fig. 8.36.

Decoded sequence of information bits is $i_1 = 0$, $i_2 = 1$ and $i_3 = 0$, where the second bit is decoded with smaller reliability than the other two. It should be noted that this code is in fact a block code (although realized by the convolution encoder)

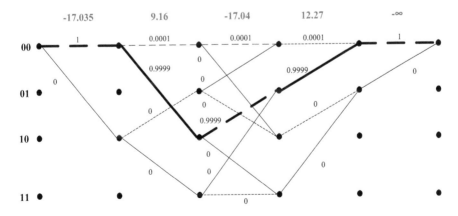

Fig. 8.36 The result of BCJR procedure

because at the encoder input is the three bits block to which two bits (tail bits) are added for the trellis termination, to which at the encoder output the corresponding sequence has ten bits. To the next information bits 001 correspond terminating bits 11, while from the line encoder in this case the following sequence is emitted −1, −1, +1, +1, +1, −1, +1, −1, +1, +1.

(b) It can be noticed that the branch metrics depend on the channel signal-to-noise ratio and on the symbol probabilities at the encoder input, what can be easily seen if they are written in a developed form

$$\gamma_t(S^{(t-1)}, S^{(t)}) = C_t \exp\left(\frac{u_t}{2} \operatorname{ld} \frac{P(u_t = +1)}{P(u_t = -1)}\right) \exp\left(2a \frac{E_c}{N_0} \sum_{l=1}^{n} x_{tl} y_{tl}\right).$$

When $P(u_t = +1) = P(u_t = -1) = 0.5$, the first exponential term is equal to one, but the branch metrics are still dependent on E_c/N_0.

In Table 8.14 it is clearly shown that in considered example the correct decision is made in all analyzed cases, where for smaller values of parameter E_c/N_0 the estimation becomes very unreliable. For example, when $E_c/N_0 = -20$ [dB], normalized transition probabilities corresponding to the correct path are (0.504, 0.2673, 0.1839, 0.2191, 1) showing a small reliability of the decision, if compared to the values from Fig. 8.36.

Here it is essential to notice that for Viterbi decoding (either for hard outputs, either for soft outputs, i.e. SOVA) the decision in any way does not depend on the ratio E_c/N_0 neither on the signal-to-noise ratio!

(c) For a case when the symbols at the encoder input are not equiprobable, BCJR algorithm can take it into account. In Fig. 8.37 are shown LLR estimations corresponding to the probabilities $P(i_t) = 0.1$ and $P(i_t = 1) = 0.9$, as well as the paths corresponding to decoded information words (including the terminating bits as well), for the case $E_c/N_0 = 0$ [dB]. BCJR decoder favors these information sequences which are more probable, i.e., for a difference to SOVA algorithm, it takes into account the source characteristics.

Problem 8.9 The system described in the previous problem is considered, the source parameters, convolutional encoder and modulator, as well as all numerical values are the same as in the first part of the previous problem ($P_s = 1$ [µW], $V_b = 1$

Table 8.14 BCJR procedure results for various estimation of ratio E_c/N_0 in the channel

E_c/N_0 (dB)	L_c	LLR estimations				
		$t = 1$	$t = 2$	$t = 3$	$t = 4$	$t = 5$
10	40	−172.4	91.6	−172.4	91.6	−710
0	4	−17.035	9.16	17.035	9.16	−70.99
−10	0.4	−1.4372	0.7533	1.4356	0.6696	−6.79
−20	0.04	−0.1021	−0.0238	−0.0756	−0.7199	−0.6582

Problems

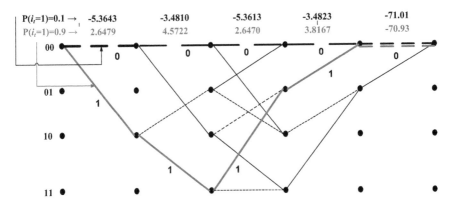

Fig. 8.37 BCJR procedure results for different symbol probabilities at the encoder input

[Mb/s], $N_0 = 10^{-12}$ [W/Hz], equally probable symbols at the encoder input). Decode the emitted information sequence for a received sequence

$$y = (-0.95, -1.2, -0.1, -0.05, -1.2, 1.01, 0.3, 1.13, -0.4, -16),$$

if in one block three information bits are transmitted, and the decoding is performed by using

(a) Log-MAP algorithm
(b) Max-log-MAP algorithm

Solution
BCJR algorithm, although optimum, has some drawbacks:

- the overflow easily happens during the calculation forward and backward metrics,
- a great number of multiplications should be performed.

In previous problem it was shown that the first drawback can be lessened by normalization of a calculated metrics after every step. However, both problems can be simultaneously solved by calculating logarithmic branch metrics values

$$\Gamma_t(S^{(t-1)}, S^{(t)}) = \ln \gamma_t(S^{(t-1)}, S^{(t)}) = \ln C_t + \frac{u_t}{2}L(u_t) + \frac{L_c}{2}\sum_{l=1}^{n}x_{tl}y_{tl},$$

providing that metrics forward and backward are obtained by a series of successive additions

$$A_t(S^{(t)}) = \ln \alpha_t(S^{(t)}) = \max_{S^{(t-1)}} * \left\{ A_{t-1}(S^{(t-1)}) + \Gamma_t(S^{(t-1)}, S^{(t)}) \right\}$$

$$B_{t-1}(S^{(t-1)}) = \ln \beta_t(S^{(t)}) = \max_{S^{(t)}} * \left\{ \Gamma_t(S^{(t-1)}, S^{(t)}) + B_t(S^{(t)}) \right\}$$

where the maximizing in the first case is performed over a set of all input paths into a state $S^{(t)}$, while in the second case the choice is performed over these paths leaving the considered state. A procedure for maximization performed in above equalities is just the point where these two algorithms differ.

(a) For **log-MAP algorithm** [74] this operator can be changed by relation

$$\max^*\{a, b\} = \max\{a, b\} + \ln(1 + \exp(-|a - b|)),$$

and the numerical values of corresponding forward and backward metrics are given in Table 8.15. When calculating metrics, operation max* is always used for the case when there are two arguments and the above equality can be directly applied.

When the metrics forward and backward are found, they can be used for finding the estimations for LLR because

$$L(u_t|y) = \max_{R_1}^* \left\{ A_{t-1}(S^{(t-1)}) + \Gamma_t(S^{(t-1)}, S^{(t)}) + B_t(S^{(t)}) \right\}$$
$$- \max_{R_0}^* \left\{ A_{t-1}(S^{(t-1)}) + \Gamma_t(S^{(t-1)}, S^{(t)}) + B_t(S^{(t)}) \right\},$$

and because log-MAP algorithm is a logarithmic equivalent of MAP algorithm, LLR values are the same as in the first part of a previous problem

$$L(u_1|y) = -17.035, \ L(u_2|y) = 9.16, \ L(u_3|y)$$
$$= -17.035, \ L(u_4|y) = 9.16, \ L(u_5|y) = -70.99.$$

Table 8.15 Node metrics forward and backward for log-MAP algorithm

		t = 0	t = 1	t = 2	t = 3	t = 4	t = 5
Forward	$A_t(0)$	0.0000	4.3000	4.6000	4.9801	11.2801	44.0801
	$A_t(1)$	$-\infty$	$-\infty$	-4.2000	8.4200	5.8812	$-\infty$
	$A_t(2)$	$-\infty$	-4.3000	4.0000	4.2203	$-\infty$	$-\infty$
	$A_t(3)$	$-\infty$	$-\infty$	-4.4000	0.5172	$-\infty$	$-\infty$
Backward	$B_t(0)$	44.0801	39.7801	30.3200	29.9400	32.8000	0.0000
	$B_t(1)$	0	0	29.5600	35.6600	-32.8000	$-\infty$
	$B_t(2)$	0	31.3451	40.0800	-31.1400	$-\infty$	$-\infty$
	$B_t(3)$	0	0	31.2400	-34.4600	$-\infty$	$-\infty$

Problems

It should be noticed that maximization in the above relation can be performed and for more than two arguments, this problem can be overcomed easily by a recursive by using of the corresponding operation (in every step two inputs are represented as one output, becoming the input for the next step).

(b) For **max-log-MAP algorithm** [75] the approximation is used

$$\max{}^*\{a,b\} = \max\{a,b\},$$

the corresponding metric numerical values are given in Table 8.16. When the signal-to-noise ratio is sufficiently large (here it is $E_c/N_0 = 10$ [dB]) the differences in metric values in respect to BCJR algorithm are very small, while for the worse channel conditions these differences grow.

LLR estimations in this case can be found directly by using of relation

$$L(u_t|\mathbf{y}) = \max_{R_1}\left\{A_{t-1}(\mathbf{S}^{(t-1)}) + \Gamma_t(\mathbf{S}^{(t-1)}, \mathbf{S}^{(t)}) + B_t(\mathbf{S}^{(t)})\right\}$$
$$- \max_{R_0}\left\{A_{t-1}(\mathbf{S}^{(t-1)}) + \Gamma_t(\mathbf{S}^{(t-1)}, \mathbf{S}^{(t)}) + B_t(\mathbf{S}^{(t)})\right\}$$

yielding

$$L(u_1|\mathbf{y}) = -17.24,\ L(u_2|\mathbf{y}) = 9.16,\ L(u_3|\mathbf{y})$$
$$= -17.24,\ L(u_4|\mathbf{y}) = 9.16,\ L(u_5|\mathbf{y}) = -44.08.$$

It is obvious that and log-MAP and max-log-MAP take into account the channel quality influence (by using parameter L_c depending on E_c/N_0), as well as the binary symbols probabilities at the encoder input (by using parameter $L(u_t)$).

$$L(u_t|\mathbf{y}) = F\{L(u_t), L_c y_t\}$$

Table 8.16 Node metrics forward and backward for max-log-MAP algorithm

		$t=0$	$t=1$	$t=2$	$t=3$	$t=4$	$t=5$
Forward	$A_t(0)$	0.0000	4.3000	4.6000	4.9800	11.2800	44.0800
	$A_t(1)$	$-\infty$	$-\infty$	-4.2000	8.4200	5.8800	$-\infty$
	$A_t(2)$	$-\infty$	-4.3000	4.0000	4.2200	$-\infty$	$-\infty$
	$A_t(3)$	$-\infty$	$-\infty$	-4.4000	0.0002	$-\infty$	$-\infty$
Backward	$B_t(0)$	44.0800	39.7800	30.3200	29.9400	32.8000	0.0000
	$B_t(1)$	0	0	29.5600	35.6600	-32.8000	$-\infty$
	$B_t(2)$	0	31.1400	40.0800	-31.1400	$-\infty$	$-\infty$
	$B_t(3)$	0	0	31.2400	-34.4600	$-\infty$	$-\infty$

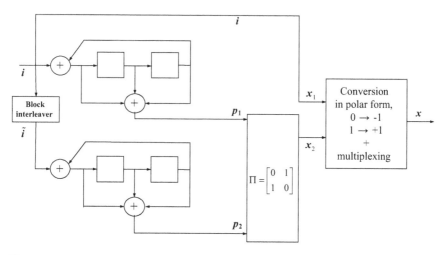

Fig. 8.38 Turbo encoder block diagram

which is their important advantage in respect to SOVA algorithm. By this procedure, for a priori estimation LLR of information bits values at the encoder input, an improved estimation is obtained, after the receiving the input sequence *y* at the decoder input (more information is obtained!).

Problem 8.10 Consider turbo encoder having the structure shown in Fig. 8.38, where the interleaving is performed over a block of $N = 5$ bits, according to the scheme

$$I\{.\} = \begin{pmatrix} 1 & 2 & 3 & 4 & 5 \\ 5 & 1 & 4 & 2 & 3 \end{pmatrix}$$

where the second row denotes the bit positions in a sequence obtained at the interleaver output, while in the first row are ordinal numbers of bits in the original sequence. Encoder output is led into the puncturing block (described by a puncturing matrix Π), represented by polar pulses and after multiplexing symbol-by-symbol is transmitted over a channel where is AWGN, resulting in $E_c/N_0 = 0$ [dB].

(a) Find sequence *x* emitted into the channel if at the encoder input is the information bits sequence $i = (10101)$. In particular, explain the procedure for terminating bits obtaining and at trellis show the paths corresponding to code sequences at the outputs of component encoders.
(b) Decode the transmitted information sequence if log-MAP algorithm is used and at the turbo decoder input the received sequence is

Problems

$$y = (-2, 1.5, -0.5, 0.4, 1.8, 0.3, 0.2, -0.13, 0.1, 0.1, -0.2, 0.15, 0.1, 0.08).$$

Solution

(a) In this case **turbo encoder** is realized as a parallel cascade combination of two recursive systematic convolutional encoders (RSC), separated by a block interleaver. At the beginning it is supposed that both encoders are reset and for the input sequence i (determining output x_1) firstly are found the outputs of the first and the second RSC where are the parity-check bits and after that, they are punctured, so as that at output x_2 alternatively are sent odd bits of sequence p_1 and even bits of sequence p_2. All sequences are given in details in Table 8.17, and the sequence at the encoder output is

$$x = (+1, +1, -1, -1, +1, +1, -1, -1, +1, +1, -1, +1, +1, +1).$$

Sequences corresponding to two RSC encoder outputs are shown as well in Fig. 8.39, by heavy lines. It is obvious that to a bit sequence at the first RSC input, being $i = (10101)$, correspond the terminating bits 01, while to the sequence of information bits at the second RSC (i.e. at the interleaver output $i' = (11001)$) correspond the terminating bits 10. In Table 8.16 in the last two columns the sequences are given corresponding to the trellis termination that must be sent to the

Table 8.17 Sequences in some points of the encoder from Fig. 8.38

t	1	2	3	4	5	6	7
i	1	0	1	0	1	0	1
i'	0	0	1	1	1	1	0
p_1	1	1	1	0	1	1	1
p_2	0	0	1	0	0	1	0
x_1	1	0	1	0	1	0	1
x_2	1	0	1	0	1	1	1

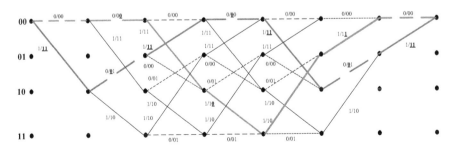

Fig. 8.39 Codewords forming in two RSC

encoder output to return (reset) it in the initial state and to be ready to transmit a new information block.

Decoding procedure is performed *iteratively* [68, 79]:

1. At the first decoder input:
   - sequence y_1 is led, corresponding to a transmitted sequence x_1, containing information about sent information symbols. On this basis the series of symbols $L_c\, y_1$ is calculated.
   - by combination of y_1 and $y_{2,p1}$ (corresponding to the sequence x_2, to the parity-check bits from the first RSC decoder output, while the rest of sequence is filled with zeros), a sequence $y^{(I)}$ is formed and from it by using log-MAP algorithm, the improved estimation $L_1(u_t|y)$ is found, and further

$$L_{e1}(u_t|y) = L_1(u_t|y) - L_2(u_t) - L_c y_1.$$

   where at the beginning it is set usually $L_2(u_t) = 0(\forall t)$, i.e. it is determined by the source characteristics.

2. at the second decoder input
   - sequence y_1' is led, corresponding to the transmitted sequence x_1 after the interleaver $\rightarrow L_c\, y_1'$.
   - from sequences y_1' and $y_{2,p2}$ the sequence $y^{(II)}$ is formed, and the second decoder finds

$$L_{e2}(u_t|y) = L_2(u_t|y) - L_1(u_t) - L_c y_1',$$

   where $L_1(u_t) = I\{L_{e1}(u_t|y)\}$ is determined by a value obtained in the previous step.

As shown in Fig. 8.40, in the next step $L_{e2}(u_t)$ is led through deinterleaver and a priori LLR are found for the first decoder in the second iteration, because

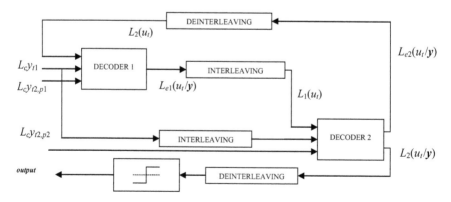

Fig. 8.40 Turbo decoding principle

$L_2(u_t) = I^{-1}\{L_{e2}(u_t|y)\}$. This procedure is repeated successively until the codeword is decoded or until a given maximum number of iterations is achieved. In this case, the path corresponding to the calculated values $L_1(u_t|y)$ is shown in Fig. 8.41. To these values corresponds the procedure of determining $L_{e1}(u_t)$, shown below

$y_1 =$	-2	-0.5	1.8	-0.2	0.1	-0.2	0.1
$L_1(u_t|y) =$	-2.9806	-3.4712	7.9516	-1.4854	-0.0031	-1.4803	0.1193
$L_2(u_t) =$	0	0	0	0	0	0	0
$L_c y_1 =$	-8	-2	7.2	-0.8	0.4	-0.8	0.4
$L_{e1}(u_t) =$	5.0194	-1.4712	0.7516	-0.6854	-0.4031	-0.6803	-0.2807

Of course, the receiver does not know whether the heavy line in the Fig. 8.41 really corresponds to the emitted information sequence. However, even here it can be concluded that the first bit of information sequence is $i_1 = 1$ (highly reliable!) although the corresponding received symbol is negative (-2). The rest of the bits cannot be yet reliable decoded and a procedure is continued by leading $L_{e1}(u_t)$ through the interleaver and sent to the other RSC decoder as an estimation of a priori RLL for sequences y_1'.

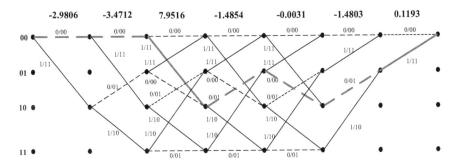

Fig. 8.41 The decoding procedure for the first RSC during the first iteration

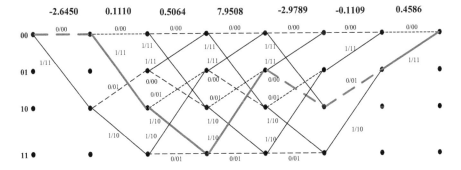

Fig. 8.42 The decoding procedure for the second RSC during the first iteration

Table 8.18 Turbo decoders outputs during the first four iterations, $L_1(u_t|y)$ and $L_2(u_t|y)$

		$t=1$	$t=2$	$t=3$	$t=4$	$t=5$	$t=6$	$t=7$
The first RSC decoder	iter = 1	−2.9806	−3.4712	7.9516	−1.4854	−0.0031	−1.4803	0.1193
	iter = 2	2.0350	−3.7935	9.1071	−0.8265	1.1469	−0.8789	1.1729
	iter = 5	17.5989	−6.7464	14.9302	−0.9586	4.4211	−0.9622	4.4210
	iter = 10	44.8247	−13.4653	26.7319	−1.9421	11.2654	−1.9421	11.2654
The second RSC decoder	iter = 1	−2.6450	0.1110	0.5064	7.9508	−2.9789	−0.1109	0.4586
	iter = 2	−3.5272	0.4030	1.4530	9.2648	−1.8882	−0.4028	−0.9194
	iter = 5	−6.8240	0.0487	4.7568	15.2578	17.4199	−0.0487	−4.6431
	iter = 10	−13.2631	−0.9421	11.8841	27.3506	44.9234	0.9421	−11.6595

Finding $L_2(u_t|y)$ is shown in Fig. 8.42 and the further procedure is

$$
\begin{aligned}
y'_1 &= \quad -0.5 \quad\quad -0.2 \quad\quad 0.1 \quad\quad 1.8 \quad\quad -2 \quad\quad 0 \quad\quad 0 \\
L_2(u_t|y) &= -2.6450 \quad 0.1110 \quad 0.5064 \quad 7.9508 \quad -2.9789 \quad -0.1109 \quad 0.4586 \\
L_1(u_t) &= -1.4712 \quad -0.6854 \quad -0.4031 \quad 0.7516 \quad 5.0194 \quad 0 \quad 0 \\
L_c y'_1 &= \quad -2 \quad\quad\quad -0.8 \quad\quad 0.4 \quad\quad 7.2 \quad\quad -8.0 \quad\quad 0 \quad\quad 0 \\
L_{e2}(u_t) &= 10.0350 \quad -1.7935 \quad 1.9071 \quad -0.0265 \quad 0.7469 \quad -0.0789 \quad 0.7729
\end{aligned}
$$

and $L_{e2}(u_t)$ is a set of a priori LLR estimations which are led to the input of the first decoder in the second iteration.

Estimations $L_1(u_t|y)$ and $L_2(u_t|y)$ after a few first iterations are given in Table 8.18. Erroneously decoded bits after every particular iteration are shadowed in table and it is clear that already after the second iteration, the first RSC decoder successfully decodes full information sequence, while the second component decoder needs more than five iterations to eliminate the errors influence. It is obvious that the estimations are more and more reliable after additional iterations. It is easy to verify that a decoded sequence is $i = (10101)$ and it just corresponds to the sequence at the decoder input in the first part of this problem (and, as earlier shown, to it corresponds the sequence $i' = (11001)$ at the interleaver output).

Chapter 9
Low Density Parity Check Codes

Brief Theoretical Overview

Besides the turbo codes, there is one more class of linear block codes that makes possible to approach to the Shannon bound. These are *Low Density Parity Check* (LDPC) codes. They were proposed by Gallager [83]. In principle they have a *sparse parity-check matrix*. In such a way the corresponding parity-check equations have a small number of terms providing significantly smaller complexity compared to standard linear block codes. They provide the *iterative decoding with a linear complexity*. Even then it seemed that this class of codes has a good performance, but the contemporary hardware and software were not suitable for their practical implementation. Using graph theory Tanner in 1981 [84] proposed an original interpretation of LDPC codes, but his work was practically ignored for the next 15 years. Not until mid-nineties some researches started to consider these codes and the decoding using graphs. Probably the most influential was work of David McKay [85] who demonstrated that, the performance very near to the Shannon limit can be achieved by using iterative decoding.

LDPC codes are linear block codes constructed by using control matrix H where the number of nonzero components is small. The corresponding parity-check equations, even for a long codeword have a small number of terms yielding substantially smaller complexity than the corresponding equations for typical linear block code which has the same parameters. For binary codes, control matrix has a great number of zeros and a few ones.

According to Gallager definition, the following conditions should be satisfied yielding a code which has low density parity-checks:

1. Control matrix H has a fixed number of ones in every row, denoted by ρ.
2. Control matrix H has a fixed number of ones in every column. This number is denoted by γ (usually $\gamma \geq 3$).

3. To avoid the cycles in the corresponding bipartite graph (will be explained later) in any two columns (or rows) the coincidence of ones should not be more than at one position.
4. Parameters s and v should be as small as possible in relation to codeword length n.

The code satisfying above conditions is denoted as C_{LDPC} (n, γ, ρ). Control matrix has n columns and $n-k$ rows, and the following must holds $\gamma n = \rho(n-k)$. If the columns of matrix H are linearly independent code rate of LDPC code is

$$R = 1 - \gamma/\rho,$$

and matrix density is

$$r = \rho/n = \gamma/(n-k),$$

being for typical LDPC codes $r \approx 0.02$ and **regular LDPC code** (Problems 9.1, 9.2, and 9.6) is obtained. However, it is very difficult to construct such code, especially because of condition 3. It was shown that for efficient LDPC codes the cycles cannot be avoided. Therefore, it is allowed that a number of ones in some rows (columns) differs from prescribed values and in this case γ and ρ are average values of these parameters. The corresponding code is called *irregular LDPC code* (Problem 9.1).

As an example, consider control matrix

$$H = \begin{bmatrix} 1 & 1 & 0 & 0 & 0 & 0 \\ 0 & 0 & 1 & 1 & 0 & 0 \\ 0 & 0 & 0 & 0 & 1 & 1 \end{bmatrix}.$$

Number of ones in every column is $\gamma = 1$, and number of ones in every row is $\rho = 2$. Therefore it is regular LDPC code having the matrix density $r = 1/3 = 2/6 = 0.333$ and the code rate $R = 1-\gamma/\rho = 1/3$. The matrix has $n = 6$ columns and $n-k = 3$ rows. It is code $(n, k) = (6, 3)$.

Consider now control matrix

$$H = \begin{bmatrix} 0 & 1 & 0 & 1 & 0 & 1 & 1 & 1 & 0 & 0 & 0 & 1 \\ 1 & 0 & 1 & 1 & 0 & 0 & 0 & 0 & 1 & 0 & 0 & 0 \\ 0 & 1 & 0 & 0 & 1 & 0 & 1 & 0 & 0 & 0 & 0 & 1 \\ 1 & 0 & 0 & 1 & 0 & 0 & 0 & 0 & 0 & 1 & 1 & 0 \\ 0 & 0 & 1 & 0 & 1 & 1 & 0 & 0 & 0 & 1 & 0 & 0 \\ 1 & 0 & 1 & 0 & 0 & 0 & 1 & 1 & 0 & 0 & 1 & 0 \\ 0 & 1 & 0 & 0 & 0 & 1 & 0 & 1 & 1 & 1 & 0 & 0 \\ 0 & 0 & 0 & 0 & 1 & 0 & 0 & 0 & 1 & 0 & 1 & 1 \end{bmatrix}.$$

Number of ones in every column is $\gamma = 3$, while the number of ones differs in rows and has the average value $\rho = 4.5$. It is irregular LDPC code which has matrix

Brief Theoretical Overview

density $r = 3/8 = 4.5/12 = 0.375$ and the code rate $R = 1 - \gamma/\rho = 1/3$. The matrix has $n = 12$ columns and $n-k = 8$ rows, basic code parameters are $(n, k) = (12, 4)$. Therefore, a code rate can be found and if the number of ones in every column is not the same.

At the end consider now square matrix having $\gamma = \rho = 3$ ones in every column and row

$$H = \begin{bmatrix} 1 & 1 & 0 & 1 & 0 & 0 & 0 \\ 0 & 1 & 1 & 0 & 1 & 0 & 0 \\ 0 & 0 & 1 & 1 & 0 & 1 & 0 \\ 0 & 0 & 0 & 1 & 1 & 0 & 1 \\ 1 & 0 & 0 & 0 & 1 & 1 & 0 \\ 0 & 1 & 0 & 0 & 0 & 1 & 1 \\ 1 & 0 & 1 & 0 & 0 & 0 & 1 \end{bmatrix}.$$

In any two columns (or rows) the ones do not coincide at more than one position and the matrix density is relatively small ($r = 3/7 = 0.4286$), this matrix seems to satisfy the condition to be LDPC code control matrix. However, it should be noticed that the code rate is here $R = 1-3/3 = 0$!

Furthermore, the rows in this matrix are not linearly independent and it cannot be control matrix of a linear block code. By eliminating the linearly dependent rows, the following is obtained

$$H_{\text{LN}} = \begin{bmatrix} 1 & 1 & 0 & 1 & 0 & 0 & 0 \\ 0 & 1 & 1 & 0 & 1 & 0 & 0 \\ 0 & 0 & 1 & 1 & 0 & 1 & 0 \\ 0 & 0 & 0 & 1 & 1 & 0 & 1 \end{bmatrix}.$$

However, although the number of ones in every rows is fixed ($\gamma' = 3$), the number of ones in every column varies from 1 to 3 (the average value is $\gamma' = 1.71$). It may be seen that the number of matrix rows is $(n-k) = 4$ and a linear block code $(7, 3)$ is obtained having the code rate $R = 1 - \gamma'/\rho' = 3/7$.

For the last example, there is one more question—do this code is regular or not? Obviously, code which has the control matrix H_{LN} (linearly independent rows) is irregular. But it can be obtained from matrix H, which satisfies conditions for a regular code. Let instead control matrix H_{LN} for decoding control matrix H is used and **syndrome decoding** is performed (it can be applied for every linear block code). If r is received vector, the corresponding syndrome is $S = rH^{\text{T}}$. In this case to the received vectors will correspond seven-bit syndrome vectors because their length is defined by the number of matrix H rows. However, there are three linearly dependent rows in this matrix and to all possible words would not correspond $2^n = 128$ but only $2^{(n-3)} = 16$ possible syndromes. Therefore, by using matrix H the same results will be obtained as by using matrix H_{LN} for decoding, where the rows are linearly independent. The conclusion is that a code can be defined by matrix having linearly independent rows, and this code can be defined and by matrix having linearly dependent rows as well and it can be said that the code is regular!

Having the above in view, the construction of regular codes can be substantially simplified if linear dependence of control matrix rows is allowed. Using this approach the first (and relatively exact) method for construction of regular LDPC codes was proposed by Gallager in his pioneering work [83]. The steps in control matrix construction are:

– The wished codeword length (n) is chosen. Number of ones in rows and columns is fixed being v and s, respectively.
– One positive integer is chosen $m = n/\rho$, and a matrix to be constructed is divided into s submatrices of dimensions $m \times m\,\rho$, where

$$H((j-1) \times m + 1 : jm, 1 : n) = H_j(1 : m, 1 : n), j = 1, 2, \ldots, s.$$

– Now, submatrix H_1 is formed, in such way that in the ith row, by ones the positions from $(i-1)\rho + 1$ to $i\,\rho$ ($i = 1, 2, \ldots m$) are filled, providing that rows of H_1 do not have common ones, and columns have not more than one.
– The other submatrices are obtained by permutations of columns of matrix H_1. Columns permutations are chosen so as that rows of the full matrix do not have common ones, and columns have not more than one. Total number of ones in matrix H is $m\rho\gamma$, and matrix density is $r = 1/m$.

For example, consider control matrix which has parameters $n = 20$, $\rho = 4$, $\gamma = 3$ and $m = 5$. Firstly, submatrix H_1 dimensions 5×20 is formed, and later using search other submatrices (H_2 and H_3) are found, fulfilling the above conditions. One possible control matrix H obtained using such procedure is

$$H = \begin{bmatrix} H_1 \\ -- \\ H_3 \\ -- \\ H_3 \end{bmatrix} = \begin{bmatrix}
1 & 1 & 1 & 1 & 0 & 0 & 0 & 0 & 0 & 0 & 0 & 0 & 0 & 0 & 0 & 0 & 0 & 0 & 0 & 0 \\
0 & 0 & 0 & 0 & 1 & 1 & 1 & 1 & 0 & 0 & 0 & 0 & 0 & 0 & 0 & 0 & 0 & 0 & 0 & 0 \\
0 & 0 & 0 & 0 & 0 & 0 & 0 & 0 & 1 & 1 & 1 & 1 & 0 & 0 & 0 & 0 & 0 & 0 & 0 & 0 \\
0 & 0 & 0 & 0 & 0 & 0 & 0 & 0 & 0 & 0 & 0 & 0 & 1 & 1 & 1 & 1 & 0 & 0 & 0 & 0 \\
0 & 0 & 0 & 0 & 0 & 0 & 0 & 0 & 0 & 0 & 0 & 0 & 0 & 0 & 0 & 0 & 1 & 1 & 1 & 1 \\
\hline
1 & 0 & 0 & 0 & 1 & 0 & 0 & 0 & 1 & 0 & 0 & 0 & 1 & 0 & 0 & 0 & 0 & 0 & 0 & 0 \\
0 & 1 & 0 & 0 & 0 & 1 & 0 & 0 & 0 & 1 & 0 & 0 & 0 & 0 & 0 & 1 & 0 & 0 & 0 & 0 \\
0 & 0 & 1 & 0 & 0 & 0 & 1 & 0 & 0 & 0 & 0 & 0 & 0 & 1 & 0 & 0 & 1 & 0 & 0 & 0 \\
0 & 0 & 0 & 1 & 0 & 0 & 0 & 0 & 0 & 0 & 1 & 0 & 0 & 0 & 1 & 0 & 0 & 0 & 1 & 0 \\
0 & 0 & 0 & 0 & 0 & 0 & 0 & 1 & 0 & 0 & 0 & 1 & 0 & 0 & 0 & 1 & 0 & 0 & 0 & 1 \\
\hline
1 & 0 & 0 & 0 & 0 & 1 & 0 & 0 & 0 & 0 & 1 & 0 & 0 & 0 & 0 & 1 & 0 & 0 & 0 & 0 \\
0 & 1 & 0 & 0 & 0 & 0 & 1 & 0 & 0 & 1 & 0 & 0 & 0 & 1 & 0 & 0 & 0 & 0 & 0 & 0 \\
0 & 0 & 1 & 0 & 0 & 0 & 0 & 1 & 0 & 0 & 0 & 1 & 0 & 0 & 0 & 0 & 0 & 1 & 0 & 0 \\
0 & 0 & 0 & 1 & 0 & 0 & 0 & 0 & 1 & 0 & 0 & 0 & 1 & 0 & 0 & 1 & 0 & 0 & 0 & 0 \\
0 & 0 & 0 & 0 & 1 & 0 & 0 & 0 & 0 & 1 & 0 & 0 & 0 & 1 & 0 & 0 & 0 & 0 & 0 & 1
\end{bmatrix}$$

The matrix rang is 13, and from 15 rows only 13 are linearly independent. The obtained code is (20, 7), parameters are $n-k = 13$, code rate $R = 0.35$ and matrix density $r = 0.2$ (Problem 9.3).

On the basis of explained procedure, the conclusion can be drawn that Gallager method for regular LDPC code construction is not fully deterministic. In fact, it is not clearly defined how to perform the permutations in submatrices $H_2, H_3, \ldots, H_s$, but only the needed features. Such an approach usually results in codes having good performance, but it is relatively difficult from all possible combinations to choose LDPC code with a given codeword length n and code rate $R = k/n$, i.e. (n, k) code having good characteristics. However, later the structured LDPC codes were proposed where the control matrix structure can be found using a deterministic algorithm. Some of these algorithms are based on Euclidean and projective geometry [86, 87], cyclic permutations [88] or combined construction [89]. These problems are also considered in papers and monographs [90, 91, 92].

Tanner 1981 proposed an alternative way for considering control matrix of LDPC code by using the **bipartite graphs** [84] (Problems 9.1, 9.2, 9.4, 9.5, and 9.6). This approach provides advanced technique for decoding. To explain this method better, firstly we will consider two classical ways for linear block codes decoding. The decoding of any linear block code can be performed using **syndrome** (Problem 9.2). Using such approach firstly on the basis of matrix H and relation $vH^T = 0$, where $v = [v_1, v_2, \ldots, v_n]$ is a valid code word, a system of $n-k$ equations with n unknown variables is written. Every row of control matrix defines one equation for parity-check, and the position of one in that row defines the position of symbol in equation.

On the base of this system, the bits of code word can be found. Tanner graph is bipartite graph visualizing the relation between two types of nodes—**variable nodes** (v_j) denoting the sent symbols, and **(parity-)check nodes** $(c_i,)$—nodes corresponding to parity-checks which relate the emitted symbols (bits). In such way, for any linear block code, if (i, j)th element of matrix H equals one, at the corresponding bipartite graph, there is a line between variable node v_j and (parity-)check node c_i. The state of a check node depends on the values of variable nodes to which it is connected. For some check node it is said that it is **children node** of variable nodes to which is connected, and a variable node is **parent node** for all check nodes connected to it.

As an example consider Hamming code (7, 4) where the control matrix is (Problem 5.3)

$$H = \begin{bmatrix} 1 & 0 & 1 & 0 & 1 & 0 & 1 \\ 0 & 1 & 1 & 0 & 1 & (1) & (1) \\ 0 & 0 & 0 & 1 & 1 & (1) & (1) \end{bmatrix} \quad \begin{array}{l} c_1 : v_1 + v_3 + v_5 + v_7 = 0 \\ c_2 : v_2 + v_3 + v_6 + v_7 = 0 \\ c_3 : v_4 + v_5 + v_6 + v_7 = 0 \end{array},$$

where the corresponding equations are at the right side. The corresponding bipartite graph is given in Fig. 9.1.

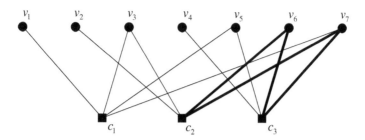

Fig. 9.1 Bipartite graph for Hamming code (7, 4)

If the received word is $y = (0000011)$, syndrome is formed on the base of rows of control matrix, but instead of the code word bits, the received bits are written

$$s_1 = y_1 + y_3 + y_5 + y_7 = 1$$
$$s_2 = y_2 + y_3 + y_6 + y_7 = 0,$$
$$s_3 = y_4 + y_5 + y_6 + y_7 = 0$$

and it is concluded that the error happened at position $S = [s_1 s_2 s_3] = 1$.

From graph shown in Fig. 9.1 it can be concluded that the variable node v_6 is connected with check nodes c_2 and c_3. Variable node v_7 is connected (among others) with the same check nodes—c_2 i c_3. These connections are the consequences of binary ones written in control matrix H in parentheses. These connections in bipartite graphs are denoted by heavy lines forming **4-girth-cycle**.

The described procedure can be applied to any linear block code, and to the LDPC codes as well. Of course, generally, decimal equivalent of syndrome does not give directly the error position, i.e. the process of error position finding is not finished. For longer code words (and practically, interesting LDPC codes have control matrix of large dimensions to provide for small density), a smaller decoding complexity is obtained by **majority logic decoding** (MLD) (mentioned in Chap. 7) (Problem 9.2). Let A_l is a set of all rows of matrix having one at lth position. Then, on the base of positions of ones in these rows, for every such set of matrix A_l rows, a set of s equations (chosen from n–k) can be written **orthogonal** to bit v_l (l = 1, 2, ...,n), i.e. in this set every symbol occurs only once except v_l. If in these equations code word bits are changed to received word bits, MLD can be used and it is possible to correct up to $\lfloor \gamma/2 \rfloor$ errors and minimal Hamming distance is $d_{\min} > s + 1$. However, for small s this method is not especially effective, but the value of s for LDPC codes cannot be large.

Structural feature of the matrix H generated by Gallager method is that in rows corresponding to one code word, the other code word bits except v_l appear only once. Because of that, if the regular LDPC code is constructed on the above rules, in the graph there would be no girths having cycles length four. Besides, during the code construction, the aim is to eliminate all girths having short cycles from the corresponding bipartite Tanner graph.

Brief Theoretical Overview

Consider control matrix of regular LDPC code

$$H = \begin{bmatrix} 1 & 1 & 0 & 1 & 0 & 0 & 0 \\ 0 & 1 & 1 & 0 & 1 & 0 & 0 \\ 0 & 0 & 1 & 1 & 0 & 1 & 0 \\ 0 & 0 & 0 & 1 & 1 & 0 & 1 \\ 1 & 0 & 0 & 0 & 1 & 1 & 0 \\ 0 & 1 & 0 & 0 & 0 & 1 & 1 \\ 1 & 0 & 1 & 0 & 0 & 0 & 1 \end{bmatrix}.$$

Set of equations containing the matrix rows having one in the first column, denoted with A_1, is

$$\begin{aligned} c_1 &: v_1 + v_2 + v_4 = 0 \\ c_5 &: v_1 + v_5 + v_6 = 0, \\ c_7 &: v_1 + v_3 + v_7 = 0 \end{aligned}$$

and it should be noticed that in this set every symbol occurs once except v_1. This set of equations is orthogonal to v_1. Set of equations containing the matrix rows having one in the second column, denoted with A_2, is

$$\begin{aligned} c_1 &: v_1 + v_2 + v_4 = 0 \\ c_2 &: v_2 + v_3 + v_5 = 0. \\ c_6 &: v_2 + v_6 + v_7 = 0 \end{aligned}$$

This set is orthogonal to v_2.

If the sequence $r = (0011100)$ is at the decoder input, parity-checks for set A_1 will be $(1, 1, 1)$ and the first estimated bit of the code word is $x_1 = 1$ (at that bit the error happened!). Parity-checks for A_2 will be (100) and the first estimated bit of the code word is $x_1 = 0$ because of the majority of zeros (there is no error at the second bit). Further procedure would show that the error happened only at the first transmitted bit, i.e. that the emitted code word is $x = (1011100)$.

Tanner bipartite graph corresponding to this regular LDPC code is shown in Fig. 9.2. Weight of every node is defined by the number of entering edges. For bipartite graph where all nodes at one its side have the same weight it is said the code is regular. It can be easily noticed that all variable nodes have the same weight $\gamma = 3$, and that all check nodes have as well the same weight $\rho = 3$. Also, in this case Tanner graph does not have 4-girth-cycles, but has 6-girth-cycles. As stressed earlier, during construction, the aim is that such cycles are as long as possible.

A simple procedure of iterative decoding LDPC codes proposed Gallager himself in his pioneering work [63]. As shown above, the information about "successful" transmission is in the syndrome. If some of syndrome components equal one, it means that some of control sums defined by matrix columns are not satisfied. **Iterative decoding** of LDPC codes based on **hard decision** consists of

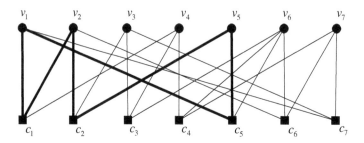

Fig. 9.2 Tanner graph for considered example

complementing some syndrome bits, performed in iterations. The result is a decoded code word corresponding to the syndrome having all zeros.

The procedure to find which bits should be complemented is usually called **bit-flipping** (BF) algorithm (Problems 9.3, 9.4, and 9.10) and consists of the next steps:

1. By above explained procedure, on the basis of known matrix H the sums of parity-checks are formed. For the bits of received word y, using

$$S = yH^T$$

the syndrome is found. If all syndrome components are zeros, it is concluded that the decoder code word is just the received vector y and the decoding is finished.
2. If the syndrome differs from zero, all control sums are calculated corresponding to nonzero syndrome bits. Let these sums form the set Z. Let ith bit of received word is denoted by y_i. This bit appears only in some parity-checks of Z. Parity-check equations where y_i appears, which are not satisfied form subset $Z^{(y_i)} \in Z$.
3. From all received word bits (y_i, $i = 1, 2, \ldots , n$), the bits are found for which subset $Z^{(y_i)}$ has the biggest number of elements. Otherwise speaking, the received bits are found which take part in the largest number of subset elements. If there are more such bits all of them should be taken into account.
4. Received bit(s) found according the previous step should be complemented (flipped). By this procedure, the modified code word is obtained $y^{(1)}$, being the result of the first iteration.
5. The procedure from steps 1–4 is repeated iteratively for vectors $y^{(1)}, y^{(2)}, \ldots y^{(N)}$, until all syndrome bits are not equal zero. Vector $y^{(N)}$ satisfying this condition corresponds to the received code word on the basis *bit-flipping* procedure.

It is considered today that by using BF algorithm a good performance can be achieved with relatively small complexity of realization. Low-complex parallel bit-flipping algorithm with fixed threshold is especially important, as well as the simplest *massage-passing* algorithm, known as Gallager-B (Problems 9.4 and 9.5).

Brief Theoretical Overview

Very popular are modifications of this algorithm, especially weighted BF algorithms (*weighted bit-flipping* [93]), modified weighted BF (*modified weighted bit-flipping* [94]) and weighted BF algorithm based on reliability ratio (*reliability ratio weighted bit-flipping* [95]).

As an example consider the control matrix of regular LDPC code (only positions of ones are denoted, while the zero positions are omitted):

$$H = \begin{array}{c} \\ I \\ II \\ III \\ IV \\ \\ V \\ VI \\ VII \\ VIII \\ \\ IX \\ X \\ XI \\ XII \end{array} \begin{bmatrix} 1 & 2 & 3 & 4 & 5 & 6 & 7 & 8 & 9 & 10 & 11 & 12 & 13 & 14 & 15 & 16 \\ 1 & 1 & 1 & 1 & & & & & & & & & & & & \\ & & & & 1 & 1 & 1 & 1 & & & & & & & & \\ & & & & & & & & 1 & 1 & 1 & 1 & & & & \\ & & & & & & & & & & & & 1 & 1 & 1 & 1 \\ \hline 1 & & & & 1 & & & & 1 & & & & 1 & & & \\ & 1 & & & & 1 & & & & 1 & & & & 1 & & \\ & & 1 & & & & 1 & & & & 1 & & & & 1 & \\ & & & 1 & & & & 1 & & & & 1 & & & & 1 \\ \hline 1 & & & & & 1 & & & & & 1 & & & & & 1 \\ & 1 & & & & & 1 & & & & & 1 & 1 & & & \\ & & 1 & & & & & & 1 & 1 & & & & 1 & & \\ & & & 1 & 1 & & & & & 1 & & & & & 1 & \end{bmatrix}$$

It is obvious that the code word length is $n = 16$, the number of ones over rows and columns is, respectively, $\gamma = 4$ and $\gamma = 3$, and control matrix density is $r = 0.25$.

If the received word is $y = (1100101000000000)$, then $yH^T = (000001101001)$. In the syndrome at the positions VI, VII, IX and XII are ones. Therefore, the following sums are not satisfied:

- VI control sum (where are 1, 2, 3 and 4 received word bits)
- VII control sum (where are 3, 7, 11 and 15 received word bits)
- XI control sum (where are 1, 6, 11 and 16 received word bits)
- XII control sum (where are 4, 5, 10 and 15 received word bits)

In the above control sums bits 1, 3, 4, 11 and 15 appear the most frequently (two times) and the received word bits at these places are inverted yielding the word $y' = (0111101000000010)$, to which now the syndrome $y'H^T = (010101101001)$ corresponds. Now ones in syndrome are at the II, IV, VI, VII, IX and XII position. It is obvious that the parity-checks are not satisfied. Now, in control sums bits 6 and 15 appear the most frequently (three times). Bit 6 appears in sums II, VI and IX, and bit 15 in sums IV, VII and XII, while the other bits appear at most two times. By inverting bits 6 and 15, the second estimation of the transmitted code word is

$\mathbf{y}'' = (0111111000000000)$ yielding $\mathbf{y}''\mathbf{H}^T = (000000000000000)$, and $\mathbf{y}''$ is the decoded code word.

Besides the above explained algorithm, Gallager in his work also exposed the basic idea for LDPC *iterative decoding* based on *soft decision*. He has shown that by using such algorithms, substantially better performance can be obtained, where the number of operations per bit in every iteration is not dependent of the block length. The proposed algorithm is based on MAP, later known as *belief-propagation* [83, 96] (Problem 9.6). This method is based on solution of equations on graph, yielding as a result the exact values of a posteriori probabilities if the graph has not cycles. The graphs corresponding to LDPC codes usually have cycles and this algorithm does not guarantee a correct decoding. In spite of this, Gallager iterative algorithm in praxis achieves the good results because the correct decoding most frequent can be performed even if the estimated a posteriori probabilities are not exact. Iterative version of *belief propagation* algorithm suitable for practical implementation is here usually called *sum-product algorithm* (SPA) [25, 96] (Problems 9.6 and 9.10). This algorithm is based on the *message passing* between Tanner graph nodes. During *sum-product* algorithm, every connected variable node d_j and check node h_i change information according to the following iterative procedure:

1. The initial values of probabilities are chosen Q_{ij}^x. For these values a priori estimations of received symbols f_j^x are chosen. These probabilities in principle depend on the type of the channel used. For a channel with only additive white Gaussian noise, they are chosen as

$$f_j^0 = \frac{1}{\sqrt{2\pi}\sigma} e^{-(r_j+1)^2/(2\sigma^2)}, \quad f_j^1 = \frac{1}{\sqrt{2\pi}\sigma} e^{-(r_j-1)^2/(2\sigma^2)} \quad j = 1, 2, \ldots, n,$$

where r_j denotes the jth bit of the received word. In fact, f_j^0 denotes a priori probability that (logical) zero ($d_j = 0$) was emitted, and f_j^1 denotes a priori probability that (logical) one ($d_j = 1$) was emitted. These values are the initial probability estimations that variable nodes are in state 0 or 1. These estimation are sent to check nodes where the symbol v_j appears.

2. Other the check node c_i receives from all connected variable nodes v_j the probability estimations Q_{ij}^x (for $x = 0$ and $x = 1$), the probabilities are calculated that ith parity-check equation is satisfied if the v_j is in the state x (corresponding to the symbol v_j value and can be 0 or 1). These probabilities are

$$R_{ij}^x = \sum_{\mathbf{v}:v_j=x} P(c_i/\mathbf{v}) \prod_{k \in N(i)/j} Q_{ik}^{v_k}$$

where $N(i)$ is a set of indexes of all parent nodes connected to check node c_i, and $N(i)/j$ is the same set but without node v_j to which the information is sent. Term $P(c_i/\mathbf{v})$ denotes the probability that parity-check equation is satisfied, and

Brief Theoretical Overview

summing is performed over all possible decoded vectors v for which the parity-check equation is satisfied when the informed node is in the state x.

3. At the variable node d_j after receiving from all connected nodes the values of probabilities R_{ij}^x (for $x = 0$ and $x = 1$), the correction of probabilities that this variable node is in state x (it can be 0 or 1) is performed. These probabilities are

$$Q_{ij}^x = \alpha_{ij} f_j^x \prod_{k \in M(j)/i} R_{kj}^x,$$

where $M(j)/i$ is a set of indexes of all check nodes connected to node v_j, without node c_i to which the information is sent. Coefficient α_{ij} is a normalization constant chosen so as the condition $\sum_x Q_{ij}^x = 1$ is satisfied.

This process, shown in Fig. 9.3 is repeated iteratively and stops after the information yielding vector d for which the corresponding syndrome is zero vector. If after some number of iterations zero syndrome is not obtained, the process stops when some, predefined, number of iterations was made. In both cases the decoder yields optimally decoded symbols, in the maximum a posteriori probability sense, but if syndrome is not equal to zero, the emitted code word is not decoded.

Consider regular LDPC code and the corresponding Tanner graph from the above example. The channel is with additive white Gaussian noise only. The procedure starts by calculating a priori probabilities $f_1^0, f_1^1, f_2^0, f_2^1, \ldots, f_7^0, f_7^1$ sent to the check nodes. To the first check node the initial information from the first, second and fourth variable node $\left(Q_{ij}^x = f_j^x\right)$ are sent, concerning the probability that the corresponding node is in the state 0 or 1. Now, the first check node has to return some information to every of connected variable nodes. They are different.

Parity-check equation for the first check node is $v_1 + v_2 + v_4 = 0$. Coefficient R_{11}^0 is the estimation sent by the check node 1 to the variable node 1. It is calculated

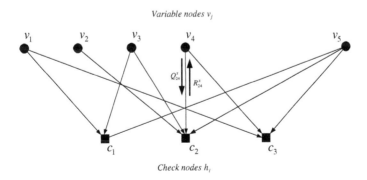

Fig. 9.3 The change of information between variable and check nodes

under assumption that the corresponding parity-check equation is satisfied if the first bit in code word, v_1, is zero. In this case the number of ones in equation should be even and there are only two combinations of other bits entering into parity-check for the first check node satisfying simultaneously this condition yielding

$$R_{12}^0 = Q_{12}^0 Q_{14}^0 + Q_{12}^1 Q_{14}^1.$$

Coefficient R_{11}^1 is also the estimation which v_1 sends to c_1 but it is calculated by supposing that the parity-check equation is satisfied when $v_1 = 1$. In this case the number of ones should be odd and there are two combinations of bits v_2 and v_4 satisfying this condition, yielding

$$R_{11}^1 = Q_{12}^0 Q_{14}^1 + Q_{12}^1 Q_{14}^1.$$

This step of iteration process is shown in Fig. 9.4. When all values of coefficients R_{ij}^0 and R_{ij}^1 are calculated, the first estimation of message symbols is made. The corresponding vector $\widehat{v}$ is calculated as follows

$$\widehat{v}_j = \arg\max_x f_j^x \prod_{k \in M(j)} R_{kj}^x$$

and if the syndrome for this vector is not zero, procedure continues by correcting probabilities Q_{ij}^x, joined to the corresponding variable nodes.

As said earlier, decoding of LDPC codes converges to the original message if there are no cycles in the corresponding bipartite graph. However, the relatively short cycles are unavoidable as well and when the corresponding LDPC code has good performance. But, the degrading effect of small cycles is diminished as the code length is longer, and substantially is small for code words long over 1000 bits. Also, there exist special procedures to eliminate the short cycles or to reduce their number [25].

The realization of *sum-product* algorithm in a *logarithmic domain* (Problems 9.7 and 9.10) can be further considered. The motives to use the logarithmic domain are the same as for turbo codes decoding—avoiding the products calculation always

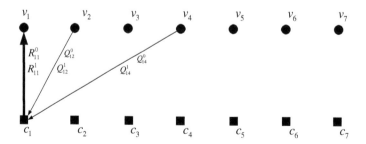

Fig. 9.4 One step in *sum-product* algorithm

when it is possible and lessening the overflow effect. Here logarithmic likelihood ratio is used. It can be used as well as for hard decoder inputs (with smaller correcting capability). However, this algorithm has one serious drawback. To find the information which the check node send to it adjoined variable node j one has to determine

$$\beta_{i,j} = 2 \tanh^{-1} \left\{ \prod_{k \in N(i) \backslash j} \tanh(\alpha_{i,k}/2) \right\},$$

and still more serious problems appear when the hardware block should be implemented corresponding to this relation. That is the reason to use the SPA procedure versions being much easier to be implemented. The simplest (and still rather exact) approximation of the above equality is

$$\beta_{i,j} = \left[\prod_{k \in N(i) \backslash j} \mathrm{sgn}(\alpha_{i,k}) \right] \times \left[\min_{k \in N(i) \backslash j} |\alpha_{i,k}| \right],$$

it is called *min-sum* algorithm (Problem 9.8) having still possible further modifications as *self-correcting min-sum algorithm* (Problem 9.8). The quasi-cyclic codes (Problem 9.9) can be considered as the LDPC codes as well, but the obtaining of the parity-check matrix is not always easy. The class of *Progressive Edge Growth* (PEG) LDPC codes (Problem 9.10) can be constructed as well, where the parity-check matrix is formed by a constructive method (it is a structured, and not a random approach with the limitations as for Gallager codes). For such a code Monte Carlo simulation of signal transmission over BSC with AWGN was carried out (Problem 9.10). For the case when sum-product algorithm is applied, the error probability 10^{-6} is achieved for substantially smaller values of parameter E_b/N_0 in respect to the case when bit-flipping algorithm is applied. Some modification of bit-flipping algorithm (e.g. gradient multi-bit flipping—GDFB) allows to increase the code gain from 0.4 to 2.5 dB (for $P_{e,rez} = 10^{-6}$). On the other hand, even a significant simplification of sum-product algorithm (min-sum is just such one) in the considered case does not give a significant performance degradation—coding gain decreases from 6.1 to 5.8 dB. It is just a reason that contemporary decoding algorithms are based on *message-passing* principle, with smaller or greater simplifications.

Problems

Problem 9.1 At the input of a decoder, corresponding to a linear block code (7, 4), a code word r transmitted through the erasure channel is led. It is known that the first four bits are not transmitted correctly (but their values are not known), while it is supposed that the last three bits are received correctly being $y_5 = 1$, $y_6 = 0$, $y_7 = 1$. Find the values of erased (damaged) bits:

(a) If a code used is obtained by shortening the Hamming systematic code (8, 4), its generator is given in Problem 5.4, and a shortening was performed by the elimination of the first information bit;
(b) If the code was used which has the parity-check matrix

$$H = \begin{bmatrix} 1 & 1 & 0 & 1 & 0 & 0 & 0 \\ 0 & 1 & 1 & 0 & 1 & 0 & 0 \\ 0 & 0 & 1 & 1 & 0 & 1 & 0 \\ 0 & 0 & 0 & 1 & 1 & 0 & 1 \end{bmatrix};$$

(c) For both codes draw a Tanner graph and on its basis explain the difference in the decoding complexity of the above codes.

Solution

In this problem it is supposed that the erasure channel is used, where the emitted symbol can be transmitted correctly, or it can be "erased", meaning that there is an indication that the symbol (denoted by E) was not transmitted correctly, but there is not the information do it originate from the emitted zero or one (illustrated in Fig. 9.5). Here, as a difference to Problem 4.8, it is not permitted that an emitted symbol during transmission was inverted, i.e. that the error occurred and that the receiver has not information about that (i.e. the transitions $0 \to 1$ and $1 \to 0$ are not permitted).

The case is considered when code (7, 3) is used and the emitted code word $x = (x_1, x_2, x_3, x_4, x_5, x_6, x_7)$ should be found, if the received word is $y = $ (EEEE101). Although this code can detect two errors and correct only one error in the code word at the BSC (it is a shortened Hamming code with one added parity-check bit), in the following it will be shown that here all four erased symbols can be corrected.

(a) In Problem 5.4 the construction of code (8, 4) is explained, which has the generator matrix

$$G_{(8,4)} = \begin{bmatrix} 1 & 0 & 0 & 0 & | & 1 & 1 & 0 & 1 \\ 0 & 1 & 0 & 0 & | & 1 & 0 & 1 & 1 \\ 0 & 0 & 1 & 0 & | & 0 & 1 & 1 & 1 \\ 0 & 0 & 0 & 1 & | & 1 & 1 & 1 & 0 \end{bmatrix}$$

Code (7, 3), obtained by shortening of the previous code by elimination of the first information bit has the following generator matrix

$$G_1 = \begin{bmatrix} 1 & 0 & 0 & 0 & | & 1 & 1 & 0 & 1 \\ 0 & 1 & 0 & 0 & | & 1 & 0 & 1 & 1 \\ 0 & 0 & 1 & 0 & | & 0 & 1 & 1 & 1 \\ 0 & 0 & 0 & 1 & | & 1 & 1 & 1 & 0 \end{bmatrix} = \begin{bmatrix} 1 & 0 & 0 & 1 & 0 & 1 & 1 \\ 0 & 1 & 0 & 0 & 1 & 1 & 1 \\ 0 & 0 & 1 & 1 & 1 & 1 & 0 \end{bmatrix},$$

Fig. 9.5 BEC (Binary Erasure Channel) graph

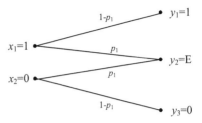

and the corresponding parity-check matrix is

$$H_1 = \begin{bmatrix} 1 & 0 & 1 & 1 & 0 & 0 & 0 \\ 0 & 1 & 1 & 0 & 1 & 0 & 0 \\ 1 & 1 & 1 & 0 & 0 & 1 & 0 \\ 1 & 1 & 0 & 0 & 0 & 0 & 1 \end{bmatrix}.$$

The code word must fulfill the condition $xH_1^T = 0$, i.e. in a developed form

$$x_1 \oplus x_3 \oplus x_4 = 0$$
$$x_2 \oplus x_3 \oplus x_5 = 0$$
$$x_1 \oplus x_2 \oplus x_3 \oplus x_6 = 0$$
$$x_1 \oplus x_2 \oplus x_7 = 0$$

By supposing that the last three symbols are transmitted correctly, then $x_5 = y_5 = 1$, $x_6 = y_6 = 0$, $x_7 = y_7 = 1$ and a system of four equations with four unknown variables is obtained

$$x_1 \oplus x_3 \oplus x_4 = 0$$
$$x_2 \oplus x_3 = 1$$
$$x_1 \oplus x_2 \oplus x_3 = 0$$
$$x_1 \oplus x_2 = 1$$

the solutions are $x_1 = 1$, $x_2 = 0$, $x_3 = 1$ and $x_4 = 0$, the reconstructed word is $x = (1010101)$.

For any linear block code, on the parity-check matrix basis, a system with $n-k$ equations can be written, and in such way it is possible to reconstruct no more than $n-k$ erased bits. It should be noticed that for the solving of this system some method of linear algebra should be applied (e.g. the substitution method).

(b) For the code (7, 3) which has the parity-check matrix

$$H_2 = \begin{bmatrix} 1 & 1 & 0 & 1 & 0 & 0 & 0 \\ 0 & 1 & 1 & 0 & 1 & 0 & 0 \\ 0 & 0 & 1 & 1 & 0 & 1 & 0 \\ 0 & 0 & 0 & 1 & 1 & 0 & 1 \end{bmatrix},$$

the corresponding equation system on the basis of the relation $xH_2^T = 0$ is

$$x_1 \oplus x_2 \oplus x_4 = 0, \; x_2 \oplus x_3 \oplus x_5 = 0, \; x_3 \oplus x_4 \oplus x_6 = 0, \; x_4 \oplus x_5 \oplus x_7 = 0.$$

This last three symbols are correctly transmitted ($x_5 = y_5 = 1$, $x_6 = y_6 = 0$, $x_7 = y_7 = 1$), and above equations make possible a successive code word bits reconstruction

$$\begin{aligned} x_4 \oplus x_5 \oplus x_7 = 0 &\Rightarrow x_4 = x_5 \oplus x_7 = 0, \\ x_3 \oplus x_4 \oplus x_6 = 0 &\Rightarrow x_3 = x_4 \oplus x_6 = 0, \\ x_2 \oplus x_3 \oplus x_5 = 0 &\Rightarrow x_2 = x_3 \oplus x_5 = 1, \\ x_1 \oplus x_2 \oplus x_4 = 0 &\Rightarrow x_1 = x_2 \oplus x_4 = 1, \end{aligned}$$

and the reconstructed code word is $x = (1100101)$. The proposed decoding method is in fact an *iterative* method—in the first step the first three equations (every considered for itself) have not a solution, while from the last one x_4 is found. In the next step, the first two equations have not a solution, but the third one has it. In the third step x_2 is found using the second equation and the previously found values for x_3 and x_4. In the last step the first code word bit is reconstructed.

The following could be noticed:

- Generally, for a code (n, k), if from every equation one variable can be found, the corresponding decoder complexity is linear, i.e. $O((n-k))$.
- In the first case the number of binary ones in the rows is three or four (on the average $\rho_1 = 13/4$) and a number of ones in the columns varies from one to three, but on the average it is $\gamma_1 = 13/7$, and a code rate can be found from the relation $R_1 = 1 - \gamma_1/\rho_1 = 1 - 4/7 = 3/7$.
- In the second case a number of ones in every row is fixed ($\rho_2 = 3$), but a number of ones in matrix columns is not the same (from one to three, on the average $\gamma_2 = 12/7$), and the code rate is here the same as well $R_2 = 1 - \gamma_2/\rho_2 = 3/7$.
- The matrix density is relatively high for both cases ($r_1 = \rho_1/n = 0.4643$, $r_2 = \rho_2/n = 0.4286$), both codes are irregular.
- An optimum method for code which has parity-check matrix H_2 decoding is iterative—it is optimal to decode firstly the fourth bit, then the third, the second, and at the end the first one. In the same time a decoding procedure has a linear complexity!
- To provide for *iterative decoding with a linear complexity*, it is fundamental to achieve that a parity-check matrix allows the decoding of one bit from one equation in every iteration. If a matrix has a small number of ones in every row,

it will provide (with some good luck concerning the error sequence!) that in every row there is no more than one bit with error.
- Therefore, a necessary condition is that a parity-check matrix is a *sparse matrix*, the codes fulfilling this condition are *Low Density Parity Check* (LDPC) *codes* [83].

(c) Tanner graph [84] is a bipartite graph to visualize the connections between the two types of nodes, the *variable nodes*, usually denoted by v_j, representing the code word emitted symbols and the *(parity-check nodes)*, denoted by c_i, representing the parity-check equations.

Tanner graph for a code having parity-check matrix H_1 is shown in Fig. 9.6, with especially marked cycle having a length four (4-girth-cycle), corresponding to the variable nodes v_1 and v_2, i.e. to the check nodes c_1 and c_2 (there is one more cycle of the same length, the reader should find it).

It can be noticed that just this cycle makes impossible to decode the symbols v_1 and v_2 directly, from the last four equation of the system

$$c_1: \quad v_1 \oplus v_3 \oplus v_4 = 0$$
$$c_2: \quad v_2 \oplus v_3 \oplus v_5 = 0$$
$$c_3: \quad v_1 \oplus v_2 \oplus v_3 \oplus v_6 = 0$$
$$c_4: \quad v_1 \oplus v_2 \oplus v_7 = 0$$

At the *Tanner bipartite graph*, corresponding to matrix H_2, shown in Fig. 9.7, it can be noticed that the cycle of the length four does not exist. It just makes possible to decode successively a code word bits with the linear complexity. The shortest cycle at this graph has the length six and it is drawn by heavy lines. Therefore, to provide the iterative decoding with the linear complexity, the parity-check matrix graph should not have the short length cycles!

Problem 9.2 Parity-check matrix of one linear block code is

$$H = \begin{bmatrix} 1 & 1 & 1 & 1 & 0 & 0 & 0 & 0 & 0 \\ 1 & 0 & 0 & 0 & 1 & 1 & 1 & 0 & 0 \\ 0 & 1 & 0 & 0 & 1 & 0 & 0 & 1 & 1 & 0 \\ 0 & 0 & 1 & 0 & 0 & 1 & 0 & 1 & 0 & 1 \\ 0 & 0 & 0 & 1 & 0 & 0 & 1 & 0 & 1 & 1 \end{bmatrix}.$$

(a) Find the parity-check matrix density and verify whether the code is regular,
(b) Draw the Tanner graph corresponding to the code and find the minimum cycle length,
(c) Decode the received word

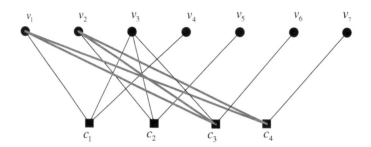

Fig. 9.6 Tanner graph corresponding to the first considered code

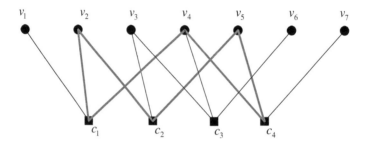

Fig. 9.7 Tanner graph corresponding to the second considered code

$$y = (0001000000)$$

using majority logic decision.

Solution

(a) The number of ones in every column is $\gamma = 2$, and in every row is $\rho = 4$, the code is **regular**. Matrix density is $r = 2/4 = 0.5$ and code rate $R = 1-\gamma/\rho = 1/2$. Parity-check matrix has $n = 10$ columns and $n-k = 5$ rows, it is the code $(n, k) = (10, 5)$.

(b) The corresponding Tanner graph is shown in Fig. 9.8. In the parity-check matrix is not possible to find a rectangle at which corners are binary ones and it is obvious that in the graph there are no cycles of the length four. One cycle of a length six is shown by heavy lines. The cycle length must be an even number and it is obvious that it is a **minimum cycle length** for this graph.

(c) **Majority logic decoding** (majority voting) starts by forming a set of equations orthogonal to a single code word bit (similarly to the procedure for Reed-Muller codes decoding):

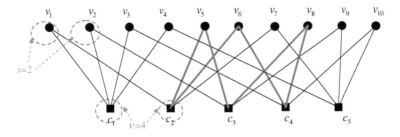

Fig. 9.8 Tanner graph of the code (10, 5)

1. equations set orthogonal to v_1:

$$A_1: c_1: v_1 \oplus v_2 \oplus v_3 \oplus v_4 = 0, \quad c_2: v_1 \oplus v_5 \oplus v_6 \oplus v_7 = 0,$$

2. equations set orthogonal to v_2:

$$A_2: c_1: v_1 \oplus v_2 \oplus v_3 \oplus v_4 = 0, \quad c_3: v_2 \oplus v_5 \oplus v_8 \oplus v_9 = 0,$$

3. equations set orthogonal to v_3:

$$A_3: c_1: v_1 \oplus v_2 \oplus v_3 \oplus v_4 = 0, \quad c_4: v_3 \oplus v_6 \oplus v_8 \oplus v_{10} = 0,$$

4. equations set orthogonal to v_4:

$$A_4: c_1: v_1 \oplus v_2 \oplus v_3 \oplus v_4 = 0, \quad c_5: v_4 \oplus v_7 \oplus v_9 \oplus v_{10} = 0,$$

5. equations set orthogonal to v_5:

$$A_5: c_2: v_1 \oplus v_5 \oplus v_6 \oplus v_7 = 0, \quad c_3: v_2 \oplus v_5 \oplus v_8 \oplus v_9 = 0,$$

6. equations set orthogonal to v_6:

$$A_6: c_2: v_1 \oplus v_5 \oplus v_6 \oplus v_7 = 0, \quad c_4: v_3 \oplus v_6 \oplus v_8 \oplus v_{10} = 0,$$

7. equations set orthogonal to v_7:

$$A_7: c_2: v_1 \oplus v_5 \oplus v_6 \oplus v_7 = 0, \quad c_5: v_4 \oplus v_7 \oplus v_9 \oplus v_{10} = 0,$$

8. equations set orthogonal to v_8:

$$A_8: c_3: v_2 \oplus v_5 \oplus v_8 \oplus v_9 = 0, \quad c_4: v_3 \oplus v_6 \oplus v_8 \oplus v_{10} = 0,$$

9. equations set orthogonal to v_9:

$$A_9 : c_3 : v_2 \oplus v_5 \oplus v_8 \oplus v_9 = 0, \quad c_5 : v_4 \oplus v_7 \oplus v_9 \oplus v_{10} = 0,$$

10. equations set orthogonal to v_{10}:

$$A_{10} : c_4 : v_3 \oplus v_6 \oplus v_8 \oplus v_{10} = 0, \quad c_5 : v_4 \oplus v_7 \oplus v_9 \oplus v_{10} = 0.$$

If at the decoder input is the sequence $\mathbf{y} = (0001000000)$, the syndrome can be found

$$\mathbf{S} = \mathbf{y}\mathbf{H}^T = (10001),$$

and syndrome bits determine the parity-check sums, while the parity-check values for single bits of a code word, corresponding to the sets A_1–A_{10} provide for the code word bits reconstruction

$$A_1 : (1, 0) \to v_1 = ?, \quad A_2 : (1, 0) \to v_2 = ?, \quad A_3 : (1, 0) \to v_3 = ?,$$
$$A_4 : (1, 1) \to v_4 = 1, \quad A_5 : (0, 0) \to v_5 = 0, \, A_6 : (0, 0) \to v_6 = ?,$$
$$A_7 : (0, 1) \to v_7 = ?, \quad A_8 : (0, 0) \to v_8 = 0,$$
$$A_9 : (0, 1) \to v_9 = ?, \quad A_{10} : (0, 1) \to v_{10} = ?$$

From the above it is reliably determined that at the 5th, 6th and the 8th code word bits there were no errors, and at the 4th bit almost surely the error occurred, while for the rest of the bits the reliable decision cannot be made. As a difference from the example in introductory part of this chapter, in this case the number of parity-check sums is even (it is not desirable) and small as well, i.e. only two parity-check sums are formed for every code word bit. Supposing that no more than one error occurred at the code word, the reconstructed code word is

$$x' = (0000000000).$$

The decoding procedure would be highly more efficient if the number of ones is greater per column and odd if possible. But, on the other hand, the matrix density would grow yielding probably the occurrence of the shorter cycles as well, i.e. the code quality would be degraded. Therefore, a majority decoding is not a procedure providing for an efficient LDPC codes decoding, i.e. it is relatively successful only for codes which have very long code words and a relatively dense parity-check matrix.

Problem 9.3 In a communication system Gallager code (20, 7) is used, the parity-check matrix given in in introductory part of this chapter (it is repeated here for convenience!). By bit-flipping algorithm decode the following received words

(a) $y = (11000000000000000000)$, (b) $y = (00011000000000011000)$.

Solution
Bit-flipping algorithm is an iterative procedure using as inputs the received bits and a parity-check matrix structure. For Gallager code (20, 7) it is

$$H = \begin{bmatrix} 1 & 1 & 1 & 1 & 0 & 0 & 0 & 0 & 0 & 0 & 0 & 0 & 0 & 0 & 0 & 0 & 0 & 0 & 0 & 0 \\ 0 & 0 & 0 & 0 & 1 & 1 & 1 & 1 & 0 & 0 & 0 & 0 & 0 & 0 & 0 & 0 & 0 & 0 & 0 & 0 \\ 0 & 0 & 0 & 0 & 0 & 0 & 0 & 0 & 1 & 1 & 1 & 1 & 0 & 0 & 0 & 0 & 0 & 0 & 0 & 0 \\ 0 & 0 & 0 & 0 & 0 & 0 & 0 & 0 & 0 & 0 & 0 & 1 & 1 & 1 & 1 & 0 & 0 & 0 & 0 \\ 0 & 0 & 0 & 0 & 0 & 0 & 0 & 0 & 0 & 0 & 0 & 0 & 0 & 0 & 0 & 1 & 1 & 1 & 1 \\ \hline 1 & 0 & 0 & 0 & 1 & 0 & 0 & 0 & 1 & 0 & 0 & 0 & 1 & 0 & 0 & 0 & 0 & 0 & 0 & 0 \\ 0 & 1 & 0 & 0 & 0 & 1 & 0 & 0 & 0 & 1 & 0 & 0 & 0 & 0 & 0 & 1 & 0 & 0 & 0 & 0 \\ 0 & 0 & 1 & 0 & 0 & 0 & 1 & 0 & 0 & 0 & 0 & 0 & 1 & 0 & 0 & 0 & 1 & 0 & 0 & 0 \\ 0 & 0 & 0 & 1 & 0 & 0 & 0 & 0 & 0 & 1 & 0 & 0 & 0 & 1 & 0 & 0 & 0 & 0 & 1 & 0 \\ 0 & 0 & 0 & 0 & 0 & 0 & 0 & 1 & 0 & 0 & 0 & 1 & 0 & 0 & 0 & 1 & 0 & 0 & 0 & 1 \\ \hline 1 & 0 & 0 & 0 & 0 & 1 & 0 & 0 & 0 & 0 & 1 & 0 & 0 & 0 & 0 & 0 & 1 & 0 & 0 \\ 0 & 1 & 0 & 0 & 0 & 0 & 1 & 0 & 0 & 0 & 1 & 0 & 0 & 0 & 1 & 0 & 0 & 0 & 0 \\ 0 & 0 & 1 & 0 & 0 & 0 & 0 & 1 & 0 & 0 & 0 & 1 & 0 & 0 & 0 & 0 & 0 & 1 & 0 \\ 0 & 0 & 0 & 1 & 0 & 0 & 0 & 0 & 1 & 0 & 0 & 0 & 1 & 0 & 0 & 1 & 0 & 0 & 0 \\ 0 & 0 & 0 & 0 & 1 & 0 & 0 & 0 & 0 & 1 & 0 & 0 & 0 & 0 & 1 & 0 & 0 & 0 & 0 & 1 \end{bmatrix}.$$

(a) For the received word $y = (11000000000000000000)$ the syndrome is $yH^T = (000001100011000)$ showing that the following sums are not satisfied:

- VI control sum (where are 1, 5, 9 and 13 received word bits)
- VII control sum (where are 2, 6, 10 and 17 received word bits)
- XI control sum (where are 1, 6, 12 and 18 received word bits)
- XII control sum (where are 2, 7, 11 and 16 received word bits)

In the above control sums bits 1, 2 and 6 appear the most frequently (two times) and the received word bits at these places are inverted yielding the word $y' = (00000100000000000000)$, to which the syndrome $y'H^T = (010000100010000)$ corresponds. Now, the II, VI and XI parity-check rows are considered where one appears only in the sixth column in every of these rows. By inverting this bit, the second estimation of the code word y'' is obtained consisting of zeros only, yielding the following $y''H^T = (000000000000000)$, and the vector y'' is a decoded code word.

(b) For the received word $y = (00011000000000011000)$ the syndrome is $yH^T = (110111101101001)$, where every syndrome element corresponds to the sum of corresponding elements in 4, 5, 16 and 17 column of parity-check matrix, denoted by dot-line-dot in Fig. 9.9. The syndrome shows that the parity-check sums I, II, IV, V, VI, VII, IX, X, XII and XV are not satisfied, corresponding to the shadowed rows of the parity-check matrix in the Fig. 9.9. If only these rows are considered, columns denoted by ordinal numbers 2, 5, 15, 16 and 20 contain the highest number (three) of ones.

After the first correction (carried out by inverting bits 2, 5, 15, 16 and 20) the first estimation of the codeword is found $y' = (01010000000000101001)$. To this code word (having ones at the positions 2, 4, 15, 17 and 20, determining the columns of interest in Fig. 9.10) corresponds the syndrome $y'H^T = (000100000101000)$, and in IV, X and XII control sums the 16th bit the most frequently appears (three times). It can be easily seen from the shadowed rows in Fig. 9.10. Therefore, in the second iteration only this bit is inverted and second code word estimation is formed as $y'' = (01010000000000111001)$.

Because $r''H^T = (000000000000000)$, i.e. the syndrome equals zero, the last estimation is a valid code word. Therefore, the result of decoding is

$$x = y'' = (0\underline{1}01\underline{0}000000000\underline{1}1100\underline{1}),$$

where the bits inverted in the received word are underlined and in bold.

Of course, it may not be the transmitted code word. For example, it was possible that the transmitted word is an all zeros word and the four bits were inverted (by errors) yielding the received word r. Hamming distance of these two words is four, while the distance of the received word to the decoded word r'' is four as well. Therefore, it can be said that a decoded word is one from the nearest to the received word in a Hamming distance sense, but is not necessarily the right solution.

Problem 9.4 LDPC code (10, 5) is defined with the parity-check matrix from the formulation of Problem 9.2.

(a) Write the parity-check equations and explain the parallel realization of bit-flipping algorithm with the fixed threshold $T = 2$,
(b) By using the explained algorithm decode the following received words

$$y_1 = (0000100000), \quad y_2 = (1000100000).$$

Solution
As it was shown in Problem 9.2, the corresponding parity-check equations are:

$$c_1 = v_1 \oplus v_2 \oplus v_3 \oplus v_4, \quad c_2 = v_1 \oplus v_5 \oplus v_6 \oplus v_7, \quad c_3 = v_2 \oplus v_5 \oplus v_8 \oplus v_9,$$
$$c_4 = v_3 \oplus v_6 \oplus v_8 \oplus v_{10}, \quad c_5 = v_4 \oplus v_7 \oplus v_9 \oplus v_{10}.$$

and all these parity-checks are satisfied (the result is equal to zero) if vector v is a valid codeword.

Problems

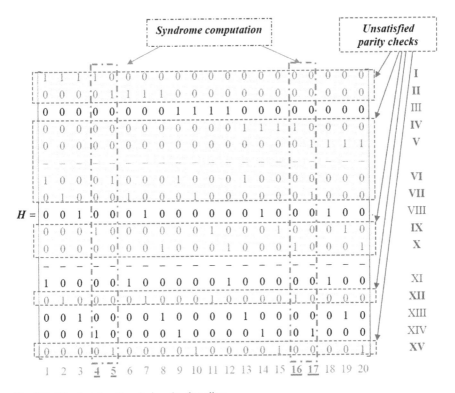

Fig. 9.9 The first iteration during the decoding

For the updating of the jth bit in the received word, it is important to identify the parity-check equations that include this bit. Let $c_i^{(j)}$ denote the ith parity-check equation containing the jth bit in the received word, which is satisfied only if the corresponding bipartite graph (shown in Fig. 9.8) has an edge that connects the jth variable node and the ith check node.

For regular LDPC codes, there are exactly ρ inputs into every check node (in this case $\rho = 4$), and it is easy to understand that the operation in check nodes can be realized by using ρ-input exclusive or (XOR) gates. On the other hand, there are exactly γ parity-checks $c_i^{(j)}$ that are important to decide whether the jth bit will be flipped or not (in this case $\gamma = 2$). If the number of unsatisfied parity-checks is greater or equal than the predefined threshold $T \leq \gamma$, the jth bit will be inverted (flipped) in the current iteration

$$D_j = \begin{cases} 1, & \sum h_i^{(j)} \geq T, \\ 0, & \sum h_i^{(j)} < T. \end{cases}$$

Therefore, operations in variable nodes can be realized by using ν-input majority logic (ML) gates with the decision threshold T. The updated bit is then calculated as

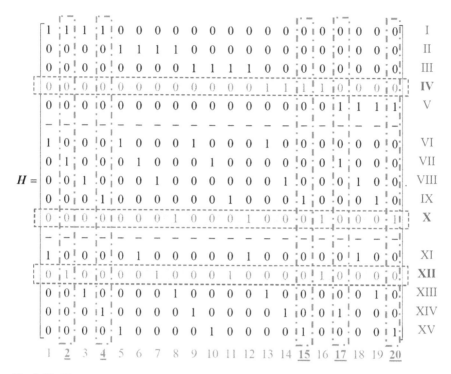

Fig. 9.10 The second iteration during the decoding

$v_j^* = v_j \oplus D_j$. In the parallel decoder realization, every symbol ($j = 1, 2, \ldots, n$) is updated at the same time, in parallel.

According to the formulation of the problem, in this case it is assumed that the threshold has the fixed value for all symbol nodes in all iterations, i.e. $T = 2$. The corresponding hardware realization (on the logic gate level) is given in Fig. 9.11, where the updating of the fifth bit is illustrated.

(b) When the received word is $y_1 = (0000100000)$ only the fifth received bit is not equal to zero, and therefore only the second and the third parity-check are unsatisfied. As it is shown in Table 9.1, two parity-checks are satisfied only for the fifth symbol of the received word and this bit is flipped. It is easy to verify that vector with all zeros represents the valid codeword and the decoding process is finished.

If the threshold is set to the typical value $T = \lceil \gamma/2 \rceil = 1$, it is easy to see that seven bits should be flipped. It is recommended to reader to check do the obtained word after the first iteration is a codeword in this case.

If the received word is $y_2 = (1000100000)$, the decoding procedure is illustrated in Table 9.2. As decoder structure is the same, only the values of the parity-checks are changed. As only the first and the fifth bit in the received word have non-zero values, only the first and the third parity-check are unsatisfied. If the threshold is $T = 2$, only the second bit in the received bit should be flipped and the estimated word after the first iteration is $\mathbf{y_2'} = (1100100000)$. For this word, all parity-checks are satisfied (h_1–h_3 as a summation of two binary ones, c_4 and c_5 as a summation of all zeros), i.e. it is the valid codeword and the decoding process is finished.

Now two cases will be considered when (1000100000) can be received:

- codeword (1100100000) is transmitted, the channel introduced the error at the second bit and the decoding process was successful
- codeword (0000000000) is transmitted, and the channel introduced the errors at the first and the fifth bit. In such case, the decoder produced one additional error at the second position of the estimated vector and the decoding process was unsuccessful.

Now, consider why the errors at positions 1 and 5 cannot be corrected if the codeword all-zeros is transmitted. The Tanner graph for the case when r_2 is received is given in Fig. 9.12. One can see that errors are located inside the cycle with length 6. As this code is girth-6 code, the errors are located within the cycle with the minimum length that is not desirable.

As the zero-valued variable nodes do not have any impact to the parity-checks, it is enough to analyze the impact of variable nodes v_1, v_5, check nodes connected to them (c_1, c_2, c_3) and the variable nodes that are connected to these parity-checks at least T times (could be flipped in some scenario). In this case, besides v_1 and v_5, only v_2 is connected to c_1 and c_3, and the other variable nodes are connected only to one check node from set $\{c_1, c_2, c_3\}$. The structure consisted of these nodes in bipartite graph is presented in Fig. 9.12, where full circles represents the non-zero

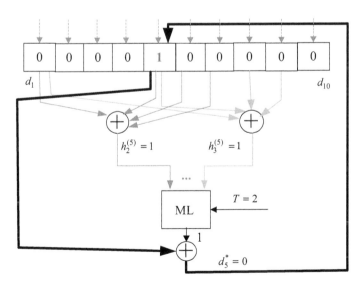

Fig. 9.11 Parallel bit flipping, updating of the fifth bit

Table 9.1 The decoding process during the first iteration, received word y_1

j	1	2	3	4	5	6	7	8	9	10
$c_i^{(j)}$	$c_1^{(1)}=0$ $c_2^{(1)}=1$	$c_1^{(2)}=0$ $c_3^{(2)}=1$	$c_1^{(3)}=0$ $c_4^{(3)}=0$	$c_1^{(4)}=0$ $c_5^{(4)}=0$	$c_2^{(5)}=1$ $c_3^{(5)}=1$	$c_2^{(6)}=1$ $c_4^{(6)}=0$	$c_2^{(7)}=1$ $c_5^{(7)}=0$	$c_3^{(8)}=1$ $c_4^{(8)}=0$	$c_3^{(9)}=1$ $c_5^{(9)}=0$	$c_4^{(10)}=0$ $c_5^{(10)}=0$
$\sum c_i^{(j)}$	1	1	0	0	2	1	1	1	1	0
D_j	0	0	0	0	1	0	0	0	0	0
v_j^*	0	0	0	0	0	0	0	0	0	0

Table 9.2 The decoding process during the first iteration, received word y_2

j	1	2	3	4	5	6	7	8	9	10
$c_i^{(j)}$	$c_1^{(1)}=1$ $c_2^{(1)}=0$	$c_1^{(2)}=1$ $c_3^{(2)}=1$	$c_1^{(3)}=1$ $c_4^{(3)}=0$	$c_1^{(4)}=1$ $c_5^{(4)}=0$	$c_2^{(5)}=0$ $c_3^{(5)}=1$	$c_2^{(6)}=0$ $c_4^{(6)}=0$	$c_2^{(7)}=0$ $c_5^{(7)}=0$	$c_3^{(8)}=1$ $c_4^{(8)}=0$	$c_3^{(9)}=1$ $c_5^{(9)}=0$	$c_4^{(10)}=0$ $c_5^{(10)}=0$
$\sum c_i^{(j)}$	1	2	1	1	1	0	0	1	1	0
D_j	0	1	0	0	0	0	0	0	0	0
v_j^*	1	1	0	0	1	0	0	0	0	0

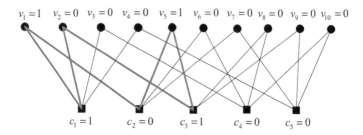

Fig. 9.12 Tanner graph of code (10, 5) for received word r_2

symbols (errors), empty circles zero-symbols, full squares unsatisfied parity-checks and empty squares satisfied parity-checks. It is clear that maximum number of two unsatisfied parity-checks is for v_2, as two full squares are connected to them.

The same structure is presented in Fig. 9.13, after the flipping of the second bit in the received word. In this case all parity-checks are satisfied, and the codeword (1100100000) is decoded.

Finally, as is was shown that at least one weight-three vector satisfy all parity-check equations, it is clear that the minimum Hamming distance of this code is not greater than three. Therefore, it is not possible to construct the decoder that could correct more than one error per codeword. So, it can be concluded that the described bit-flipping decoder is the best possible for the described scenario.

Problem 9.5 LDPC code (10, 5) is defined with the parity-check matrix from the formulation of Problem 9.2.

(a) Shortly explain Gallager-B decoding algorithm,
(b) By using the explained algorithm with fixed threshold $T = \lceil \gamma/2 \rceil$ decode the following code word

$$y = (0000100000).$$

Fig. 9.13 The error patterns (**a**) in the received word, (**b**) in the decoded word

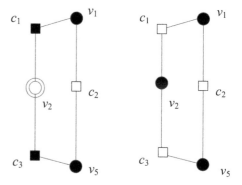

Solution

(a) The Gallager-B algorithm represents the simplest possible *message passing* algorithm, where the information is transmitted across the edges of the bipartite graph, from variable nodes to the check nodes and vice versa. Let assume that the variable nodes that are connected to one check node are neighbor variable nodes, and the check nodes connected to one variable node are neighbor parity nodes as well. In message passing algorithms:

- variable nodes send *reports* about their current status to the check nodes,
- by using the reports from the neighbor variable nodes, check nodes make an *estimation* of the value of the current variable node (in the next iteration, the current state of variable node is not of direct interest, it is more important how their neighbors see it!)
- by using majority of the estimations from the neighbor check nodes, new value of the variable node is obtained.

Gallager-B algorithm represents iterative decoding procedure operating in a binary field where, during the iteration, binary messages are sent along the edges of Tanner graph. Let $E(x)$ represent a set of edges incident on a node x (x can be either symbol or parity-check node). Let $m_i(e)$ and $m'_i(e)$ denote the messages sent on edge e from variable node to check node and check node to variable node at iteration i, respectively. If the initial value of a bit at symbol node v is denoted as $r(v)$, the Gallager-B algorithm can be summarized as follows:

Initialization ($i = 1$): For each variable node v, and each set $E(v)$, messages sent to check nodes are computed as follows

$$m_1(e) = r(v).$$

Step (i) (*check node update*): For each check node c and each set $E(c)$, the update rule for the ith iteration, $i > 1$, is defined as follows

$$m'_i(e) = \left(\sum_{e' \in E(c) \setminus \{e\}} m_{i-1}(e) \right) \bmod 2.$$

Step (ii) (*variable node update*): For each variable node v and each set $E(v)$, the update rule for the ith iteration, $i > 1$, is defined as follows

$$m_i(e) = \begin{cases} 1, & \text{if } \sum_{e' \in E(v) \setminus \{e\}} m'_i(e) \geq \lceil \gamma/2 \rceil \\ 0, & \text{if } \sum_{e' \in E(v) \setminus \{e\}} m'_i(e) \leq \gamma - 1 - \lceil \gamma/2 \rceil, \\ r(v), & \text{otherwise.} \end{cases}$$

(b) Decoding of the fifth bit will be explained step by step:

(1) *Initialization*

For variable node v_5 incident edges are $v_5 \to c_2$ and $v_5 \to c_3$ and messages that are initially sent are:

$$m_1(v_5 \to c_2) = m_1(v_5 \to c_3) = 1,$$

while the messages sent over all other edges are equal to binary zero (as the other bits in received vector are equal to zero), as shown in Fig. 9.14.

(2) *Step (i): update of check nodes* (illustrated in Fig. 9.15):

- Update for check node c_2

$$m'_2(c_2 \to v_5) = (m_1(v_1 \to c_2) + m_1(v_6 \to c_2) + m_1(v_7 \to c_2)) \bmod 2$$
$$= (0 + 0 + 0) \bmod 2 = 0.$$

- Update for check node c_3

$$m'_2(c_3 \to v_5) = (m_1(v_2 \to c_3) + m_1(v_8 \to c_3) + m_1(v_9 \to c_3)) \bmod 2$$
$$= (0 + 0 + 0) \bmod 2 = 0.$$

(c) *Step (ii): variable node update* (illustrated in Fig. 9.16):

Under the assumption that all parity-checks are satisfied, new estimation of the observed variable node is equal to message that is sent from the check node.

As different check nodes can make the different estimations, the final estimation in the current iteration can be obtained by a majority voting. On the other hand, to the edge on the graph to the ith parity-check node, the majority voting of the estimations from the other check nodes is sent, excluding the ith node.

For variable node $j = 5$ and set $E(12) = \{c_2 \to v_5, c_3 \to v_5\}$ one obtains the estimations according to Fig. 9.16:

1. For the edge $v_5 \to c_2$ one obtains

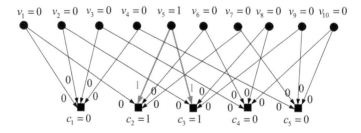

Fig. 9.14 Initialization, *red bold lines* corresponds to send binary ones

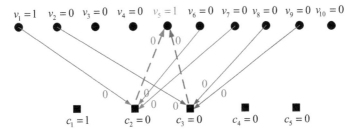

Fig. 9.15 Update of check nodes, *red lines* correspond c_2, *blue* to c_3, *solid lines* report from variable nodes to check nodes, *bold dashed lines* report the estimations of symbol v_{12} by using parity-checks

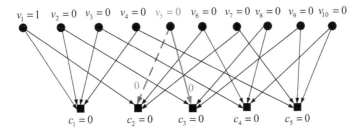

Fig. 9.16 Update of check nodes, *red lines* correspond c_2, *blue* to c_3, *solid lines* report from variable nodes to check nodes, *bold dashed lines* report the estimations of symbol v_{12} by using parity-checks

$$\sum_{e' \in E(5) \setminus \{v_5 \to c_2\}} m'_2(e') = m'_2(c_3 \to v_5) = 0 \Rightarrow m_2(v_5 \to c_2) = 0$$

2. For the edge $v_5 \to c_3$ one obtains

$$\sum_{e' \in E(5) \setminus \{v_5 \to c_3\}} m'_2(e') = m'_2(c_2 \to v_5) = 0 \Rightarrow m_2(v_5 \to c_3) = 0$$

Therefore, the updated word is $\mathbf{y}' = (0000100000)$ and it can be easily checked that $\mathbf{y}'$ is valid codeword as $\mathbf{y}'\mathbf{H}^T = 0$. Therefore, the decoding procedure is finished.

Problem 9.6 LDPC code parity-check matrix is

$$H = \begin{bmatrix} 0 & 1 & 0 & 1 & 1 & 0 & 0 & 1 \\ 1 & 1 & 1 & 0 & 0 & 1 & 0 & 0 \\ 0 & 0 & 1 & 0 & 0 & 1 & 1 & 1 \\ 1 & 0 & 0 & 1 & 1 & 0 & 1 & 0 \end{bmatrix}.$$

(a) Find the code basic parameters and draw the corresponding Tanner graph.
(b) Find all code words corresponding to this matrix and find the code weight spectrum.
(c) If the code words are transmitted by the unit power polar pulses, find the samples at the decoder input if the corresponding sample values of additive Gaussian noise are

$$n = (-1.1, 0.2, -0.3, -1.5, 0, 1, -0.4, -0.15, -0.25).$$

(d) Supposing the channel noise variance $\sigma^2 = 0.49$, decode the words found above by using *sum-product* algorithm.

Solution

(a) Number of ones in every column here is $s = 2$, and number of ones in every row is $\rho = 4$. It is a **regular LDPC** code, matrix density is $r = 2/4 = 0.5$ and the code rate $R = 1 - \gamma/\rho = 1/2$. The matrix has $n = 8$ columns and $n-k = 4$ rows, the code basic parameters are $(n, k) = (8, 4)$.

Tanner graph of the code is shown in Fig. 9.17. It can be noticed that there are no cycles of minimum length. A rule for its forming is shown as well—connections between the second check node and the corresponding variable nodes (the second matrix row) are denoted by heavy lines, while the connections of the seventh variable node and the nodes corresponding to parity-check sums where they appear (the seventh matrix column) are denoted by dashed lines.

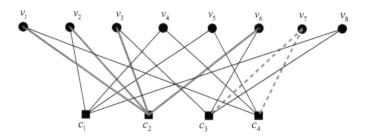

Fig. 9.17 Tanner graph for code (8, 4)

Problems

$$H = \begin{bmatrix} 0 & 1 & 0 & 1 & 1 & 0 & 0 & 1 \\ 1 & 1 & 1 & 0 & 0 & 1 & 0 & 0 \\ 0 & 0 & 1 & 0 & 0 & 1 & 1 & 1 \\ 1 & 0 & 0 & 1 & 1 & 0 & 1 & 0 \end{bmatrix} \begin{matrix} 1 \\ 2 \\ 3 \\ 4 \end{matrix}$$
$$\begin{matrix} 1 & 2 & 3 & 4 & 5 & 6 & 7 & 8 \end{matrix}$$

(b) On the basis of the parity-check matrix, vector v (eight bits) is a code vector if the following is satisfied

$$c_1 : v_2 \oplus v_4 \oplus v_5 \oplus v_8 = 0,$$
$$c_2 : v_1 \oplus v_2 \oplus v_3 \oplus v_6 = 0,$$
$$c_3 : v_3 \oplus v_6 \oplus v_7 \oplus v_8 = 0,$$
$$c_4 : v_1 \oplus v_4 \oplus v_5 \oplus v_7 = 0.$$

From these four linearly independent equations a maximum of four unknown variables can be determined, all code words can be found if the values for four bits are fixed, while the other four are obtained by solving this system. For example, if bits v_3, v_4, v_5, v_6, are fixed, the other bits are found by solving the following system of equations

$$\begin{matrix} v_2 \oplus v_8 = v_4 \oplus v_5, \\ v_1 \oplus v_2 = v_3 \oplus v_6, \\ v_7 \oplus v_8 = v_3 \oplus v_6, \\ v_1 \oplus v_7 = v_4 \oplus v_5. \end{matrix} \Rightarrow (v_1, v_2, v_7, v_8) \otimes \begin{bmatrix} 0 & 1 & 0 & 1 \\ 1 & 1 & 0 & 0 \\ 0 & 0 & 1 & 1 \\ 1 & 0 & 1 & 0 \end{bmatrix}$$
$$= (v_4 \oplus v_5, v_3 \oplus v_6, v_3 \oplus v_6, v_4 \oplus v_5),$$

but this system has not a unique solution in spite that the matrix H rank is 4!

Not until five bits are fixed, e.g. v_1, v_2, v_3, v_4, v_5, the others bits can be found from relations

$$c_2 : v_6 = v_1 \oplus v_2 \oplus v_3,$$
$$c_4 : v_7 = v_1 \oplus v_4 \oplus v_5,$$
$$c_1 : v_8 = v_2 \oplus v_4 \oplus v_5.$$

and it is clear that one possible generator matrix is

$$G = \begin{bmatrix} 1 & 0 & 0 & 0 & 0 & 1 & 1 & 0 \\ 0 & 1 & 0 & 0 & 0 & 1 & 0 & 1 \\ 0 & 0 & 1 & 0 & 0 & 1 & 0 & 0 \\ 0 & 0 & 0 & 1 & 0 & 0 & 1 & 1 \\ 0 & 0 & 0 & 0 & 1 & 0 & 1 & 1 \end{bmatrix}$$

The matrix has a full rank, it is the code (8, 5) and the corresponding code words are

(00000000), (01000101), (10000110), (11000011)
(00001011), (01001110), (10001101), (11001000)
(00010011), (01010110), (10010101), (11010000)
(00011000), (01011101), (10011110), (11011011)
(00100100), (01100001), (10100010), (11100111)
(00101111), (01101010), (10101001), (11101100)
(00110111), (01110010), (10110001), (11110100)
(00111100), (01111001), (10111010), (11111111)

The code spectrum is

$a(0) = 1, a(2) = 2, a(3) = 8, a(4) = 10, a(5) = 8, a(6) = 2, a(8) = 1, a(1) = a(7) = 0$,

minimum Hamming distance is $d_{min} = 2$ and it cannot be guaranteed even that this code will always could correct one error in a code word. However, further it will be shown that in some situations, using a suitable decoding procedure, it could correct more!

(c) Consider the transmission of code word $x = (11111111)$. This vector is emitted as a polar pulse train $x^* = (+1, +1, +1, +1, +1, +1, +1, +1)$. The transmission is through the channel with AWGN which has a known samples values, and at its output the vector $y^* = (-0.1, 1.2, 0.7, -0.5, 1.1, 0.6, 0.85, 0.75)$ is obtained.

As shown in Fig. 9.18 this signal can be decoded using various procedures—after hard decision majority decoding can be applied or iterative decoding on the *bit-flipping* algorithm basis. The received word in this case would be $y = (01101111)$, where the errors occurred at the positions 1 and 4. It is recommended to the reader to find the corresponding result if the iterative decoding with hard decisions is applied.

In the following the procedure for decoding of the received word using *sum-product* algorithm (SPA), known as well as *belief-propagation* algorithm, proposed

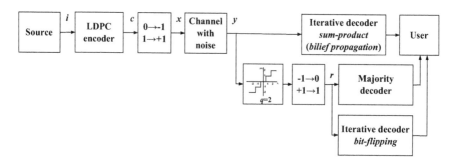

Fig. 9.18 Block scheme of a transmission system using LDPC codes

Problems

firstly in [90] and rediscovered in [93] will be described. It is supposed the AWGN in channel has a standard deviation $\sigma = 0.7$, the coefficients f_j^c are calculated firstly on the basis of probability density functions according to the transmitted signal values

$$f_j^0 = \frac{1}{\sqrt{2\pi}\sigma} e^{\frac{-(y_j+1)^2}{2\sigma^2}}, \quad f_j^1 = \frac{1}{\sqrt{2\pi}\sigma} e^{\frac{-(y_j-1)^2}{2\sigma^2}}$$

the corresponding numerical values are given in Table 9.3. In such a way the initial metrics are found showing the likelihood that single samples originate from the one of the two possible emitted symbols.

Further, two sets of coefficients are calculated sent from the variable to the check nodes in the first iteration, Q_{ij}^0 and Q_{ij}^1. As presented in Fig. 9.19, the initial values of these coefficients are found from

$$Q_{ij}^c = \begin{cases} f_j^c, & H(i,j) = 1, \\ 0, & H(i,j) = 0. \end{cases}$$

After the check node c_i receives from all connected variable nodes v_j the probability values Q_{ij}^c (for $c = 0$ and $c = 1$), the probabilities are calculated that ith parity-check equation is satisfied, if the variable node has the value c, using the relation

$$R_{ij}^c = \sum_{d_j=c} P(c_i/v) \prod_{k \in N(i)/j} Q_{ik}^{d_k}$$

where $N(i)$ is a set of parent (variable) nodes indexes connected to the check node c_i, and $N(i)/j$ is the same set, but without the node v_j to which the information is sent.

The term $P(c_i/v)$ denotes the probability that jth parity-check equation is satisfied, and summing is carried out over all possible decoded vectors v for which the parity-check equation is satisfied when the informed variable node is in the state c. Therefore, the coefficient R_{23}^1 is the estimation that the check node 2 sends to the variable node 2 (shown in Fig. 9.20) and it is calculated under the condition that the corresponding parity-check equation $v_2 \oplus v_4 \oplus v_5 \oplus v_7 = 0$ is satisfied and when the second variable node is in the state $v_2 = 1$, being

Table 9.3 The received vector values and the corresponding coefficients f_j^x

j	1	2	3	4	5	6	7	8
y_j	−0.1	1.2	0.3	−0.3	1.1	0.6	0.85	0.75
f_j^0	0.2494	0.0041	0.1016	0.3457	0.0063	0.0418	0.0173	0.0250
f_j^1	0.1658	0.5471	0.3457	0.1016	0.5641	0.4841	0.5570	0.5347

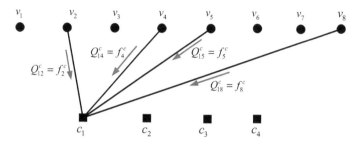

Fig. 9.19 Initial information transmission from the variable nodes to the check nodes

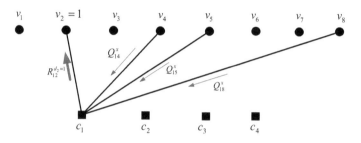

Fig. 9.20 Transmission of initial information from check node to variable nodes

$$R_{12}^1 = Q_{14}^0 Q_{15}^0 Q_{17}^1 + Q_{14}^0 Q_{15}^1 Q_{17}^0 + Q_{14}^1 Q_{15}^0 Q_{17}^0 + Q_{14}^1 Q_{15}^1 Q_{17}^1,$$

because for $v_2 = 1$, the combinations (v_4, v_5, v_7): $(0, 0, 1)$, $(0, 1, 0)$, $(1, 0, 0)$, $(1, 1, 1)$ satisfy the parity-check.

Values R_{ij}^0 and R_{ij}^1 make possible to find the first estimation of the received vector.

$$\left.\begin{array}{l} v_1^0 = f_1^0 R_{21}^0 R_{41}^0 = 7.25 \times 10^{-4} \\ v_1^1 = f_1^1 R_{21}^1 R_{41}^1 = 5.49 \times 10^{-4} \end{array}\right\} \Rightarrow v_1 = 0$$

$$\left.\begin{array}{l} v_2^0 = f_2^0 R_{12}^0 R_{22}^0 = 3.79 \times 10^{-5} \\ v_2^1 = f_2^1 R_{12}^1 R_{22}^1 = 6.98 \times 10^{-4} \end{array}\right\} \Rightarrow v_2 = 1$$

$$\left.\begin{array}{l} v_3^0 = f_3^0 R_{23}^0 R_{33}^0 = 4.97 \times 10^{-5} \\ v_3^1 = f_3^1 R_{23}^1 R_{33}^1 = 3.79 \times 10^{-3} \end{array}\right\} \Rightarrow v_3 = 1$$

$$\left.\begin{array}{l} v_4^0 = f_4^0 R_{14}^1 R_{44}^1 = 3.85 \times 10^{-4} \\ v_4^1 = f_4^1 R_{14}^1 R_{44}^1 = 5.25 \times 10^{-4} \end{array}\right\} \Rightarrow v_4 = 1$$

$$\left.\begin{array}{l} v_5^0 = f_5^0 R_{15}^0 R_{45}^0 = 4.19 \times 10^{-5} \\ v_5^1 = f_5^1 R_{15}^1 R_{45}^1 = 9.15 \times 10^{-4} \end{array}\right\} \Rightarrow v_5 = 1$$

$$\left.\begin{array}{l} v_6^0 = f_6^0 R_{26}^0 R_{36}^0 = 6.49 \times 10^{-5} \\ v_6^1 = f_6^0 R_{26}^1 R_{36}^1 = 3.90 \times 10^{-4} \end{array}\right\} \Rightarrow v_6 = 1$$

$$\left.\begin{array}{l} v_7^0 = f_7^0 R_{37}^0 R_{47}^0 = 2.23 \times 10^{-5} \\ v_7^1 = f_7^1 R_{37}^1 R_{47}^1 = 5.16 \times 10^{-3} \end{array}\right\} \Rightarrow v_7 = 1$$

$$\left.\begin{array}{l} v_8^0 = f_8^0 R_{18}^0 R_{38}^0 = 8.39 \times 10^{-5} \\ v_8^1 = f_8^1 R_{18}^1 R_{38}^1 = 1.53 \times 10^{-3} \end{array}\right\} \Rightarrow v_8 = 1$$

By forming the first estimation of the codeword $v' = (01111111)$ the first iteration is ended. Because this vector does not satisfy the relation $v' \otimes H^T = 0$, the procedure should be continued (as a sent word is $x = (1111110)$, it is obvious that

Problems

there is still one more error at the first bit, but the decoder, of course, does not "know" it!).

The next iteration starts by calculating coefficients Q_{ij}^0 and Q_{ij}^1. The variable node v_j from connected check nodes receives the probabilities R_{ij}^c values (for $c = 0$ and $c = 1$), the probability is corrected that this node is in the state c

$$Q_{ij}^c = \alpha_{ij} f_j^c \prod_{k \in M(j)/i} R_{kj}^c, \quad \alpha_{ij} = \left(\sum_c f_j^c \prod_{k \in M(j)/i} R_{kj}^c \right)^{-1}$$

where α_{ij} is a normalization constant and $M(j)/i$ is the set of indexes of check nodes connected to the node v_j, but without the check node c_i to which the information is sent.

On the basis of Fig. 9.21 it is obvious that except the check node to which the considered variable node sends the information, only one more node is connected to it. For the variable node $j = 1$, it can be written [93]

$$Q_{21}^0 = \alpha_{21} f_1^0 R_{41}^0, \quad Q_{21}^1 = \alpha_{21} f_1^1 R_{41}^1, \quad \alpha_{21} = (f_1^0 R_{41}^0 + f_1^1 R_{41}^1)^{-1},$$
$$Q_{41}^0 = \alpha_{41} f_1^0 R_{21}^0, \quad Q_{21}^1 = \alpha_{41} f_1^1 R_{21}^1, \quad \alpha_{41} = (f_1^0 R_{21}^0 + f_1^1 R_{21}^1)^{-1},$$

while, e.g. for the sixth variable node, the corresponding equations are

$$Q_{26}^0 = \alpha_{26} f_1^0 R_{36}^0, \quad Q_{26}^1 = \alpha_{26} f_1^1 R_{36}^1, \quad \alpha_{21} = (f_1^0 R_{36}^0 + f_1^1 R_{36}^1)^{-1},$$
$$Q_{36}^0 = \alpha_{36} f_1^0 R_{26}^0, \quad Q_{36}^1 = \alpha_{36} f_1^1 R_{26}^1, \quad \alpha_{36} = (f_1^0 R_{26}^0 + f_1^1 R_{26}^1)^{-1},$$

After that, the calculating procedure for coefficients R_{ij}^0 and R_{ij}^1 is repeated, as described earlier.

Numerical values of coefficients Q_{ij}^0, Q_{ij}^1, R_{ij}^0 and R_{ij}^1, for all combinations of check and variable nodes, are given in Tables 9.4 9.5 9.6 9.7 9.8 and 9.9, respectively.

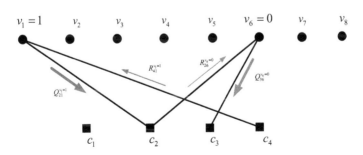

Fig. 9.21 Information transmission from variable nodes to the check nodes, start of the next iteration

Table 9.4 Coefficients R_{ij}^0 values in the first iteration

i/j	1	2	3	4	5	6	7	8
1	0	0.1343	0	0.0108	0.1301	0	0	0.1366
2	0.0208	0.0691	0.0702	0	0	0.0740	0	0
3	0	0	0.0237	0	0	0.0210	0.0257	0.0245
4	0.1396	0	0	0.0806	0.0508	0	0.0501	0

Table 9.5 Coefficients R_{ij}^1 values in the first iteration

i/j	1	2	3	4	5	6	7	8
1	0	0.0250	0	0.1652	0.0238	0	0	0.0203
2	0.1385	0.0510	0.0501	0	0	0.0518	0	0
3	0	0	0.1453	0	0	0.1558	0.1361	0.1415
4	0.0239	0	0	0.0554	0.0682	0	0.0680	0

Table 9.6 Values of coefficients Q_{ij}^0 in the second iteration

i/j	1	2	3	4	5	6	7	8
1	0	0.01	0	0.918	0.0083	0	0	0.0081
2	0.8977	0.0385	0.0093	0	0	0.0082	0	0
3	0	0	0.0744	0	0	0.1099	0.0224	0.2401
4	0.1845	0	0	0.335	0.0578	0	0.0058	0

Table 9.7 Values of coefficients Q_{ij}^1 in the second iteration

i/j	1	2	3	4	5	6	7	8
1	0	0.99	0	0.082	0.9917	0	0	0.9919
2	0.1023	0.9615	0.9907	0	0	0.9885	0	0
3	0	0	0.9256	0	0	0.8901	0.9776	0.7599
4	0.8155	0	0	0.6650	0.9422	0	0.9942	0

Table 9.8 Values of coefficients R_{ij}^0 in the second iteration

i/j	1	2	3	4	5	6	7	8
1	0	0.9044	0	0.0259	0.9030	0	0	0.9028
2	0.0574	0.8814	0.8587	0	0	0.8603	0	0
3	0	0	0.3063	0	0	0.2887	0.3274	0.1829
4	0.3558	0	0	0.2242	0.3971	0	0.4079	0

Table 9.9 Values of coefficients R_{ij}^1 in the second iteration

i/j	1	2	3	4	5	6	7	8
1	0	0.0956	0	0.9741	0.0970	0	0	0.0972
2	0.9426	0.1186	0.1413	0	0	0.1397	0	0
3	0	0	0.6937	0	0	0.7113	0.6726	0.8171
4	0.6442	0	0	0.7758	0.6029	0	0.5921	0

The second estimation of the received vector is found from

$$\left.\begin{array}{l}5.10\times 10^{-3}\\ 1.01\times 10^{-1}\end{array}\right\}\Rightarrow v_1=1,\quad \left.\begin{array}{l}3.25\times 10^{-3}\\ 6.2\times 10^{-3}\end{array}\right\}\Rightarrow v_2=1,\quad \left.\begin{array}{l}7.85\times 10^{-3}\\ 5.10\times 10^{-2}\end{array}\right\}\Rightarrow v_3=1,\quad \left.\begin{array}{l}2.57\times 10^{-3}\\ 4.34\times 10^{-2}\end{array}\right\}\Rightarrow v_4=1$$

$$\left.\begin{array}{l}2.27\times 10^{-3}\\ 3.3\times 10^{-2}\end{array}\right\}\Rightarrow v_5=1,\quad \left.\begin{array}{l}1.04\times 10^{-2}\\ 4.81\times 10^{-2}\end{array}\right\}\Rightarrow v_6=1,\quad \left.\begin{array}{l}2.32\times 10^{-3}\\ 2.22\times 10^{-1}\end{array}\right\}\Rightarrow v_7=1,\quad \left.\begin{array}{l}4.13\times 10^{-3}\\ 4.25\times 10^{-2}\end{array}\right\}\Rightarrow v_8=1$$

and $v''=(11111111)$, this vector satisfies the relation $v''\otimes H^T=0$, a valid code word is reconstructed and the decoding procedure is ended. It should be noticed that the node for which the metrics are recalculated (it reports or is informed) does not take part in the corresponding estimation calculation, providing that a procedure converge to the correct solution. Algorithm is simpler if the number of ones in every row is smaller, what is one of the reasons that it is desirable for the LDPC parity-check matrix to have a small density.

Furthermore, during the decision whether the zero or one is decoded, the relative ratio of values $d_{j,0}$ i $d_{j,1}$ is only important, and these values can be presented in a normalized form. These normalized values are the estimation of likelihood for a decision made after the iteration having the ordinal number *iter*, which, similarly as for turbo codes decoding can be denoted as

$$L_j^{0,(iter)}=\frac{v_j^0}{v_j^0+v_j^1},\quad L_j^{1,(iter)}=\frac{v_j^1}{v_j^0+v_j^1},\quad j=1,2,\ldots,n$$

corresponding to the probability that in the *j*th symbol interval the bit 0/1 is emitted, when the received word is known. It is clear that a priori probabilities are 0.5 (if it is not said that the symbols are not equiprobable) and initial estimation of likelihoods is

$$L_j^{0,(0)}=\frac{f_j^0}{f_j^0+f_j^1},\quad L_j^{0,(0)}(v_j)=\frac{f_j^1}{f_j^0+f_j^1},\quad j=1,2,\ldots,n$$

corresponding to the estimation before the decoding started (after the zeroth iteration), i.e. to reliability estimation if only hard decision of the receiving bits was made, without any processing by the decoder.

In Table 9.10 is given how the reliability estimations for single symbols change during the iterative decoding (shadowed fields in the table correspond to wrongly reconstructed bits):

Table 9.10 Reliability estimations for symbols during the iterations for $\sigma = 0.7$

Iter	$L_j^{s,(iter)}/j$	1	2	3	4	5	6	7	8
0	$L_j^{0,(0)}$	0.6006	0.0074	0.0543	0.8850	0.0111	0.0795	0.0302	0.0447
	$L_j^{1,(0)}$	0.3994	0.9926	0.9457	0.1150	0.9889	0.9205	0.9698	0.9553
1	$L_j^{0,(1)}$	0.5690	0.0514	0.0130	0.4228	0.0438	0.0164	0.0043	0.0520
	$L_j^{1,(1)}$	0.4310	0.9486	0.9870	0.5772	0.9562	0.9836	0.9957	0.9480
2	$L_j^{0,(2)}$	0.0482	0.3440	0.1335	0.0559	0.0644	0.1776	0.0103	0.0887
	$L_j^{1,(2)}$	0.9518	0.6560	0.8665	0.9441	0.9356	0.8224	0.9897	0.9113
3	$L_j^{0,(3)}$	0.0584	0.0135	0.0377	0.1053	0.0068	0.0488	0.0105	0.0661
	$L_j^{1,(3)}$	0.9416	0.9865	0.9623	0.8947	0.9932	0.9512	0.9895	0.9339
5	$L_j^{0,(5)}$	0.0190	0.0123	0.0072	0.0280	0.0103	0.0097	0.0024	0.0083
	$L_j^{1,(5)}$	0.9810	0.9877	0.9928	0.9720	0.9897	0.9903	0.9976	0.9917
10	$L_j^{0,(10)}$	0.0008	0.0003	0.0003	0.0008	0.0002	0.0004	0.0001	0.0004
	$L_j^{1,(10)}$	0.9992	0.9997	0.9997	0.9992	0.9998	0.9996	0.9999	0.9996

- before the decoding, the estimation is wrong at the first and the fourth symbol, but these estimations are highly unreliable, while the values for the other bits are very reliable,
- after the first iteration the error at the fourth bit is corrected, but its estimation is unreliable, while the error at the first bit is not corrected, but this estimation is not reliable as well,
- after the third iteration all errors are corrected and a valid code word is decoded, but the second bit estimation has still a moderate reliability,
- after the third estimation all bits are estimated with the reliability higher than 90% and the reliability further grows with the increasing iteration number and after the tenth iteration it is greater than 99.9%!

It is shown that the decoding can be stopped always in that iteration where a valid code word is reconstructed, because the algorithm must converge to a unique solution and the estimation reliability only grows with the number of iterations. Besides, the reliability estimation of the obtained result depends on the estimated channel signal-to-noise ratio (f_j^c depends on y_j and σ^2), and not only on the received word. In Table 9.11 the values of these estimations are given before the decoding and during the first and the second iteration. The reader should draw the corresponding conclusion on the basis of the obtained results.

Problem 9.7 Communication system uses for error control coding LDPC code described by the parity-check matrix from the previous problem. For transmission the unit power polar pulses are used and the signal at the channel output is

Table 9.11 Reliability estimations for symbols during the iterations for $\sigma = 1$

Iter	$L_j^{\cdot,(Iter)}/j$	1	2	3	4	5	6	7	8
0	$L_j^{0,(0)}$	0.5498	0.0832	0.1978	0.7311	0.0998	0.2315	0.1545	0.1824
	$L_j^{1,(0)}$	0.4502	0.9168	0.8022	0.2689	0.9002	0.7685	0.8455	0.8176
1	$L_j^{0,(1)}$	0.5419	0.1351	0.1429	0.5512	0.1463	0.1621	0.1005	0.2110
	$L_j^{1,(1)}$	0.4581	0.8649	0.8571	0.4488	0.8537	0.8379	0.8995	0.7890
2	$L_j^{0,(2)}$	0.3638	0.1976	0.2102	0.4262	0.1765	0.2403	0.1275	0.2292
	$L_j^{1,(2)}$	0.6362	0.8024	0.7899	0.5738	0.8235	0.7597	0.8725	0.7708

$$\mathbf{y} = (-0.1, 1.2, 0.7, -0.5, 1.1, 0.6, 0.85, 0.75).$$

Decode the signal by applying a logarithmic version of *sum-product* algorithm if the channel noise is AWGN with variance $\sigma^2 = 0.49$ for two following cases:

(a) Channel output is continuous,
(b) Before the decoder input, the decision block is inserted which has a zero threshold.

Solution
The same code and the same sequence at the channel output are considered as in the previous problem, but the decoding procedure differs, i.e. here the realization of *sum-product* algorithm in **logarithmic domain** is considered. The motives to use a logarithmic domain are the same as for turbo codes decoding—avoiding the products calculation always when it is possible (i.e. their change by summations) and lessening the overflow effect.

In this algorithm version on the basis of initial estimations of a posteriori probabilities $P(x_n = +1/y_n)$ and $P(x_n = -1/y_n)$ for every symbol a priori logarithmic likelihood ratio is defined

$$\Lambda_j^{(0)} = \log(P(x_n = +1/y_n)/P(x_n = -1/y_n)).$$

Similarly, a logarithmic likelihood ratio can be defined corresponding to information sent by a variable nodes to the connected check node Q_{ij}^c (corresponding to the probability that this variable node is in c state, on the basis of the variable node estimation itself), and information sent by the check node to the adjoined variable node j, R_{ij}^0 (corresponding to the probability that this variable node is in the state c, on the basis of the estimation reporting check node).

$$\alpha_{i,j} = \log(Q_{i,j}^1/Q_{i,j}^0), \quad \beta_{i,j} = \log(R_{i,j}^1/R_{i,j}^0).$$

Sum-product algorithm in logarithmic domain can now be described by the following steps [95]:

1. For all check nodes connected with *j*th variable node is set initially

$$\alpha_{i,j} = \Lambda_j^{(0)}.$$

2. For every iteration the following is calculated

$$\beta_{i,j} = 2\tanh^{-1}\left\{\prod_{k\in N(i)\setminus j} \tanh(\alpha_{i,k}/2)\right\}, \quad \alpha_{i,j} = \Lambda_j^{(0)} + \sum_{k\in M(j)\setminus i} \beta_{k,j},$$

where $\tanh(x)$ denotes hyperbolic tangent. A posteriori information concerning symbols after iteration, denoted by ordinal number *iter*, are

$$\Lambda_j^{(iter)} = \Lambda_j^{(0)} + \sum_{i\in M(j)} \beta_{i,j},$$

and the estimation of a corresponding code word symbol after current iteration is found from $v_j^{(iter)} = \text{sgn}(\Lambda_j^{(0)})$, where $\text{sgn}(x)$ is the sign of the argument x.

3. Procedure from the previous step is repeated iteratively until $vH^T = 0$ is obtained or if a fixed (given in advance) number of iterations is achieved.

In this problem two different cases will be considered, as shown in Fig. 9.22. In the first case the initial estimations are obtained on the "soft inputs" basis, while in the second case they are obtained from the quantized samples (equivalent to the received binary word). As it will be shown, the output in both cases is "soft", i.e. besides the send code word estimation a logarithmic likelihood ratio will be found for every code word symbol, estimated on the channel output sequence basis.

(a) When the estimation is based on the non quantized samples, the initial estimations are formed using the same relations as in a previous problem

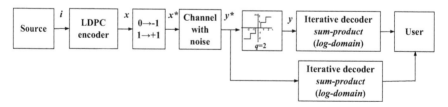

Fig. 9.22 Block scheme of a transmission system using LDPC codes when the decoding is performed using *sum-product* algorithm in a logarithmic domain

$$f_j^0 = \frac{1}{\sqrt{2\pi}\sigma} e^{\frac{-(y_j+1)^2}{2\sigma^2}}, \quad f_j^1 = \frac{1}{\sqrt{2\pi}\sigma} e^{\frac{-(y_j-1)^2}{2\sigma^2}} \Rightarrow \Lambda_j^{(0)} = \log\left(f_j^1/f_j^0\right).$$

The variables $\alpha_{i,j}$, $\beta_{i,j}$ values are further calculated by the above expressions and the numerical values, concluding with the second iteration, are given in Table 9.12.

On the basis of numerical values given in Table 9.13 the same conclusions can be drawn as in the previous problem. It is easy to verify that between the logarithms of likelihood ratio and a posteriori estimations that zero or one were sent (when "non-logarithmic" SPA logarithm version is used) the connection can be established

$$\Lambda_j^{(iter)} = \log(L_{j,iter}^{(1)}/L_{j,iter}^{(0)}),$$

and it is left to a reader to verify whether this relation is satisfied for the numerical values given in Tables 9.10 and 9.13.

(b) When the estimation is based on the quantized samples, system part from the encoder output to the decoder input can be described as a binary symmetric channel where the crossover probability is

$$p = \frac{1}{2}\text{erfc}\left(\frac{1}{\sqrt{2}\sigma}\right) = 0.0766.$$

In this case, the quantizer output sequence is

$$y_q = (-1, +1, +1, -1, +1, +1, +1, +1),$$

and the initial estimations are the same for all "positive" symbols (i.e. for all "negative" symbols) given by

$$\Lambda_j^{(0)} = y_{qj}\log((1-p)/p).$$

Numerical values of variables $\alpha_{i,j}$, $\beta_{i,j}$ are calculated by this relation and given in Table 9.14, while the corresponding likelihood ratios are given in Table 9.15. It can be noticed that a large number of symbols has the same likelihood ratio and that the relation between the likelihood ratios does not changes significantly during a few first iterations.

Already after the first iteration the algorithm reconstructs the word

$$v = (00101111),$$

which satisfies the relation $v\mathbf{H}^T = 0$, i.e. it is a valid code word. However, it is clear that it is not a sent code word (the sent code word x is composed from a binary ones

Table 9.12 Estimations of symbols reliability during the iterations, log-SPA, soft inputs

Iter		i|j	1	2	3	4	5	6	7	8
0	$\Lambda_j^{(0)}$		−0.4082	4.8980	2.8571	−2.0408	4.4898	2.4490	3.4694	3.0612
1	α_{ij}	1	0	4.8980	0	−2.0408	4.4898	0	0	3.0612
		2	−0.4082	4.8980	2.8571	0	0	2.4490	0	0
		3	0	0	2.8571	0	0	2.4490	3.4694	3.0612
		4	−0.4082	0	0	−2.0408	4.4898	0	3.4694	0
	β_{ij}	1	0	−1.6791	0	2.7264	−1.6988	0	0	−1.9089
		2	1.8944	−0.3041	−0.3367	0	0	−0.3573	0	0
		3	0	0	1.8132	0	0	2.0048	1.6678	1.7518
		4	−1.7642	0	0	−0.3742	0.2934	0	0.3055	0
2	α_{ij}	1	0	4.5938	0	−2.4150	4.7832	0	0	4.8130
		2	−2.1724	3.2188	4.6703	0	0	4.4538	0	0
		3	0	0	2.5204	0	0	2.0917	3.7749	1.1523
		4	1.4863	0	0	0.6856	2.7910	0	5.1371	0
	β_{ij}	1	0	−2.2471	0	3.6268	−2.2309	0	0	−2.2287
		2	2.7977	−2.0055	−1.8045	0	0	−1.8180	0	0
		3	0	0	0.8175	0	0	0.9018	0.7200	1.4970
		4	0.5936	0	0	1.2412	0.4175	0	0.3725	0

Problems

Table 9.13 Logarithmic likelihood ratio during the iterations, log-SPA, soft inputs

Iter	$\Lambda_j^{(Iter)}/j$	1	2	3	4	5	6	7	8
0	$\Lambda_j^{(0)}$	−0.4082	4.8980	2.8571	−2.0408	4.4898	2.4490	3.4694	3.0612
1	$\Lambda_j^{(1)}$	−0.2779	2.9147	4.3336	0.3114	3.0844	4.0965	5.4426	2.9041
2	$\Lambda_j^{(2)}$	2.9832	0.6453	1.8701	2.8272	2.6764	1.5328	4.5619	2.3295
3	$\Lambda_j^{(3)}$	2.7796	4.2936	3.2386	2.1395	4.9860	2.9701	4.5433	2.6488
5	$\Lambda_j^{(5)}$	3.9420	4.3898	4.9216	3.5455	4.5636	4.6213	6.0449	4.7795
10	$\Lambda_j^{(10)}$	7.0962	8.2109	8.1903	7.0840	8.3884	7.9194	9.1831	7.7597

only), and besides the two errors occurred in a channel, the decoder introduced an additional error at the second code word bit.

It is interesting to notice that the Hamming distance of the reconstructed code word $x' = (00101111)$ to the received word $y = (01101111)$ equals one, while the distance the received word to the emitted word $x = (11101111)$ equals two. Therefore, the algorithm made the best possible decision on the received word basis and the parity-check matrix structure knowledge.

On the other hand, squared Euclidean distance of sequence $y^* = (-0.1, 1.2, 0.7, -0.5, 1.1, 0.6, 0.85, 0.75)$ to the concurrent emitting sequences $(-1, -1, 1, -1, 1, 1, 1, 1)$ and $(+1, +1, +1, +1, +1, +1, +1, +1)$ respectively is 6.245 i 3.845, and the sequence corresponding to the emitted code word is nearer to the received sequence, explaining the result from the first part of this problem.

Now it is obvious that the decoding error in the second part of solution is not the consequence of the non-optimum decoding algorithm (it is the same in both cases), but of the quantization error. Of course, if the quantization was carried out using a sufficient number of levels, this effect would become negligible.

Problem 9.8 Transmission system uses LDPC error control code described by parity-check matrix given in the previous problems, for transmission the unit power polar pulses are used and the signal at the channel output is

$$y = (-0.1, 1.2, 0.7, -0.5, 1.1, 0.6, 0.85, 0.75),$$

Decode the signal if the channel noise is AWGN with variance $\sigma^2 = 0.49$, when the signal is led directly to the decoder input by applying the following:

(a) *Min-sum* algorithm,
(b) *Min-sum* algorithm supposing that the initial LLR-s are determined by signal samples,
(c) Self-correcting *min-sum* algorithm supposing that the initial LLR-s are determined by signal samples.

Table 9.14 Estimations of symbols reliability during the iterations, log-SPA, hard inputs

Iter		i/j	1	2	3	4	5	6	7	8
0	L_j		−2.49	2.49	2.49	−2.49	2.49	2.49	2.49	2.49
1	α_{ij}	1	0	2.49	0	−2.49	2.49	0	0	2.49
		2	−2.49	2.49	2.49	0	0	2.49	0	0
		3	0	0	2.49	0	0	2.49	2.49	2.49
		4	−2.49	0	0	−2.49	2.49	0	2.49	0
	β_{ij}	1	0	−1.4095	0	1.4095	−1.4095	0	0	−1.4095
		2	1.4095	−1.4095	−1.4095	0	0	−1.4095	0	0
		3	0	0	1.4095	0	0	1.4095	1.4095	1.4095
		4	−1.4095	0	0	−1.4095	1.4095	0	1.4095	0
2	α_{ij}	1	0	1.0805	0	−3.8995	3.8995	0	0	3.8995
		2	−3.8995	1.0805	3.8995	0	0	3.8995	0	0
		3	0	0	1.0805	0	0	1.0805	3.8995	1.0805
		4	−1.0805	0	0	−1.0805	1.0805	0	3.8995	0
	β_{ij}	1	0	−2.8019	0	0.9814	−0.9814	0	0	−0.9814
		2	0.9814	−2.8019	−0.9814	0	0	−0.9814	0	0
		3	0	0	0.4759	0	0	0.4759	0.2411	0.4759
		4	−0.4759	0	0	−0.4759	0.4759	0	0.2411	0

Table 9.15 Logarithmic likelihood ratio during the iterations, log-SPA, hard inputs

Iter	$\Lambda_j^{(iter)}/j$	1	2	3	4	5	6	7	8
0	$\Lambda_j^{(0)}$	-2.4900	2.4900	2.4900	-2.4900	2.4900	2.4900	2.4900	2.4900
1	$\Lambda_j^{(1)}$	-2.4900	-0.3290	2.4900	-2.4900	2.4900	2.4900	5.3090	2.4900
2	$\Lambda_j^{(2)}$	-1.9844	-3.1139	1.9844	-1.9844	1.9844	1.9844	2.9721	1.9844
3	$\Lambda_j^{(3)}$	-3.4898	-1.2587	3.4898	-3.4898	3.4898	3.4898	3.5515	3.4898
5	$\Lambda_j^{(5)}$	-3.4409	-3.7699	3.4049	-3.4049	3.4049	3.4049	4.3304	3.4049
10	$\Lambda_j^{(10)}$	-5.7482	-4.1346	5.7482	-5.7482	5.7482	5.7482	7.1994	5.7482

Solution

As known from the previous problem, logarithmic version of a *sum-product* algorithm has the substantial advantages in respect to its original version:

- the number of multiplications is decreased in favor the additions
- the overflow probability is decreased
- the number of relations whose values should be found is two times smaller—instead of Q_{ij}^0 and Q_{ij}^1 only $\alpha_{i,j}$ is calculated, and instead of R_{ij}^0 and R_{ij}^1 only $\beta_{i,j}$ is calculated,
- a number of mathematical operations in every step is smaller, there is no need for normalization etc.

However, the algorithm explained in a previous problem has one serious drawback. To find the information which the check node sends to it adjoined variable node j one has to determine

$$\beta_{i,j} = 2 \tan h^{-1} \left\{ \prod_{k \in N(i) \backslash j} \tan h(\alpha_{i,k}/2) \right\},$$

it is obvious that the corresponding numerical value cannot be easily found, and still more serious problems appear when the hardware block should be implemented corresponding to this relation.

That is the reason to use the SPA procedure versions much easier to be implemented:

1. The simplest (and still rather exact) approximation of the above equality is

$$\beta_{i,j} = \left[\prod_{k \in N(i) \backslash j} \text{sgn}(\alpha_{i,k}) \right] \times \left[\min_{k \in N(i) \backslash j} |\alpha_{i,k}| \right],$$

Table 9.16 Estimations of symbols reliability by the check nodes, *min-sum*, soft inputs

Iter		i/j	1	2	3	4	5	6	7	8
0	$\Lambda_j^{(0)}$		-0.4082	4.8980	2.8571	-2.0408	4.4898	2.4490	3.4694	3.0612
1	β_{ij}	1	0	-2.0408	0	3.0612	-2.0408	0	0	-2.0408
		2	2.4490	-0.4082	-0.4082	0	0	-0.4082	0	0
		3	0	0	2.4490	0	0	2.8571	2.4490	2.4490
		4	-2.0408	0	0	-0.4082	0.4082	0	0.4082	0
2	β_{ij}	1	0	-2.4490	0	4.4898	-2.4490	0	0	-2.4490
		2	2.8571	-2.4490	-2.4490	0	0	-2.4490	0	0
		3	0	0	1.0204	0	0	1.0204	1.0204	2.0408
		4	1.0204	0	0	2.0408	1.0204	0	1.0204	0

being the key modification defining the difference between SPA and ***min-sum algorithm*** [97].

2. The modification concerning the information which the variable node i sends to check node j connected to it is possible as well, and the corresponding value is now calculated in two steps

$$a_{i,j} = \Lambda_j^{(0)} + \sum_{k \in M(j) \setminus i} \beta_{k,j}, \quad \alpha_{i,j} = \begin{cases} a_{i,j}, & \text{sgn}(\alpha_{i,j}) = \text{sgn}(a_{i,j}), \\ 0, & \text{sgn}(\alpha_{i,j}) \neq \text{sgn}(a_{i,j}), \end{cases}$$

and this correction defines ***self-correcting min-sum algorithm*** [98].

(a) The modification introduced in *min-sum* algorithm directly influences the coefficient β_{ij} values, these variable values during the first two iterations are given in Table 9.16. Of course, during next iterations, these coefficient values changes will influence at the other relevant parameters, what can be easily followed from a posteriori LLR values, given in Table 9.17.

Although *min-sum* is an approximate version of *sum-product* algorithm, the approximation here allowed the code word decoding already after the first iteration. However, the estimation of the first bit after the first iteration and the estimation of the second bit after the second iteration are extremely unreliable! Because of that, not until the third iteration one can be relatively reliable asserted that a code word was successfully decoded.

(b) The logarithmic likelihood ratios obtained by using self-correcting *min-sum* algorithm after some number of iterations are given in Table 9.18. Firstly, it can be noticed that in this case LLR symbol values grow substantially slower with the number of iterations increase, from which the conclusion could be drawn that the estimations are less reliable. However, in the case of self-correcting *min-sum* algorithm, LLR-s converge substantially faster to the stable values, and in this case, by the rule, a smaller number of iteration is needed to make a decision. On the other hand, the small coefficients values

Table 9.17 Logarithmic likelihood ratio during iterations, *min-sum*, soft inputs

Iter	$\Lambda_j^{(iter)}/j$	1	2	3	4	5	6	7	8
0	$\Lambda_j^{(0)}$	−0.4082	4.8980	2.8571	−2.0408	4.4898	2.4490	3.4694	3.0612
1	$\Lambda_j^{(1)}$	1.1×10^{-16}	2.4490	4.8980	0.6122	2.8571	4.8980	6.3265	3.4694
2	$\Lambda_j^{(2)}$	3.4694	8.9×10^{-16}	1.4286	4.4898	3.0612	1.0204	5.5102	2.6531
3	$\Lambda_j^{(3)}$	4.0816	5.5102	3.4694	2.4490	6.9388	3.4694	5.5102	3.0612
5	$\Lambda_j^{(5)}$	5.5102	4.8980	5.9184	5.3061	5.5102	5.5102	7.5510	6.9388
10	$\Lambda_j^{(10)}$	12.6531	12.0408	12.0408	12.8571	13.2653	12.0408	14.2857	12.0408

Table 9.18 Logarithmic likelihood ratio during iterations, self-correcting *min-sum*, soft inputs

Iter	$\Lambda_j^{(iter)}/j$	1	2	3	4	5	6	7	8
0	$\Lambda_j^{(0)}$	-0,4082	4.8980	2.8571	-2.0408	4.4898	2.4490	3.4694	3.0612
1	$\Lambda_j^{(1)}$	1.1×10^{-16}	2.4490	4.8980	0.6122	2.8571	4.8980	6.3265	3.4694
2	$\Lambda_j^{(2)}$	2.4490	8.9×10^{-16}	1.4286	2.4490	2.0408	1.0204	4.4898	2.6531
3	$\Lambda_j^{(3)}$	2.0408	2.4490	2.4490	0.4082	2.4490	2.4490	3.4694	1.0204
5	$\Lambda_j^{(5)}$	2.4490	2.4490	2.4490	1.0204	2.4490	3.0612	3.4694	1.0204
10	$\Lambda_j^{(10)}$	2.4490	2.4490	2.4490	1.0204	2.4490	3.0612	3.4694	1.0204

provide good protection against the overflow, being of importance in the practical implementation. It is interesting to note that in a case when at the decoder input, the values obtained by a hard decisions were led, this procedure resulted in an exact reconstruction of the sent code sequence (SPA did not succeeded, see the second part of the previous problem).

Finally, the case is considered when for the initial LLR values the signal samples are used, i.e. when

$$\Lambda_j^{(0)} = y_j.$$

Although it is clear that this approach is not optimum and that the LLR for some symbols by a rule will have a substantially smaller values, numerical results (given in Table 9.19) show that the convergence rate in respect to the previously considered case did not changed. Because of that, such an approach is often used for the implementation of simplified *sum-product* algorithm versions.

Table 9.19 Logarithmic likelihood ratio during iterations, self-correcting *min-sum*, input samples are used as the initial LLR values

Iter	$\Lambda_j^{(iter)}/j$	1	2	3	4	5	6	7	8
0	$\Lambda_j^{(0)}$	-0.1	1.2	0.7	-0.5	1.1	0.6	0.85	0.75
1	$\Lambda_j^{(1)}$	-1.1×10^{-16}	0.6	1.2	0.15	0.7	1.2	1.55	0.85
2	$\Lambda_j^{(2)}$	-0.6	-2.2×10^{-16}	0.35	0.6	0.5	0.25	1.1	0.65
3	$\Lambda_j^{(3)}$	0.5	0.6	0.6	0.1	0.6	0.6	0.85	0.25
5	$\Lambda_j^{(5)}$	0.6	0.6	0.6	0.25	0.6	0.75	0.85	0.25
10	$\Lambda_j^{(10)}$	0.6	0.6	0.6	0.25	0.6	0.75	0.85	0.25

Problems

Problem 9.9

(a) Explain the construction of parity-check matrix of a *quasi-cyclic* code where $n-k = 4$, the first circulants is $a_1(x) = 1 + x$ and the second $a_2(x) = 1 + x^2 + x^4$. Draw a corresponding bipartite graph.

(b) Find this code generator matrix starting from the fact that its second circulant is invertible. Write all code words and verify whether the code is cyclic.

(c) If it is not known that a code is quasi-cyclic, find its matrix by parity-check matrix reduction in a systematic form to a reduced standard row-echelon form.

(d) Find a generator matrix by transforming the corresponding parity-check matrix in approximate lower diagonal form.

Solution

(a) Parity-check matrix of *quasi-cyclic* (QC) LDPC code in a general case has a form [99,100]

$$H = [A_1, A_2, \ldots, A_l]$$

where $A_1, A_2, \ldots, A_l$ are binary circulant matrices having dimensions $u \times u$, where $u = n-k$.

In the case when $l = 2$ and $u = 5$, the first row of every **circulant matrix** is determined by the corresponding polynomial coefficients, while the other rows are obtained by their successive cyclic shifts

$$A_1 = \begin{bmatrix} 1 & 1 & 0 & 0 & 0 \\ 0 & 1 & 1 & 0 & 0 \\ 0 & 0 & 1 & 1 & 0 \\ 0 & 0 & 0 & 1 & 1 \\ 1 & 0 & 0 & 0 & 1 \end{bmatrix}, \quad A_2 = \begin{bmatrix} 1 & 0 & 1 & 0 & 1 \\ 1 & 1 & 0 & 1 & 0 \\ 0 & 1 & 1 & 0 & 1 \\ 1 & 0 & 1 & 1 & 0 \\ 0 & 1 & 0 & 1 & 1 \end{bmatrix},$$

and the parity-check matrix of the corresponding LDPC code finally is

$$H = \begin{bmatrix} 1 & 1 & 0 & 0 & 0 & 1 & 0 & 1 & 0 & 1 \\ 0 & 1 & 1 & 0 & 0 & 1 & 1 & 0 & 1 & 0 \\ 0 & 0 & 1 & 1 & 0 & 0 & 1 & 1 & 0 & 1 \\ 0 & 0 & 0 & 1 & 1 & 1 & 0 & 1 & 1 & 0 \\ 1 & 0 & 0 & 0 & 1 & 0 & 1 & 0 & 1 & 1 \end{bmatrix}.$$

Tanner bipartite graph for this code is shown in Fig. 9.23. Although the groups of symbol nodes, corresponding to every circulant do not overlap, it does not guarantee that the small length cycles will be avoided.

(b) It is known that generator matrix of quasi-cyclic code generally has the form

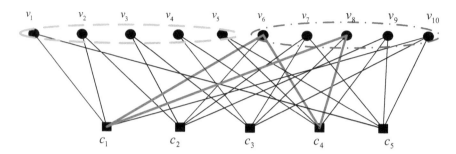

Fig. 9.23 Bipartite quasi-cyclic code graph

$$G = \begin{bmatrix} I_{u(l-1)} & \begin{matrix} (A_l^{-1}A_1)^T \\ (A_l^{-1}A_2)^T \\ \ldots \\ (A_l^{-1}A_{l-1})^T \end{matrix} \end{bmatrix},$$

where matrix A_1 is invertible (it is not necessarily the last right square submatrix in H!). It is obvious that a corresponding code has the code word length $n = ul$, while the information word length is $k = u(l-1)$.

In this example circulant $a_2(x)$ is invertible yielding $a_2^{-1}(x) = x^2 + x^3 + x^4$, and it can be easily verified that the product of corresponding matrices modulo-2 yields the unity matrix

$$A_2^{-1} = \begin{bmatrix} 0 & 0 & 1 & 1 & 1 \\ 1 & 0 & 0 & 1 & 1 \\ 1 & 1 & 0 & 0 & 1 \\ 1 & 1 & 1 & 0 & 0 \\ 0 & 1 & 1 & 1 & 0 \end{bmatrix} \Rightarrow A_2 A_2^{-1} = \begin{bmatrix} 1 & 0 & 0 & 0 & 0 \\ 0 & 1 & 0 & 0 & 0 \\ 0 & 0 & 1 & 0 & 0 \\ 0 & 0 & 0 & 1 & 0 \\ 0 & 0 & 0 & 0 & 1 \end{bmatrix} = I_5$$

and generator matrix becomes

$$G = \begin{bmatrix} I_5 & (A_2^{-1}A_1)^T \end{bmatrix} = \begin{bmatrix} 1 & 0 & 0 & 0 & 0 & 1 & 0 & 0 & 1 & 0 \\ 0 & 1 & 0 & 0 & 0 & 0 & 1 & 0 & 0 & 1 \\ 0 & 0 & 1 & 0 & 0 & 1 & 0 & 1 & 0 & 0 \\ 0 & 0 & 0 & 1 & 0 & 0 & 1 & 0 & 1 & 0 \\ 0 & 0 & 0 & 0 & 1 & 0 & 0 & 1 & 0 & 1 \end{bmatrix}.$$

Code corresponding to this generator matrix is systematic and the first five bits are completely determined by information word at the encoder input. The corresponding transformation is

$$i = (00001) \rightarrow x = (0000100101)$$

From the generator matrix it can be easily noticed that the successive cyclic shifts of this word correspond to information words

$$x' = (1000010010) \rightarrow i' = (10000)$$
$$x'' = (0100001001) \rightarrow i'' = (01000)$$
$$x''' = (1010000100) \rightarrow i''' = (10100)$$

here first two relations are really satisfied (to these code words correspond written information sequences), but the third equality is not satisfied because to information sequence (10100) corresponds code word

$$i''' = (10100) \rightarrow x''' = (1010000110).$$

Because of the fact that to one information sequence cannot correspond two different code words, it is obvious that a code is not cyclic, because at least one cyclic code word shift is not a code word. This code is only quasi-cyclic, because the parity-check bits are given by matrix corresponding to the polynomial

$$\left(a_2^{-1}(x) * a_2^{-1}(x)\right)^T == 1 + x^3.$$

(c) **Standard Parity-check matrix row-echelon form** satisfies the condition that in the neighboring rows of parity-check matrix (being not all zeros rows!) a leading one of the lower row appears in the one column to the right regarding to the upper row. Reduced standard form includes an additional condition—the column comprising a leading one in some row has not ones at the other positions, i.e. the parity-check matrix begins by unit matrix having dimensions $(n - k) \times (n - k)$.

It is obvious that the first four rows of the found parity-check matrix satisfy a standard echelon form, but the fifth row does not satisfy it. In this case it is difficult by using standard operations over the rows (reordering of rows, change of one row by sum modulo-2 of two rows) to form complete matrix satisfying standard echelon form. Therefore, step-by-step, the try will be made to form parity-check matrix in a systematic form. In the first column, one should be only at the first position, the fifth row is changed by the sum of the first and the fifth row, forming the matrix H_1. In the second column, one should be only at the second position, and the first row of matrix H_1 is changed by the sum of the first and the fifth row, and the fifth row is changed by the sum of the second and the fifth row yielding the matrix H_2. Besides, one should be careful not to decrease the matrix rang by these transformations.

$$H = \begin{bmatrix} 1 & 1 & 0 & 0 & 0 & 1 & 0 & 1 & 0 & 1 \\ 0 & 1 & 1 & 0 & 0 & 1 & 1 & 0 & 1 & 0 \\ 0 & 0 & 1 & 1 & 0 & 0 & 1 & 1 & 0 & 1 \\ 0 & 0 & 0 & 1 & 1 & 1 & 0 & 1 & 1 & 0 \\ 1 & 0 & 0 & 0 & 1 & 0 & 1 & 0 & 1 & 1 \end{bmatrix} \to H_1 = \begin{bmatrix} 1 & 1 & 0 & 0 & 0 & 1 & 0 & 1 & 0 & 1 \\ 0 & 1 & 1 & 0 & 0 & 1 & 1 & 0 & 1 & 0 \\ 0 & 0 & 1 & 1 & 0 & 0 & 1 & 1 & 0 & 1 \\ 0 & 0 & 0 & 1 & 1 & 1 & 0 & 1 & 1 & 0 \\ 0 & 1 & 0 & 0 & 1 & 1 & 1 & 1 & 1 & 0 \end{bmatrix},$$

$$H_1 = \begin{bmatrix} 1 & 1 & 0 & 0 & 0 & 1 & 0 & 1 & 0 & 1 \\ 0 & 1 & 1 & 0 & 0 & 1 & 1 & 0 & 1 & 0 \\ 0 & 0 & 1 & 1 & 0 & 0 & 1 & 1 & 0 & 1 \\ 0 & 0 & 0 & 1 & 1 & 1 & 0 & 1 & 1 & 0 \\ 0 & 1 & 0 & 0 & 1 & 1 & 1 & 1 & 1 & 0 \end{bmatrix} \to H_2 = \begin{bmatrix} 1 & 0 & 0 & 0 & 1 & 0 & 1 & 0 & 1 & 1 \\ 0 & 1 & 1 & 0 & 0 & 1 & 1 & 0 & 1 & 0 \\ 0 & 0 & 1 & 1 & 0 & 0 & 1 & 1 & 0 & 1 \\ 0 & 0 & 0 & 1 & 1 & 1 & 0 & 1 & 1 & 0 \\ 0 & 0 & 1 & 0 & 1 & 0 & 0 & 1 & 0 & 0 \end{bmatrix}.$$

In the third column now one should be eliminated in the second row (changing by the sum of the second and fifth row of matrix H_2). Further in the new obtained matrix a similar procedure is applied for elimination of ones in the fourth column—in the third row (sum 3rd and 5th) and after in the fifth row (sum 4th and 5th). As the zero in fifth row and fifth column of matrix H_3 cannot be eliminated and it cannot be written in systematic form, this approach does not yield the results!

$$H_2 = \begin{bmatrix} 1 & 0 & 0 & 0 & 1 & 0 & 1 & 0 & 1 & 1 \\ 0 & 1 & 1 & 0 & 0 & 1 & 1 & 0 & 1 & 0 \\ 0 & 0 & 1 & 1 & 0 & 0 & 1 & 1 & 0 & 1 \\ 0 & 0 & 0 & 1 & 1 & 1 & 0 & 1 & 1 & 0 \\ 0 & 0 & 1 & 0 & 1 & 0 & 0 & 1 & 0 & 0 \end{bmatrix} \to H_3 = \begin{bmatrix} 1 & 0 & 0 & 0 & 1 & 0 & 1 & 0 & 1 & 1 \\ 0 & 1 & 0 & 0 & 1 & 1 & 1 & 1 & 1 & 0 \\ 0 & 0 & 1 & 1 & 0 & 0 & 1 & 1 & 0 & 1 \\ 0 & 0 & 0 & 1 & 1 & 1 & 0 & 1 & 1 & 0 \\ 0 & 0 & 0 & 1 & 1 & 0 & 1 & 0 & 0 & 1 \end{bmatrix},$$

$$H_3 = \begin{bmatrix} 1 & 0 & 0 & 0 & 1 & 0 & 1 & 0 & 1 & 1 \\ 0 & 1 & 0 & 0 & 1 & 1 & 1 & 1 & 1 & 0 \\ 0 & 0 & 1 & 1 & 0 & 0 & 1 & 1 & 0 & 1 \\ 0 & 0 & 0 & 1 & 1 & 1 & 0 & 1 & 1 & 0 \\ 0 & 0 & 0 & 1 & 1 & 0 & 1 & 0 & 0 & 1 \end{bmatrix} \to H_4 = \begin{bmatrix} 1 & 0 & 0 & 0 & 1 & 0 & 1 & 0 & 1 & 1 \\ 0 & 1 & 0 & 0 & 1 & 1 & 1 & 1 & 1 & 0 \\ 0 & 0 & 1 & 0 & 1 & 0 & 0 & 1 & 0 & 0 \\ 0 & 0 & 0 & 1 & 1 & 1 & 0 & 1 & 1 & 0 \\ 0 & 0 & 0 & 0 & 0 & 1 & 1 & 1 & 1 & 1 \end{bmatrix}.$$

Of course, it does not mean that the parity-check matrix cannot be written in a systematic form, but only means that it cannot be obtained by reducing it to reduced standard echelon form. Knowing that a code is quasi-cyclic, from generator matrix it is easily obtained

$$H_s = \begin{bmatrix} A_2^{-1}A_1 & I_5 \end{bmatrix} = \begin{bmatrix} 1 & 0 & 1 & 0 & 0 & 1 & 0 & 0 & 0 & 0 \\ 0 & 1 & 0 & 1 & 0 & 0 & 1 & 0 & 0 & 0 \\ 0 & 0 & 1 & 0 & 1 & 0 & 0 & 1 & 0 & 0 \\ 1 & 0 & 0 & 1 & 0 & 0 & 0 & 0 & 1 & 0 \\ 0 & 1 & 0 & 0 & 1 & 0 & 0 & 0 & 0 & 1 \end{bmatrix}.$$

(d) Parity-check matrix has an ***approximate lower triangular form*** if it is in the form [101]

Problems

$$H = \begin{bmatrix} A & B & T \\ C & D & E \end{bmatrix}$$

where matrix T is square, dimensions $(n-k-g) \times (n-k-g)$, where g denotes *gap* and if this value is smaller, the encoding procedure complexity is smaller. Matrix B then has dimensions $(n-k-g) \times g$, while dimensions of A are $(n-k-g) \times k$.

This matrix further by **Gauss-Jordan** elimination is reduced to a form

$$\tilde{H}_T = \begin{bmatrix} I_{n-k-g} & 0 \\ -ET^{-1} & I_g \end{bmatrix} \otimes \tilde{H}_T = \begin{bmatrix} A & B & T \\ \tilde{C} & \tilde{D} & 0 \end{bmatrix}$$

where the matrix E is "erased", and the following is valid

$$\tilde{C} = -ET^{-1}A + C, \quad \tilde{D} = -ET^{-1}B + D,$$

On the basis of this matrix form, the code word can be obtained in a systematic form, where it has the form

$$x = (i, p_1, p_2),$$

where i is the information word and the other component vectors are formed by using relations

$$p_1 = \tilde{D}^{-1}\tilde{C}i^T, \quad p_2 = -T^{-1}(Ai^T + Bp_1).$$

In this example $n = 10$ and $k = 5$. The matrix firstly should be obtained in a suitable form. By changing the first row by the sum of the first and the third row the matrix H' is obtained, and then by changing the second row of this matrix by sum of the second and the fourth row, the matrix H'' is formed

$$H = \begin{bmatrix} 1 & 1 & 0 & 0 & 0 & 1 & 0 & 1 & 0 & 1 \\ 0 & 1 & 1 & 0 & 0 & 1 & 1 & 0 & 1 & 0 \\ 0 & 0 & 1 & 1 & 0 & 0 & 1 & 1 & 0 & 1 \\ 0 & 0 & 0 & 1 & 1 & 1 & 0 & 1 & 1 & 0 \\ 1 & 0 & 0 & 0 & 1 & 0 & 1 & 0 & 1 & 1 \end{bmatrix} \rightarrow$$

$$H' = \begin{bmatrix} 1 & 1 & 1 & 1 & 0 & 1 & 1 & 0 & 0 & 0 \\ 0 & 1 & 1 & 0 & 0 & 1 & 1 & 0 & 1 & 0 \\ 0 & 0 & 1 & 1 & 0 & 0 & 1 & 1 & 0 & 1 \\ 0 & 0 & 0 & 1 & 1 & 1 & 0 & 1 & 1 & 0 \\ 1 & 0 & 0 & 0 & 1 & 0 & 1 & 0 & 1 & 1 \end{bmatrix}$$

$$\mathbf{H}' = \begin{bmatrix} 1 & 1 & 1 & 1 & 0 & 1 & 1 & 0 & 0 & 0 \\ 0 & 1 & 1 & 0 & 0 & 1 & 1 & 0 & 1 & 0 \\ 0 & 0 & 1 & 1 & 0 & 0 & 1 & 1 & 0 & 1 \\ 0 & 0 & 0 & 1 & 1 & 1 & 0 & 1 & 1 & 0 \\ 1 & 0 & 0 & 0 & 1 & 0 & 1 & 0 & 1 & 1 \end{bmatrix} \rightarrow$$

$$\mathbf{H}'' = \begin{bmatrix} 1 & 1 & 1 & 1 & 0 & 1 & 1 & 0 & 0 & 0 \\ 0 & 1 & 1 & 1 & 1 & 0 & 1 & 1 & 0 & 0 \\ 0 & 0 & 1 & 1 & 0 & 0 & 1 & 1 & 0 & 1 \\ 0 & 0 & 0 & 1 & 1 & 1 & 0 & 1 & 1 & 0 \\ 1 & 0 & 0 & 0 & 1 & 0 & 1 & 0 & 1 & 1 \end{bmatrix}$$

Finally, by reordering the third and the fourth row in the above matrix, the parity-check matrix is formed in an approximate lower triangular form

$$\mathbf{H}_T = \begin{bmatrix} 1 & 1 & 1 & 1 & 0 & | & 1 & | & 1 & 0 & 0 & 0 \\ 0 & 1 & 1 & 1 & 1 & | & 0 & | & 1 & 1 & 0 & 0 \\ 0 & 0 & 0 & 1 & 1 & | & 1 & | & 0 & 1 & 1 & 0 \\ 0 & 0 & 1 & 1 & 0 & | & 0 & | & 1 & 1 & 0 & 1 \\ 1 & 0 & 0 & 0 & 1 & | & 0 & | & 1 & 0 & 1 & 1 \end{bmatrix},$$

where

$$\mathbf{A} = \begin{bmatrix} 1 & 1 & 1 & 1 & 0 \\ 0 & 1 & 1 & 1 & 1 \\ 0 & 0 & 0 & 1 & 1 \\ 0 & 0 & 1 & 1 & 0 \end{bmatrix}, \quad \mathbf{B} = \begin{bmatrix} 1 \\ 0 \\ 1 \\ 0 \end{bmatrix}, \quad \mathbf{T} = \begin{bmatrix} 1 & 0 & 0 & 0 \\ 1 & 1 & 0 & 0 \\ 0 & 1 & 1 & 0 \\ 1 & 1 & 0 & 1 \end{bmatrix},$$

$$\mathbf{C} = \begin{bmatrix} 1 & 0 & 0 & 0 & 1 \end{bmatrix}, \quad \mathbf{D} = [0], \quad \mathbf{E} = \begin{bmatrix} 1 & 0 & 1 & 1 \end{bmatrix}.$$

Here is $n - k = 5$ and it is obvious that $g = 1$, and the desired parity-check matrix form is

$$\tilde{\mathbf{H}}_T = \begin{bmatrix} 1 & 1 & 1 & 1 & 0 & | & 1 & | & 1 & 0 & 0 & 0 \\ 0 & 1 & 1 & 1 & 1 & | & 0 & | & 1 & 1 & 0 & 0 \\ 0 & 0 & 0 & 1 & 1 & | & 1 & | & 0 & 1 & 1 & 0 \\ 0 & 0 & 1 & 1 & 0 & | & 0 & | & 1 & 1 & 0 & 1 \\ 1 & 0 & 1 & 0 & 0 & | & 1 & | & 0 & 0 & 0 & 0 \end{bmatrix},$$

where

$$\tilde{\mathbf{C}} = \begin{bmatrix} 1 & 0 & 1 & 0 & 0 \end{bmatrix}, \quad \tilde{\mathbf{D}} = [1].$$

For two previously considered information sequences now it is obtained

… Problems

$$i = (10000) \to p_1 = (1), p_2 = (0010) \to x = (i, p_1, p_2) = (1000010010),$$
$$i = (10100) \to p_1 = (0), p_2 = (0110) \to x = (i, p_1, p_2) = (1010000110)$$

and it can be noticed that the following is always valid

$$p_1 = i \otimes \left(\tilde{D}^{-1}\tilde{C}^T\right)^T = i \otimes \begin{bmatrix} 1 \\ 0 \\ 1 \\ 0 \\ 0 \end{bmatrix},$$

$$p_2 = i \otimes \left(-T^{-1}(A + B\tilde{D}^{-1}\tilde{C})\right)^T = i \otimes \begin{bmatrix} 0 & 0 & 1 & 0 \\ 1 & 0 & 0 & 1 \\ 0 & 1 & 0 & 0 \\ 1 & 0 & 1 & 1 \\ 0 & 1 & 0 & 0 \end{bmatrix}.$$

and matrices multiplied from the left by information vector just correspond to the sixth, i.e. to 7–10 columns of the generator matrix obtained in the second part of the problem solution using a totally independent way. The procedure described in this part of solution is general (not limited to quasi-cyclic codes only), guaranteeing that the systematic code obtaining even in a case if it is applied by successive vectors p_1 and p_2 calculations, has approximately linear complexity.

Problem 9.10

(a) Explain the calculation of sphere packing bound and using this relation find the probability that the decoding is unsuccessful if the transmission is over BSC, if the crossover probability is $p < 0.14$, and a used code has code rate $R = 1/2$, code word length being $10 \leq n \leq 1000$.

(b) Draw the probability that a code word after transmitting over BSC is decoded unsuccessfully. Give the results for regular PEG (*Progressive Edge Growth*) LDPC code, code word length $n = 200$, if for decoding are applied bit-flipping, min-sum, self-correcting min-sum and sum-product algorithm. Compare the obtained results to the limiting values found in a previous part.

(c) For the same code as above draw the dependence of the error probability after decoding on the crossover probability, if the transmission is through AWGN channel for $E_b/N_0 < 10$ dB. Compare the obtained results to the corresponding Shannon bound and find the code gain for $P_e = 10^{-4}$.

(d) In the case of the channel with AWGN, when there is no a quantizing block, the channel can be considered as having a discrete input (a binary one!) while the output is continuous. For such case use PEG code which has the code word length $n = 256$, code rate $R = 1/2$, and calculate the residual bit error probability (BER) if for decoding are applied bit-flipping, gradient multi-bit flipping (GDBF), min-sum, and sum-product algorithm.

Solution

a) **Sphere packing bound** in general case can be written in a form

$$A_q(n,d) \le \frac{q^n}{\sum_{l=0}^{e_c}\binom{n}{l}(q-1)^l}, \quad e_c = \left\lfloor \frac{d-1}{2} \right\rfloor,$$

where $A_q(n,d)$ is a maximum number of words of code basis q and the minimum Hamming distance is at least d. It is easy to notice that sphere packing bound is only the second name for a Hamming bound (defined in Problem 5.8), and for a case of binary linear block codes $A_q(n,d) = 2^k$ and the above relation reduces to

$$2^k \le \frac{2^n}{\sum_{l=0}^{e_c}\binom{n}{l}} \Rightarrow 2^{n(1-R)} \ge \sum_{l=0}^{e_c}\binom{n}{l}.$$

From the above one can find a number of errors correctable in the optimum case, in the code word of the length n and for the code rate R (because $k = Rn$).

(b) The lower bound for the unsuccessful decoding is

$$P_{e,\min}(p) \le 1 - \sum_{l=0}^{e_c}\binom{n}{l}p^l(1-p)^{n-l},$$

and as to one code word usually corresponds one transmission frame, the above relation in effect gives the *frame error probability* (*rate*) (FER). Besides, the equality is satisfied only if the code is perfect, while in a general case (and for non perfect codes) the exact expression for the probability that at least one error is not corrected depends on the weight code spectrum given in fourth part of the Problem 5.8. The corresponding numerical values are shown in Fig. 9.24.

(c) Firstly, the case will be considered for binary symmetric channel, having crossover probability p. The class of *Progressive Edge Growth* (PEG) LDPC codes is considered, where the parity-check matrix is formed by a constructive method (it is a structured, and not a random approach with the limitations as for Gallager codes) described in [102]. To analyze these codes, a Monte Carlo simulation method was used. Basics of that procedure are shown in Fig. 9.25 —at the coded sequence the uncorrelated sequence is superimposed where the probability of ones equals p and the decoding is performed. The residual bit error probability is estimated by comparing the obtained sequence to the sequence at the encoder input and the corresponding results for the code PEG (200,100) and various decoding algorithms are shown in Fig. 9.26.

Problems

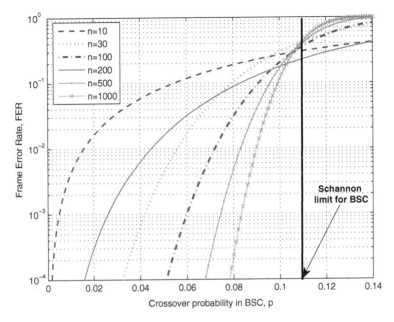

Fig. 9.24 FER in BSC, sphere packing bounds

From the obtained diagram it can be seen that from all the considered algorithms SPA provides the minimum and bit-flipping the maximum error probability. Simplified variants of SPA decoding procedure have the smaller complexity, but the decoder performances are degraded as well. It is obvious that by min-sum algorithm correction these degradations can be lessened substantially. However, even when the optimum decoding algorithm is used, there is a significant difference to the performances foreseen by sphere packing bound, being the consequence of "non perfectness" of the considered PEG code.

(d) For the case of the channel with AWGN, when there is no a quantizing block, the channel can be considered as a channel which has a discrete input (a binary one!) while the output is continuous. In this case the Monte Carlo simulation is

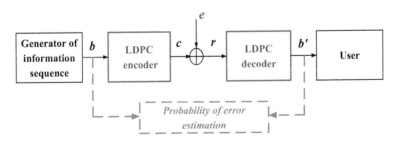

Fig. 9.25 Block scheme of a simulation procedure, BSC

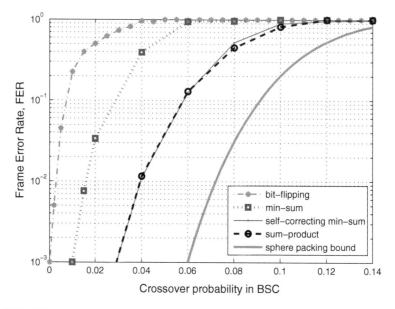

Fig. 9.26 FER in BSC

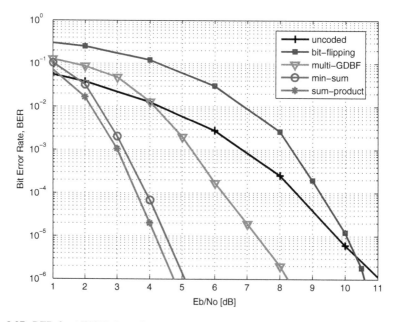

Fig. 9.27 BER for AWGN channel

performed on the modulation channel level and the results are shown in Fig. 9.27 for PEG code having the code word length $n = 256$ and code rate $R = 1/2$. The detailed description of code construction and the simulation procedure for this case are given in [103].

For the case when sum-product algorithm is applied, the error probability 10^{-6} is achieved for the substantially smaller values of parameter E_b/N_0 in respect to the case when bit-flipping algorithm is applied. Some modification of bit-flipping algorithm (e.g. gradient multi-bit flipping—GDBF) allows to increase the code gain from 0.4 to 2.5 dB (for $P_{e,res} = 10^{-6}$). On the other hand, even a significant simplification of sum-product algorithm (min-sum is just such one) in the considered case does not gives a significant performances degradation—coding gain decreases from 6.1 to 5.8 dB. It is just a reason that contemporary decoding algorithms are based mainly on *message-passing* principle, with smaller or greater simplifications.

The Shannon bound for code rate $R = 1/2$ is 0.19 dB [1], it is clear that even SPA algorithm for $P_{e,res} = 10^{-6}$ provides performances being about 4.5 dB worse than that limiting value. Of course, this difference can be substantially decreased if the code word length is longer (luckily, PEG procedure allows the codes construction having practically arbitrary code words length). On the Shannon theory basis, providing the negligible small probability of error for $E_b/N_0 < 0.19$ dB is possible only if the code rate is additionally reduced, and even when $R \to 0$, a reliable transmission can be provided only if $E_b/N_0 > -1, 59$ dB.

References

1. D. Drajić, P. Ivaniš, *Uvod u teoriju informacija i kodovanje (An Introduction into Information Theory and Coding)* (Akademska misao, Beograd, 2009). (in Serbian)
2. T.M. Cover, J.A. Thomas, *Elements of Information Theory* (Wiley, New York, 1991)
3. N. Abramson, *Information Theory and Coding* (McGraw-Hill Book Company, New York, 1963)
4. D.B. Drajić, *Uvod u statističku teoriju telekomunikacija* (Akademska misao, Beograd, 2003)
5. G. Lukatela, D. Drajić, G. Petrović, R. Petrović, Digitalne telekomunikacije, II izd. (Građevinska knjiga, Beograd 1984)
6. I.S. Gradshteyn, I.M. Ryzhik, *Table of Integrals, Series and Products*, 5th edn. (Academic Press Inc, Cambridge, 1994)
7. P.M. Fenwick, Huffman Code Efficiencies for Extensions of Sources. IEEE Trans. Commun. **43**, 163–165 (1995)
8. G. Yule, On a method of investigating periodicities in disturbed series, with special reference to Wolfer's sunspot numbers. Philos. Trans. R. Soc. London Ser. A **226**, 267–298 (1927)
9. S. Haykin, *Adaptive Filter Theory*, 3rd edn. (Prentice Hall, Englewood Cliffs, New Jersey, 1996)
10. C.E. Shannon, A mathematical theory of communication. Bell Syst. Tech. J. **27**, 379–423 (1948). (623–656)
11. R. Fano, *Transmission of Information* (Wiley, New York, 1961)
12. D.A. Huffman, A method for the construction of minimum redundancy codes. Proc. IRE **40**, 1098–1101 (1952)
13. N. Faller, An adaptive system for data compression, in *Record of the 7th Asilomar Conference on Circuits, Systems and Computers* (1973), pp. 593–597
14. R.G. Gallager, Variations on a theme by Huffman. IEEE Trans. Inf. Theor. **24**, 668–674 (1978)
15. D.E. Knuth, Dynamic Huffman coding. J. Algorithms **6**, 163–180 (1985)
16. J.S. Vitter, Dynamic Huffman coding. ACM Trans. Math. Softw. **15**, 158–167 (1989)
17. J. Ziv, A. Lempel, A universal algorithm for sequential data compression. IEEE Trans. Inform. Theor. **IT-23**, 337–343 (1977)
18. J. Ziv, A. Lempel, Compression of individual sequences via variable-rate coding. IEEE Trans. Inform. Theor. **IT-24**, 530–536 (1978)
19. T.A. Welch, A technique for high performance data compression. Computer **17**(6), 8–19 (1984)
20. E.N. Gilbert, Capacity of a burst-noise channel. Bell Syst. Tech. J. **39**, 1253–1266 (1960)
21. E.O. Elliott, Estimates of error rates for codes on burst-noise channels. Bell Syst. Tech. J. **42**, 1977–1998 (1963)

22. H. Nyquist, Certain topics in telegraph transmission theory. Trans. AIEE **47**, 617–644 (1928)
23. R.W. Hamming, Error detecting and error correcting codes. Bell Sys. Tech J. **29**, 147–160 (1950)
24. M.J.E. Golay, notes on digital coding. Proc. IRE **37**, 657 (1949)
25. S. Lin, D.J. Costello Jr., *Error Control Coding—Fundamentals and Applications*, 2nd edn. (Prentice Hall, Englewood Cliffs, NJ, 2004), Prentice Hall, Englewood Cliffs, N.J., 1983
26. R.H. Morelos-Zaragoza, *The Art of Error Correcting Coding* (Wiley, Hoboken, 2002)
27. R.E. Blahut, *Theory and Practice of Error Control Codes* (Addison-Wesley Publishing Company, Reading, Massachusetts, 1983)
28. I.S. Reed, A class of multiple-error-correcting codes and the decoding scheme. IRE Trans. Inform. Theor. **IT-4**, 38–49 (1954)
29. D.E. Muller, Application of Boolean algebra to switching circuit design and to error detection. IRE Trans. Electr. Comp. **EC-3**, 6–12 (1954)
30. D. Brown, Error detecting and correcting binary codes for arithmetic operations. IRE Trans. Electron. Comput. **EC-9**, 333–337 (1960)
31. R.R. Varshamov, G.M. Tenengolz, One asymetrical error-correction codes. Avtomatika i Telematika **26**, 288–292 (1965)
32. R.R. Varshamov, A class of codes for asymmetric channels and a problem from the additive theory of numbers. IEEE Trans. Inform. Theor. **IT-19**, 92–95 (1973)
33. A.J.H. Vinck, H. Morita, Codes over the ring of integer modulo m. IEICE Trans. Fundam. **E-81-A**, 2013–2018 (1998)
34. W.W. Peterson, E.J. Weldon Jr., *Error-Correcting Codes*, 2nd edn. (The MIT Press, Cambridge, 1972)
35. W.W. Peterson, D.T. Brown, Cyclic codes for error detection. Proc. IRE **49**, 228–235 (1961)
36. G. Castagnoli, J. Ganz, P. Graber, Optimum cyclic redundancy-check codes with 16-bit redundancy. IEEE Trans. Commun. **38**, 111–114 (1990)
37. A. Hocquenghem, Codes correcteurs d'erreurs. Chiffres **2**, 147–156 (1959)
38. R.C. Bose, D.K. Ray-Chaudhuri, On a class of error correcting binary group codes. Inform. Control **3**, 68–79 (1960)
39. S.B. Wicker, *Error control systems for digital communication and storage* (Prentice Hall Inc, New Jersey, 1995)
40. W.W. Peterson, Encoding end error-correction procedures for the Bose-Chaudhuri codes. IRE Trans. Inform. Theor. **6**, 459–470 (1960)
41. E.R. Berlekamp, Nonbinary BCH decoding, in *Proceedings of the International Symposium on Information Theory* (1967). San Remo, Italy
42. I.S. Reed, G. Solomon, Polynomial codes over certain finite fields. SIAM J. Appl. Math. **8**, 300–304 (1960)
43. D.C. Gorenstein, N. Zierler, A class of error-correcting codes in p^m symbols. J. Soc. Industr. Appl. Math. **9**, 207–214 (1961)
44. J. Massey, Shift-register synthesis and BCH decoding. IEEE Trans. Inform. Theor. **15**, 122–127 (1969)
45. R.E. Blahut, *Theory and practice of error control codes* (Addison-Wesley Publishing Company, Reading, Massachusetts, 1983)
46. P. Elias, Coding for noisy channels. IRE Convention Rec. **3**, 37–46 (1955). (Pt. 4)
47. G.D. Forney Jr, Convolutional codes I: algebraic structure. IEEE Trans. Inform. Theor. **IT-16**, 720–738 (1970), **IT-17**, 360 (1971)
48. J.L. Massey, *Threshold Decoding* (MIT Press, Cambridge, 1963)
49. D.E. Muller, Application of Boolean algebra to switching circuit design and to error detection. IRE Trans. Electron. Comp. **EC-3**, 6–12 (1954)
50. I.S. Reed, A class of multiple-error-correcting codes and the decoding Sheme. IRE Trans. Inform. Theor. **PGIT-4**, 38–49 (1954)

References

51. J.M. Wozencraft, Sequential decoding for reliable communication. National IRE Convention Rec. **5**(2), 11–25 (1957)
52. R.M. Fano, A Heuristic discussion of probabilistic decoding. IEEE Trans. Inform. Theor. **IT-19**, 64–74 (1963)
53. К. Ш. Зигангиров, "Некоторые последовательные процедуры декодирования", *Проблемы передачи информации*, Т.2. (1966), стр. 13–25
54. F. Jelinek, Fast sequential decoding algorithm using a stack. IBM J. Res. Dev. **13**, 675–678 (1969)
55. R.W. Chang, J.C. Hancock, On receiver structures for channels having memory. IEEE Trans. Inform. Theor. **IT-12**, 463–468 (1966)
56. J.L. Massey, Coding and modulation in digital communications, in *International Zurich Seminar on Digital Communications* (Zurich, Switzerland, 1974), pp. E2(1)–E2(4)
57. G. Ungerboeck, I. Csajka, On improving data-link performance by increasing the channel alphabet and introducing sequence coding, in *International Symposon on Information Theory* (Ronneby, Sweden, 1976)
58. G. Ungerboeck, Channel coding with multilevel/phase signals. IEEE Trans. Inform. Theor. **IT-28**, 55–67 (1982)
59. G. Ungerboeck, Trellis-coded modulation with redundant signal sets Part I: introduction. IEEE Commun. Mag. **25**, 5–11 (1987)
60. G. Ungerboeck, Trellis-coded modulation with redundant signal sets Part II: State of the art. IEEE Commun. Mag. **25**, 12–21 (1987)
61. R. Johanneson, K.S. Zigangirov, *Fundamentals of convolutional coding* (IEEE press, New York, 1999)
62. A.J. Viterbi, Error bounds for convolutional codes and an asymptotically optimum decoding algorithm. IEEE Trans. Inform. Theor. **IT-13**, 260–269 (1967).
63. A.J. Viterbi, J.K. Omura, *Principles of Digital Communication and Coding* (McGraw-Hill, New York, 1979)
64. J.B. Cain, G.C. Clark, J.M. Geist, Punctured convolutional codes of rate $(n-1)/n$ and simplified maximum likelihood decoding, IEEE Trans. Inform. Theor. **IT-25**, 97–100 (1979)
65. G. Ungerboeck, Channel coding with multilevel/phase signals. IEEE Trans. Inform. Theor. **IT-28**, 55–67 (1982)
66. L.R. Bahl, J. Cocke, F. Jelinek, J. Raviv, Optimum decoding of linear codes for minimizing symbol error rate. IEEE Trans. Inform. Theor. **IT-20**, 284–287 (1974)
67. J.K. Wolf, Efficient maximum likelihood decoding of linear block codes. IEEE Trans. Inform. Theor. **IT-24**, 76–80 (1978)
68. B. Vucetic, J. Yuan, *Turbo Codes—Principles and Applications* (Kluwer Academic Publishers, Boston, 2000)
69. B. Honary, G. Markarian, *Trellis Decoding of Block Codes* (Kluwer Academic Publishers, Boston, 1997)
70. D.J. Costello Jr., J. Hagenauer, H. Imai, S.B. Wicker, Applications of error-control coding. IEEE Trans. Inform. Theor. **44**, 2531–2560 (1998)
71. S. Lin, T. Kasami, T. Fujiwara, M. Fossorier, *Trellises and Trellis-Based Decoding Algorithms for Linear Block Codes* (Kluwer Academic Publishers, Boston, 1998)
72. J. Hagenauer et al., Variable-rate sub-band speech coding and matched channel coding for mobile radio channels, in *Proceedings of the 38th IEEE Vehicular Technology Conference* (1988), pp. 139–146
73. L.R. Bahl, J. Cocke, F. Jelinek, J. Raviv, Optimum decoding of linear codes for minimizing symbol error rate, IEEE Trans. Inform. Theor. **IT-20**, 284–287 (1974).
74. P. Robertson, E. Villebrun, P. Hoeher, A comparison of optimal and sub-optimal MAP decoding algorithms operating in the log domain, in *Proceedings of the IEEE ICC '95* (Seattle, June 1995), pp. 1009–1013

75. W. Koch, A. Baier, Optimum and sub-optimum detection of coded data disturbed by time-varying intersymbol interference, in *Proceedings of the IEEE GLOBECOM '90*, November 1990, Vol. II, pp. 1679–1684
76. M.C. Valenti, An efficient software radio implementation of the UMTS turbo codec, in *Proceedings of the IEEE International Symposium on Personal, Indoor and Mobile Radio Communications* (San Diego, CA, Sept 2001), pp. G-108–G-113
77. J. Hagenauer, P. Hoeher, A Viterbi algorithm with soft-decision outputs and its applications, in *GLOBECOM '89*, Nov, pp. 1680–1686
78. Y. Li, B. Vucetic, Y. Sato, Optimum Soft-Output Detection for Channels with Inter-symbol Interference. IEEE Trans. Inform. Theory **41**, 704–713 (1995)
79. C. Berrou, A. Glavieux, P. Thitimajshima, Near optimum error correcting coding and decoding: turbo codes, in *Proceedings of the ICC '93* (Geneva, 1993), pp. 1064–1070
80. C. Berrou, The ten-year old turbo codes are entering into service. IEEE Commun. Mag. **41**(7), 110–116 (2003)
81. S. Benedetto, G. Montorsi, "Unveiling turbo codes: some results on parallel concatenated coding, IEEE Trans. Inform. Theor. **IT-43**, 591–600 (1996)
82. B. Sklar, A primer on turbo code concepts. IEEE Commun. Mag. **35**, 94–102 (1997)
83. R. Gallager, Low-density parity-check codes. IRE Trans. Inform. Theor. **7**, 21–28 (1962)
84. L.M. Tanner, A recursive approach to low complexity codes. IEEE Trans. Inform. Theor. **27**, 533–547 (1981)
85. D.J.C. MacKay, R.M. Neal, Near Shannon limit performance of low density parity check codes. Electron. Lett. **32**, 1645–1646 (1996)
86. Y.Y. Tai, L. Lan, L. Zeng, S. Lin, K.A.S. Abdel-Ghaffar, Algebraic construction of quasi-cyclic LDPC codes for the AWGN and erasure channels, IEEE Trans. Commun. **54**, 1765–1774 (2006)
87. J. Xu, L. Chen, I. Djurdjevic, S. Lin, K.A.S. Abdel-Ghaffar, Construction of regular and irregular LDPC codes: geometry decomposition and masking. IEEE Trans. Inform. Theor. **53**, 121–134 (2007)
88. M.P.C. Fossorier, Quasi-cyclic low-density parity-check codes from circulant permutation matrices. IEEE Trans. Inform. Theor. **50**, 1788–1793 (2004)
89. B. Vasic, O. Milenkovic, Combinatorial constructions of low-density parity-check codes for iterative decoding. IEEE Trans. Inform. Theor. **50**, 1156–1176 (2004)
90. T.J. Richardson, R. Urbanke, Efficient encoding of low-density parity-check codes. IEEE Trans. Inform. Theor. **47**, 638–656 (2001)
91. T. Richardson, R. Urbanke, The renaissance of Gallager's low-density parity-check codes. IEEE Commun. Mag. **41**(8), 126–131 (2003)
92. T. Richardson, R. Urbanke, Modern Coding Theory, 2007, online: http://lthcwww.epfl.ch/mct/index.php
93. Y. Kou, S. Lin, M. Fossorier, Low-density parity-check codes based on finite geometries: a rediscovery and new results. IEEE Trans. Inform. Theor. **47**, 2711–2736 (2001)
94. J. Zhang, M. Fossorier, A modified weighted bit-flipping decoding of low density parity-check codes. IEEE Comm. Lett. **8**, 165–167 (2004)
95. F. Guo, L. Hanzo, Reliability ratio based weighted bit-flipping decoding for low-density parity-check codes. Electron. Lett. **40**, 1356–1358 (2004)
96. D.J.C. MacKay, Good error correcting codes based on very sparse matrices. IEEE Trans. Inform. Theor. **45**, 399–431 (1999)
97. J. Chen, M. Fossorier, Density evolution for two improved BP-based decoding algorithms of LDPC codes. IEEE Commun. Lett. **6**, 208–210 (2002)
98. V. Savin, Self-corrected min-sum decoding of LDPC codes, in *Proceedings of the IEEE International Symposium on Information Theory, ISIT 2008* (Toronto, July 2008), pp. 146–150
99. M.P.C. Fossorier, Quasi-cyclic low-density parity-check codes from circulant permutation matrices. IEEE Trans. Inform. Theor. **50**, 1788–1793 (2004)

References

100. B. Vasic, O. Milenkovic, Combinatorial constructions of low-density parity-check codes for iterative decoding. IEEE Trans. Inform. Theor. **50**, 1156–1176 (2004)
101. S.J. Johnson, *Introducing Low-density Parity-Check Codes*, Published Internal Technical Report, Department of Electrical and Computer Engineering, University of Newcastle, Australia
102. T. Richardson, R. Urbanke, *Modern Coding Theory* (Cambridge University Press, New York, 2008)
103. Omran Al Rasheed, Dajana Radović, Predrag Ivaniš, Performance analysis of iterative decoding algorithms for PEG LDPC codes in Nakagami fading channels, Telfor J. **5**, 97–102 (2013)

Index

A
Algebraic coding theory, 158, 385

B
Block codes
  bound
    Singleton, 156
    Varshamov-Gilbert, 156, 215
  cyclic code, 239, 240
  dual code, 162
  equivalent code, 159, 172
  error control procedure
    ARQ, 154, 155, 163, 165, 188
    FEC, 154, 155, 163, 165
    hybrid, 154, 155
  extended code, 159
  generator matrix, 159–162, 168, 169, 172, 173, 178, 184, 194, 201, 202, 208
  hamming codes, 156, 157, 159, 173–175, 177, 178, 180, 181, 183–186
  hamming distance, 154, 155, 157, 158, 162, 164, 170, 172, 179, 194, 212, 213, 247
  hamming weight, 158, 159, 170, 179, 183, 203, 219
  linear block code, 153, 157, 158, 160, 168, 169, 170, 178, 194, 199, 201, 202, 215, 216
  MDS codes, 156, 197
  parity-check, 157, 162, 174
  parity-check matrix, 162, 164, 179, 184, 196
  perfect code, 156
  Reed-Muller codes, 164
  repetition codes, 165, 166, 168
  syndrome, 156, 163, 164, 171, 172, 174, 176, 177, 182, 186, 194, 196, 220
  weight spectrum, 158, 159, 160, 189, 195, 203, 207, 212, 215

C
Channel
  alphabet
    input, 2, 91, 95, 104, 109, 113, 115, 124, 126–128, 131, 132, 141
    output, 2, 91, 95, 98, 102, 109, 113, 133, 136, 140, 143, 147, 364, 391, 392, 420, 432, 487, 488, 491
  binary (BC), 92, 93, 107, 120, 123, 124, 215, 391
  binary erasure (BEC), 94, 136, 137, 155, 461
  binary symmetric (BSC), 93, 120, 121, 122, 123, 124, 128, 133, 151, 165, 183, 194, 197, 198, 366, 489, 504
  burst noise channel
    Gilbert model, 110
    Gilbert-Elliott model, 110
  capacity, 101, 102, 106, 111, 123, 124, 139, 142, 143, 145
  continuous, 91, 103, 104, 123, 124, 139, 142, 143, 145
  decision
    hard, 400, 404, 415, 453, 480, 485, 496
    soft, 2, 414, 417, 427, 456
  decision rule, 106
    MAP, 108, 114, 116, 117, 441
    ML, 108, 117, 391, 394
  discrete, 2, 91, 92, 95, 103, 109, 121, 136, 139–143, 145, 391–393, 399
  probabilities
    input (a posteriori, a priori), 95
    output (a posteriori, a priori), 95
    transitional, 11
  second Shannon theorem, 2, 111, 167
  with memory, 93, 109, 151
  without memory, 2, 95, 109, 392
Convolutional codes

Convolutional codes (*cont.*)
  catastrophic error propagation, 343
  code rate, 327, 328, 350, 374, 402
  constrain length, 3, 328
  encoder, 153
  error event, 337
  error path, 344
  free distance, 3, 334, 345, 374, 376, 377
  nonsystematic, 327, 329, 342
  polynomial description, 330
  punctured convolutional codes, 340, 374, 376
  state diagram, 3, 329, 334, 349, 351, 378
  systematic, 329, 342, 378, 402, 443
  TCM, 340, 378
  transfer function matrix, 334, 344
  tree code, 327
Cyclic codes
  algorithm
    Berlekamp-Massey, 250, 309, 310, 313
    Euclidian, 317
    Gorenstein-Zierler, 306–308, 313
  BCH codes, 244, 245, 247, 248, 282, 285, 290, 291, 295, 297, 299, 310
  CRC codes, 244, 273, 275, 277, 279, 280
  polynomial
    erasure locator, 250
    error magnitude, 248, 297, 309, 310
    error locator, 248, 250, 292–294, 297–299, 306, 309, 310
    erasure locator, 250
    generator, 239, 240, 242–255, 258–260, 262, 265, 266, 268–286, 288–291, 295, 296, 299–304, 319
    irreducible, 238, 239, 317–319, 321–323
    parity-check, 246, 251, 252, 255, 256, 262
    primitive, 238, 242, 246, 282, 286, 287, 289, 295, 299, 302, 303, 305, 308, 311, 322, 323
    syndrome, 248, 297, 299, 310, 312
  Reed-Solomon codes, 248, 286, 299, 301, 312

D

Data compression (source encoding)
  average code word length, 2, 49, 53–56, 59, 60, 62–64, 67, 69, 71, 74, 75
  binary code, 45, 46, 48, 49, 54, 58, 60, 69, 78
  block code, 45, 46, 153
  code base, 45, 48
  code efficiency, 50, 54, 55, 58, 77
  code tree, 47, 48, 50–54, 60, 62–65, 67, 78, 83
  code word length, 45, 49, 50, 53–56, 59, 60, 62–64, 67, 69, 71, 74, 75, 84
  compact code, 52, 54, 62, 64, 67
  complete code, 46
  compression ratio, 50, 54, 56–59, 62, 71–75, 77, 84–86, 88–90
  FGH algorithm
    Vitter modification, 54, 83
  First Shannon theorem, 49, 71, 73
  Huffman algorithm
    adaptive, 78, 81, 83
  instantaneous code, 48, 49, 59, 62, 66
  Kraft inequality, 48, 49, 59, 60, 63, 66, 67
  nonsingular code, 45
  perfect code, 55, 57, 58
  sibling property, 54, 62, 64, 78, 80, 82, 83
  Shannon Fano coding, 50, 54–57, 63–66
  singular code, 45
  unique decodability, 46

E

Encoding
  Channel (error control)
    block, 153, 327
    convolutional, 327, 328, 334
    cyclic, 237, 240, 385
    linear, 237, 385
    turbo, 385
  source (data compression), 1, 2, 45–90, 153

I

Information
  mutual, 5, 7, 8, 99–105, 107, 108, 112, 113
  rate, 11, 14, 15, 16, 102, 111–113, 121, 122, 124–129, 131, 133–135, 139, 166, 167, 341, 350
  source, 2, 5, 6, 8–10, 12, 15, 113, 121, 122, 124, 130, 131, 135, 138, 139
  unit (Shannon), 10, 11, 127

L

LDPC
  belief propagation, 3, 456, 480
  bit flipping, 3, 454, 455, 467, 468, 471, 474, 480, 503, 505, 507
  decoding
    hard, 2, 199, 200, 364, 365, 414, 417, 459
    iterative, 3, 385, 399, 447, 453, 456, 462, 463, 475, 480, 486, 488

Index 517

soft, 2, 3, 109, 142, 147, 199, 200, 201, 250, 334, 339, 364, 366, 367, 391, 414, 417, 427, 456
irregular, 448, 449, 462
nodes
   parity(-check), 451–453, 456–459, 463, 469, 471, 475–478, 481–483, 487, 488, 493–495
   variable, 451–453, 456–459, 463, 469, 471, 475–478, 481–483, 485, 487, 488, 493, 495
   children, 451
   parent, 451, 456, 481
regular, 448, 449, 450, 451, 452, 453, 455, 457, 463, 464, 469, 478, 503
sum-product algorithm, 456, 458, 459, 478, 487, 488, 495, 497, 503, 507
Tanner bipartite graph
   girth cycle, 452, 453, 463

M

Matrix
   channel, 2, 91
   doubly stochastic, 24
   generator, 3, 159–163, 168, 169, 172, 173, 178, 179, 184, 194, 201–203, 206–209, 211, 214, 217, 218, 220, 221, 240, 241, 242, 249, 250, 251, 253–256, 258–261, 267, 302, 303, 386, 388–390, 405, 406, 408–410, 414, 419, 420, 460, 479, 497–500, 503
   parity-check, 3, 162, 164, 171, 174, 178–180, 184, 194, 196, 202, 205, 207, 208, 216–218, 241, 256, 258, 389, 406, 408–410, 447, 459, 460–464, 466–468, 474, 477, 479, 485, 487, 491, 497, 499–502, 504
   stochastic, 6, 21, 92
   transition, 6, 19–25, 73, 92–95, 106–110, 112, 133, 137, 145

R

Random process
   autocorrelation function, 13
   discrete, 12, 13
   discrete autocorrelation function, 31
   PN generator, 31
   power density spectrum, 14, 143, 147, 339, 432

pseudorandom (PN), 13, 31
stationary, 13
wide sense stationary, 13

S

Source
   binary, 6, 10, 18–20, 34, 35, 73, 84, 89, 111, 112, 115, 118, 122, 128, 133, 136, 139, 141, 143, 145, 147
   continuous, 2, 13, 14, 15, 39, 41, 42
   discrete
      adjoined, 12, 143
      alphabet, 5, 7, 19, 45, 77
      encoding, 45, 77
      entropy, 15, 32, 34, 136
      extension, 2, 17
   with memory (Markov)
      ergodic, 6
      homogeneous, 22
      joined source, 55
      state, 6, 20, 21, 22, 25, 27, 29
         absorbing, 6
         diagram, 6–8, 20–27, 29–31, 34, 35, 110, 329, 330, 334, 335, 345–351, 353, 358, 359, 363, 377, 378, 410
         stationary, 34, 110
         stationary probabilities, 8, 20–22, 74, 75, 149
         trellis, 6, 9, 26, 29, 31, 34, 35, 407, 408
   without memory, 5, 41, 52

T

Trellis decoding of linear block codes
   algorithms
      generalized VA, 3, 385, 390, 392, 394, 419–422, 426, 428
      BCJR, 3, 394, 398–400, 419, 421, 426, 432, 438, 439, 441
      Max-Log-MAP, 3, 399, 439, 441
      Constant-Log-MAP, 3, 399
      SOVA, 3, 399–401, 426–428, 431, 432, 438, 442
      trellis oriented generator matrix, 3, 399–401, 426–428, 431, 432, 438, 442
Turbo codes
   encoder, 402, 405, 442, 443
   decoder, 401, 405, 447, 458, 487